INDUSTRIAL AIR POLLUTION ENGINEERING

CONTENTS

Foreword

Of all the areas in which chemical engineers work, none has captured the public's attention more than pollution control, for nothing is more important than protecting the precious environment that sustains our life. This important and politically sensitive issue has given birth to a broad range of technological developments for the abatement and disposal of air and water pollutants and solid wastes. Nowhere has this been more evident than in the chemical process industries, where the potential for toxic pollution is greater than in perhaps any other industrial segment, and where the total annual expenditure for pollution control has hovered around the $2-billion mark.

This volume "Industrial Air Pollution Engineering," opens with a short introduction, followed by a section on air-pollution instrumentation. Particulate collectors are spotlighted next, with separate sections devoted to particulate scrubbers, dust collectors, and electrostatic precipitators. The final section covers removal of gaseous pollutants, with the emphasis on SO_2 removal.

The companion volume, "Industrial Wastewater and Solid Waste Engineering," is a collection of articles on cleaning up liquid industrial effluents and determining the best methods for dealing with hazardous and non-hazardous chemical plant wastes.

The presentation of material in these two volumes naturally reflects the approach taken in *Chemical Engineering*. Thus, they are a source of practical chemical engineering technology, written by a multitude of authors, that is virtually unmatched in the literature. The articles contained in these books will be of value not only to process and operating engineers, who will find much useful information on bringing chemical processes to the highest level of environmental compatibility at the lowest cost, but also to design engineers, who can profit from the valuable experience embodied herein in designing environmentally sound plants from the grassroots up.

One caution: Since changing governmental regulations have played a significant role in the development of environmental technology, the date of original publication is included with each article. The reader is advised to take this date into account when reading an article that cites regulations. Many articles that were published when different standards were in effect have been included because of the value of the technical information they contain.

Introduction

How to attack air-pollution control problems

Cleaning up emissions that contain toxic or hazardous substances entails much more than simply ordering some treatment units. Careful study of the process and its emissions, as well as accurate definition of performance requirements, is crucial to the success of any cleanup effort.

W. L. OConnell, Hydroscience Environmental Systems

☐ Toxic materials used in the chemical industry may pose a threat to human and animal health, and to vegetation and crops, when present in the environment at even very low concentrations. In devising controls for atmospheric emissions of these substances, engineers at Hydroscience have developed a strategy for attacking air-pollution control problems, which they have been applying successfully for more than five years. They have found that most difficulties with treatment systems stem not from improper specification or design of equipment, but rather from oversights during one of the essential planning steps that should precede these activities.

Although there is some overlap between them, these five steps include (1) setting up an emission limit, (2) identifying all emission sources, (3) investigating process modification, (4) defining the control problem, and (5) selecting a control system.

Setting up an emission limit

The first problem in specifying air pollution controls for toxic materials is to set an acceptable emission level. In some cases, this has already been done by a regulatory agency; but more often, the agency will require that the owner choose an emission level and submit a request for a permit. The agency then approves or rejects the application. In order to forestall the procedure's evolving into a guessing game that could undesirably affect your company's budget, schedule, and image with the agency, you will need a reliable method of establishing reasonable emission levels.

Originally published October 18, 1976

Reverse-pulse baghouses equipped with acid-resistant, fiberglass fabric filters remove particles from high-sulfur, coal-fired boiler emissions.

Baghouse Accessories Co.

If the emission is one of the materials listed in Table I, the work of selecting a level has already been done. The EPA's Emission Standards for Hazardous Air Pollutants [1] sets the maximum-permissible emission levels for new and existing sources of the 4 materials now designated as "hazardous"; the agency's Standards of Performance for New Stationary Sources [2] specifies the maximum permissible emission levels of the 5 criteria pollutants listed in the table for 23 categories of new sources (Table II). Standards for existing sources emitting these pollutants are set by the various states. These standards, however, do not cover most of the situations occurring in the chemical industry.

For other than these materials, the engineer will probably have to devise his own emission levels. It usually proves easier to identify toxic materials than to set their acceptable emission levels. Four lists that have been published by governmental agencies can be of some help. These lists are not directed at atmospheric emissions, however, so their applicability is limited and care must be taken in their use.

The first of these is the list of 15 cancer-suspect materials (Table III) contained in the Occupational Safety and Health (OSHA) standards [3]. Although the standard specifies precautions to be used in handling the compounds and leaves little doubt that they are considered quite toxic, it is of little help in setting emission levels. The only reference to emission controls is a statement requiring "decontamination" of exhaust air before discharge. Decontamination is frequently, but not always, interpreted to mean that the material in question cannot be detected in the airstream by the best commonly available analytical technique. Most of these materials—except for vinyl chloride, 4,4'-methylenebis (2-chloroaniline), and ethyleneimine—have limited use in the chemical industry.

A more-helpful listing is contained in EPA's proposed regulations designating substances that are hazardous in natural waters [4]. To identify likely toxic air pollutants from this list of 306 materials, it is convenient to divide it into three groups: inorganics, economic poisons, and other organics. Over 50% are inorganics, with

heavy-metal salts predominating (Table IV). Most heavy metals are toxic to animal life at low levels; and since many of their compounds—particularly the oxides—have low solubilities in water, the metals tend to accumulate in the environment at the point of discharge. *Recent experiences with mercury, cadmium, and lead indicate that it is prudent to consider most heavy metal compounds toxic enough to require very efficient emission controls.*

About 15% of the materials on the list are either pesticides or herbicides. The absence of a pesticide or herbicide from the list is no proof that it is nontoxic; it is entirely possible that it is not produced in large enough quantities to be included. Almost all pesticides or herbicides have the potential for causing environmental problems, and they should be classed with materials requiring very efficient controls.

It is harder to identify from the EPA list organics that will present emission problems, since the list was intended only for water-pollution control. Useful, though, are criteria EPA has used to identify toxic materials. The most helpful of these is that the material have an inhalation LC 50* of less than 20 cm^3/m^3 (ppm, volume per volume). The criteria of an oral LD 50† of less than 50 mg/kg of body weight and a dermal LD 50** of less than 200 mg/kg have also been used in identifying toxic materials.

Unfortunately, in the EPA list, no materials have been picked under the first criterion, and only 35 under the latter two criteria; most materials were picked because of their toxicity to aquatic animals. A good source for locating LC 50's and LD 50's is the National Institute for Occupational Safety and Health's (NIOSH) "Registry of Toxic Effects of Chemical Substances—1975" [5]. The registry contains toxicity data for over 13,000 substances and is extensively cross-referenced.

A third grouping of materials purported to be toxic is contained in the recent consent agreement between EPA

New stationary sources whose emissions are regulated	Table II

Asphalt concrete plants	Primary zinc smelters
Basic-oxygen steel furnaces	Secondary-brass and bronze-ingot plants
Coal preparation plants	Secondary lead smelters
Diammonium phosphate plants	Sewage sludge incinerators
Electric-arc steel furnaces	Steam generators (power plants)
Fluid cat-cracker regenerators and waste-heat boilers	Sulfuric acid plants
Incinerators (municipal waste)	Superphosphoric acid plants
Nitric acid plants	Triple superphosphate plants
Petroleum storage tanks	Triple superphosphate storage facilities
Portland cement plants	
Primary-aluminum-reduction plants	Wet-process phosphoric acid plants
Primary copper smelters	
Primary lead smelters	

Maximum permissible emission levels have been set by EPA for new stationary sources of certain substances	Table I

Hazardous air pollutants— regulated for all sources	Criteria pollutants— regulated for 23 categories of new sources
Asbestos	
Beryllium	Carbon monoxide
Mercury	Fluorides
Vinyl Chloride*	Nitrogen oxides
	Particulates
	Sulfur dioxide
	(Hydrocarbons)†
	(Photochemical oxidants)†

*Proposed
†Criteria pollutants for which EPA has set no emission limits

---

*That concentration of material in air which will kill ½ the test population within 14 d after being inhaled for 1 h.
†That amount of material which will kill ½ the test population within 14 d after a single oral dose.
**That amount of material which will kill ½ the test population within 14 d after a 24-h skin exposure.

and four environmental action groups which sued for promulgation of regulations to control the discharge of toxic pollutants into waterways. The 65 materials listed are not specifically named as toxic; nevertheless, EPA has agreed to set effluent limitations and treatment standards for them at a large number of point sources unless the materials are demonstrated to be nontoxic or, for other reasons, do not require regulation. The list—contained in Appendix A [6] of the agreement—partially duplicates the "Hazardous Substances" list, since it contains heavy-metal compounds and pesticides. Additional organics are included, many of which have widespread application. Some of the more prominent organics are shown in Table V. It should be noted that the list refers to toxicity in water, which is not necessarily the same as toxicity in an atmospheric emission.

A fourth valuable listing is the American Conference of Governmental Industrial Hygienists' (ACGIH) tabulation of Threshold Limit Values [7]. The TLV is that time-weighted average concentration to which nearly all workers—based on present knowledge—can be exposed indefinitely for eight hours a day, five days a week without adverse effects. An earlier edition of this list was included in the OSHA standards [3].

TLV standards were derived for the control of workers' exposures and should not be used to determine relative toxicities, evaluate or control air-pollution problems, or assess the effects of continuous exposures. Extrapolation from the data is limited, since the TLV may be set to eliminate an irritation or a nuisance problem, as well as a chronic or acute toxicity problem. Moreover, the working population is much more healthy and resilient than the total population, which contains many very young, very old, and infirm members.

In spite of disclaimers by ACGIH, the TLV's are sometimes used for the purposes mentioned, because better information is just not available. When TLV's are so used, safety factors of 10 to 100 are generally applied. Choosing this safety factor is risky and usually requires knowledge and judgment that is possessed only by toxicologists and industrial hygienists, or others with similar training. Such decisions are not recommended for most engineers.

Various other information sources on toxicity are available. EPA has published quite a few studies on the toxicity of specific materials; a recent example is the report on the environmental effects of several metal compounds [8]. NIOSH has published a series of documents proposing criteria for occupational exposures to industrial contaminants. Sources that are not generally open to the public, but sometimes available to purchasers, are manufacturers' privately collected toxicity data.

Setting an emission level is not the hopeless task it might seem to be at this point. Any material will fall into one of four categories with respect to the five lists cited: (1) contained in a list for which emission limits are set, (2) contained in, or meets the criteria of one of the other lists, (3) not contained in a list, but similar to a listed material, and (4) not similar to any material contained in a list.

If the material falls in the fourth category, special air-pollution control precautions are seldom required except in unusual situations. The only cases where there should be a question as to the toxicity and allowable emission levels are in categories (2) and (3). In such cases, the engineer should seek the advice of a toxicologist or industrial hygienist. Specialists can frequently estimate a permissible exposure-level for the general population that accounts for the stability of the material and possible chronic effects from environmental accumulation or biomagnification.

Cancer-suspect agents [3]   **Table III**

2-Acetylaminofluorene
4-Aminodiphenyl
Benzidine
Bis-chloromethyl ether
3,3'-Dichlorobenzidine (and its salts)
4-Dimethylaminoazobenzene
Ethyleneimine
Methyl chloromethyl ether
4,4'-Methylenebis(2-chloroaniline) *
α-Naphthylamine
β-Naphthylamine
4-Nitrobiphenyl
N-Nitrosodimethylamine
β-Propiolactone
Vinyl chloride

* Being repromulgated in revised form

Predominating types of compounds in the "hazardous substances" list [4]   **Table IV**

Ammonium	Cobalt	Mercury
Antimony	Copper	Nickel
Arsenic	Cyanide	Uranium
Beryllium	Fluorine	Zinc
Cadmium	Iron	Zirconium
Chromium	Lead	

Some organics listed in the consent order obtained by the Natural Resources Defense Council from EPA [6] *   **Table V**

Benzene	Isophorone
Chlorinated benzenes	Nitrophenols
Chlorinated ethanes	Nitrosamines
Chlorinated phenols	Phenol
Chloroalkyl ethers	Phenolate esters
Dichloroethylenes	Polychlorinated biphenyls
Ethylbenzene	Polynuclear aromatics
Haloethers	Tetrachloroethylene
Halomethanes	Toluene
Hexachlorobutadiene	Trichloroethylene
Hexachlorocyclohexane	Vinyl compounds
Hexachlorocyclopentadiene	

* Abridged; see original for details

Data needed for designing air-pollution control systems	Table VI

Physical form (gas, liquid, solid)
Flowrate of stream
Concentration of material to be collected
Particle size distribution
Variations in flow or concentration (extent and time)
Physical properties (solubility, resistivity, ease of handling)
Chemical properties (reactivity, corrosivity, toxicity)
Temperature and pressure of stream

The influence of other sources of related materials, as well as any naturally occurring background concentrations, should also be considered. How much the source should be allowed to contribute to the overall concentration is then calculated from a dispersion model, based on local topography and meteorology.

Identifying emission sources

Once an emission limit has been set, attention shifts to the process envelope. It is normally not enough just to attach a well-designed control device to the vent emitting the pollutant. With the low emission levels permitted for the more toxic materials, emissions from sources other than the main process vent figure highly, and may sometimes overshadow even those from the main vent. It is good practice any time, and essential when dealing with toxic pollutants, to study the entire process and identify all emission points and all possible solutions to the control problem. Some frequently overlooked emission points which have been found to contribute heavily to a process's emissions are:

Accidental releases
a. Spills
b. Relief valve operation

Uncollected emissions
a. Tank breathing
b. Packing-gland or rotary-seal leakage
c. Vacuum pump discharges
d. Sampling station emissions
e. Flange leaks

Re-emission of collected material
a. Vaporization from water wastes in ditches or canals
b. Vaporization from aeration basins
c. Re-entrainment or vaporization from landfills
d. Losses during transfer operations

Process modification

As the process is studied, the possibility of modifying it to reduce or eliminate a pollutant should be considered. This type of control usually offers the most economical way to reduce emissions, since little capital is normally needed to implement it. Often there are improvements in operating efficiency to be had by reducing losses, and the cost of a terminal control system is also cut, since it has to handle less material.

Nonetheless, modifications of this type are frequently opposed by operating personnel, who resist even looking at inconveniencing changes to the process if another control method exists. This resistance lasts until the completed study reveals the cost of add-on controls for the unmodified process to be totally unreasonable. If process modification has been ignored up to this point, redoing the study will be costly in both time and money. Thus, it is wise to insist on carrying out both types of studies from the start. Some techniques that are efficacious in reducing emissions are:

1. Substitution of a less-toxic or less-volatile solvent for the one being used.
2. Replacement of a raw material with a purer grade in order to reduce the amount of inerts vented from the process, or the formation of undesirable impurities and byproducts.
3. Changing the process operating conditions to cut down the formation of undesirable byproducts.
4. Recycling process streams to recover waste products, conserve materials, or diminish the formation of an undesirable byproduct by the law of mass action.
5. Enclosing certain process steps to lessen contact of volatile materials with air.

This type of study need not be laborious. Once an engineer is acquainted with the technique, the proper questions will occur to him automatically as he studies the process. Since the process design engineer is highly familiar with the process, it is often most economical for him to perform the study.

As the study proceeds, a running count of the cumulative effects of all process modifications should be kept. When emission sources are eliminated, the remaining ones should be compared with the allowable emissions level, so that the required control efficiency can be determined. A comparison of the required control efficiency with the efficiencies of the control devices likely to be used will indicate how far the process study should be carried.

Often a fairly simple means of control will be uncovered. In one case, a light hydrocarbon being released from a reaction ran afoul of a hydrocarbon emission regulation. Much effort was expended in finding ways to collect, purify, and sell the material before a process modification study was permitted. In a short time, it was discovered that the cause of the emission was a side reaction stemming from too high a temperature, and an insufficient excess of one reactant. Laboratory work was needed to set new process conditions, but the resulting changes increased product yield, decreased consumption of one reactant, and cut the hydrocarbon emission rate to a point where no further controls were required. The project netted the operator a substantial profit.

Occasionally a complex solution will be necessary. In another case, an herbicide facility required an extremely efficient control system to prevent damage to vegetation around the plant. Problems with wastewater treatment demanded the minimizing of aqueous wastes. As a solution, the strategy shown in Fig. 1 was devised. The pollutants were removed from the vent gases at the wet end of the process by a packed tower, which also condensed the water given off by the dryer and recycled it to the reactor. The effluent from the scrubber and

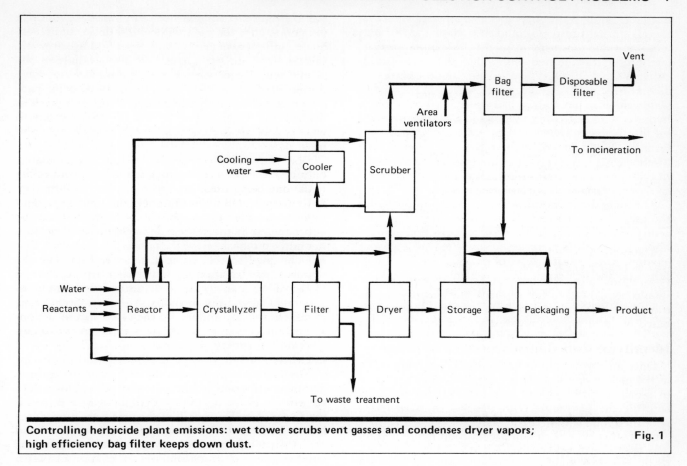

Controlling herbicide plant emissions: wet tower scrubs vent gasses and condenses dryer vapors; high efficiency bag filter keeps down dust.

Fig. 1

vents at the dry end of the process were passed through a bag filter to remove the bulk of the dust, and then through a high-efficiency disposable filter to remove any bleed-through from the bags. The dry area was enclosed, air-conditioned, and vented through the bag filter to avert the escape of fugitive dust through open doors and windows.

If the emitted pollutant tends to accumulate in the environment, it is wise to sample soil, water, and plant and animal life in the area of the plant site for a background level of the material. This can serve both as a defense against future accusations of contamination, and as a starting point for an ongoing monitoring program that gauges the efficiency of existing controls in preventing undesirable accumulations.

Defining the control problem

The final step before choosing a control device is to define the properties of the vent stream. The properties of interest will depend upon the control device being considered as well as the emission conditions.

Basic data needed are shown in Table VI. At this point in the study, a good deal should already be known about the properties of the emission; however, most of the information will likely be based on theoretical calculations, or the results of small-scale tests. Do exhibit a certain amount of skepticism about such data. All of it should be verified, and the conditions under which it was taken should be checked against those in the plant. Even laboratory data may be inaccurate,

since it is not always possible to simulate actual plant conditions in the lab, and differences often occur in agitation rates, flow velocities, and heat-loss rates. If the process has been implemented at the pilot-plant scale, sample the emission and determine its properties directly. In selected cases, conditions may be ideal enough for control equipment to be designed safely from theory, but care should still be taken to ensure that all variables have been accounted for.

Plan the sampling procedures carefully so that all relevant information is obtained. Stack-gas sampling is not as simple as it sometimes appears, and it is easy to get inaccurate results that give no inkling of their inaccuracies. When planning a sampling program, those not well versed in sampling should seek the advice of an expert.

There are three circumstances in which inaccurate data commonly lead to major problems in the design of a control system.

First is the failure to recognize the presence of another phase. If particulates, especially fines, are present when only a gas is expected—or vice versa—serious control-efficiency problems can follow, since most devices designed for removing gaseous pollutants are far less efficient on particulates, and most particulate control-devices are inefficient at removing gases. Only a few devices do both jobs well.

Where the particulate is an aerosol resulting from the condensation of a relatively nonvolatile material, the problem is especially difficult, since a sample at the only available sample point may give one answer, whereas a sample further upstream could show more gaseous material and one downstream might show more particulate. Sulfur trioxide frequently offers this type of difficulty. When dry, SO_3 is gaseous and can be absorbed in 98% H_2SO_4 or 80% isopropanol. When SO_3 contacts water, either in moist air or in a water-filled absorber, it hydrates to H_2SO_4 and condenses to a submicron aerosol that is not collected well by either reagent. In this form, it should be collected with a dry filter. EPA's Method 8 [2] employs both an absorber and a filter, but in the plant, dual control devices are usually prohibitively expensive.

The second circumstance is the failure to recognize the presence of fine particles in an emission. Until recently, this was a minor problem, but now fine particulates in the atmosphere are being cited as a serious environmental hazard. A lot of material has been published on the fine particulate problem and its control [9,10,11], and more work is in progress [12].

Fine particles are particularly important in the treatment of toxic pollutants for two reasons. First, fine particles stay airborne longer than coarse ones and increase the chances of the surrounding population being exposed to the pollutant. Second, the low emission levels normally required for toxic materials will generally force removal of most fine material from the emission.

Qualitatively, the presence of fine particles can sometimes be detected by the hazy appearance of a plume, but this rule does not hold when coarse particulates or steam are present. Unless you have experience with a similar emission, you should check particle size distribution with one of the many cascade impactors or optical devices on the market. Take care to extend the test into sufficiently fine sizes, so that all particles that may have to be removed by the control device will be accounted for.

The third problem-causing circumstance is failure to recognize variations in the characteristics of the emission. Virtually all emissions vary from season to season, as the temperatures of both the air and the cooling water change. Most emissions—even from continuous processes—vary from day to day, and some, particularly from batch processes, vary from hour to hour. The ranges of these variations must be estimated in order to determine their effect on the control device. If the controls are designed for the worst case, they will usually handle less-severe cases; even so, you should be aware of problems peculiar to the control devices, such as the sensitivity of the venturi scrubber to flowrate, or of the electrostatic precipitator to dust resistivity.

Selecting a control system

Once the emission rate and the properties of the pollutant stream have been defined, the job of specifying the control device can begin. Controls for toxic materials must normally have very high collection efficiencies. Furthermore, they must be reliable enough so that the protection they offer will always be there when the process is operating. Since most such devices are subject to one malfunction or another that destroys their efficiency, some type of alarm or interlock is essential to ensure that the emission source does not continue to operate after the pollution-control device has failed. Some units, particularly baghouses and electrostatic precipitators, are sectionalized so that a failure in one section will not jeopardize complete control and force shutdown of the process. When the material being handled is so toxic that small emission increments (such as might result from a slow decline in efficiency due to wear or fouling) are critical, a stack monitor may also be in order.

Of the large number of control devices now available, only a few types are commonly used for controlling toxic materials. These are included here because they have a high collection efficiency and reliability, and return the collected material in a form that can be either reused in the process or safely disposed of. It is not implied that these are the only types of controls that may be used for collecting toxic materials. The demands of special situations may match the capabilities of one of the many controls not listed here better than those that have been included. There are few, if any, blanket rules that apply, and each case should be considered on its own merits. More comprehensive discussions on the variety of controls available can be found elsewhere [13,14].

Although it may at first seem that choosing from among the various systems for controlling toxic substances would require detailed economic studies, this is rarely the case. The details of the problem—in particular the form in which the material is to be collected and the desired control efficiency—generally narrow the choice to one or two possibilities. A decision can often be made from a rough knowledge of the comparative economics of the various systems. A quick guide to costs and areas of application for several controls is given in Table VII.

Fabric filters

Fabric filters (see photo) have perhaps the broadest applicability of any particulate control device. They can be used on virtually any range of particle concentration and size, so long as the dust can be suspended and delivered to the unit. They are less affected by variations in particle size than any other common dust collector. In fact, it is usually not necessary to have more than a rough idea of the particle size distribution to select an acceptable unit. The dust is recovered in a dry form suitable for recycle or disposal in a landfill. Commercial units are available for a wide variety of gas flows, ranging from a few hundred to over a million cfm.

The advantages of the fabric filter are restricted mainly by the properties of available fabrics. Most fabrics have temperature limits substantially below 250°F, although a few such as Nomex, Teflon and glass withstand up to 450°F. If gas cooling is required, direct evaporation of water must be done with care because the filters are quite sensitive to blinding by wet dust. Hygroscopic, sticky or tarry dust can cause the same problem. In addition, some fabrics are attacked by acidic, alkaline, or solvent materials in the dust or gas.

In fabric filters, the dust-laden gas passes through a cloth layer so that the particles are at first removed from the gas by the fabric and ultimately by a layer of dust

Quick comparison guide for several air-pollution control systems   **Table VII**

Control	Capital cost	Operating cost	Normal Gas flow	Particle size	Concentration
Baghouse	Medium	Medium	Any	Any	Any
Carbon adsorber	High	Low	Any	—	Low-medium
Disposable filter	Low	High	Low	Any	Low
Electrostatic precipitator	High	Low	High	Fine	Low-medium
Incinerator	Low	High	Low-medium	—	Medium-high
Packed tower	Low	Low	Low-medium	—	Any
Thick fiber bed	High	Medium	Low-medium	Fine	Low-medium
Thin fiber bed	High	High	Low-medium	Fine	Low-medium
Venturi scrubber	Low	High	Low-medium	Medium	Any

buildup. Most of the filtration is done by the dust layer, especially with the woven fabrics. Much of the art of fabric filtration arises from the need to remove the dust layer efficiently enough so that the flow resistance is kept low without compromising filtration efficiency. Two basic cleaning systems are commonly used. One employs a mechanical shaker to dislodge most of the accumulated dust cake from woven fabric bags. The other uses a short, high-pressure air pulse that opposes the normal gas flow and vibrates the bag to dislodge all of the cake; bags for this system are generally made of a felted fabric.

Fabric filters are usually sized according to the ratio of the gas flow (cfm) to the cloth area (ft^2), which may range from 1:1 up to 20:1. A good analytical method of determining the best air/cloth ratio has not been discovered, although a lot of work has been devoted to finding one [15]. Generally, woven fabrics are operated below 8 cfm/ft^2, and felted fabrics above 8 cfm/ft^2. Sizing is estimated from experience, or from the results of small-scale tests that measure the effects of the fabric type and the frequency and energy of cleaning, as well as the flowrate.

The efficiencies of fabric filters are normally above 98% and can go to 99.9%. Most material losses occur during or shortly after cleaning, and vary directly with the frequency of cleaning and the air-to-cloth ratio.

Electrostatic precipitators

There are three types of electrostatic precipitators available, each with its own preferred area of application. The conventional precipitator is widely used in many basic industries for high-efficiency collection of relatively fine particles from large volumes of gas; it is commonly used on toxic pollutants in the metallurgical industry, but not in the chemical industry. The two-stage or positive-corona precipitator, used for indoor air-cleaning, is being supplied in small modular units for collecting aerosols and very dilute dusts. Wet electrostatic precipitators, which have a film of water flowing over the collection electrode, are designed for applications where the resistivity or stickiness of the dust prohibits the use of a conventional unit.

Electrostatic precipitators all function by inducing a charge on the dust particle with a high-voltage corona

discharge. The electrostatic field between the discharge electrode and a grounded collection electrode influences the particle to drift toward the collection plate. Velocities and plate areas are chosen so as to keep the particle within the collector until it reaches the plate. The plate is cleared of collected particulates by periodic mechanical vibration, continuous water flushing, or gravity drainage of a liquid. Inefficient charging of the particles, maldistribution of gas, turbulence in the gas, and re-entrainment during cleaning of dust from the plates all serve to reduce the efficiency of the unit.

All three precipitators can be designed to give greater than 99% collection efficiencies on particles well into the submicron range, although efficiencies are somewhat lower in the 0.1 to 1.0 μ size range. Mechanical pre-collectors are frequently used to reduce the amount of coarse dust reaching the precipitator.

For the proper design of a precipitator, the size distribution of the dust should be well defined. Temperatures are generally not restricting, since standard precipitators can operate up to 1,000°F. Liquid particulates can be handled easily if the liquid drains off the plate freely. A tubular collection electrode is frequently used to collect sulfuric- and phosphoric-acid mists, instead of the standard flat plates. Sticky and tarry dusts can be handled by wet precipitators but not by the other types. A big advantage of precipitators is that for large flows, a very low pressure drop is needed.

The major drawback to using precipitators is their sensitivity to particle resistivity. Too high or too low a resistivity will cause collection problems. Where possible, the resistivity of the dust can be adjusted by changing the temperature or by adding a conditioning agent such as water, NH_3, or SO_3. Wet precipitators have been designed to overcome some of the problems associated with particle resistivity; however, they create in turn a problem of dust slurry disposal. Precipitators do not adapt well to changes in operating conditions and are generally used only on continuous processes.

The operational theory of electrostatic precipitators is well established, and much work has been done recently on mathematical modeling [16,17]. Sizing is generally done with the classical Deutsch equation (1922), which states that efficiency is a power function of the drift velocity of the particle toward the collecting electrode.

The drift velocity can be calculated, but is frequently derived from tests; it usually ranges from 0.1 to 0.7 ft/s. Specific collecting areas range from 0.1 to 0.4 ft^2/ft^3/min. Standard precipitators are rarely used below 10,000 cfm. Two-stage precipitators are used from a few hundred up to a few thousand cfm. Precipitators are high-capital, low-operating-cost units and are usually preferred for large flowrates.

Venturi scrubbers

The venturi scrubber is the only control device in common use that can collect fine particulates efficiently while cleaning up gaseous pollutants. It removes water-soluble gases with efficiencies up to 98%, depending primarily upon the liquid to gas ratio employed.

The venturi's efficiency in collecting particulates is influenced by pressure drop and the number of submicron particles in the gas stream. The scrubber can be designed for almost any control efficiency on particle diameters above 0.3 μ (microns), which is the collection limit for most venturis operating at a pressure drop of 80 in. The theoretical minimum in particle size is reached at 0.2 μ, when the required throat velocity approaches the speed of sound. At higher pressure drops, multistage or positive-displacement blowers may have to be substituted for the usual single-stage centrifugal ones. Noise from these types of blowers may require provisions for abatement.

The venturi scrubber collects particles by inertial impaction. Dust-laden gas is pulled through a constriction at the entrance duct so that its velocity is boosted to 200–500 ft/s at the throat. Scrubbing liquor, usually water, is injected by low-velocity jets at a rate of 3 to 10 gal/1000 ft^3 of gas. The high gas velocities atomize the water into droplets whose size varies with the velocity, but usually ranges from 50 to 500 μ diameter. During the period in which the droplets are accelerating to the gas velocity, the dust particles impact on them, because the inertia of the particles prevents them from flowing around the droplets. The efficiency of particle collection is proportional to the square of the particle diameter, the approach velocity, the particle density, and the reciprocal of the droplet diameter.

The dust-laden water droplets must then be recovered from the gas stream with a separator—cyclonic, vane, and packed separators being the most popular. The separator should be chosen carefully, since an inefficient separator can undo the work of a good venturi. Unless the dust-laden effluent can be exploited in the process, it is usually necessary to remove the dust, adjust the pH, and recycle the water, since scrubber water carrying a toxic material can rarely be discharged untreated. Treatment facilities for this effluent can add considerable cost to the control system and take up a large amount of space.

Specifying a venturi hinges on the choice of pressure drop to achieve a required efficiency. This can be done theoretically if the size distribution and density of the dust are known; methods of calculation have been published by Calvert [18,19] and others.

The efficiency of venturi scrubbers is very sensitive to the pressure drop, and those with adjustable throat openings are preferred if the gas flow varies. Gases to be scrubbed should be presaturated with water in order to maintain efficiency. Saturation and cooling, if required, is simpler with a venturi than with a baghouse or electrostatic precipitator, since carryover of water droplets from a direct-quench system does not raise problems. Venturi scrubbers occupy less space than these other dust collection devices, and costs for corrosion-resistant construction are generally lower. Venturis are matched most frequently with small gas flows that do not contain a large number of extremely fine particles. They are especially suited to variable or intermittent flows, highly corrosive conditions, and services in which both a gas and a particulate must be controlled.

Fiber-bed aerosol collectors

Collecting liquid particulates (aerosols or mists) which are flammable or corrosive, or flowing at a low rate, is often uneconomical with electrostatic precipitators. Doing the job with scrubbers may also be precluded if mixing with water is undesirable. For such cases, a fiber-bed aerosol collector may be appropriate.

There are two basic types of fiber-bed aerosol collectors on the market—one with a thick, low-velocity bed such as the Brink mist eliminator, and the other with a relatively thin, high-velocity bed such as the Wheelabrator Ultra-dyne. The thick-bed type can only be used on materials that will drain freely from the bed, or will flush out with a spray of solvent—usually water; very viscous or insoluble materials will plug the bed. The thin-bed type uses a renewable media and handles materials that would plug the thick-bed types. Thin-bed collectors are not as efficient on very fine aerosols, though, and they present the added problem of disposing of a large roll or mat loaded with the recovered toxic material.

In thin-bed units, the contaminated air is pulled through a relatively dense bed of fine fiberglass at the high velocity of 1,000 to 2,000 ft/min. When the aerosol particles impact on the glass fibers, those with high viscosity stick to the fibers, whereas those with low viscosity coalesce and emerge from the bed to be collected by another mist-eliminator. Compared with those in venturi scrubbers, gas velocities through thin-bed collectors are three to four times higher, and targets ten times smaller; hence, this type of unit handles fine particulates more efficiently than high-energy venturis, which draw on the same collection principle.

Thick-bed collection units operate in the laminar flow region with a much more densely packed bed, about ten times as deep. The same impaction principle is used to collect the coarser particles, and the deeper, denser bed helps to offset the lower velocity. With dense packing, direct interception will also be appreciable. Fine particles, however, are collected by Brownian motion and diffusion-controlled mechanisms. Constant bombardment by gas molecules will eventually cause fine particles to be knocked into a fiber if sufficient residence time and closely spaced targets are provided. With this type of unit, a low velocity will not compromise efficiency as it will with the thin-bed type.

Fiber-bed aerosol collectors are sized empirically, either from experience with similar applications or from direct test data. Deep-bed units normally operate with superficial velocities of 5 to 30 ft/min at about 5 to 15 in H$_2$O pressure drop, and generally incorporate several tubular elements in a single case. Trade-off exists between the size of the unit (and hence the velocity or pressure drop) and its operating efficiency.

Disposable filters

Disposable filters are used to remove small amounts of particulates from moderate to small volumes of air. They are available in a wide variety of styles and efficiencies, ranging from the conventional furnace-type filter that may hold up about two-thirds of a typical dust, to a HEPA (High Efficiency Particulate Air) filter that will stop more than 99.99% of a 0.3-micron aerosol. They are normally supplied in 24-in-square modules that have rated capacities between 1,000 and 2,000 cfm, although other sizes are available. They are frequently used for recovering radioactive materials and pathogenic organisms, and for maintaining "clean" rooms. Disposables are also relied upon as backup filters for collecting leakage from other dust collectors that handle toxic materials such as beryllium or the solid carcinogens [3,20].

Filtration works by inertial impaction and interception, but involves surface phenomena in filters of high efficiency. Most of these units use folded or corrugated media to increase the surface area available for a given physical size. The highest-efficiency filters available are HEPA filters which were developed for collecting radioactive particulates. A popular size contains 200 ft^2 of thin media folded within a 2 ft $\times$ 2 ft $\times$ 1 ft housing; the folds are set off by corrugated spacers. The pressure drop for this kind of filter rises rapidly as dust builds up, and filter life is short for all but very low dust-loadings. Low efficiency filters rely upon a more porous mat, so that depth filtration takes place and the filters have a greater dust-storage capacity.

Since high-efficiency filters are much more expensive than low-efficiency units, the life of the high-efficiency unit is usually extended with a prefilter. In a properly designed system, the high-efficiency filter should last from several months to over a year. One major operating difficulty with very-high-efficiency filters is the bypassing of air around the filter; for this reason HEPA filters are constructed with rigid frames and rely on a resilient gasket to provide a good seal. Many sealing devices have been designed to maintain a tight seal around the filter assembly and prevent bypassing from occurring.

Packed towers

The most prevalent apparatus for controlling gaseous emissions is the scrubber tower. With the right scrubbing liquid, it is an extremely versatile device that can be adapted to treat almost any gas in any concentration. Its only serious limitation is its tendency to plug when large quantities of sticky particulates are present. A tower will remove a moderate quantity of coarse ($>20\ \mu$) particulates if they are removed from the liquid before it is recirculated.

A packed tower that contacts countercurrently a gas with a reactive scrubbing liquid is probably the most widely adopted scrubber for controlling toxic gases, although sieve towers are sometimes coupled with noncorrosive systems having large flows. Caustic soda, sulfuric acid and sodium hypochlorite are often used as scrubbing liquids when water proves inadequate.

With a reactive scrubbing liquid, tower operation approximates a gas-film controlled system with zero back-pressure, so that the number of transfer units (N_{og}) required to reduce the concentration of a gas to any given value Y_2 from an inlet value Y_1 is given by:

$$N_{og} \cong \ln(Y_1/Y_2)$$

Since the height of a transfer unit (HTU) is known for most common packings, the height of the tower can be estimated. For rough approximations, an HTU of 2 ± 1 ft may be used for many situations. The ratio of liquid to gas flow (moles) is chosen to ensure adequate wetting of the packing, and furnish an excess of reagent to react with the maximum anticipated emission rate. Scrubbing columns are generally sized conservatively to allow for transient overloads. Care should be taken to ensure that proper distribution and redistribution of the scrubbing liquid is accomplished. Scrubbing columns are subject to several mechanical problems that can lead to a loss in efficiency that may not be apparent to the operator. When toxic materials are being collected, it is important that reliable instrumentation be installed to ensure that malfunctions are detected and corrected.

Carbon adsorption

Adsorption onto activated carbon is sometimes used to control emissions of toxic or odorous organics, as well as to collect and recycle organics valuable enough to make recovery economically attractive. In air-pollution control applications, carbon is pressed into service for collecting organics that cannot be picked up by scrubbers, and cannot be incinerated economically. These organics are normally nonpolar substances boiling above 150°F, concentrated at less than 1,000 ppm in a dust-free gas stream below 100°F. Organics with boiling points above 350°F resist removal from the carbon except by thermal regeneration.

Organics are adsorbed onto carbon in the large volume of pores (up to 25%) whose diameters range from 1–5 $\times$ 10^{-7} cm. The mechanisms involved include Van der Waals surface attraction, followed by capillary condensation.

The capacity of a carbon for adsorbing an organic is a complex function of the properties of the carbon and the organic, and the operating conditions. Prediction of an organic's affinity for carbon is difficult, except by analogy to data for a homologue. At saturation, most carbons retain 20 to 40 lb of organic per 100 lb of carbon; under actual operating conditions, this value is rarely obtained, and working capacities run about 10 to 20 lb per 100 lb of carbon. A more-detailed discussion of carbon adsorption can be found elsewhere [21].

Several adsorber systems are in general use. If the organic can be removed from the carbon by heat, a fixed bed with steam or hot air regeneration is most commonly resorted to, although a moving bed with external regeneration is occasionally employed. The regenerated organic is normally concentrated enough to be recovered economically by condensation or incineration. But, if the organic concentration proves too low ($<$1 ppm), a thin-bed adsorber may be deployed; saturated carbon from this adsorber is withdrawn periodically and either discarded or returned for regeneration to one of several firms specializing in such treatment. When the carbon holds toxic materials, you should make certain that the method of disposal eliminates the emission problem, and does not just transfer it to another location.

Incineration

Incinerating emissions of flammable organic vapors is favored when the heat generated can be put to use, or when no other control system is practical. Difficult-to-treat odors and solvent emissions from ovens lend themselves to this type of treatment. However, organics containing halogens, sulfur, or phosphorus generate acid gases that can be scrubbed out of flue gases only at great cost; these organics are rarely burned if another method of control exists. Normally, safety and insurance regulations limit the maximum concentration of organics to 25% of the lower explosive limit. This concentration will only supply about half of the heat required to sustain combustion, and even with heat exchange, auxiliary fuel is usually necessary.

Part of the waste gas mixes with the auxiliary fuel and ignites in a burner to form a stable flame. The remainder is then combined with hot combustion gases so that the entire mixture is raised to 1,400–1,500°F. The hot gases must be thoroughly mixed so that all of the gas stays above 1,400°F for 0.3–0.5 s. These conditions result in the complete conversion of organics to CO_2. Too-rapid flame quenching or inadequate combustion time and temperature will generate substantial quantities of CO or aldehydes. Liquid particulates can be consumed under these conditions, but solid particulates will demand additional residence time, up to about 1.0 s; combustion of soot requires higher temperatures as well. More details of fume incineration can be found in an EPA report on the subject [22].

In some cases, the fuel costs can be cut by carrying out the combustion over a platinum catalyst at 800–900°F. The worst drawback to catalytic combustion is the sensitivity of the catalyst to deactivation or poisoning by a variety of metals, phosphorus, sulfur, and the halogens. If these materials are excluded, the choice between a catalytic and a thermal incinerator is largely a matter of balancing the initial and replacement costs of the catalyst against the savings in fuel that will be realized.

References

1. 40 CFR [Code of Federal Regulations] 61; 38 F.R. [Federal Register] 8820, Apr. 6, 1973, et seq.
2. 40 CFR 60; 36 F.R. 24876, Dec. 23, 1971, et seq.
3. 29 CFR 1910.93; 39 F.R. 23540, June 27, 1976.
4. 40 F.R. 59960, Dec. 30, 1975.
5. "Registry of Toxic Effects of Chemical Substances—1975", (Toxic Substances List—1973 and 1974), National Institute for Occupational Safety and Health.
6. "Natural Resources Defense Council v. Train," U.S. District Court, District of Columbia, No. 73-2153 Consent Agreement, *Environment Reporter Decisions*, 8 ERC, June 26, 1976, p. 2,120.
7. "Threshold Limit Values for Chemical Substances and Physical Agents in the Workroom Environment with Intended Changes for 1974," American Conference of Governmental Industrial Hygienists, Cincinnati, Ohio, 1975.
8. "Preliminary Investigation of Effects on the Environment of Boron, Indium, Nickel, Selenium, Tin, Vanadium, and Their Compounds", Versar Inc., Vol. 6, Vanadium, EPA-560-2-75-005f, PB-245 989, Aug. 1975.
9. Vandergrift, A. E., Shannon, L. J., and Gorman, P. G., Controlling Fine Particles, *Chemical Engineering*, Vol. 80, No. 14, June 18, 1973, p. 107.
10. Shannon, L. J., Control Technology for Fine Particulate Emissions, EPA-650-2-74-027, NTIS-PB-236 646, May 1974.
11. Calvert, S., others, Fine Particle Scrubber Performance Tests, EPA-650-2-74-093, NTIS-PB-240 324, Oct. 1974.
12. Iammartino, N. R., Gaging the Potential for Fine Particle Control, *Chemical Engineering*, Vol. 82, No. 15, June 23, 1975, p. 86.
13. Rymarz, T. M., and Klepstein, D. H., Removing Particulates from Gases, *Chemical Engineering*, Vol. 82, No. 21, Oct. 6, 1975, p. 113.
14. Parekh, R., Equipment for Controlling Gaseous Pollutants, *Chemical Engineering*, Vol 82, No. 21, Oct. 6, 1975, p. 129.
15. Billings, C. E., and Wilder, J., "Handbook of Fabric Filter Technology" (4 vols.), EPA-APTD-0690-3, PB-200 648 to 651, Dec. 1970.
16. Oglesby, S., others, "A Manual of Electrostatic Precipitation Technology." EPA-APTD-0611, PB-196 381, Aug. 1970.
17. Gouch, J. P., others, "A Mathematical Model of Electrostatic Precipitation," EPA-650-2-75-037, PB-246 188, Apr. 1975.
18. Stern, A. C., ed., "Air Pollution," 2nd Academic Press, New York, N.Y., 1968, p. 483.
19. Calvert, S., others, Wet Scrubber System Study, Vol. 1, "Scrubber Handbook," EPA-R2-72-118a, PB-213 016, Aug. 1972, p. 5–112.
20. Control Techniques for Beryllium Air Pollutants, U.S. Environmental Protection Agency, Publication AP-116, Feb. 1973.
21. Package Sorption Device System Study, MSA Research Co., EPA-R2-73-202, PB-221 138, Apr. 1973.
22. Rolke, R. W., Hawthorne, R. D., Garbett, L. R., others, Afterburner Systems Study, EPA-R2-72-062, PB-212 560, Aug. 1972.

The author

W. L. OConnell is a Senior Environmental Specialist with Hydroscience Associates, Inc., 9041 Executive Park Drive, Knoxville, TN 37919, where he is responsible for air-pollution control and stack sampling activities. Before transferring to Hydroscience, he was responsible for air-pollution control research activities at the Environmental Research Laboratory of The Dow Chemical Co. at Midland, Mich. He has a B.S. in chemical engineering from Drexel University and is a member of AIChE and the Air Pollution Control Assn.

Air Pollution Control

If your control systems meet today's standards, they're probably not good enough. Systems must be designed with tomorrow's requirements in mind. Also, don't be guided by widely-held misconceptions about the relationships between power input and pollutant recovery.

AARON J. TELLER, Teller Environmental Systems, Inc.

Air pollution control has entered a new phase in its evolution. No longer can a company gain time simply by putting an extra piece of equipment on the line. This is true because the regulations are more restrictive and the authorities are conducting their own effluent measurements. The economics of pollution control are becoming more significant and the death knell is ringing for redundant systems.

For example, soon we will no longer be able to afford the "successful system" consisting of:

> An afterburner . . .
> followed by a cooling system . . .
> followed by a baghouse . . .
> followed by a reactor . . .
> followed by a venturi.

The evolution to complexity is being terminated. The new direction is toward systems with little or no redundancy that are compatible with the manufacturing operation and that are based on an understanding of effluent behavior. Above all, we need systems that are more economical.

General Design Factors

Factors affecting system design are: 1) characteristics of the effluent; 2) level of reduction required; 3) technology of component equipment; 4) potential for recovering the value of the pollutant.

Effluent characteristics are critical and are not subject to generalizations. For example, one system consisted of particulates and an acid gas, with the gas present in relatively low concentration. In this case, reduction of particulate matter by 99.5% did not have a significant effect on opacity. Thus the problem did not lie with the particulates, but with a changing characteristic of the low

Originally published May 8, 1972

concentration acid gas. The answer was to install a nucleator, which reduced opacity to less than 5%.

This example points out that it is critical in the design of control systems to understand the physical and chemical characteristics of the stream. This is especially true because of the synergies that can occur when the stream's composition is changed.

Plan For Future Regulations

The degree of emission reduction should not be based on existing regulations, but rather on those that can be reasonably anticipated. Few things are as inevitable as increased taxes and lower emission limits. Here are some rules of thumb on what to expect:

- Opacity objectives should be lower than 10%.
- Plan for particulate loadings of less than 0.02 gr./scf.
- Gaseous emissions should be limited to 1 ppm., except where high toxicity will require even lower values.
- Concentration of gaseous components at plant boundaries should be on the order of 1 to 2 ppb.
- Liquid wastes will be prohibited.

These restrictions may appear excessive based on today's standards. But if the present legislative trend continues, they will be realistic within five years and, in fact, they have been reached in some communities. Furthermore, technology does exist that is capable of achieving these limits.

An engineering comparison of component control technology is difficult, to say the least. One may consider: mechanism, costs, size, power, performance and safety. Thus, to attempt a rigorous comparison of equipment applicability requires a five dimensional grid, which may be a more difficult problem than control of pollution itself.

For instance, a comparison of cost and size relationships is hazardous because the cost element for corrosion requirements is small for some types of equipment and extreme for others. The mechanism of pollutant recovery is significant and may determine whether or not a particular control device is suitable. An understanding of mechanism can also dispel the old wives' tale that increased recovery requires increased power.

The mechanisms of pollutant recovery are presented in Table I for particulate collection only and Table II for both particulate collection and gas absorption.

Power Misconception

The misconception that has pervaded industry is that there is always a relationship between power expended and pollutant recovery. This belief was valid when only one mechanism was used for recovery. In the case of particulates, for example, inertial impact was the basis for the relationship. This mechanism applies to baffles, turbulent contactors, tray towers, wet cyclones and venturis. In these devices, a power-separation relationship is obvious.

Methods For Particulate Collection—Table I

Brownian Separator

The Brownian Separator removes low-viscosity liquid particulates in the particle size range from 0.01 to 0.05 microns. This patented device consists of filaments, usually glass, arranged so that the space between them is less than the mean free path of the particle to be collected. Thus, the particle cannot escape; it must collide with one of the filaments where it is captured. Eventually, liquid builds up on the filaments and runs down to a collection point.

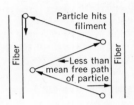

Filtration

The panel filter or baghouse will remove solid particles down to the 0.1 micron level. The panel-type filter is limited to relatively light dust loadings because cleaning is time-consuming and costly. The baghouse, on the other hand, can be used with heavy grain loadings. These units are made up of a large number of cylindrical bag filters, which may be cleaned automatically to provide continuous operation. The baghouse should not be used for oily, hygroscopic or explosive dusts. Gas temperature is limited by the composition of the filter medium.

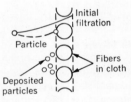

Electrostatic Precipitator

In this common collecting device, a high voltage is imposed on a relatively slow-moving gas stream. Charged particles are attracted to the grounded electrode, where they are periodically removed by rapping, vibration or washing. Conventional precipitators are quite bulky because of low air velocities. They have the advantage of providing dry collection, generally down to the 0.2 to 0.5 micron range. Theoretically there is no minimum limit to the size of particles that can be collected.

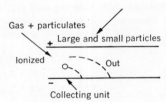

Centrifugal Collectors

Available in a number of different designs, cyclone separators rely on centrifugal force to drive particles to the wall of the chamber where they drop out of the gas stream into a collector. They are useful for relatively coarse separations. A large-diameter cyclone will remove particles 15 microns and larger. Small-diameter units, usually connected in parallel, are effective down to the 10-micron range.

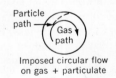

Imposed circular flow on gas + particulate

Inertial Separators

A variety of dry mechanical collectors are based on inertial impaction of particles on baffles arranged in the gas stream. Because there is a practical limit on how closely together the baffles can be placed, these devices are best for separations above 20 microns. However, some new designs can achieve 5 to 8 micron separations.

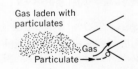

Gravity Settling

The oldest and simplest means of particle separation is the gravity settling chamber, in which particles fall onto collecting plates. These are useful for removing coarse dusts above 40 microns, and may be placed ahead of other separation equipment.

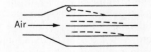

Equipment For Removing Gases and Dusts—Table II

Spray Towers

In spray towers the gas stream passes at low velocity through water sprays created by pressure nozzles. These units are simple but only moderately effective. They will remove particles down to the 10 to 20 micron range and absorb very soluble gases.

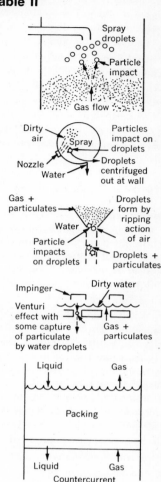

Wet Cyclones

In this version of the cyclone separator, swirling gas flows through water sprays. Droplets containing dusts and absorbed gas are separated by centrifugal force and collected at the bottom of the chamber. Wet cyclones are more effective than spray towers. They will absorb fairly soluble gases and remove dusts down to the 3 to 5 micron range.

Venturi Collectors

The venturi design relies on high gas velocities on the order of 100 to 500 ft./sec. through a constriction where water is added. The impact breaks the water into droplets, with the fineness of spray determined by gas velocity. Venturis collect particles to the 0.1 micron level and recover soluble gases.

Perforated Impingement Trays

Here, gas flows through small orifices at 40 to 80 ft./sec. and hits a liquid layer to form spray. The liquid captures particles and gases and is collected on impingement baffles. Performance is relatively good with particle collection possible down to 1 micron and absorption of fairly soluble gases.

Packed Towers

Packed towers were primarily created for gas absorption. With some new designs, they can also be used for dust removal. Gas-liquid flow may be concurrent, cocurrent or crossflow. Two kinds of mass and particulate transfer are possible, depending on the packing. Extended surface packings provide absorption by spreading the liquid surface, and collect particles by cyclonic action. Recovery is limited to 10-micron sizes and larger. With filament packing, absorption by surface renewal is twice as effective and particles to 3 microns are separated by inertial impact. A cross-flow tower gives the highest solids handling capability with the lowest pressure drop.

Nucleation Scrubber

This process grows submicron particles, down to 0.01 micron, by condensation. It collects grown particles on filament-type packing by inertial impact at low energy. At the same time, absorption occurs as in a conventional cross-flow packed tower.

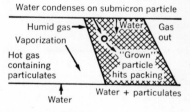

Turbulent Contactor

With this device gas flows through spherical packing made up of 0.5 to 2-in. balls, which oscillate or bounce in liquid. Particles down to the 1 to 5 micron range collect by baffle impact on the spheres. Gases will absorb in the turbulent liquid and absorption efficiency will depend on the number of stages.

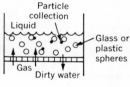

Wet Filters

This is an open filter whose efficiency is improved by spraying with water. Its absorption ability is limited to those gases that are very soluble.

Chromatographic Recovery

The chromatographic unit (proprietary) contains inexpensive extended surface solids with a mono-to-termolecular coating of an active reagent. Thus the solute gas need only diffuse to the surface of the solid where it reacts either by ionic or molecular reaction. Where the reaction is ionic, the absorption is generally irreversible. Where the reaction is molecular, the absorption can be reversible and the solute gas can be stripped off essentially pure.

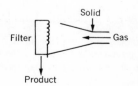

The misconception arises when considering filtration, electrostatic precipitation, nucleation and Brownian impact. These devices circumvent the inertial impact mechanism, or use it only as one component of the process.

As shown in Fig. 1, inertial impact systems do conform to the energy relationships, while those using other mechanisms deviate widely. In many cases, they may use as little as 5% of the energy needed to achieve the same degree of collection with an inertial impact unit.

Of course, the power requirement by itself is only one criterion for selection. There are disadvantages in all equipment and typical problem areas are listed in Table III.

Absorption Rate Vs. Power

Misconceptions also exist in the technology of removing gases by absorption. Many assume that there is a relationship between recovery and power, although the data in Fig. 2 indicate that no such relationship exists. Recovery is primarily a function of the availability of young liquid or solid surface. Please note *young surface,* not total surface.

Absorption is rapid on fresh liquid surface. In the case of a soluble gas, the transfer rate approaches zero after 0.003 sec., mostly because of the slow rate of diffusion away from the surface to the body of the liquid. With low solubility gases, absorption ceases even more quickly. Thus, the controlling factor is not the surface itself, but rather the rate at which the surface is renewed.

Spray towers, spray chambers, wet cyclones and venturis all provide large surface areas for contact, but have little capability for renewal of the liquid surface. For this reason, these units have relatively low efficiency with high power inputs. The tray tower, packed tower and turbulent contactor offer surface renewal, but with different relationships between absorption rate and power requirements.

In order to prevent misunderstanding about product literature claims of 99% recovery, it should be noted that the comparisons in Fig. 2 are based on 90 F. scrubbing liquid and emission levels no less than 1 ppm. Liquor temperature is critical, not only from the aspect of solubility equilibrium, but also because it affects the kinetics of dissolution in the liquid boundary layer.

Limitations of Collection Equipment—Table III

Equipment	Disadvantage
Settling chamber	Size
Baffle chamber	Reentrainment and dust buildup
Packed crossflow (T)	Irrigation minima
Packed crossflow (S)	Plugging
Nucleation	Irrigation minima
Precipitator	Size and maintenance
Baghouse	Bag life
Chromatographic baghouse	Bag life
Dry cyclone	Plugging with damp conditions
Wet cyclone	Plugged nozzles and misting
Packed countercurrent	Plugging
Thermal venturi	Erosion, minimum gas temp. 1,200 F.
Venturi	Mist carryover
Brownian separator	Cannot handle solids

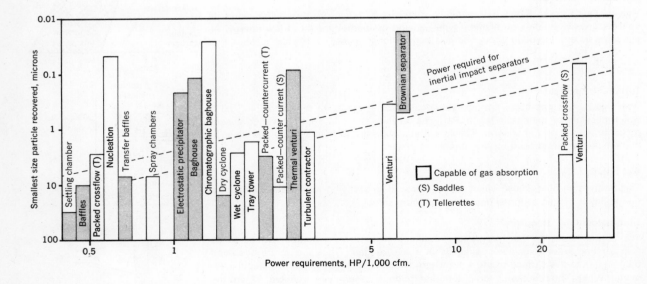

DIRECT RELATIONSHIP between power input and particle size recovered is valid only for inertial impact collectors—Fig. 1.

The dry scrubber (chromatographic separator) is effective for low concentration streams and can serve for both particulate and gas removal. Because there is effectively no depth to the liquid phase, the rate of transfer is on the order of 500 times that of liquid-gas absorption. Concentrations up to 500 ppm. have been treated with a transfer depth of 0.05 in. of chromatographic material, giving an emission level to 1 to 2 ppm.

System Selection

A comparison of equipment operation and capabilities only serves to help establish what components may go into the system. In achieving economic solutions, the system is of prime concern and will often control the choice of equipment.

One example of restriction by the system is control of emissions from secondary aluminum manufacture. The process itself consists entirely of solids handling and the skill level of the operators is not high. Stack emissions contain chlorine, hydrogen chloride, hydrogen fluoride, carbonaceous particulates and inorganic particulates.

A conventional control system might be made up of an afterburner, cooling system, baghouse and scrubber. Carbonaceous matter would be removed in the afterburner, inorganics in the baghouse and acid gases in the scrubber. Such a system could cut emissions to an acceptable level. But it would be complex, would call for new operating labor, and operating costs exclusive of labor would be on the order of $70,000/yr. per furnace.

One actual solution to the above situation is based on a chromatographic dry absorber that collects solids and absorbs acid gases, producing a neutral, dry product. No special labor is required and operating costs, exclusive

of labor, are around $15,000/yr. per furnace. Opacity of the exhaust stream is less than 5% with solids loading under 0.01 gr./scf. Maximum gas concentration is 10 ppm. and the average 1 ppm.

Another systems example involves a jet engine test cell emitting some 1 million cu.ft./min. The exhaust contains 0.01 to 1.0 micron carbonaceous particulates plus oil droplets in the 20 to 50 micron range. Dry collection was ruled out because of the potential explosive hazard from static buildup. Additional ducting was not possible due to excessive size and vibration problems.

A scrubber was built on the exhaust stack, in which the nucleation process utilizes the thermal energy of the system to give high collection efficiency—5% opacity and 0.004 gr./scf. solids. Solids can be separated and burned, while the scrubber water is cooled and recycled.

In the two cases outlined here, the total system parameters rather than mere separation characteristics controlled the design. This is the direction that the new regulations are forcing on us. It may truely be a blessing in disguise.

Meet the Author

Aaron J. Teller is president of Teller Environmental Systems, Inc., 295 Fifth Ave., N.Y., NY 10016. His career has involved both industry and academic experience, and before founding his own company in 1970 he was dean of The Cooper Union. He holds a doctorate in chemical engineering from Case Institute of Technology. Dr. Teller was instrumental in founding the Metropolitan Engineer's Council of Air Resources and the National Engineering Council on Air Resources, of which he serves as chairman.

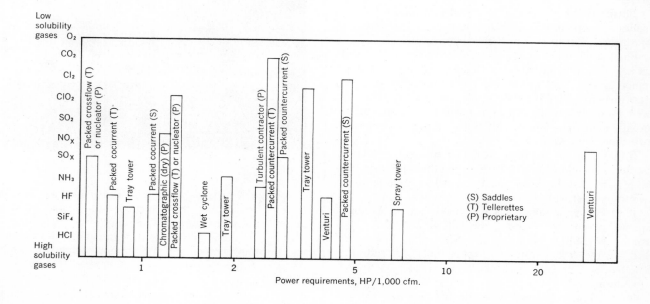

POWER REQUIREMENTS for absorption equipment vary widely—Fig. 2.

Using chemical-resistant masonry in air-pollution-control equipment

Bricks, mortars, cements and a whole range of other masonries are used to line chimneys and associated air-pollution-control equipment. This article outlines how to select materials for best corrosion resistance in various industrial applications.

Walter Lee Sheppard Jr., Consultant.

☐ Stack gases generated by smelting of ores, combustion of hydrocarbon fuels, and incineration of industrial wastes vary substantially in makeup and exit temperature. Their corrosiveness—particularly when dissolved in water—makes them especially troublesome to handle in pollution-control units such as scrubbers.

Let us look first of all at the corrosive conditions which exist in each of the three applications before going into details about which masonry linings should be selected.

Gases given off by smelting operations usually fall into the lower-temperature ranges, and contain sulfur oxides as their principal contaminants, although others are often present.

Then too, the burning of coal and oil produces sulfur oxides—in amounts considerably less than those released by smelters, but usually at higher temperatures.

Finally, the incineration of liquid and solid wastes, and industrial waste gases, is carried out at much higher temperatures. Although wastes given off by incineration may include oxides of nitrogen, phosphorus and sulfur as well as hydrogen fluoride, the primary contaminant is usually hydrogen chloride. All of these gases produce acid in contact with water.

If the firing is done in an oxidizing atmosphere (excess oxygen), the following gases are among those present: SO_2 or SO_3, HCl, HF, NO_X, and P_2O_5. All of these produce acid in contact with water.

If the firing is done in a completely reducing atmosphere (absence of oxygen), the following residues will result: free sulfur, HCl, N_2 and NH_3, phosphides, and HF. If the free sulfur is stripped out, the resulting solution, will, in the absence of HCl or HF, be alkaline or close to neutral.

Since combustion atmospheres are unlikely to be completely reducing (i.e., absolutely no oxygen present), in most cases some SO_2 and NO_x will be formed. If traces of vanadium, nickel or other metallic catalysts are present some SO_3 may also be formed. If sulfur and nitrogen oxides are present, the scrubbing liquor will be acidic.

Handling these three types of waste gases (from smelting, burning of fuels and incineration) follows basically the same pattern: scrubbing the gas, which is then vented, followed by treatment of the scrubbing liquor. However, due to variations in the quantity and type of contaminant, and the temperature of the exit gas, the procedure in each case is modified to some degree.

Originally published November 20, 1978

Photo: Swemco, N.Y.

Scrubber handling waste gases from an ore smelter

Fig. 1

Smelting

Because there are great quantities of sulfur in many ores, the designer is often faced with the disposal or conversion of large amounts of byproducts. Generally, one of three methods is followed:

1. Manufacture of sulfuric acid from the spent scrubbing liquor.
2. Reduction of the sulfur oxides to elemental sulfur.
3. Simple scrubbing of the gases, followed by neutralizing of the scrubbing liquid. (If lime or limestone is used, calcium sulfate is formed. This method poses an additional disposal problem.)

Acid-resistant masonry need not be used in the second method, but it is frequently required in the first and third methods.

From the smelters, offgases at temperatures in the 600 to 900°F range are carried by ducts to a scrubber-absorber, in which the liquor is recirculated water, whose acid content can range from 3% to as much as 15%, depending on the design and the process. Wet scrubbers cannot be expected to attain removal efficiencies in excess of 70% SO_3, 95% SO_2, and 80% HCl.

Fig. 1 is a photograph of a scrubber handling waste gases from an ore smelter at Hurley, N.M. Fig. 2 is a vertical sectional diagram of the same scrubber.

Since exit gases will be at temperatures in the 125 to 180°F range, condensation of acid droplets may be anticipated in the exit ducts carrying the scrubbed gases to the stacks, and in the stacks as well. If a reheater is installed to raise the exit gases above the dew point (usually consid-

ered to be in the 350 to 450°F range) the acid in the exit gases remains a vapor (provided the ducts are insulated) and discharges with them, except, of course, during shutdowns. For those parts of the system that are heated or are subject to possible chemical attack from acid condensate during idle periods (this includes the scrubber entry and exit ducts), a recently developed 100% closed-cell, foamed borosilicate-glass block can be installed as a lining—the mortar most often used with this block is silica sol binder and a borosilicate glass filler. This construction has several advantages:

1. Chemical protection is provided against acid condensation.
2. Lower skin temperature reduces heat losses. External insulation, with its attendant high maintenance cost, can be eliminated.
3. At 12 lb/ft³, the lining is lighter than other types, permitting a saving in structural support.
4. The contoured block is quick and inexpensive to install, and is easily repaired.

Fig. 3 shows the effect of using foamed glass block on thermal profiles in masonry-lined equipment.

When the 960°F top temperature of the borosilicate glass foam is exceeded, its surface may be protected with refractory or acid brick layers, which provide sufficient insulation to bring the surface temperature of the foamed glass below its upper limit. (Acid brick in North America is normally composed of either red shale or fireclay and is made to conform with ASTM Standard C-279.)

If mostly sulfur oxides are involved, and there are little or no chlorides or fluorides present, alloys may be more convenient to use in the high-temperature entry duct. However, the exit duct, which operates at a far lower temperature, can usually be constructed more economically using carbon steel and a 2-in. layer of borosilicate glass foam.

The scrubbers that handle off-gases from smelters have in the past been largely of the low-pressure variety—packed vessels or even open bodies—in which gases rise countercurrent to the scrubbing liquid. (Venturis and other high-pressure scrubbers are discussed further on.) Common designs call for a membrane (typically, rubber applied as a sheet or fluid) protected by 4 in. of acid brick; the latter is most often composed of red shale bedded and bonded with a furan resin mortar.

Depending on the service conditions, alternative selections may be made for the membrane, the brick (fireclay rather than shale) or the mortar (perhaps a phenolic). However, the designer should avoid using a silicate or silica mortar in such scrubbers, since they can be washed out by large volumes of water, which often occurs when water circulates through the unit during startup, or when the operation of the smelter is interrupted.

Since silicates cannot long survive in alkaline conditions, they should likewise be avoided if alkaline groundwater is used, or if soda ash or other neutralizers are added to the scrubber solutions. Polyester and epoxy mortars are occasionally used in such service, but are not normally recommended, since their maximum service temperatures are in the 225 to 250°F range. An interruption in the supply of water to the scrubber, with the consequent overheating of the lining, would severely damage a brick structure laid with these mortars.

If fluorides contained in the off-gases are sufficient to build up an HF (or acid fluoride) content in the scrubbing liquid exceeding 50 ppm, excessive damage will occur within a relatively short period to any masonry lining containing silica or alumina. Under such conditions, the lining should be faced on all exposed surfaces with carbon brick laid in a furan (or phenolic) resin mortar filled with carbon, graphite or barytes. As HF content increases, the use of any siliceous or aluminous materials may have to be eliminated, even in the protected rear courses, due to the penetration of HF through the carbon brick.

If the designer selects lead (or a ferrous alloy) as the membrane, then carbon-bearing material cannot be employed in direct contact between the masonry and the membrane, since such contact creates a galvanic couple in which the carbon (or graphite) is positive to the metal. This couple causes the metal to waste gradually until it is penetrated or punctured. In such units, it has sometimes been observed that the pattern of the brick and mortar has been incised into the lead membrane. The use of a barytes filler (which is a nonconductor) in the resin mortar is recommended.

The port at which the hot gases enter the scrubber and first come in contact with sprayed scrubbing liquid may require special design measures. In the dry zone, where the temperature is beyond the maximum operating limit for resin mortars, the mortar used to bond the acid-brick bed to the masonry lining should contain an acid-resistant silicate base and a refractory filler.

Since exit gases are usually saturated with water vapor bearing highly corrosive, entrained sulfur oxides, they often pass through a mist eliminator before entering the duct to the chimney. The body of the eliminator is usually above and resting on the scrubber body so that it can drain directly into it.

Because the walls are wet and are usually at moderate temperatures in the 140 to 160°F range, they can be protected in a variety of ways. Some designers have opted for FRP (fiberglass-reinforced plastic), whereas others have applied flake-glass-filled resin coatings, and similar heavy organic materials. Since all these materials have low thermal limits, pump failure or an interruption in the water supply, with the concomitant overheating, may damage the coating. To protect against such a possibility, some designers prefer a ceramic lining of a silicate mortar, applied by trowel or gun to a 1/2 to 3/4-in. thickness directly on the steel. Several such units are operating satisfactorily after three or more years of service.

From the eliminator, the gases pass to the chimney. If they are not reheated, the chimney will operate at a low (positive) pressure and a temperature of 130 to 150°F at the most—well within the condensation range of any remaining acid contaminants in the gases. However, if the water in the scrubber fails, temperatures within the chimney can reach the 500°F range or higher.

Low-temperature chimneys have been fabricated with FRP or flake-glass resin-lined steel, and generally have operated satisfactorily. However, if one wishes to protect against a possible runaway, or if hot smelter gases will be vented during maintenance or replacement of the scrubbing equipment, the safest material is acid-resistant brick, bedded and bonded in a potassium silicate mortar, which should be free of any sodium.

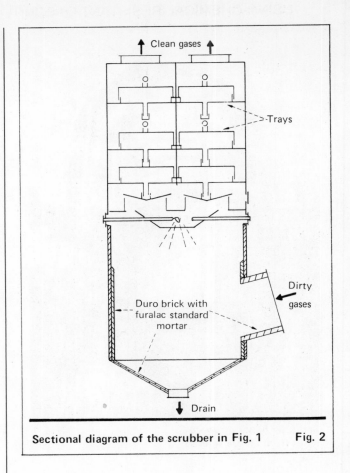

Sectional diagram of the scrubber in Fig. 1   Fig. 2

If any sodium is present in the mortar, sulfuric acid condensing on the brickwork will react with it to form sodium sulfate, a hydrating material that crystallizes in the brick joints and slowly picks up moisture from the air until its volume increases by 50%. This gradually creates a disrupting internal pressure of up to 9,000 psi in the brickwork. Eventually, after six to ten years, this will cause severe damage and require expensive repairs. Although potassium silicate mortars containing no sodium are somewhat more expensive than sodium silicate ones, they add little to the initial cost of a chimney, and help to avert extensive maintenance expenditures later on.

The liquid employed in the scrubbing of such smelter off-gases is collected either for manufacture of byproducts or disposal, as indicated above. The lining of equipment handling the liquid in the manufacture of sulfuric acid has been discussed in an earlier paper and will not be repeated here. [2].

Steam and power generation

Generating steam and power by burning fossil fuels also yields oxides of sulfur as waste-gas contaminants, though in considerably lower concentrations than does a smelter. Here, too, the off-gases must usually be scrubbed before venting, though the treatment methods differ from those noted under smelting.

The gases leave the boilers in the 700-1,200°F range through ducts that carry them to a high-pressure scrubbing system, through which a scrubbing medium (often, simply water, but sometimes containing a mild neutralizing alkali) is recirculated. From the scrubber, the off-gases

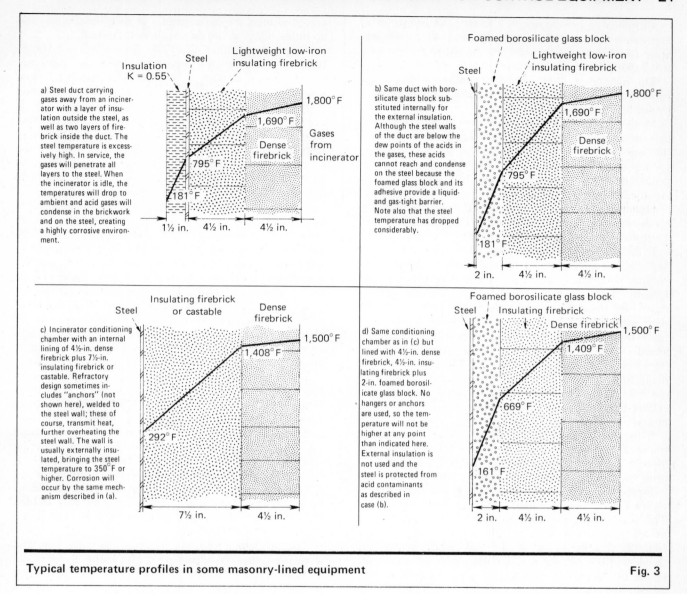

Typical temperature profiles in some masonry-lined equipment   Fig. 3

are carried through a duct to the stack, usually without the mist eliminator mentioned under smelting.

Comments made earlier with regard to the two sections of duct are largely applicable here as well. Certainly the exit duct's conditions are much the same. However, the entry duct is somewhat hotter and contains far fewer contaminants. Frequent shutdowns in steam plants cannot be tolerated, so there is far less exposure of the steel walls to the condensation of contaminants, and little need for masonry linings.

The scrubber is usually of a venturi type, the entry portion being a cone with wet walls, and the scrubbing liquid being flooded down the walls and through the venturi throat. The gases travel at high speed as they pass down the cone and into the throat, atomizing the scrubbing liquid and carrying it against a target—such as the bottom of a receiver vessel below the venturi—and then into a separator vessel where the gases expand and rise more slowly into the exit duct and chimney.

In this system, the high-velocity gases flow in considerable volume and contain not only sulfur oxides (in small

but troublesome amounts) but also fly ash. To the anticipated damage caused by chemical corrosion at the cone and venturi, we therefore have to add the abrasive action of the fly ash, which will scrub off any corrosion product that might otherwise remain as a protective layer. For this reason, acid-resistant brick is often used on the surface of the cone and the venturi, and on the walls directly above the cone. The mortar used here is normally a furan. Because of the high speed of the gases, and the rapid rise in temperature that could occur were the pumps of the scrubbing system to fail, a high-temperature formulation (one that can endure temperatures in the range of 425-450°F for an extended period) is strongly recommended.

Brick selection is influenced by the venturi's design. If its construction permits the temperature of the steel to rise to that of the internal brick lining, a brick with a coefficient of expansion a little higher than either fireclay or shale, such as silicon carbide, should be chosen. Otherwise, the brick lining may loosen with temperature cycling and break up due to the severe vibration typical of this service. Acid-resistant brick linings have a coefficient of expansion of approximately 4.1 x 10^{-6}, as compared to approximately 6.8 x 10^{-6} for carbon steel. As the temperature rises from ambient (usually about 140°F), the steel will expand more than the brick.

The membrane used with these linings is very often rubber, as in scrubbers that treat waste gases discharged by smelting operations.

Many venturi scrubbers handling boiler off-gases contain no other masonry linings. However, in some cases, the bottom of the receiver, or the target against which the gas-liquid mixture leaving the venturi impinges, is also protected with a lining of brick laid with furan or modified phenolic resin mortar.

Handling of the exit gases follows the pattern described above, and is subject more or less to the same conditions. Duct and chimney design are similar.

If the acid wastes from the scrubber are to be treated and discarded, they may be piped to an "equalizer" or holding tank, often partially or completely below grade. Usually made of concrete, the tank is lined with a membrane (usually asphaltic) and one or more courses of shale acid-resistant brick that is bedded and bonded in a furan resin mortar. Lime or another neutralizing agent is added, either in this tank or in a neutralizer vessel immediately downstream from it. The lining of the neutralizer vessel is usually acid brick, similar to that in the holding tank. If steel tanks are used, they may be rubber-lined, with an inner course of brick and furan mortar.

If the waste scrubbing liquids are borne by gutters or trenches, they are usually lined with a $1/4$ to $3/8$-in.-thick asphalt membrane (applied hot), inner-lined with acid brick bedded and bonded with a furan resin mortar.

Piping to and between these vessels may be made of vitrified clay, manufactured with bell-and-spigot ends. The procedure for joining the pipes is as follows: (1) trowel furan mortar onto the bell shoulder and the spigot end; (2) seat the spigot in the bell tightly against the shoulder; (3) caulk the annular space between the inside of the bell wall and the outside of the pipe shaft with asbestos (or other ceramic) fiber that has been saturated in the same furan mortar; (4) pour into the balance of the annular space molten sulfur mortar plasticized with 1.2% Thiokol. As an alternative to step (4), the balance of the annular space may be packed with a nonshrinking high-bond epoxy mortar. This latter approach is recommended if zero leakage is essential. It should be noted that preformed, pressure-fitted urethane, neoprene and other elastomeric joints can never be made liquid-tight; this is due to (a) out-of-roundness, (b) wall-thickness variations that are still within ASTM tolerances, and (c) the pressures (although slight) to which the line is subjected.

Manholes on waste pipelines are constructed and lined in the same manner as are the holding and neutralizer tanks, out of poured reinforced concrete; they are membrane-lined, and inner-lined with 4 in. of acid-resistant brick bedded and bonded with furan mortar.

Incinerators

Incinerator operating temperatures vary considerably, depending on the types of material to be incinerated, the fuel employed, and the needs of the user. Consequently, the associated pollution-control equipment must be engineered to meet the requirements of each installation, and no lining material can be considered standard.

Off-gas temperatures can range from 1,100 to 2,100°F. Because of this high temperature level, the gases are usually passed through a "conditioning chamber," from which they emerge at a temperature between 500 and 600°F. The gases then pass to the scrubber.

In many installations, such as municipal incinerators, an electrostatic precipitator or other device for removing large particles follows the conditioner and precedes the scrubber. In some designs, the precipitator will follow the scrubber, though a precipitator will, of course, eliminate only particulate matter, whether liquid or solid.

Most precipitators are designed without chemical-resistant linings. However, if one is to be installed in an incineration system immediately following the "conditioning" chamber, liquid droplets will certainly be found in some areas. Gases entering the precipitator are saturated with water vapor, and some cooling during residence must be anticipated. Probably the best internal lining for such a vessel is the foamed borosilicate-glass block described earlier. If the body of the precipitator is to bear an electrical charge, however, a different type of lining will have to be considered, perhaps one employing carbon or graphite staves, joined with a high-temperature mortar.

The scrubber can be of any type, but the venturi is very popular. As described previously, it consists of entry cone, throat, receiver, transition piece and separator, from which the scrubbed gases pass to the chimney.

The difference between this third system and those discussed in the previous two sections lies primarily in the entry duct and the additional "conditioner" vessel. It should be noted, however, that the liner construction for the scrubber may also vary slightly. Basically, conditions for exit duct and chimney will be similar to those in both the foregoing systems, and so the design and the lining selection are much the same.

The primary differences between this entry duct and those previously considered are the far higher temperature and the cyclical nature of operation. Although some incinerators operate continuously and others may be "down" only a few times a year, the majority will be idle perhaps once a week. This means, of course, gas condensation in the entry duct, with accompanying corrosion.

Under high-temperature operation, the use of special alloys, which can be quite costly, is indicated, unless a refractory insulating lining is provided to drop the temperature into the range acceptable for a less-expensive alloy or carbon steel. In addition, the presence of chlorides (HCl) is almost certain in incinerators, and while this is of little consequence with acid-resistant brick linings, it can have a serious effect on many of the less-expensive alloys. Lastly, the present interest in energy saving and the desirability of eliminating external insulation (which requires frequent maintenance) make the use of the combination of refractory brick, insulating brick and foamed borosilicate-glass block particularly attractive.

The "conditioner" could be one of many designs. One type is a large open chamber, with the gases entering under (or together with) a finely divided spray of water. Another is an extra-large, vertical section of duct directly over (and perhaps attached to) the entry cone of a venturi. In either case, the function of the equipment is only to reduce the temperature of the gas and not to scrub it. Therefore, the designer does not usually consider that this chamber will have wet walls or bottoms, though he usually provides a drain leading into the industrial sewer to provide for emergencies.

From what this author has seen of municipal incinerators, it is these chambers, usually designed by refractory engineers without the advice of those experienced in the uses of chemical-resistant masonry, that require the most frequent and continuous repairs in the system.

In most municipal operations, it appears difficult to assure inspection, to keep the spray heads in proper order and to ensure uniform pressure and operation of the water supply. As a result, the walls of these chambers are either seriously overheated and dry, or are flooded, or a combination of both, side-by-side. After a few months of operation, the bottoms are almost always wet, often to a depth of several inches of water.

These chambers should be redesigned to provide that the walls be protected by either a positive membrane such as a rubber sheet or a multilayer ceramic, as described earlier. In order to create an all-chemically-resistant structure, this multilayer ceramic might be covered by foamed borosilicate-glass block faced with layers of insulating and refractory brick that are bonded and bedded in high-temperature resin mortars in the covered layers, and in a potassium silicate mortar in the exposed layer (which, due to the mentioned poor control of the water supply, may require maintenance). The bottoms must be protected by the same membrane system, continuous from walls to floor, and this in turn covered by foamed borosilicate glass, and faced with acid brick that is bedded and bonded in high-temperature furan mortar.

When the conditioner is part of the scrubber complex and is directly over the entry cone, there is no bottom, and instead the streams of water from faulty spray heads run down the walls and merge with the scrubbing liquid, which enters at this point. Experience has shown that under these conditions the potassium silicate mortars require frequent maintenance, and a high-temperature furan will outperform them. In such designs, sheet rubber membranes, faced with a layer of foamed borosilicate glass protected with a single layer of acid brick, have performed well.

The same membrane system has been continued through the scrubber cone and venturi throat, protected by one or two layers of acid brick bonded with furan and omitting the foamed glass. Because of the variety of contaminants in the off-gases, the same protection has often been continued through the receiver and the separator. The use of troweled or gunned silicate-bonded ceramic linings alone has also been tried in such units, with mixed results. One unit is still operating successfully in its fourth year, but a second required considerable maintenance and subsequent facing with acid brick after only two months of service.

Expansion joints are usually necessary in chemical-resistant masonry linings in pits, trenches and sumps. Their design and detailing depends on the differences between thermal expansion coefficients of the various elements in the structure. They are also required so as to provide for the swelling that exists in all North-American-produced shale or fireclay acid-brick. Such swelling,

which is a non-uniform factor in design, occurs less in cool, dry installations, develops rather quickly in hot, wet, cycling exposures, and can account for as much as 0.16% growth of any dimension.

The designer must be particularly careful in mating brick internal linings to inlets and outlets. These should all be sleeved, with sleeves cemented into the linings. Differential movement within a vessel lining can shear the sleeves and cause lining failure. External movement can rupture the joint between sleeve and lining.

In open-body or packed scrubbers, where hot gases rise countercurrent to the scrubbing liquid, and the walls are continually wet, the gases are brought into the unit at the base and through a port in the wet wall. Care should be taken to angle the gas entry port slightly downward to prevent liquids from running into the port.

It is also important to form a shelf or "eyebrow" over the entry port to prevent cool scrubbing liquid from running down onto the hot bricks that line the entry port, causing them to spall under the thermal shock.

[When bricks are laid with the long edge parallel to the circumference of the vessel, this is called the "stretcher" position. When they are laid with the width parallel, this is called the "header" position. Large inlets are surrounded by a circle of bricks known as a "bullseye." To form a shelf or "eyebrow," the bricks surrounding the upper half of the bullseye are placed in the header position extending outward, while those surrounding the lower half of the bullseye are placed in the stretcher position.]

The designer should bear in mind that before making a firm material selection it is wise to contemplate any operating problems that may develop, and consider how these problems may affect the performance of the materials being evaluated. Also remember that some contaminants, such as HF, the salts of which are very water-soluble, are difficult or impossible to remove from the scrubbing liquid by simple neutralization. They will continue to concentrate as the scrubbing liquid is recirculated, until what started as a trace material of little consequence creates a serious hazard and causes extensive damage to some of the equipment.

References

1. Sheppard, Jr., W. L. 'Handbook of Chemically Resistant Masonry," Published by author, 923 Old Manoa Road, Havertown, PA 19083, 1977.
2. Paper by McDowell, Jr., D. W. and Sheppard, W. L., Jr., Use of Non-Metallics in Mineral Acid Plant Construction, presented at the Toronto meeting of NACE, April 1975.

The Author:

Walter Lee Sheppard, Jr., is a specialist and consultant in the selection and use of chemical-resistant masonry and is president of CCRM, Inc., 923 Old Manoa Road, Havertown, PA 19083, telephone (215) 449-2167. He has spent over 30 years in this specialized field, after receiving a B. Chem. from Cornell University and an M.S. in Chemistry from the University of Pennsylvania.

A member of the Natl. Soc. of Professional Engineers, the Natl. Soc. of Corrosion Engineers, the Assoc. of Consulting Chemists and Chemical Engineers and the Amer. Soc. of Testing and Materials, Mr. Sheppard is a registered professional engineer in Delaware and California.

Controlling fugitive emissions

EPA is now in the process of developing standards for the control of these emissions. Here is a summary of the agency's draft standards, along with some alternative methods proposed by industry. These standards include air-quality rules, as well as a control program of monitoring, reporting and maintenance.

Michael J. Wallace, Sandoz, Inc.

☐ Federal air-quality regulations have always implicitly included control measures for fugitive emissions, but, until recently, enforcement has been directed only at point-source emissions.

Now, however, the U.S. Environmental Protection Agency (EPA) has begun to concern itself with fugitive emissions.

EPA has written draft standards for controlling such emissions, and the agency is soliciting comments in order to issue a set of proposed rules and, eventually, standards. Here, we shall summarize the control methods described in the draft standards, as well as review alternative strategies for the control of fugitive emissions. Let us now:

1. Define fugitive emissions.
2. Estimate the degree to which these emissions contribute to the U.S.'s air pollution.
3. Review existing federal air-pollution regulations.
4. Summarize EPA's draft standards on monitoring, reporting and maintenance, along with some alternative methods.
5. Estimate industry's cost of a fugitive-emissions program.

Originally published August 27, 1979

What are fugitive emissions?

In contrast to a point-source emission such as that from a reactor vent or a boiler exhaust stack, a fugitive emission results from an equipment leak and is characterized by a diffuse release of volatile organic carbon (VOC) compounds or hydrocarbons into the atmosphere. These hydrocarbons are non-methanes (in the C_2 through C_6 range) and are generally photochemically reactive—that is, they are precursors to oxidants.

According to EPA's proposed generic standard, a fugitive emission is a leak that registers over 1,000 ppm on a hydrocarbon analyzer at the emission source. Point-source emissions can be measured quantitatively, since both concentration and volumetric flowrate are readily monitored in a confined system, while fugitive emissions, in contrast, being essentially leaks, are not confined in space. These leaks result in a diffuse discharge that is difficult to measure quantitatively.

(Methods of monitoring fugitive emissions will be discussed later on. We will not consider fugitive emissions of particulates, because EPA's main thrust is on controlling hydrocarbons.)

Amount of fugitive emissions

Table I shows sources of hydrocarbon emissions (including fugitive emissions), compiled from EPA [1] data.

For 1975, petroleum processes (refineries) accounted

U.S. hydrocarbon emissions declined in the early 1970s, probably due to pollution rules		Table I

Source	Emission rate million metric tons/yr	
	1974	1975
Transportation	11.3	10.6
Stationary fuel combustion	1.6	1.3
Petroleum processes	4.5	4.6
Chemical processes	9.7	9.0
Other industries	0.9	0.9
Solid waste	0.9	0.8
Forest and agricultural burns	0.9	0.9
Totals	**29.8**	**28.1**

Totals	1970	1971	1972	1973
Million metric tons/yr	30.7	30.2	30.9	30.8

Compiled from Ref. [1].

for about 16% of all hydrocarbon emissions. Chemical processes—petrochemical plants and synthetic-organic-chemical manufacturing industry (SOCMI) plants—accounted for some 32% of total emissions. And transportation (vehicles) produced approximately 38%.

There was about a 9% decline in total emissions from a peak of 30.9 million metric tons/yr in 1972, to 28.1 million m.t./yr in 1975. (Data for 1975 through 1978 were not available to the author at the time of writing this article.) The relatively large decrease of 9% is more than a standard statistical deviation, and it appears that a net decrease in hydrocarbon emissions has been brought about by air-pollution-control efforts.

As a comparison, EPA [2] reported that the total yearly hydrocarbon emission rate due to the natural degradation of vegetation and leaf litter is about 11 million m.t./yr. This equals nearly 40% of total yearly manmade hydrocarbon emissions as reported for 1975. How much of these manmade emissions are fugitive emissions?

There are no direct measurements or estimates of total fugitive-hydrocarbon emissions in the U.S. However, we can make a rough order-of-magnitude guess. Kremer [3] found that, depending on the type of plant, about one third of all air-pollutant emissions are due to leakage from sealing elements such as valves, pump seals and flanges. Assuming that these data apply universally, then, using EPA's 1975 figures, about 4.5 million m.t./yr of hydrocarbons result from fugitive emissions in the petroleum and chemical process industries (CPI). This corresponds to approximately 16% of all non-natural hydrocarbon emissions in the U.S., or, as stated above, about 33% of all hydrocarbons in the petroleum refining and chemical process industries. Even as only an order-of-magnitude estimate, this relatively high percentage of total emissions underscores the growing concern over fugitive emissions.

U.S. air-quality rules

The 1977 Amendments to the Clean Air Act of 1970 have given EPA power to implement broad control over all aspects of industrial air pollution. Most critical to chemical engineers are those restrictions placed on

modification or expansion of existing facilities, or on construction of new ones.

The 1977 Amendments require that any industry proposing a new or modified major source of air pollutants must file for a New Source Review, so as to obtain a construction permit. A major source is defined as one that, without any control device, could potentially emit 100 tons/yr of pollutant. This is a small quantity, amounting to 23 lb/h for a source in continuous operation 365 days a year. Under this regulation, many facilities that had once escaped air-pollution control regulations are now required to apply for a New Source Review, and it may require from 6 to 42 mo to obtain a permit.

To determine the amount of uncontrolled emissions, one must consider all air-pollutant emissions upstream of any control device, such as a baghouse, scrubber or carbon-absorption unit. Any device, such as a reflux condenser, that is essential to the operation of an industrial process is not considered a control device. In this case, uncontrolled emissions are determined downstream of the device.

Fig. 1 shows a reactor with reflux- and after-condensers. There are two possible points, A or B, for evaluating potential air contaminants. Point B is between the reflux- and after-condensers, and Point A is downstream of the after-condenser. During normal reflux operation, the reflux-condenser is necessary in order to return solvent to the reactor, and the after-condenser is used primarily as an air-pollution control device to remove any remaining condensables from the reflux-condenser vent line. In this case, Point B would be used in determining the potential uncontrolled air-pollutant emission rate.

However, the point at which potential emissions are determined is now being decided in a federal court. The court may rule that the point should be downstream of any control device. Such a decision would, of course, benefit industry, especially the smaller-volume SOCMI firms. This is because this decision would significantly raise the level that defines a major source.

If a plant modification, expansion or new facility is taken as one source, then fugitive emissions and the applicable control methods for them will have to be considered in determining the potential air-pollutant emissions.

If a New Source Review is required, then it must be determined whether the facility is in what is called a nonattainment area. Such an area is one that does not meet the federal National Ambient Air Quality Standards (NAAQS) for ambient levels of either sulfur dioxide, carbon monoxide, hydrocarbons, nitrogen dioxide, particulate matter or photochemical oxidants. In such areas the planned production facility must proceed with what is termed the nonattainment process, the first step of which is application of control technology known as the Lowest Achievable Emission Rate (LAER).

Guidelines for LAER have not yet been developed. It assumes pollutant control by the best available technology, without considering the cost. Again, this refers to both point-source and fugitive air-pollutant emissions.

After estimating the possible reduction in emissions by applying LAER, it is determined whether a detailed

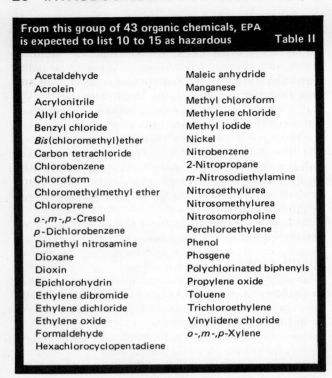

From this group of 43 organic chemicals, EPA is expected to list 10 to 15 as hazardous   Table II

Acetaldehyde	Maleic anhydride
Acrolein	Manganese
Acrylonitrile	Methyl chloroform
Allyl chloride	Methylene chloride
Benzyl chloride	Methyl iodide
Bis(chloromethyl)ether	Nickel
Carbon tetrachloride	Nitrobenzene
Chlorobenzene	2-Nitropropane
Chloroform	*m*-Nitrosodiethylamine
Chloromethylmethyl ether	Nitrosoethylurea
Chloroprene	Nitrosomethylurea
o-,*m*-,*p*-Cresol	Nitrosomorpholine
p-Dichlorobenzene	Perchloroethylene
Dimethyl nitrosamine	Phenol
Dioxane	Phosgene
Dioxin	Polychlorinated biphenyls
Epichlorohydrin	Propylene oxide
Ethylene dibromide	Toluene
Ethylene dichloride	Trichloroethylene
Ethylene oxide	Vinylidene chloride
Formaldehyde	*o*-,*m*-,*p*-Xylene
Hexachlorocyclopentadiene	

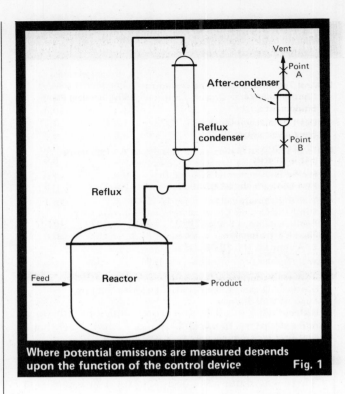

Where potential emissions are measured depends upon the function of the control device   Fig. 1

New Source Review is necessary. If not, then the proposed facility may apply for a permit to construct. If a detailed review is required, then the facility must conduct a program of ambient and background air-quality monitoring and mathematical modelling to see if NAAQS are met. If so, then the facility may proceed with a Prevention of Significant Deterioration (PSD) review, which is outlined below. If not, the facility must apply control-technology offsets and proceed with additional mathematical modelling.

If the facility is to be located in an attainment area, then a PSD review is required. In this program, one applies control technology to meet New Source Performance Standards (NSPS). These standards can be specified as emission limitations, control-equipment specifications, or work-practice standards. Next, it must be determined whether applying NSPS reduces potential emissions to less than 50 tons/yr. If so, then a permit can be sought. If not, several additional steps may be required, including the monitoring of local air quality for one year.

The procedures for permit application are more complex than indicated here. A detailed analysis can sometimes be obtained from an in-house environmental staff, or, if not, from an environmental consultant.

Clean Air Act

Three sections of the Clean Air Act provide EPA with a basis to make rules for the control of fugitive emissions:

■ *Sec. 110* deals with *control technology guidelines*. Under this section, EPA is required to develop pollutant emission-control technologies for typical industrial facilities such as refineries or petrochemical plants. Development is based, in large part, upon data collected from monitoring surveys conducted by EPA or its designated con-

sultants/contractors. Methods of monitoring and results of some specific major fugitive-emission studies will be discussed later on.

■ *Sec. 111* deals with *work-practice standards* for pollutant emission control. The theory and application of workplace standards will be discussed in detail in this report. This section also deals with the development of New Source Performance Standards. EPA has issued a 59-item list for the selection of NSPS projects by priority and by industry type. Synthetic organic chemical manufacturing is first on this list, followed by: No. 2—industrial surface coatings: cans; No. 3—petroleum refineries: fugitive sources; No. 4—industrial surface coatings: paper; and No. 5—dry cleaning.

■ *Sec. 112* deals with the development of *hazardous air-pollutant standards*. All regulations under Sec. 111 and 112 are scheduled to be issued by Aug. 7, 1982, as required by the Clean Air Act. Sec. 112 calls for the development of two regulations in regard to fugitive emissions: One deals with hazardous organic pollutants and the other with general organic pollutants. EPA is required to list, as necessary, such hazardous pollutants and to establish standards that provide an ample margin of safety to protect public health. Generic standards (treated next in this article) will be proposed when organic chemicals are listed as hazardous air pollutants. A list of 43 chemicals is being considered (see Table II). From this list, 10 to 15 chemicals will probably be listed as hazardous.

Proposed generic standard

We will now consider the generic standard that EPA is proposing for fugitive-emission control and compare it with the work practice standard being proposed by the Chemical Manufacturers Assn. (CMA), formerly the Manufacturing Chemists Assn., a Washington, D.C.,

EPA has set this schedule for issuing proposed fugitive-emissions guidelines and regulations   **Table III**

Control Techniques Guideline Document
    Control of Volatile Organic Emissions from Chemical Plant Equipment Leaks
    Draft (industry review)—May 1979
    Final—September 1979

National Emission Standards for Hazardous Air Pollutants
1. Refinery Fugitive Benzene Sources
    Pacific Environmental Services, Inc.
    EPA contract No. 68-02-3060
Federal Register promulgation—early 1981
2. Synthetic Organic Chemical Manufacturing Industry Fugitive Benzene Sources
    Pacific Environmental Services, Inc.
    EPA contract No. 68-02-3060
Federal Register promulgation—late 1981

New Source Performance Standards
    Synthetic Organic Chemical Manufacturing Industry
    Fugitive VOC Sources
    Radian Corp.
    EPA contract No. 68-02-3058
Federal Register promulgation—mid-1981

trade organization composed of major manufacturers of industrial chemicals.

CMA's proposed standard is summarized because it will influence EPA in setting a final standard. This is because the agency has been listening more to industry lately in setting policy.

EPA's projected schedule for fugitive-emissions guidelines and regulations is presented in Table III. In addition to the draft control-techniques-guideline document for the control of volatile organic emissions (VOC) from chemical-plant equipment leaks, it is assumed that draft documents for emissions and performance standards will also be available for comment prior to final promulgation of the regulations.

As stated in the December 1978 EPA draft standard, "Generic standards are standards which may be applied universally to similar emissions sources." These are independent of chemical type or process technology and are based on the similarity of processing equipment, particularly secondary equipment, such as valves and pump seals. As proposed by EPA, compliance would be affected by (1) a program of fugitive-emission detection, (2) correction of fugitive emissions by closer maintenance controls, and (3) good housekeeping practices.

The generic standards have to be implemented quickly, and with a minimum of capital. Also the standards have to remain essentially unchanged, even if additional other standards are promulgated. This is not only to keep costs down, but to expedite implementation and minimize retrofitting. Therefore, since direct quantitative monitoring of fugitive emissions is impractical, the new standard has to be centered on equipment or work-practice standards.

In fact, the draft generic standard contains an effective leak-detection program, with provisions for subsequent repair and maintenance. The generic standard requires different types of emission sources to be in-spected at various predetermined intervals for subsequent repair. The intervals will be established from data collected in surveys of refineries and petrochemical plants, and will vary from weekly to annually, depending on the frequency of the particular types of leaks and their projected emission rates. Inspections will be the most frequent for sources having a high probability of leakage and a likelihood of a large emission rate. The EPA proposed inspection intervals for compressor seals, pipelines, and pressure-relief valves are 1 to 3 mo. For pump seals, valves in liquid service, process drains and flanges, the rate is 3 to 12 mo. These ranges vary, depending upon the particular situation. So far, EPA has not defined these situations. More-specific inspection criteria will be treated later in this article when we discuss individual emission sources.

Cutoff range

To reduce the potential number of pieces of equipment that require inspection, EPA has proposed a lower cutoff range that excludes process streams carrying trace quantities of hazardous pollutants. The proposed lower limit is 1 to 10% by volume, depending upon the nature of the hazardous chemical constituent. This range is based upon results of recent engineering studies of refineries and petrochemical plants.

Using EPA's definition of a fugitive emission as one that produces a hydrocarbon concentration of at least 1,000 ppm at the source, the lower hazardous-pollutant concentration of 1% in the process stream would mean that the leak concentration would be limited to 10 ppm. Similarly, a lower cutoff of 10% would permit a hazardous-pollutant concentration of 100 ppm.

As stated in the draft generic standard, processing equipment containing concentrations at or above the cutoff levels would be inspected to determine whether gaseous or liquid leaks were occurring. If a gaseous leak occurred and the concentration were greater than 1,000 ppm (converted to an equivalent basis of hexane) when using a standard portable hydrocarbon detector, then repair would be required within a specified period of time.

This proposed time limit is between 5 and 15 days, depending upon the nature of the leaks. For example, more time is required when certain repairs are needed, as is the case of a damaged single mechanical seal. If repair of a fugitive-emission source prior to a plant shutdown would result in still greater emissions, then repair could be delayed until the shutdown. The delay, however, would be subject to review and approval by EPA.

The above summary noted the more salient features of the draft generic standard. The actual draft goes into much more detail, including extensive test methods and instrument-calibration techniques.

CMA proposed standard

Major portions of industry have taken exception to the proposed draft generic standard. Recently, the Chemical Manufacturers Assn. has proposed its own standard—a work-practice standard—for the chemical industry. This includes all the fugitive-emission sources listed in EPA's draft generic standard, but differs in the particulars of leak detection, monitoring, repair, recordkeeping and reporting. In general, CMA has proposed that:

■ Each facility develop its own leak detection and repair plan based upon EPA's criteria.

■ Each facility maintain a log of when leaks are detected and repaired.

■ All leaks should be repaired as soon as feasible.

■ An annual report should be submitted to EPA.

■ Weekly visual checks and operational checks must be made every time there is a transfer process.

■ Quarterly monitoring checks should be made on equipment most liable to leak.

■ Measures should be included to allow modification of the plan, as required, once results based on new studies or experience are available.

CMA intends to use EPA's proposed standard as a basis for its own standard. However, CMA feels that each facility should be allowed to develop its own cost-effective detection and control program, based on its own experience. Whatever form the EPA guidelines finally take, they should be flexible enough to allow a company to depart from them as long as such a departure is justified by the facility's past experience.

CMA's proposed work-practice standard applies to hazardous air pollutants as treated under Sec. 112 of the Clean Air Act. The standard will be applied for fugitive emissions of a hazardous air pollutant in cases where the EPA administrator, after a thorough study, determines that the work-practice standard is necessary to control the fugitive emission.

Further, the standard will be limited to facilities that process, store or transfer fluid mixtures containing 10% or more, by volume, of a hazardous air pollutant. The standard is not applicable to any fugitive-emission source that does not emit a hazardous air pollutant into the ambient air, nor does it apply to research and development facilities that process less than 5,000 lb/d of a hazardous chemical.

An essential part of the proposed CMA standard is the development of a Leak Detection and Repair Plan (LDRP) by each facility that processes hazardous air pollutants as described under Sec. 112. The facility shall adhere to the LDRP and keep a copy of it. The LDRP must also be certified by the owner of the facility as meeting the fugitive-emission criteria as established by EPA under the Clean Air Act.

The plan is to be reviewed and updated, as required, once every three years, or within 90 days of a major modification of the facility.

The LDRP shall, at the minimum: (1) develop a schedule and recordkeeping program for routine surveillance and/or monitoring of fugitive emissions; (2) establish a written plan for detection and repair of leaks, and a reasonable schedule for repair; (3) provide a written plan for sampling procedures, housekeeping (e.g., spill cleanup) and onsite waste handling; (4) develop recordkeeping procedures for all aspects of the LDRP and save these records for one year; and (5) establish a written plan for specifying sufficient personnel to fulfill the LDRP, and provide a training program with a written manual.

CMA's proposed standard also contains a detailed

Portable organic-vapor analyzer used extensively in making individual source surveys for EPA   Fig. 2

Century Systems Corp.

description of how an LDRP should be prepared, and includes discussions of housekeeping and sampling, chemical storage, chemical transfer and handling, waste handling, process vents, air-pollution control devices, and administrative procedures.

Having dealt exclusively thus far with the regulatory aspects of fugitive-emission control, we now turn to the more technical areas of concern. We will first look at monitoring and reporting, then at individual sources of fugitive emissions, and then at costs of these programs.

Monitoring fugitive emissions

There are several methods for detecting fugitive emissions and thereby enabling their control. These include source monitoring, area monitoring, and fixed-point monitoring. Only source monitoring has been investigated enough to establish its effectiveness. EPA has suggested that the use of a combination of monitoring techniques may be effective in locating large leaks that result in the more-significant releases of fugitive emissions in a plant.

Methods other than source monitoring are presented here, in part, in case there is a desire to develop data and to compare these with EPA's.

In evaluating the various monitoring techniques, a fundamental axiom of fugitive-emission detection and control is that a small number of sources contribute the majority of fugitive emissions. This will be discussed in greater detail later when we cover the sources of such emissions.

Area monitoring requires that an instrument operator, using a portable VOC analyzer with a stripchart recorder, follow a predetermined path through a process unit. VOC emissions are measured at a given distance, for example, 3 ft, from all equipment within the area. The operator makes a complete survey around each piece of equipment, being careful that the upwind and down-

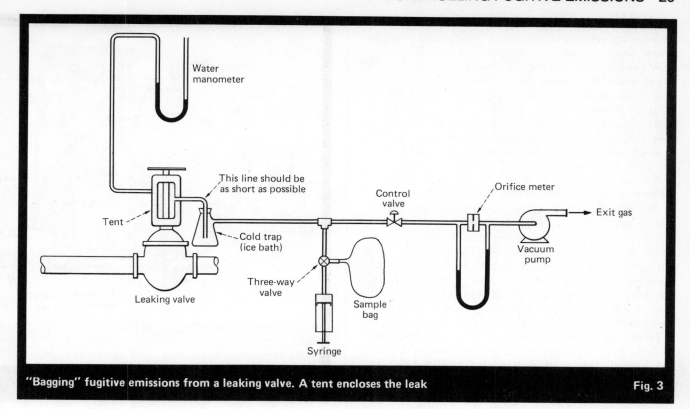

"Bagging" fugitive emissions from a leaking valve. A tent encloses the leak   Fig. 3

wind sides of the equipment are sampled. If a peak in concentration is observed, the location is noted, and a subsequent, more-detailed survey is made to locate the source.

It is estimated that as much as 50% of all significant leaks in the process unit can be detected by the walk-through unit-area survey, if all major leaks in the area have been repaired via an ongoing monitoring process, and if the plant has reduced voc background concentrations.

An advantage of this type of survey is that leaks from accessible equipment can be located quickly, and with considerably less manpower than required for an individual component survey. Disadvantages include possible detection of voc emissions from adjacent units, and low or missed voc readings due to high winds or gusts that dilute emissions. In addition, followup source monitoring is still required to determine the exact location of the source.

EPA considered continuous area monitoring to detect gaseous leaks of hazardous organic compounds—but found it impractical. The main reason for rejecting this method was that continuous monitoring equipment for measuring ambient-air concentrations of specific hazardous organic air pollutants would most likely not be available by the time that the particular air pollutant was identified as hazardous.

Fixed-point monitoring places analyzers at specific fixed points in a process area to monitor automatically for fugitive emissions. voc is monitored both locally and centrally, so that action can be taken when elevated concentrations are observed. Fixed-point monitoring is used only in those areas where voc compounds are handled.

EPA says that such monitoring may also be used as a maintenance tool to detect pending or existing equipment failures that can result in leaks. In this method, the individual samplers are placed either near specific pieces of equipment that handle vocs or in a grid pattern throughout the process area. If a concentration peak is observed, then the operator performs an individual component survey to locate the leak.

A major advantage of fixed-point monitoring is that it readily adapts to a particular processing facility's specific needs. Another advantage is that it has the lowest manpower requirement of the fugitive-emission monitoring systems.

Some of the disadvantages are that this system has the highest capital cost and still requires the use of a portable voc analyzer to locate the leak, especially if the process-area method is used. If background voc concentrations have been minimized by a proper monitoring, detection and repair program that has corrected major leaking equipment, this method is estimated to detect one third of all leaks.

Source monitoring detects leaks by checking individual sources—valves, flanges and seals, among others. A portable voc detector is used in this method. The sample probe of the instrument is placed at a point of a particular pipeline component where leakage could occur. The probe is moved along the surface with particular care that both upwind and downwind areas are sampled. The probe should traverse the source within one centimeter of the surface.

For sources having leak areas open directly to the atmosphere, such as cooling towers, process effluent drains, and pressure-relief valves, the probe is placed at both the center of the emission area and around its periphery. A more-detailed description of the fugitive-emission survey protocol for each type of source is presented later on under the discussion of individual emission sources.

For measuring the organic vapor concentration in the collected sample, EPA has made extensive use of the Century Systems Corp.'s (Arkansas City, Kans.) Model 128 organic vapor analyzer (see Fig. 2). This instrument uses a flame-ionization detector that is accurate to 1 ppm of organic vapor, and the unit gives a direct readout with a 2-s response time.

Other source-monitoring methods

Several other methods of fugitive-emission source-testing have been used in Europe and in the U.S. Kremer [3] describes four ways to evaluate such losses under laboratory-type conditions. The first is the pressure-drop method, in which an isolated and completely sealed unit is pressurized, the source gas turned off, and the drop in static pressure noted. This method is useful mainly for detection of leak rates of whole sections of lines, including flanges, valves and other fittings.

In the pressure-retention method, pressure drop due to leakage is compensated for by introducing a source gas, and the rate at which gas must be added to bring the fitting back up to pressure is recorded.

The capsule method measures leakage rates of individual components, such as flanges and seals, both in the laboratory and in the field. In this method, the component is enclosed in a globe, and a flush gas is circulated through the globe in a closed loop by means of a pump. A hydrocarbon detector measures the concentration of the fugitive gas in the flush gas. This method is reliable, but too much time is needed for setting it up.

The fourth method—in more-common use in the U.S.—is to enclose the component in a plastic bag ("bagging"). Emissions are captured and then measured volumetrically. A typical setup is shown in Fig. 3. The sampling train operates under a slight negative pressure. The fugitive emission is exhausted from a polyester-film bag or tent that has been placed around the emission source. Fig. 4 shows the tent construction.

In Fig. 3, the captured fugitive gas enters a cold trap, where heavier hydrocarbons and water condense to prevent fouling of the sampling train. From there, the gas is drawn past a sampling syringe by means of a vacuum pump. An orifice meter measures the rate of flow of the fugitive gas. The syringe is filled with a measured volume of the gas, and the water manometer measures pressure to allow conversion to standard conditions (when the sample is being analyzed for VOC concentration). By means of a three-way valve, the gas sample in the syringe is transferred to a tetrafluoroethylene sample bag. The sample VOC concentration is measured by standard analytical means and, knowing the gas flowrate in the system, the mass emission rate of the fugitive emission is calculated.

Before samples are taken, ambient air is drawn through the tent for a specified time in order to equilibrate the volume inside the tent. At the same time, the background VOC concentration is determined in the area immediately around the tented component. If the

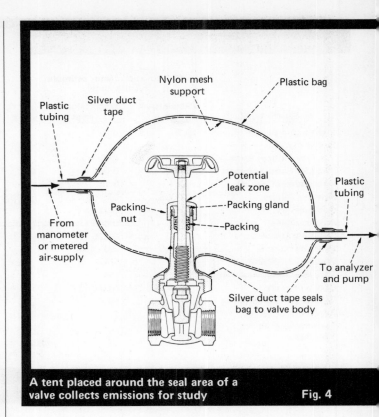

A tent placed around the seal area of a valve collects emissions for study   Fig. 4

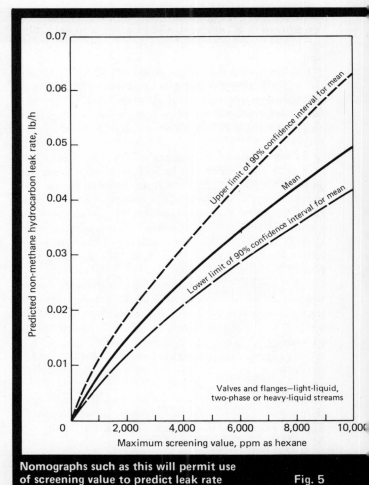

Nomographs such as this will permit use of screening value to predict leak rate   Fig. 5

EPA drafted this standard for oil refineries, based upon the use of screening values		Table IV

Source type	VOC emission factor estimate lb/(h)(source)	
	Attainment areas	Nonattainment areas
Light-liquid pumps	0.26	0.097
Pipeline valves		
a. Gas/vapor streams	0.046	0.007
b. Light-liquid streams	0.022	0.009
Pressure-relief devices	0.35	0.15
Pipeline flanges	0.00057	0.00057
Compressors	0.97	0.277
Process drains	0.070	0.046
Sampling connections	0.033	0.026
Open-ended lines	0.0070	0.0
Wastewater separators	NA	NA
Vacuum-producing systems	NA	NA
Cooling towers	Negligible	Negligible
Process-unit turnarounds	NA	NA
Accumulator vents	2.71	2.71
Pressure-relief-device discharges	NA	NA

NA—No factor available

background concentration is too high (greater than 10 ppm), a pressurized sampling train can be used.

Screening values

Since EPA intends to add fugitive emissions to point-source emissions in order to determine whether air-pollution regulations are being met, a quantitative measure of these leaks is needed. Such a measure has been developed by Radian Corp. (Austin, Tex.) [4] for EPA.

Radian studied fugitive emissions at nine petroleum refineries. One of the main objectives of its study was to establish a relationship between the screening values of fugitive-emission sources and quantitatively measured leakrates from the sources.

Screening values are defined as the maximum hydrocarbon concentration around a fugitive-emission source, as measured by a portable voc analyzer. The quantitative measurements were obtained from those same sources by "bagging" them as shown in Fig. 4, and measuring their mass fugitive-emission rates by means already described under the source-monitoring section of this article. These "baggable" sources include process valves, flanges, pump seals, compressor seals, relief valves and process drains.

The report develops nomographs that correlate mass emission rate of non-methane hydrocarbons with maximum screening values (in ppm as hexane).

Pump seals, inline valves and relief valves are among major fugitive-emission sources						Table V

Fugitive-emission source	Emission rate—lb/h					Percent leaking Ref. 4
	Ref. 6	Ref. 7	Ref. 8	Ref. 9	Ref. 4	
Valves						
Gas/vapor streams	0.02	0.062	0.082	0.046	0.047	29
Light-liquid/2-phase streams			to	0.022	0.023	37
Heavy-liquid streams			0.085	—	0.0007	7
Flanges						
Gas/vapor streams		0.0004	0.1		0.0005	3
Light-liquid/2-phase streams		0.004	to	0.0006	0.0005	5
Heavy-liquid streams		—	1.1		0.0007	2
Pump seals						
Light-liquid streams	0.11		0.16	0.26	0.26	64
Heavy-liquid streams			to		0.045	23
			0.48			
Compressor seals						
Hydrocarbon service	0.064		0.077	0.968	0.98	70
Hydrogen service			to		0.10	81
			0.65			
Relief valves						
Gas/vapor streams			0.031		0.36	46
Light-liquid/2-phase streams		0.121	to	0.352	0.013	25
Heavy-liquid streams			0.4		0.019	35
Sampling connections				0.033		
Process drains						
Light-liquid/2-phase streams			0.15	0.0704	0.085	26
Heavy-liquid streams					0.029	18
Rotary mechanical seals on pumps and compressors		0.13				
Packing seals on piston compressors		0.23				

A typical nomograph, as developed in the Radian report, is presented in Fig. 5. The nomograph was derived for the mean and for the upper and lower 90%-confidence intervals around the mean. Radian developed these nomographs not only for different fugitive-emission sources, but also for different types of process streams within a source-type category. For example, there is one nomograph for valves and flanges carrying gas or vapor streams, and another for light-liquid, two-phase or heavy-liquid streams. These nomographs approximate mass emission rate of a fugitive emission source by measuring the non-methane hydrocarbon concentration. Once this relationship is established for a given source, it may be used to determine the maximum concentration that corresponds to a maximum allowable fugitive-source mass-emission rate, as determined by an NAAQS.

Using such data, EPA will likely set some limits on mass emissions from fugitive sources, at least in guideline form. An example of this: the EPA draft baseline VOC emission factors for refineries shown in Table IV. This table gives regulatory baseline fugitive-emission factors for all refinery sources in both NAAQS attainment and nonattainment areas. For refineries in nonattainment areas, regulations following the EPA guidelines will be enforced.

Petroleum refineries in attainment areas will most likely operate with uncontrolled fugitive emissions unless a hazardous-pollutant standard is established under Sec. 112 of the Clean Air Act.

Rather than measure quantitatively the suspected fugitive-emission sources within a chemical production facility in order to determine compliance or non-compliance, it is being proposed to establish maximum concentrations by means of these nomographs.

The Radian report also establishes a set of curves correlating percent of fugitive-emission sources with maximum screening value. There are also curves correlating maximum screening values with percent of total mass emissions. These curves indicate that, in most cases, screening values above 1,000 ppm account for 50% to 90% or more of all sources of emissions. EPA is using these data to establish what it believes is a practical screening value limit that will help control the majority of emissions.

Monsanto Research Corp.'s Dayton Laboratory (Dayton, Ohio) [5] studied fugitive emissions in four chemical plants producing monochlorobenzene, butadiene, ethylene oxide/glycol, and dimethylterephthalate. Hydrocarbon fugitive-emission rates were determined for pumps, valves, flanges, process drains, compressors, agitators, sample valves, and relief devices. Monsanto used the Century Systems Corp. Model 128 organic vapor analyzer.

Their data show a significant variation in leakage rates with petrochemical process, for the same source. For example, when processing monochlorobenzene, pumps had five times the leakage rate of valves, and the flange leakage-rate was 29 times that of the pumps (0.2, 1.0, and 29 m.t./yr for valves, pumps and flanges, respectively). Yet when processing butadiene, valves had 10 times the fugitive-emission rate of pumps, and flanges showed no leakage (1,000, 96, and 0 m.t./yr for

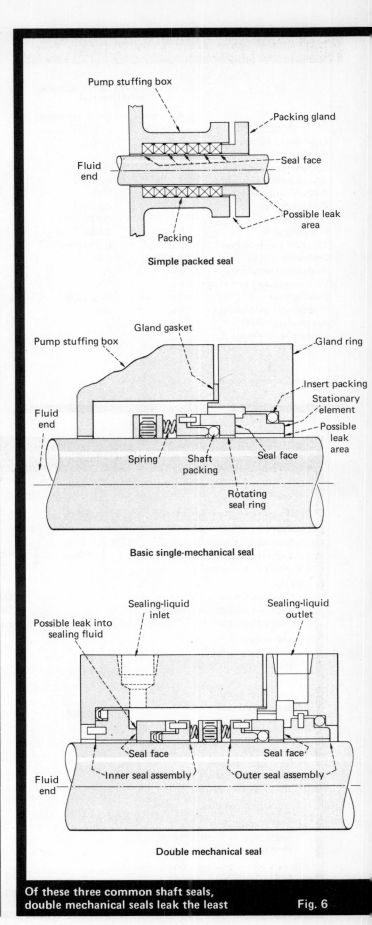

Simple packed seal

Basic single-mechanical seal

Double mechanical seal

Of these three common shaft seals, double mechanical seals leak the least

Fig. 6

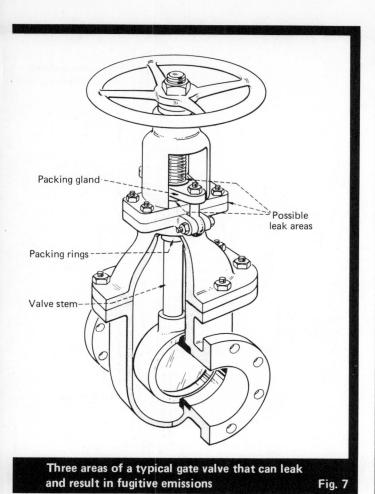

Packing gland

Possible leak areas

Packing rings

Valve stem

Three areas of a typical gate valve that can leak and result in fugitive emissions Fig. 7

this system, there are two possible leak areas: between the rotating shaft and the packing gland, and between the packing gland and the stuffing box.

The single mechanical seal in Fig. 6 reduces the possibility of leakage and limits fugitive emissions to one source point—the area between the seal's stationary element and the rotating shaft. The source of the leak is, of course, the seal face between the rotating seal ring and the seal's stationary element.

The double mechanical seal, also shown in Fig. 6, practically eliminates leakage by utilizing an externally pressurized liquid to seal the rotating and stationary seal elements. The only possibility of leakage is into the sealing liquid. With a properly chosen liquid, leakage can be controlled by absorption or reaction, with subsequent disposal of a purge stream.

It is a common fact that reciprocating pumps and compressors leak more than centrifugal units. This is because reciprocating units use packed seals, while centrifugals employ mechanical seals. Studies have shown that the leakage rate of mechanical seals is 50% less than that of packed seals.

Packed pump-seals account for approximately 5% of all fugitive emissions. Only 9% of all pumps account for this 5%-figure. The most effective method of reducing pump-seal emissions is to replace the packed seal with a mechanical seal.

According to EPA [1], the average uncontrolled emission rate, equal to 5 lb/d, can be reduced to 3 lb/d by seal replacement.

To perform an individual-source survey for detection of fugitive emissions from pump seals, EPA proposes the following: The probe of the portable VOC analyzer is traversed circumferentially around the outer periphery of the sealing element, i.e., the seal housing interface and the pump shaft. The probe should be held as closely as possible to the sealing element. When this cannot be done, because of physical restrictions, the probe should be placed within 1 cm of the interface between the shaft and seal.

In addition, any other possible points of fugitive emissions on the pump should be surveyed. When external seal fluids are employed, the seal fluid reservoir should also be surveyed.

Process valves—According to various literature sources, these account for about 75% of total fugitive emissions, or about 23% of total hydrocarbon emissions in the refinery and chemical process industries. Also, EPA surveys show that about 4% of all valves account for approximately 70% of total fugitive emissions. Clearly, then, industry should concentrate its efforts on valves to reduce fugitive emissions.

Fig. 7 shows a gate valve. The possible leakage areas are around the stem packing, the bonnet assembly and between the valve stem and packing gland.

In an analysis of leakage rates of the various valve types, Bierl, et al. [7] found that leakage rate varies with the type of packing used. In order of decreasing unit fugitive-emission rate, piston compressors (packed seals), centrifugal compressors (mechanical seals), safety-relief valves, and gate and control valves gave leakage rates of 42%, 24%, 21% and 10%, respectively, of total emissions for the source types listed. Ball valves

valves, pumps and flanges, respectively). This large variation is of concern to the CPI, because it poses a potential stumbling block to the setting of a maximum screening value for monitoring and detection of fugitive emissions.

The report also concludes that, for significant fugitive sources, average emission rates in petrochemical plants are markedly lower than those for corresponding sources for petroleum refineries.

Monsanto Research found there are also fewer fugitive-emission sources in the petrochemical plants than in petroleum refineries.

Monitoring individual sources

Having reviewed parts of the Clean Air Act that deal with fugitive emissions and gained an insight into EPA's proposed monitoring methodology, let us consider in detail the individual fugitive-emission sources. Table V presents leakage rates of various major sources. These rates are culled from various references.

Data from Ref. *4, 6* and *9* were obtained from surveys of petroleum refineries; while Refs. *7* and *8* are from petrochemical plants. Now, we will consider the individual sources:

Pumps with packed seals—Before discussing monitoring techniques for these, consider the possible leak areas of the packed seal, single mechanical seal, and double mechanical seal. Fig. 6 shows the simple packed seal. In

and flanges resulted in unit fugitive-emission rates of about 1% and 0.5%, respectively.

Referring to the influence of packing on leakage, Bierl lists the following packings in order of decreasing leakage: string, closed compressed rings, polytetrafluoroethylene (PTFE) packing, and PTFE lift rings. Regarding the effect of age on fugitive emissions from valves, Bierl pointed out that there was a five-fold increase in rate after 2 years. Also, more frequent use of a valve will result in earlier packing failure. In addition, leakage can be reduced by 90% through maintenance.

To perform an individual source survey of valves, EPA suggests placing the analyzer probe at the interface of valve stem and housing, and traversing the probe around the periphery of this interface. If other parts of the valve, such as those in a multi-piece bonnet construction, are suspected as emission sources, these should be surveyed in a similar manner.

Safety-relief valves—These are said to account for 10% of total VOC fugitive emissions.

Fig. 8 shows such a valve. Relief valves leak when subjected to repeated overpressures. In addition, the valve seat may become fouled, preventing the valve plug from closing properly, and this allows process gases and vapors to escape.

EPA proposes two methods to control fugitive emissions from relief valves: One is to vent all such devices to a flare system, which will burn up any fugitive emissions. A second method is to place a rupture disk just upstream of the relief valve to prevent its becoming fouled from normal pressure fluctuations. A pressure indicator placed between the rupture disk and the relief valve permits ready detection of a ruptured disk. EPA's suggested method for individual source surveying of these valves is to place the hydrocarbon detector probe at the center of the relief-valve exhaust horn.

Open-ended valves—These can leak from two sources: The first is the valve stem, as discussed under inline valves. The second area is at the valve seat, where leaks go directly into the atmosphere. EPA's recommended procedure for open-ended valves is to add either a second valve as backup or to place a blind flange on the atmospheric side of the valve. The suggested individual-source survey method is to place the hydrocarbon analyzer probe at the center of the uncovered opening to the atmosphere.

Sample valves—To reduce fugitive emissions from these devices, EPA proposes bleeding the sample into a closable container. The sample valve must remain closed whenever samples are not being drawn. Any samples of gases or vapors bled out of the system must be captured and either returned to the process or destroyed in an environmentally acceptable manner.

Less-significant sources

Process piping flanges—These account for only 4% of all fugitive emissions. For flanges, the emission rate increases as gasket size gets larger, and as gasket surface pressure also increases. A three- to tenfold increase in leak rate can be expected if good maintenance practices are *not* followed. To detect fugitive emissions from flanges, EPA suggests that individual-source surveys be made by placing the probe at the outer edge of the

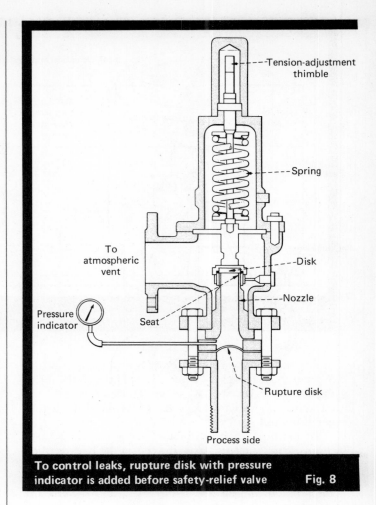

To control leaks, rupture disk with pressure indicator is added before safety-relief valve   Fig. 8

flange/gasket interface, and taking air samples around the periphery of that interface.

Compressors—As indicated in Table V, compressors have the highest leak rate, but because relatively few of them are used in the CPI, compressors account for only 2% of all fugitive emissions.

To make an individual-source survey for compressors, the same procedure is followed as is used for pumps. When the compressor has an externally applied liquid seal, the seal liquid reservoir must be surveyed by placing the probe inlet at the center of the exhaust area to the atmosphere.

Cooling towers—Though normally considered a secondary-emissions source, cooling towers should be mentioned. Generally, most of the VOC compounds in cooling water result from leaks on the process side of heat-transfer equipment. Because of the large volume of air being handled, and the resulting immediate dilution of any fugitive emission, VOC-emission-rate data for cooling towers are not accurate. Because of this high dilution rate, it has been proposed to base future emission-calculation studies on total-organic-carbon (TOC) analyses. TOC analyses are made on the cooling water itself.

Wastewater treatment plants—When considering plants as potential fugitive-emission sources, one of the first areas to look at is process drains, which account for approximately 3% of all fugitive emissions. Process

Tag number	Unit	Component	Hazardous organic chemical concentration in stream	Date leak located	Date maintenance performed	Component recheck after maintenance	
						Date	Instrument reading (ppm)

Leak detection and repair survey log

Instrument operator: _____

Recorder: _____

Survey log would record leaks and their repair for hazardous organic chemicals — Fig. 9

drains and wastewater separators contribute as much as 5 lb per thousand gal of wastewater. Through the use of vapor recovery systems and/or separator covers, EPA estimates, these emissions can be reduced to 0.2 lb per thousand gal.

To perform an individual-source survey of process drains, EPA suggests that the portable hydrocarbon probe be placed at the center of a process drain area that is open to atmosphere. For covered drains, place the probe at the surface of the cover, and traverse the periphery of the cover.

Spills—Regarding fugitive emissions due to spills, EPA has proposed housekeeping practices to reduce their amount. The recommended procedure is to clean the spill up within 8 h by siphoning it into a storage con-tainer, or by using chemical absorbents or an equivalent cleanup scheme.

Recordkeeping and reporting

EPA's draft generic standards for fugitive emissions propose requirements for recordkeeping and reporting. The draft standards state that such administrative procedures are necessary to ensure that the monitoring, detection, repair and housekeeping activities proposed by the standards are followed effectively. Any detected leaks are to be recorded in a logbook, and a notation is to be entered when the leak is repaired. EPA also wants to be notified quarterly about leaks that have not been repaired within the required interval.

The draft standard further specifies that the quar-

Tag number	Unit	Component	Hazardous organic chemical concentration in stream	Date leak located	Date maintenance performed	Dates maintenance attempted	Reasons repairs postponed or failed

EPA-proposed quarterly leak report would explain why leaks were not fixed — Fig. 10

Estimated unit capital and operating costs for fugitive-emission control devices					Table VI
Control device	Capital cost per unit, $	Annual capital-recovery cost, $	Annual miscellaneous cost, $	Annual maintenance cost, $	Total annual operating cost, $
Double mechanical seals					
New	575	90	20	30	140
Retrofitted	850	140	30	40	210
Rupture disk	610	100	23	31	154
Blinds on open-ended valves	30	5	1	2	8
Closed-loop sampling	310	50	12	15	77
Closed-vent system	6,530	1,100	300	300	1,700

terly reports include a detailed listing of the units and components that have leaked beyond the specified repair interval. In addition, the date and duration of the leak must be given, as well as concentrations of hazardous pollutants in the fugitive emission.

In the draft generic standard, EPA wants to have the option of witnessing the fugitive-emission inspections, observations and monitoring. The draft generic standard, therefore, requires that the industrial facility notify EPA of the date of any monthly, quarterly or annual inspections, observations and monitoring at least one week in advance.

Further, the standards require that once an organic chemical is listed as hazardous (under the Clean Air Act, Sec. 112), the industrial facility must send EPA an estimate of the type and amount of emissions of that pollutant. This estimate must be sent within four months after the organic has been listed as hazardous. The estimate is to be categorized by specific fugitive-emission sources, such as those included in the draft generic standards. The fugitive-emission estimates would be calculated from nominal operating capacities.

The draft generic standards require that once a fugitive emission has been detected, it must be noted in a survey log (see Fig. 9), together with information as to the hazardous-pollutant concentration, the dates that the leak was detected and repaired, and the dates and results of subsequent recheck monitoring. A weatherproof and easily seen tag containing the date the leak was detected and the component identification number must be affixed to the fugitive-emission source. After repair of the leak, the log will be completed and the tag discarded. The log must be kept for two years.

Fig. 10 shows a proposed quarterly leak report that is to be submitted to EPA. Besides listing all the leaks that had been detected since the previous report and that had not been repaired within the specified 5 to 15-d interval, a statement signed by the manager of the facility must accompany the report. This statement attests to the fact that all required weekly, monthly, quarterly and yearly inspections, observations and monitoring had been performed.

Cost of control

The cost of a fugitive-emissions control program will now be estimated for a SOCMI plant, based on a report prepared by Hydroscience, Inc. [9] for EPA. Hydroscience lists scenarios for small, medium and large plants. Here, however, we will list only unit control costs, as well as develop monitoring and manpower requirements.

This report is the first major fugitive-emission study done for EPA on a SOCMI plant. Previous work dealt with refinery emissions at first, and then, more recently, with petrochemical plants. This report comprehensively discusses individual sources of fugitive emissions in the SOCMI facilities. After providing a detailed characterization and description of each source, the report concisely discusses the applicable technologies for controlling

Annual unit manpower requirements for monitoring process equipment			Table VII
Component	Type of monitoring	Estimated time for monitoring, min	Times monitored, no/yr
Pump seals	Instrument	5	1
	Visual	0.5	52
Compressor seals	Instrument	10	4
Inline valves (gas)	Instrument	1	4
Inline valves (liquid)	Instrument	1	1
Open-ended valves	Instrument	1	1
Safety-relief valves	Instrument	8	4
Pipeline flanges	None	—	—
Cooling towers	Instrument	10	1

Annual unit manpower maintenance requirements. Pump seals take the most time to repair		Table VIII
Component	Leakage frequency, %/yr	Repair time h/leak
Pump seals	12	80
Process valves	6	1.13
Safety-relief valves	7	0
Flanges	Minor	—
Compressor seals	7	40

fugitive emissions. It also details methods that are used for leak detection.

Table VI lists unit capital and operating costs for fugitive-emission control devices. Table VII gives unit monitoring manpower requirements, and Table VIII lists fugitive-emission-source unit-maintenance manpower requirements. Tabulation of unit costs and manpower requirements allows the reader to develop a cost scenario for his own plant.

Now, we will develop costs for a typical medium-sized (100 million to 400 million lb/yr) SOCMI plant, using Hydroscience's costs. Assume such a plant has the following pieces of process equipment, which are potential sources of fugitive emissions:

Equipment	Units
Pump seals:	
Mechanical (43 single; 10 double)	53
Packed	6
Process valves:	
Inline, gas	365
Inline, liquid	670
Open-ended	415
Compressor seals	2
Safety-relief valves	50
Pipeline flanges	2,400
Cooling towers	1

To comply with EPA's proposed guidelines, it is estimated that the plant would have to install the following emission-control equipment:

Equipment	Units
Double mechanical seals (retrofitted)	43
Rupture disks (before safety-relief valves)	50
Blind flanges (on open-ended valves)	415
Closed-loop sampling trains	104
Closed-vent systems (flares)	1

For this medium-sized plant, the total estimated monitoring manpower requirement would be 177 man-hours/yr (based on a two-man crew) for individual component surveys with visual inspection. Maintenance manpower requirements are estimated to be 680 man-hours/yr. Assuming monitoring and maintenance manpower costs $15/h, the total yearly labor cost would be $12,900.

In addition, Hydroscience estimates an additional annual $5,200 for administrative costs, $2,600 for instrumentation capital-related costs (based on two portable hydrocarbon analyzers at $4,800 each), and $2,700 for instrumentation supplies, maintenance, and calibration costs. This gives a grand total fugitive-emission monitoring and repair operating cost of $23,400 per year.

In addition, one must calculate the capital and operating costs associated with the purchase, installation, maintenance and operation of the fugitive-emission control-equipment listed above, using the unit costs listed in Table VI. It is then calculated that the total installed capital cost for this equipment is $118,300.

Similarly, the following annual costs are calculated: annual capital recovery = $19,400; annual control equipment maintenance cost = $5,800; and annual miscellaneous cost = $4,800. This gives a total annual equipment operating cost of $30,000. The total fugitive-emission control operating cost is, therefore, estimated at $53,400/yr. The capital required for purchasing and installing the individual-component monitoring and fugitive-emission control equipment is $127,900.

It can thus be readily seen that the economic impact of an industrywide fugitive-emission control program can be great, particularly on the small- to medium-sized chemical producer.

It would be advantageous to chemical producers to check Hydroscience's fugitive-emission and cost data against their experience in their own facilities. Such a comparison would be invaluable in providing comments to EPA as its proposed standards proceed toward final promulgation.

References

1. U.S. EPA, Compilation of Air Pollutant Emission Factors, 3rd ed., supplement No. 8, May 1978. Pub. No. PB-288-905, National Technical Information Service, Springfield, VA 22161.
2. U.S. EPA, Air Quality Criteria for Photochemical Oxidants and Oxidant Precursors, Vol. 1—External Review Draft No. 1, Sept. 1977, U.S. EPA, Office of Research and Development, Washington, DC 20460.
3. Kremer, H., Leckraten statischer und dynamischer Dichtelemente an chemischen und petrochemischen Anlagen, Paper presented at 4th meeting OGEW/DGMK, Oct. 4–6, 1976 in Salzburg, Austria.
4. Radian Corp., Emission Factors and Frequency of Leak Occurrence for Fittings In Refinery Process Units, Feb. 1979, EPA 600/2-79-044, U.S. National Technical Information Service, Springfield, VA 22161.
5. Monsanto Research Corp., Source Assessment—Fugitive Hydrocarbon Emissions From Petrochemical Plants, EPA 600-4-78-004, Apr. 1978.
6. U.S. EPA, Guideline Document EPA-450/2-78-036, Control of Volatile Organic Compound Leaks From Petroleum Refinery Equipment, June 1978.
7. Bierl, A., et al., Leckraten von Dichtelementen, *Chemie Ingenieur Technik*, Vol. 49, No. 2 (1977), pp. 80–95.
8. Hughes, T. W., et al., Measurement of Fugitive Emissions From Petrochemical Plants, paper presented at 1979 AIChE meeting in Miami, Fla.
9. Hydroscience, Inc., Emission Control Options for the Synthetic Organic Chemicals Manufacturing Industry—Fugitive Emissions Report, report written for EPA's Emission Standards and Engineering Div., Office of Air Quality Planning and Standards, March 1979.

The author

Michael J. Wallace is senior environmental engineer for Sandoz, Inc., Route 10, East Hanover, NJ 07936. Telephone: (201) 386-7977. He has a broad range of environment-related responsibilities with regard to all of Sandoz' U.S. facilities. He holds a bachelor's degree in chemical engineering from Notre Dame University. Chairman of the Environmental Quality Committee of the Synthetic Organic Chemical Mfrs. Assn., he has lectured on environmental monitoring in several U.S. university extension programs.

Air Pollution Instrumentation

Instruments for environmental monitoring
Sampling and analyzing air pollution sources
Air-pollution instrumentation
Online instruments expedite emissions test
Analyze stack gases via sampling or optically, in place
Continuous source monitoring
Tracer-gas system determines flow volume of flue gases
CO_2 measurements can correct for stack-gas dilution

Instruments for environmental monitoring

Plants are required to monitor many pollutants in both air and water. Here is a rundown of the instruments available for the job.

D. M. Ottmers, Jr., D. C. Jones, L. H. Keith and R. C. Hall, Radian Corp.

☐ Government regulations require that industry monitor for various gaseous pollutants, particulates, lead, and the so-called "priority pollutants." Hence, monitoring of air and water qualities has become increasingly important during the past decade. Thus, it is important to know:
- The instrumental methods that are currently accepted for pollutant monitoring.
- The instrumentation available for this purpose.

This article discusses instrumentation for monitoring both air and water quality. Emphasis is placed upon how these instruments function, their approximate costs, and what systems may be used or developed in the near future.

Air-quality monitoring

Air-quality monitoring consists primarily of measuring various gaseous pollutants, particulates, and lead. Instrumentation for monitoring these atmospheric pollutants are discussed in the sections that follow. Quality assurance considerations and research in air-quality monitoring are also discussed briefly.

Gaseous pollutants

Federal regulations have established allowable limits for sulfur dioxide, nitrogen dioxide, carbon monoxide, and ozone [1,2]. To aid in attaining compliance with the oxidant standard, a standard has also been established for non-methane hydrocarbons [3].

Monitors for these gases usually are continuous, i.e., the indicated pollutant concentration may vary continuously with time. These monitors are automated, and are capable of unattended operation for days or weeks.

Also, manual samplers are available that determine an average pollutant concentration. They do this by chemically scrubbing the pollutant from a known quantity of air collected at a constant rate during the sampling period, with subsequent analysis of the captured material. However, the servicing and analysis to support these samplers is manpower intensive, and they are generally not well suited to the observation of frequent short-term averages. With improvements in the reliability and accuracy of the continuous monitors, manual samplers have gradually lost favor with most users. Consequently, this article will be limited to a discussion of continuous analyses.

Originally published October 15, 1979

The basic elements of a pollutant monitoring system are shown in Fig. 1. Air is taken in by the sample manifold, which is made of an inert material such as glass or Teflon. The sample manifold brings a fast-moving stream of air to the immediate vicinity of the monitor in order to minimize the residence time of the air sample in that portion of the system upstream of the measurement cell. Even though the sample-handling system is inert, reactions may occur between gaseous species in the air, e.g., reaction of ozone and nitric oxide to produce nitrogen dioxide. Thus, the residence time in the air transport system should be minimized.

Monitors may include filters to remove particulates that might otherwise cloud optical windows, plug orifices, and the like. Some monitors may use a scrubber to remove a gas in the air that would interfere with measurement of the one of interest.

Support gases are required for some measurement methods. These are consumed during the measurement and thus may increase operational costs and logistics problems.

Measurement cells can be based on many different methods for the pollutants of interest. Basically, the pollutant is introduced into an environment where it will undergo a chemical or physical reaction whose output can be converted to an electrical signal. Examples include:

Flame photometry—The gas of interest is pyrolyzed by a flame, with degradation products reacting to produce a particular wavelength of light. This light is detected by a photomultiplier tube and converted to an electrical signal.

Infrared—The gas of interest absorbs infrared light, which is beamed through the sample cell. This absorption results in heating, which is measured as a pressure imbalance, and converted to an electrical signal.

Chemiluminescence—The gas of interest is mixed with a gas with which it undergoes a spontaneous reaction. The reaction gives off light of a particular wavelength, which is measured via a photomultiplier tube and converted to an electrical signal.

Electrochemical—The gas of interest is dissolved in a cell with electrodes where it will react with other dissolved species. This reaction, or auxiliary reactions, generates an electric current at the electrodes that is proportional to the concentration of the gas of interest.

Fluorescence—Ultraviolet light is beamed through the

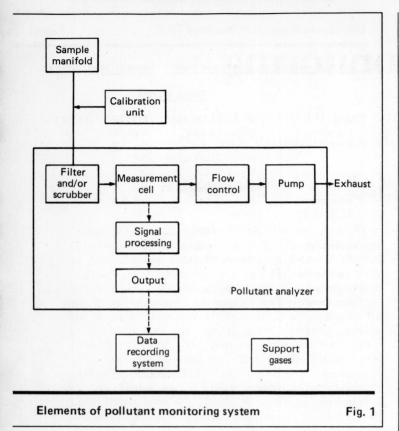

Elements of pollutant monitoring system   **Fig. 1**

sample each data channel at some frequency. Their output may be merely a voltage printed on a paper tape; more-sophisticated systems may provide a machine-readable output. Small computerized data-acquisition systems based on microprocessors can provide on-site reduction, with data storage on magnetic tape or disk. These units also may assume control functions such as automatic calibration of the instruments, calculation of zero and span drift, sequencing of high-volume samplers, and so forth.

Table I provides a listing of monitors that meet EPA's reference or equivalency specifications for the four gaseous criteria pollutants. The type of detection system used is specified for each monitor, along with remarks on support gases, scrubbers, etc.

Non-methane hydrocarbons are typically analyzed via a gas chromatograph with a flame ionization detector. No EPA-approved instruments are available for this measurement.

Particulates

Regulations also have been established for total suspended particulates [5]. Only one measurement technique is approved, the high-volume sampler. This technique uses a filter to trap particulates during a 24-h sampling period. The filter is contained in an all-weather housing, built according to a specific design. The air flowrate is measured by a chart recorder (continuously) or a rotameter (manually). This flowrate must be calibrated periodically. The filters are weighed (at a controlled humidity) before and after exposure to determine the particulate loading. The air-flow volume is converted to standard conditions (1 atm pressure and 25°C) and used to determine the particulate concentration in the sampled air.

High-volume samplers are typically run every sixth day, every third day, or daily. The sampling period is midnight to midnight.

Lead

A federal standard for ambient lead has recently been promulgated [6]. The measurement procedure involves chemical analysis of filters collected in the manner (described above) for particulates. The standard is based on a three-month average of lead on filters collected no less frequently than once every six days.

Quality-assurance considerations

Air-quality monitoring should be supported by an adequate quality-assurance program. Included are functions such as calibration method and frequency, zero/span checks, traceability of standards, analyzer audits, reports, and documentation. Quality-assurance guidelines are available from EPA [7-9]. However, EPA regional offices, or other permitting offices, may have considerable latitude in specifying particular procedures.

Research in air-quality monitoring

Research in air-quality monitoring is currently very active. For gaseous pollutants there is a major activity in remote sensing, i.e., measurement of the pollutant concentration in the air at some point far removed from

sample cell, either continuously or in pulses. Absorption of this light induces emission of fluorescent light by the species of interest. This is detected by a photo-multiplier tube and converted to an electrical signal.

Second-derivative spectroscopy—Ultraviolet light is absorbed by the gas of interest in the sample cell. Measurements are based on the shape characteristics of the absorption spectrum rather than on the change in intensity. The output signal is proportional to the second derivative of intensity with respect to wavelength.

Ultraviolet spectroscopy—Ultraviolet light passes through a sample and is measured by a photoelectric detector. The decrease in intensity—which is exponentially proportional to the concentration of the absorbing pollutant—is then converted to an electrical signal.

Generally, only one type of measurement cell is used in any particular monitor. As shown in Fig. 1, other items in the system include a flow-control system and a sample pump. Flow control may be achieved via a capillary orifice, needle valve, and so on, with suitable pressure control.

The calibration unit must supply zero air, and gases of known concentration to check or adjust instrument response. This unit may be manual or automated.

Various types of outputs may be supplied with air-quality instrumentation. Panel meters, digital displays, recorder jacks, and terminal strips are all used in various brands.

The data recording system provides a permanent record of the monitoring results. Strip-chart recorders provide a continuous record but are manpower intensive in terms of data reduction. Data loggers typically

the monitor. These systems are typically based on lasers or correlation spectroscopy. For particulates, much research has been done on size fractionating devices, which allow respirable and inhalable particulates to be quantified. New regulations will probably be required before these new techniques enjoy widespread application.

The analysis of specific organic compounds in ambient air is receiving much attention. This is currently achieved by collecting air samples and transporting them to a laboratory. However, the use of special gas-chromatograph–mass-spectrometer systems for real-time analysis shows considerable promise.

Water-quality monitoring

The chemical parameters for monitoring water pollutants are very extensive. They include traditional measurements such as pH, color, biochemical oxygen demand (BOD), chemical oxygen demand (COD), total solids (TS), nitrogen, oxygen, sulfate, sulfite, sulfide, chloride, ammonia, hardness, etc. The instruments for measuring and monitoring these parameters are common and have been the subject of many past articles, books, and reference manuals [10]. They will not be further discussed here. Instead, this discussion will be limited to a new list of chemical parameters that both industry and the EPA are required by law to begin monitoring in the near future.

These new chemical parameters are the so-called "priority pollutants," and are listed in Table II. They are distinguished from chemical parameters previously monitored for, by being specific organic compounds and isomers. (The elements are still measured as total copper, zinc, etc.)

The priority pollutants resulted from a 1976 court settlement after the EPA was sued by several environmental groups for failing to implement portions of the Federal Water Pollution Control Act. This history of the development of the list of 129 priority pollutants, and present methodology used to analyze for them is summarized in a recent review [11].

Regulations for promulgation of new-source performance standards and pretreatment standards started to be issued in June 1979, and are to be completed for 21 major industrial categories by June 1981. These regulations will require application of best available technology (BAT) economically achievable by June 1984. The EPA is spending about $90 million to gather analytical, engineering, and economic data on which to base these regulations. It is estimated that the economic impact to American industry to meet the treatment and monitoring requirements of these regulations will be between $40 and $60 billion.

Sample treatment

Before the instrumentation needed to monitor for the priority pollutants is discussed, a brief description of sample treatment is necessary to put the complexity of the monitoring requirements in perspective. There are eight different procedures required to prepare a single sample for analysis.

The organic compounds are divided into four fractions. The very-volatile organic compounds are col-

EPA-approved pollutant monitors [4]			Table I
Method and cost	Brand and model	Range(s), ppm	Notes
Sulfur dioxide			
Flame photometry $4,000-6,000	Meloy SA185-2A	0.5, 1.0	1,2
	Meloy SA285E	0.05, 0.1, 0.5, 1.0	
	Monitor Labs 8450	0.5	
	Bendix 8303	0.5, 1.0	
Fluorescence $4,000-6,000	Thermo Electron 43	0.5	3
	Beckman 953	0.5, 1.0	
	Meloy		
	Monitor Labs		
Electrochemical $7,000-8,000	Phillips PW 9755	0.5	3,2
	Phillips PW 9700	0.5	
Derivative spectroscopy $8,000	Lear-Siegler SM 1000	0.5	3
Other	Asarco 500	0.5	
Nitrogen dioxide			
Chemiluminescence $5,000-6,000	Monitor Labs 8440E	0.5	3
	Bendix 8101-C	0.5	
	CSI 1600	0.5	
	Meloy NA530R	0.1, 0.25, 0.5, 1.0	
	Beckman 952A		
	Thermo Electron 14 B/E	0.5	
	Thermo Electron 14 D/E	0.5	
Ozone			
Chemiluminescence $4,000-7,000	Meloy 0A325-2R	0.5	4
	Meloy 0A350-2R	0.5	
	Bendix 8002	0.5	
	McMillan 1100-1	0.5	
	McMillian 1100-2	0.5	
	McMillian 1100-3	0.5	
	Monitor Labs 8410E	0.5	
	Beckman 950A	0.5	
	CSI 2000	0.5	
	Phillips FP 9771	0.5	
Ultraviolet spectroscopy $3,000-4,000	Dasibi 1003-AH	0.5, 0.1	3
	Dasibi 1003-PC	0.5, 1.0	
Infrared $6,000-7,000	Bendix 8501-5CA	50	3
	Beckman 866	50	
	MSA 202S	50	
	Horiba AQM-10	50	
	Horiba AQM-11	50	
	Horiba AQM-12	50	

Notes:
1. Support gas (hydrogen) is needed.
2. Scrubber is required.
3. No support gases needed.
4. Support gas (ethylene) is needed (except Phillips 9771).

lected in a headspace-free 40-milliliter (mL) vial. They are stripped from solution by nitrogen or helium which is bubbled through a 5-mL aliquot. These compounds are trapped in a stainless-steel tube packed with Tenax and silica gel, or a similar solid adsorbent. When the purging is complete (12 min) the trap is rapidly heated to 180°C; the adsorbed compounds are flashed off the trap and retrapped on the head of a gas chromatographic column after which they are analyzed by one of the instrumental methods discussed later.

The remaining organic sample is collected in a one-gallon jug. A 1-L aliquot is extracted with a hexane/methylene-chloride mixture to remove the pesticides and polychlorinated biphenyls (PCBs). The extract is concentrated to 10 mL using a Kuderna-Danish concentrator on a steam bath. Florisil chromatography of this concentrate provides three fractions that contain all the pesticides and PCBs. Each of these fractions is concentrated again to 10 mL prior to instrumental analysis.

A 2-L aliquot of the water is made alkaline to pH 10 and extracted with methylene chloride. Then it is made strongly acidic with hydrochloric acid and reextracted with methylene chloride. The first (alkaline) extract contains both the basic and neutral priority pollutants and the second (acidic) extract contains the phenols. Each of these extracts is concentrated to about 10 mL using a Kuderna-Danish concentrator and then each is concentrated to exactly 1.0 mL using a micro-Snyder 3-ball condenser column. These concentrated extracts contain the majority of the priority pollutants.

Total cyanides are determined by reflexing 500 mL of an acidic water solution. Cyanide, as HCN, is vaporized and trapped in a sodium hydroxide solution. Buffer is added, followed by chloramine-T solution and pyridine-barbituric acid reagent. A red-blue dye is formed and the concentration is determined by using a spectrophotometer.

Total phenols are determined by distilling 500 mL of the water to remove most interferences (the phenols steam distill). Solutions of 4-aminoantipyridine and potassium ferricyanide are added to form a yellow dye. This, in turn, is extracted with chloroform and the concentration is determined by use of a spectrophotometer.

The metals are digested three different ways before they are analyzed by atomic absorption. In addition, three different techniques of atomic-absorption analysis are used: flame excitation, graphite furnace, and cold-vapor analysis. One group of metals (cadmium, chromium, copper, nickel, lead, and zinc) are reanalyzed using the graphite furnace if they are not first detected using flame excitation.

Finally, the asbestos samples are filtered through Nuclepore filters and the retained particles are carbon coated under vacuum. The organic filter is dissolved with chloroform, leaving the fibers embedded in a carbon film. A portion of the film is magnified 20,000 times with a transmission electron microscope, and the asbestos fibers are identified by selective-area electron diffraction. A representative area of the electron microscope grid is counted, and the concentration of asbestos, in millions of fibers per liter, can be calculated from the size of the water sample.

Analytical measurements

The instruments currently used to provide measurements of the priority pollutants are:

- Spectrophotometers, for total cyanides and total phenols.
- Atomic-absorption spectrophotometers, for the elements.
- Transmission electron microscopes, for asbestos.
- Gas chromatographs (GC) with electron capture detectors, for pesticides.
- Computerized gas-chromatograph–mass-spectrometers (GC-MS), for the purgeables, base/neutrals and phenols.

All of these instruments—except the spectrometer and gas chromatograph—are relatively expensive. The EPA is moving to substitute less expensive or more cost-effective instruments for monitoring purposes whenever possible. For instance, gas chromatographs with specific detectors and high-pressure liquid chromatographs will, in some cases, be able to substitute for gas-chromatograph–mass-spectrometers (GC-MS). These instruments are cheaper than the GC-MS and, if only a few of the organic priority pollutants need to be monitored, can offer significant savings. However, if many of the organic priority pollutants must be monitored, the cost of all these instruments plus the more labor-intensive sample preparation requirements may more than offset the investment in a GC-MS instrument.

Another substitution forthcoming will be inductively-coupled argon-plasma emission-spectroscopy (ICAP). Although an ICAP is many times more expensive than an atomic absorption spectrophotometer, it can analyze many of the elements at once and the labor savings, if thousands of samples are to be analyzed, will offset the extra investment in the instrument.

The instrumental methods presently in use, and those under consideration as alternative methods, are summarized in Table III.

Priority pollutant analysis

Instrumentation for measuring the priority pollutants is discussed in this section. Here, the discussions involve: the equipment types and their approximate costs, associated data systems, principles of operation, and sample detection.

Equipment types and approximate costs

Gas chromatography (GC) and gas chromatography coupled with mass spectroscopy (GC-MS) are the primary instrument systems used for analysis of the organic priority pollutants. High performance liquid chromatography (HPLC) is also used, but at present is recommended for only a few of the priority pollutants. These systems are extremely flexible and can be equipped with a variety of options to alter their characteristics and meet different analytical needs. The major features of these systems are compared in Table IV.

The prices of gas and liquid chromatographs are approximately the same. Basic instruments can be purchased for approximately $6,000. These instruments, however, include few options and cannot be used for the analysis of all the pollutants. More-flexible units

Priority pollutants as listed by EPA

Table II

Purgeable organics

Acrolein	1,1-Dichloroethylene	Bromoform
Acrylonitrile	1,1,2-Trichloroethane	Dichlorobromomethane
Benzene	1,1,2,2-Tetrachloroethane	Trichlorofluoromethane
Toluene	Chloroethane	Dichlorodifluoromethane
Ethylbenzene	2-Chloroethyl vinyl ether	Chlorodibromomethane
Carbon tetrachloride	Chloroform	Tetrachloroethylene
Chlorobenzene	1,2-Dichloropropane	Trichloroethylene
1,2-Dichloroethane	1,3-Dichloropropene	Vinyl chloride
1,1,1-Trichloroethane	Methylene chloride	1,2-*trans*-Dichloroethylene
1,1-Dichloroethane	Methyl chloride	*bis*(Chloromethyl) ether
	Methyl bromide	

Base/neutral-extractable organics

1,2-Dichlorobenzene	*bis*(2-Ethylhexyl) phthalate	Benzo(k)fluoranthene
1,3-Dichlorobenzene	Di-*n*-octyl phthalate	Benzo(a)pyrene
1,4-Dichlorobenzene	Dimethyl phthalate	Indeno(1,2,3-c,d)pyrene
Hexachloroethane	Diethyl phthalate	Dibenzo(a,h)anthracene
Hexachlorobutadiene	Di-*n*-butyl phthalate	Benzo(g,h,i)perylene
Hexachlorobenzene	Acenaphthylene	4-Chlorophenyl phenyl ether
1,2,4-Trichlorobenzene	Acenaphthene	3,3'-Dichlorobenzidine
bis(2-Chloroethoxy) methane	Butyl benzyl phthalate	Benzidine
Naphthalene	Fluorene	*bis*(2-Chloroethyl) ether
2-Chloronaphthalene	Fluoranthene	1,2-Diphenylhydrazine
Isophorone	Chrysene	Hexachlorocyclopentadiene
Nitrobenzene	Pyrene	N-Nitrosodiphenylamine
2,4-Dinitrotoluene	Phenanthrene	N-Nitrosodimethylamine
2,6-Dinitrotoluene	Anthracene	N-Nitrosodi-*n*-propylamine
4-Bromophenyl phenyl ether	Benzo(a)anthracene	*bis*(2-Chloroisopropyl) ether
	Benzo(b)fluoranthene	

Acid-extractable organics

Phenol	4,6-Dinitro-*o*-cresol	2-Chlorophenol
2-Nitrophenol	Pentachlorophenol	2,4-Dichlorophenol
4-Nitrophenol	p-Chloro-*m*-cresol	2,4,6-Trichlorophenol
2,4-Dinitrophenol		2,4-Dimethylphenol

Pesticides/PCBs

α-Endosulfan	4,4'-DDE	Aroclor 1016
β-Endosulfan	4,4'-DDD	Aroclor 1221
Endosulfan sulfate	4,4'-DDT	Aroclor 1232
α-BHC	Endrin	Aroclor 1242
β-BHC	Endrin aldehyde	Aroclor 1248
δ-BHC	Heptachlor	Aroclor 1254
γ-BHC	Heptachlor epoxide	Aroclor 1260
Aldrin	Chlordane	2,3,7,8-Tetrachlorodibenzo-
Dieldrin	Toxaphene	*p*-dioxin (TCDD)

Metals

Antimony	Chromium	Selenium
Arsenic	Copper	Silver
Beryllium	Lead	Thallium
Cadmium	Mercury	Zinc
	Nickel	

Miscellaneous

Asbestos (fibrous)	Total cyanides	Total phenols

Instrument for analyzing priority pollutants		Table III
Priority pollutants	Present instrument	Alternative instrument
Phthalate esters	GC-MS	GC with EC detector
Haloethers	GC-MS	GC with Hall detector
Chlorinated hydrocarbons	GC-MS	GC with EC detector
Nitrobenzenes	GC-MS	GC with EC detector
Nitrosamines	GC-MS	GC with alkali FI detector
Benzidines	GC-MS	HPLC with electrochemical detector
Phenols	GC-MS	GC with EC and FI detector
Polynuclear aromatics	GC-MS	HPLC with fluorescence detector
Pesticides	GC, GC-MS	– – –
Halogenated purgeables	GC-MS	GC with Hall or EC detector
Nonhalogenated purgeables	GC-MS	GC with FI detector
Nonhalogenated hydrocarbons	GC-MS	GC with FI detector
Elements	AA	ICAP
Asbestos	Electron microscope	– – –
Cyanide	Spectrophotometer	– – –
Total phenols	Spectrophotometer	– – –

GC-MS, gas-chromatograph-mass-spectrometer
GC, gas chromatograph
AA, atomic absorption spectrophotometer
EC, electron capture
FI, flame-ionization
HPLC, high-performance liquid chromatography
ICAP, inductively coupled argon-plasma emission-spectroscopy

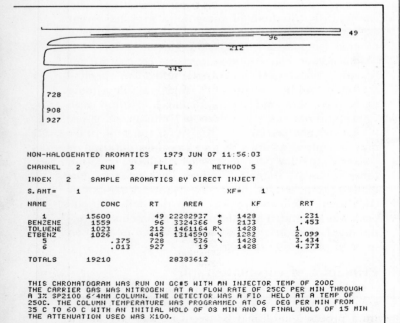

Printer/plotter output   Fig. 2

capable of multiple pollutant analysis generally cost from $10,000 to $15,000. The addition of a data system for automated data processing and reporting of results is an important consideration and adds another $5,000 to $10,000. If a data system is not purchased, a strip-chart recorder is required, which costs $750 to $1,500.

Gas-chromatograph–mass-spectrometers are considerably more expensive than the other chromatograph systems and cost approximately $80,000 to $120,000 for an instrument suitable for analyzing all the organic priority pollutants. The higher cost is primarily due to the complexity of the mass spectrometer. Not only is the mass spectrometer complicated in terms of components such as vacuum pumps, data system and the mass analyzer, but it is also quite complicated to operate and requires a skilled operator. In addition to their higher cost, GC-MS systems are usually floor-mounted and require a space two to four times that occupied by a gas or liquid chromatograph.

Although all of the present priority pollutants can be analyzed by either GC or GC-MS techniques, the choice between the two is not particularly easy. Such factors as the number and type of compounds to be analyzed, nature of the samples, type of personnel available for operating the equipment, and degree of confirmation required must be considered. After these factors have been considered, the chromatographic capability required for the analyses must be determined. In general, a minimum of two gas chromatographs equipped with a total of four different detectors is required to analyze all compounds (see Table II). The cost effectiveness of GC versus GS-MS techniques have been discussed in detail elsewhere [12,13] and will not be presented in this article.

Data systems

The data system is an essential component of all modern gas-chromatograph–mass-spectrograph instruments. Analytical capability and ease of use are directly related to the data system. Specifications related to specific ion-monitoring, background-subtraction capability, spectra searching and compound identification techniques, size of reference library, ability to use central spectra libraries via telephone communication, and computer instrument tuning ability should be thoroughly investigated.

In contrast to the gas-chromatograph–mass-spectrometer, a data system is not absolutely essential to gas chromatographs or high-performance liquid chromatographs. Nevertheless, a data system should be seriously considered as an integral part of any chromatograph system. Besides extending the analytical capability of the chromatograph, these systems can generate detailed final reports of analysis with accompanying chromatograms and be used to totally automate the analysis. In so doing they greatly reduce the amount of operator expertise required and enable most analyses to be routinely performed.

A variety of data systems are available for chromatography. Instruments range from simple digital integrators to complex computing integrators with printer/plotter outputs. The more-complex systems may be able to handle data from more than one input

Instruments available for priority pollutant analysis **Table IV**

Instrument	Use	Price range*	Major U.S.† manufacturers	Comment
Gas chromatograph	All organic priority pollutants	$6,000-25,000	Hewlett-Packard Perkin-Elmer Tracor Varian	Instrument specificity depends upon the detector used
Liquid chromatograph	Benzidines and polynuclear aromatics	$6,000-28,000	Altex Du Pont Hewlett-Packard Micromeritics Perkin-Elmer Spectra-Physics Tracor Varian Waters	Not compatible with spraying techniques of analysis. Compounds must have a high ultraviolet absorbence or be fluorescent
Gas chromatograph-mass spectrometer	All organic priority pollutants	$50,000-150,000	Finnigan Hewlett-Packard	Requires a skilled operator

*Price depends on the options chosen.
†Listings are in alphabetical order and do not include manufacturers of speciality instruments.

(multichanneled). Level of data-processing sophistication is usually not related to the number of data channels that can be handled, however. Simple digital integrators usually do no more than print peak areas, retention times, and area percents or concentrations. The more sophisticated printer/plotter systems are capable of peak identifications, multilevel calibrations based on one or more internal or external standards, flexible integration and baseline-correction algorithms, and detailed reporting with custom headings. Most of these systems can be user-programmed for total control of integration and reporting. In addition, many instruments can be equipped with floppy-disk or magnetic-tape systems for retaining programs and raw chromatographic data. This allows the data to be reprocessed using different integration algorithms and reporting formats.

A typical printer/plotter output of a computing integrator is shown in Fig. 2. The first information presented is the chromatogram with retention times of the components printed at each peak. Printing the retention times on the chromatogram greatly facilitates locating integration results for a given peak. After the chromatogram, a heading is printed that may contain information regarding the type of analysis (pesticides, etc.), date and time of analysis, details of the instrument (channel, run, file, method), and the sample analyzed. Other information such as the analyst, correction factors, etc., can also be included. Results of the analysis can be presented in a variety of formats. Information presented usually includes peak names, concentrations, retention times, integration areas, method of baseline correction, and compound response factors. The report can be terminated with a statement that identifies the chromatograph used and all operating parameters of the chromatographic analysis.

Peaks are identified in the report by either name or number. They are identified by name according to

information supplied by the user prior to the analysis. This information consists of the name to be used, the expected retention time, and the window of retention time for which a given name is to be applied to a peak. If a peak does not fall within one of these retention time windows, it is assigned a numerical value.

It should be understood that identification of a peak by name does not necessarily mean the peak is actually the compound identified, because retention times are only characteristic of a compound's identity. Usually several other compounds can be found that will have the same retention time as the compound of interest under a specific set of chromatographic conditions, and theoretically hundreds of compounds may have that same retention time simply because there are so many organic compounds (~3,000,000) known.

Information concerning component concentrations, retention times, peak areas, and method of baseline correction can be presented in various ways depending upon preference and the particular integrator used. Concentrations can be in parts per million, parts per billion, weight percent etc. Retention times of peaks may be output in units of time (seconds, minutes) or converted to other parameters such as relative retention times or Kovat's Retention Indices. Peak areas are normally output as total integrator counts. A symbol or code usually follows the peak area that tells how the peak was integrated and the method of baseline correction used. Peak heights may also be shown in addition to or in place of peak areas.

Principles of chromatography

Chromatography [14,15] is the physical method of separating a mixture into its components by an equilibrium process established between a stationary and a mobile phase. The mobile phase is a gas in gas chromatography and a liquid in liquid chromatography. As presented in Table V, gas chromatography consists of

Forms of chromatography available for pollutant analyses				Table V
Chromatography	**Mobile phase**	**Stationary phase**	**Classification**	**Major use**
Gas	Gas	Solid	Gas-solid	Separation of gases and hydrocarbons
	Gas	Liquid	Gas-liquid	Various organic separations except gases
Liquid	Liquid	Solid	Normal phase (absorption)	Nonpolar to moderately polar compounds
	Liquid	Liquid	Reverse phase (liquid-liquid)	Most nonionic compounds
	Liquid	Ion exchange resin	Ion exchange	Ionic compounds
	Liquid	Polymer (gel)	Steric exclusion (gel permeation)	Macromolecules

gas-liquid and gas-solid chromatography; whereas liquid chromatography is subdivided into normal phase, reverse phase, ion exchange, and steric exclusion.

The basic process responsible for the separation can also be used to categorize the chromatography. For instance, partition chromatography is the separation process due to equilibria established between the mobile phase and a liquid, and adsorption chromatography involves equilibria between the mobile phase and a solid.

The separation process occurs during transport of the sample through the chromatographic column. Two major types of columns have been developed for gas chromatography, packed and capillary. Packed columns, as their name suggests, contain a packing that either acts as an adsorbent or supports a stationary liquid phase. These columns are commonly coiled tubes (2 to 4 mm I.D.) made of metal or glass. Capillary columns contain the liquid phase coated on the inside wall of the tube. They are considerably smaller in diameter (0.25 to 0.5 mm) than packed columns, and vary in length from approximately 10 to 150 m. In contrast to gas chromatographs, modern columns for high-performance liquid chromatographs are almost exclusively made of metal, packed with very small particles (5 to 10 μm) and only 15 to 25 cm in length.

As shown in the block diagram of Fig. 3, the components of a basic gas-chromatograph system consist of a carrier-gas supply with pressure and flow controls, injector (injection port), column, column oven, detector, signal processor, temperature-control electronics, and output device. The carrier gas serves as the mobile phase that transports the sample through the column and into the detector. The injector is normally a heated septum inlet and forms a gas-tight seal between the column and the ambient environment.

The column and detector form the "heart" of the chromatograph system. The column performs the analytical separation, and the detector transforms the separated constituents into electrical signals. These signals are usually very small currents that require amplification by the signal processor before they are capable of driving an output device. The temperature-control electronics maintain the injector and detector at fixed temperatures, and enable the column oven to be operated at a fixed temperature (isothermal operation) or to be temperature programmed at a fixed rate from an initial to a final temperature. The great advantage of temperature programming is that it allows compounds of significantly different volatilities to be analyzed in a single chromatographic run.

The components of a liquid chromatograph system are similar to these in gas chromatography with three main differences:

1) A pump is used for driving the liquid mobile-phase through the column.

2) A solvent programmer is used instead of a temperature programmer, which is used to change the mobile phase composition during an analysis.

3) A column oven is not essential.

Two main techniques are used for introducing priority pollutant samples into a chromatograph. These are syringe injection and the use of a sample valve. Syringe injection is used to introduce small aliquots (1 to 10 μL)

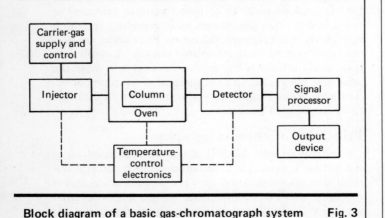

Block diagram of a basic gas-chromatograph system   Fig. 3

| Characteristics of chromatograph detectors commonly used for priority pollutant analysis | | | | | | Table VI |

Detector	Type	Classification[1]	Use	Sensitivity[2]	Specificity	Complexity[3]
Flame ionization	GC	Universal	Hydrocarbons	Nanogram	None	2
Photoionization	GC	Universal/Selective	Aromatics	High picogram	Low	1
Electron capture	GC	Selective	Pesticides, PCBs[1] chlorinated hydrocarbons, phthalate esters	Picogram	Variable	2
Nitrogen-phosphorus	GC	Specific	Nitrogen-containing compounds	High picogram	High	3
Hall detector	GC	Specific	Halogen-containing compounds	High picogram	High	3
Mass spectrometer	GC	Specific	All volatile compounds	Nanogram	High	4
Absorbance	HPLC	Selective	Polynuclear aromatics	Nanogram	Variable[4]	1
Fluorescence	HPLC	Selective	Polynuclear aromatics	High picogram	Variable	2
Electrochemical	HPLC	Selective	Benzidines	High picogram	Variable	3

[1] Universal, response is approximately the same for all compounds; selective response is greater for certain compounds; specific, response is specific for compounds that contain a given element.
[2] Quantity of material that can normally be detected in a typical analysis.
[3] Complexity: 1, requires few adjustments and little maintenance; 2, requires several adjustments; 3, requires several adjustments and routine maintenance; 4, requires numerous adjustments and routine maintenance by a skilled operator.
[4] Specificity depends upon the nature of the compounds present.

of liquid samples, primarily into gas chromatographs. Sample valves are used to introduce samples into both gas and liquid chromatographs. In gas chromatography the valve forms part of a gas sparging apparatus and directs the thermally desorbed volatile pollutants onto the head of a gas chromatography column. On the other hand, in liquid chromatography a valve is used to inject a liquid sample, normally 10 to 100 microliters (μL) in volume.

Sample detection

A wide variety of detectors exist for chromatography. Characteristics of the nine detectors commonly used for analysis of priority pollutants are summarized in Table VI. As shown in this table, the detectors used for analyzing the various categories of pollutants (see Table I) range from non-specific to highly specific. They also range from those needing little maintenance or knowledge for use to those requiring a skilled operator. Although the mass spectrometer should not be considered as a standard chromatograph detector, it is included since it is often used as such in the analysis of pollutants. A brief description of each of these nine detectors is presented below.

The *flame ionization detector* (FID) is a universal detector that responds to all organic compounds, but it is primarily used for the detection of hydrocarbons. Sensitivity is in the low nanogram-per-peak range (1×10^{-11} g/s). The detector consists of an air/hydrogen flame burning in a polarized environment (Fig. 4). During normal combustion the flame produces very few ions. However, when an organic compound enters the flame, ions are created that are collected by a collector elec-

trode to produce an ion current. The ion current, which is directly proportional to the mass of the compound entering the flame, is then amplified by an electrometer. The detector response is linear over a very wide range (approximately 10^4). The FID requires hydrogen and air burner gases at flowrates of approximately 30 and 300 cm^3/min, respectively. Detector maintenance requires occasional optimization of burner gases and cleaning of the flame jet and collector electrode.

The *photoionization detector* (PID) is similar to the FID in response characteristics. It is primarily used for the detection of organic compounds. Response is approximately ten times greater for aromatic compounds than it is for other compounds and for this reason is recommended for the detection of aromatic pollutants. The PID like the FID is a simple device that is easy to operate and maintain. The detector consists of a high energy UV light source, polarizing electrode and collector electrodes which are contained in a heated housing.

The detector signal is generated from an ion current that is formed by the high energy UV ionization of the compound. The detector is slightly more linear and sensitive than the FID, and detector operation requires no additional gases.

The *electron-capture detector* (EC) is a selective detector that responds to compounds that have a high electron affinity. The detector contains a radioactive foil (usually Ni-63) that ionizes nitrogen or argon/methane carrier gas to produce an electron cloud. This electron cloud is sampled with a collector electrode to produce a standing current. Electron negative compounds, such as the chlorinated pesticides and PCBs, that enter the detector cavity, deplete the electron population that can

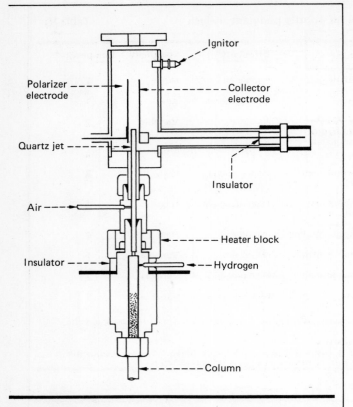

Polarizer electrode

Quartz jet

Air

Insulator

Ignitor

Collector electrode

Insulator

Heater block

Hydrogen

Column

Cross-section of the flame-ionization detector   **Fig. 4**

be sampled, thereby resulting in a detector signal. Modern electron-capture detectors are operated in a constant-current mode in order to extend their linear range to approximately 10^4. This is done by varying the sampling rate as a function of electron demand (compound concentration in the detector). The detector is extremely sensitive and will respond to sample quantities of sensitive materials as low as 10^{-12} or 10^{-13} grams. Due to its high sensitivity, the detector is also sensitive to contamination from GC column bleed and to dirty samples, but with reasonable caution is easy to use and requires little maintenance.

The *nitrogen-phosphorus detector* (N-P) is a FID that has been modified to contain an alkali source between the flame jet and the collector electrode. In recent designs the source contains a nonvolatile alkali such as rubidium silicate. The source is electrically heated to a dull red and held at a negative electrical potential with respect to the collector electrode. Hydrogen flowrates of only 1 to 5 cm^3/min are used, because such a low flow will not support a flame as classically perceived. Instead, a plasma is created around the source. The alkali acts as a catalyst and selectively ionizes certain nitrogen and phosphorus species that are created in the plasma. In this manner, an ion current similar to that in a FID is established. Selectivity to nitrogen compounds relative to hydrocarbons is usually 10^4. The N-P detector is very sensitive and can be used to detect subnanogram quantities of most nitrogen compounds. Response is approximately ten times greater for phosphorus than it is for nitrogen. Frequent adjustment of the source tempera-

ture and hydrogen flowrate may be required to maintain the same detector-response characteristics. In addition, the alkali source may need replacement every one to six months.

The *Hall electrolytic conductivity detector* (HECD) is a specific detector that can be used for the detection of halogens, sulfur, and nitrogen containing compounds. Detector components include a reactor, electrolytic conductivity cell, electrolyte system, and electronics for measuring electrolytic conductivity and heating the reactor. The relationship of these components is shown in Fig. 5. The detector operates by catalytically converting organic halogen, sulfur, or nitrogen to HX, SO_2, and NH_3, respectively. These molecules are then extracted into an electrolyte stream and the resulting change in electrolytic conductivity output is measured as the detector signal. Selectivity is achieved by either removing unwanted reaction products with a scrubber, converting them to substances that do not change the electrolyte conductivity (i.e., CH_4), or suppressing the conductivity of certain ions by using nonaqueous electrolytes. The detector can be operated in two main modes, reductive and oxidative. Halogen and nitrogen compounds are detected in the reductive mode, using hydrogen as the reaction gas. Sulfur compounds are detected in the oxidative mode with air as the reaction gas. The detector exhibits subnanogram sensitivity and selectivities of 10^5 or greater (relative to hydrocarbons). Linear range of operation is from 10^3 to 10^5, depending upon the element detected. Maintenance involves periodic replacement of the electrolyte, reaction tube, scrubber (not used in detecting halogen compounds), and the ion exchange resin.

The *quadrapole mass spectrometer* is the only system that is commonly employed as a chromatographic detector for priority pollutant analysis. This system can be considered as a mass-selective gas-chromatograph detector. Compounds entering the ion source (Fig. 6) are either directly ionized by electrons emitted from a heated filament (electron-impact mode), or indirectly by interaction of a reaction gas that has been previously ionized by electron bombardment (chemical-ionization mode). When the compound is ionized, it fragments into characteristic ions that are separated by the quadrapole filter. The separation process is very rapid and usually takes only a few milliseconds. This process results in the formation of an ion spectrum that can be used to confirm the identity of a compound. The data system can be used to record and compare spectra to those contained in a library (memory), or it can selectively monitor specific ions, resulting in a mass-specific chromatogram. Selection of characteristic ions or ratios of certain ions to be monitored results in high specificity for most compounds.

The ion information, separation, and detection must be performed under high vacuum with the aid of sophisticated electronics. Consequently, most mass spectrometers are fairly complex to operate and maintain. Sensitivity is normally in the nanogram range for most of the priority pollutants. Specificity is very high, but depends upon the ions chosen for monitoring the individual compounds.

The *ultraviolet absorption* (UV) detector is probably the

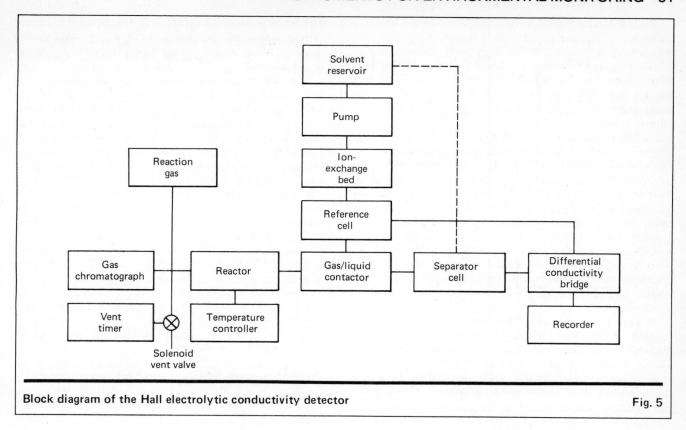

Block diagram of the Hall electrolytic conductivity detector   Fig. 5

most commonly used detector for high performance liquid chromatography (HPLC). UV detectors fall into two classes; namely, fixed wavelength and variable wavelength. Both types of UV detectors operate on the well known principles of photometry, which involve the absorption of light by certain types of organic and inorganic compounds. The amount of light absorbed by the sample compounds is directly proportional to the concentration of the compound in solution. The solvent used as the mobile phase in the chromatographic process must of course be transparent to the ultraviolet light.

Fixed-wavelength detectors operate at a single wavelength of light, which is provided by the emission of special lamps containing appropriate elements. For example, the most commonly used lamp in this type of detector is the mercury-vapor lamp, which has primary emission at 254 nm wavelength. In order to obtain other wavelengths, the lamp must be changed to another type.

Variable wavelength detectors incorporate a broad-spectrum lamp that emits a continuous band of wavelengths. The desired wavelength is then obtained by use of a grating or prism monochromator which, in effect, transmits only the selected wavelength to the sample cell.

Both types of UV detectors are moderately selective in the types of compounds they will detect. Fixed-wavelength detectors are best suited to the detection of aromatic compounds, and can be very sensitive for certain strong UV-absorbing compounds such as polynuclear aromatic hydrocarbons (PNAs). Variable wavelength detectors offer additional selectivity, and sometimes added sensitivity, for many other types of compounds.

Fluorescence detectors depend on the fact that certain types of compounds, when irradiated with light of a certain wavelength, emit light of a longer wavelength, that is, they fluoresce. Fluorescence detectors are somewhat similar to UV absorption detectors in that they require fixed or variable light sources and some type of monochromator. In fluorescence detectors, however, the amount of light *emitted* by the sample is measured rather than the amount of light absorbed.

Since relatively few types of compounds fluoresce, the degree of selectivity provided by fluorescence detectors is generally much higher than that of UV detectors. In addition, for certain types of compounds, fluorescence-detection-sensitivity is much better than by UV absorption.

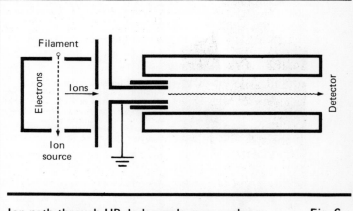

Ion path through HP dodecapole mass analyzer   Fig. 6

Both UV and fluorescence-type detectors are nondestructive of the sample. For this reason, they may be connected in series to the high-performance liquid chromatography column to obtain both types of detection almost simultaneously. By use of a dual-pen recorder, both UV and fluorescence peaks can be plotted on a single chart, providing a great deal of information about a sample.

Another type of detector for high-performance liquid chromatography that is currently gaining acceptance is the *electrochemical detector*. This type of detector is based on the fact that many compounds undergo chemical reactions that cause them to conduct a current when exposed to an electrical potential. The amount of current, or the voltage produced in the sample cell, is measured and then related to the concentration of sample. Detectors of this type may be designed to measure only a type of electrochemical reaction such as, for example, conductance or they may be multipurpose to permit measurement of oxidation and reduction as well as conductance.

A multipurpose type of detector will provide a controlled source of electrical voltage or current and be capable of measuring the current or voltage produced in the sample cell. The sample cell generally consists of two or more electrodes that are immersed in the solution flowing from the chromatographic column. The presence of sample solute in the carrier solvent then causes the electrodes to produce an electrical potential or current, which is measured.

Electrochemical detectors can be very sensitive and specific for certain types of compounds. Detection limits in the picogram range are sometimes obtainable. The use of these detectors, however, puts a number of restraints on the type of mobile phase that may be used for the chromatographic separation since the mobile phase plays an integral part in the electrochemical reactions taking place in the cell.

Summary

Instrumentation for monitoring air and water pollutants has been described. As noted, monitoring of atmospheric pollutants is more firmly established and automated. However, more-sophisticated measurement of trace organics in the atmosphere may be in the offing. Monitoring of water pollutants is in a transitional phase. Much effort is currently being expended to define suitable, yet less-expensive measurement systems. This article has described the currently used systems and presented a general description of the type of instrumentation available.

References

1. *Federal Register,* Vol. 36, Nov. 25, 1971.
2. *Federal Register,* Vol. 44, Feb. 8, 1979, for NAAQS, Ref. and Cal. Method—(Chemiluminescent was also reference method in *Fed. Reg.,* 11/25/71)
3. *Federal Register,* Vol. 36, Nov. 25, 1971.
4. *Federal Register,* Vol. 40, Apr. 25, 1975.
5. *Federal Register,* Vol. 36, Nov. 25, 1971.
6. *Federal Register,* Vol. 43, Oct. 5, 1978.
7. "Quality Assurance Handbook for Air Pollution Measurement Systems, Vol. I, Principles," EPA Environmental Monitoring and Support Laboratory, Research Triangle Park, NC 27711.
8. "Quality Assurance Handbook for Air Pollution Measurement Systems, Vol. II, Ambient Air Specific Methods," EPA-600/4-77-027a, May 1977, EPA Environmental Monitoring and Support Laboratory, Research Triangle Park, NC 27711.
9. "Guidelines for Development of a Quality Assurance Program: Reference Method for the Determination of Suspended Particulates in the Atmosphere (High Volume Method)," EPA-R4-73-0286, Office of Research and Monitoring, U.S. EPA, Washington, D.C., June 1973.
10. American Public Health Assn., American Water Works Assn., and Water Pollution Control Federation, "Standard Methods for the Examination of Water and Wastewater," American Public Health Assn., Washington, D.C. 20036, 1976, pp. 1193.
11. Keith, L. H., and Tellieard, W. A., "Priority Pollutants I—A Perspective View; *Environ. Sci. Technol.* Vol. 13, No. 4, pp. 416–423 (1979).
12. Brenner, N., others, "Gas Chromatography," Academic Press, New York, 1962.
13. Budde, W. L., and Eichelberger, N. W., *Anal. Chem.,* Vol. 51, p. 567A (1979).
14. Finnigan, R. E., others, *Environ. Sci. Technol.* Vol. 13, p. 534 (1979).
15. Snyder, L. R., and Kirkland, J. J., "Introduction to Modern Liquid Chromatography," John Wiley & Sons, New York, 1974.

The authors

Delbert M. Ottmers, Jr. is Assistant Vice-President, Engineering and Chemistry, Radian Corp., 8500 Shoal Creek Blvd., P.O. Box 9948, Austin, TX 78776, telephone (512) 454-4797. He has been involved, at Radian, in process design and development activities, with primary emphasis on air/water pollution problems. He holds B.S., M.S. and Ph.D. degrees in chemical engineering from the University of Texas. He is a member of AIChE, Tau Beta Pi, Phi Lambda Upsilon, Omega Chi Epsilon, Sigma Xi and Phi Kappa Phi.

David C. Jones is Assistant Vice-President, Technical Staff, Radian Corp. His major interest at Radian has been ambient air quality monitoring. He is an authority on instrument selection, calibration procedures, and quality assurance studies for air monitoring stations. He holds a B.S. in chemistry and a Ph.D. in physical chemistry from the University of Texas, and is a member of Alpha Chi Sigma and the National Assn. of Corrosion Engineers.

Lawrence H. Keith is Manager, Analytical Chemistry Div., Radian Corp., where he is responsible for maintaining the technical quality of the analytical staff and assuring that the laboratories keep pace with advances in instrumentation and technology. He holds a B.S. in chemistry from Stetson University, an M.S. in organic chemistry from Clemson University and a Ph.D. in natural products chemistry from the University of Georgia. He is a member of Sigma Xi, Gamma Sigma Epsilon, Kappa Kappa Psi, and ACS (Chairman, Executive Committee, Div. of Environmental Chemistry, and Past Chairman, Central Texas Section).

Randall C. Hall is Department Head/Senior Staff Scientist, Radian Corp., where he serves as head of the Organic Chemistry Dept. and as acting group leader of the Analytical Separation group within that department. He holds a B.S. in chemistry and a Ph.D. in analytical-organic and physical-organic chemistry from Texas A&M University, and is a member of ACS and Sigma Xi.

Sampling and Analyzing Air Pollution Sources

Pollution control depends on the engineer's knowing what is coming out of the stacks. Getting a representative sample and properly analyzing it is a job involving many possible pitfalls. Here's how to avoid them.

N. L. MORROW, R. S. BRIEF and R. R. BERTRAND, Esso Research and Engineering Co.

Immense emphasis is being placed on eliminating and reducing air pollution from stationary sources. So it becomes increasingly important to be able to analyze for various pollutants in these sources.

The engineer is faced with the problem of making these measurements: because he requires such analytical data for designing control equipment, because instrumentation is needed for process control, because abatement performance data are required, and, increasingly, because government regulations call for source measurements. Owing to the high cost of making source measurements, and the limitations and problems associated with current procedures, a basic understanding of the state of the measurement art is required.

In this article we will attempt to present the alternatives available today for measuring the major pollutants, and to highlight the practicalities of applying them. Due to the rapid changes occurring in this area and the limitations of time and space, emphasis will be on the most widely accepted or most promising techniques.

Getting the Sample—A Major Problem

Until recently, there has been little readily available information on stack sampling. This is slowly being corrected. The three-volume set of books by Stern[1] and the biennial reviews appearing in *Analytical Chemistry* [2] are two very useful and general references. Detailed methodologies are available from a number of different sources (e.g., Ref. 3,4,5,6), and new books and articles appear monthly. The stack-sampling methodology published by the Environmental Protection Agency is directed toward specific industries, but can be applied elsewhere.[7]

Before undertaking a source-analysis program, it is very important to decide just what information is needed and how much effort is justified. For rough estimates of emissions, emission factors[8] (average values obtained by previous source-testing of several similar processes) can be applied to determine the approximate discharge from a particular source. In many cases where published data are available, emission data can be calculated. Thus, in

Originally published January 24, 1972

combustion processes, the sulfur dioxide emission-rate can often be most easily determined by measuring the sulfur content of the fuel. If actual stack measurements are necessary, these can often be greatly simplified by careful planning.

Methodologies are always written for the worst case. But prior knowledge of process characteristics and unit construction, together with application of common and engineering sense can often result in substantial simplifications. The trap that must be avoided is the automatic extension of these simplifications from one emission source to another. Most complex chemical operations include both very simple and very complicated emission-measurement problems, and so each source must be approached as a new problem.

Sampling is the keystone of source analysis. More errors in analysis result from poor or incorrect sampling than from any other part of the measurement process. Furthermore, it is sampling that usually controls the cost of the analysis, because proper sampling strategy sets the number of samples necessary for each valid emission measurement and because it controls the locations at which the samples are obtained. Often these locations are hard-to-reach points, sometimes hazardous, and usually high aboveground.

A complete measurement requires determination of the concentration and characteristics of contaminant(s), as well as determination of the associated gas-flow. Most statutory limitations require mass rates of emission; both concentration and volumetric flowrate data are, therefore, required.

Where To Sample

The selection of a sampling site[7] and the number of sampling points needed are based on attempts to get representative samples. To accomplish this, the sampling site should be at least eight stack or duct diameters downstream and two diameters upstream from any bend, expansion, contraction, valve, fitting, or visible flame. For rectangular ducts, the equivalent diameter can be

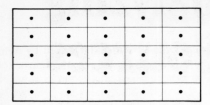

Rectangular stack
(measure at center of at least 9 equal areas)

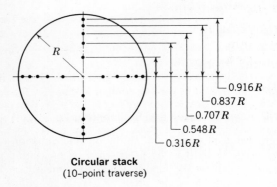

Circular stack
(10–point traverse)

TRAVERSE POINT locations for velocity measurement or for multipoint sampling—Fig. 1

calculated from the expression: Equivalent diameter = 2 (length × width)/(length + width).

After determining the sampling location(s), provision must be made to "traverse" the stack. That is, the actual sampling must be performed at a number of traverse points in the stack. These multiple samples are necessary because of the extreme gradients of flow and concentration that occur in some stacks. The concentrations of even relatively inert gases (i.e., CO_2, CH_4) have been found to vary greatly within a stack. Adding flow variations to the concentration gradients could result in ex-

treme differences in mass-emission calculations if proper traversing is not employed. The number of traverse points required on each of two perpendiculars for a particular stack may be estimated from Table I.

For rectangular stacks, the cross-section is divided into equal areas of the same shape, and the traverse points are located at the center of each equal area, as shown in Fig. 1. The ratio of the length to width of each elemental area should be between 1 and 2. A minimum of nine traverse points should be selected.

For circular stacks, the cross-section is divided into equal annular areas, and the traverse points are located at the centroids of each area, as shown in Fig. 1. When sampling circular stacks, use Table II, which gives the location of traverse points as a percentage of stack diameter from the inside wall to the traverse point.

Gas-Flow Measurements

Once the traverse points have been established, and safe access to the sampling location has been provided, velocity measurements are needed to determine gas flow. These measurements are time-consuming (since no automatic instrument exists) and so are often performed in a very perfunctory manner. This is extremely rash since the mass flow is commonly a large multiplier of the pollutant concentration and so can greatly affect the resulting pollutant mass-emission rate.

Stack-gas velocity is determined from a measurement of the velocity pressure, made by using a pitot tube. The velocity pressure is the difference between the total pressure (measured against the gas flow) and the static pressure (measured perpendicular to the gas flow). Some workers prefer a Type S (Stauscheibe or reverse type) pitot tube instead of the standard design. Both types of pitot tubes are pictured in Fig. 2.

The Type S pitot tube is designed for easy entry into small holes in the stack wall, and because of its relatively large openings does not readily plug when in the pres-

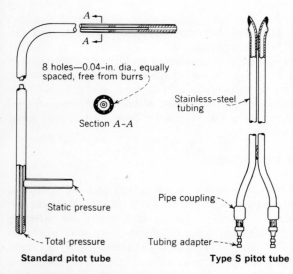

PITOT TUBE varieties—Fig. 2

Selection of Number of Traverse Points—Table I

Number of Stack Diameters Upstream and Downstream of Flow Disturbance		Number of Traverse Points on Each Diameter
Upstream	Downstream	
8+	2+	6
7.3	1.8	8
6.7	1.7	10
6.0	1.5	12
5.3	1.3	14
4.7	1.2	16
4.0	1.0	18
3.3	0.8	20
2.6	0.6	22
2.0	0.5	24

Note: If a different number of traverse points is required by disturbances upstream and downstream, choose whichever number is greater. Data are in accord with Ref. 6 (EPA).

ence of high concentrations of particulate matter. However, it requires a separate calibration for the particular velocity being measured and so does not directly read the velocity pressure.

The standard pitot tube, on the other hand (pointed tip or rounded-nose tip), reads the velocity pressure directly and, therefore, is convertible without correction factor to the velocity at the measured point. Correction factors for the Type S pitot tube ranging from 0.78 to 0.92 have been reported by the Bay Area Air Pollution Control District.[9]

Fig. 3 shows how to convert a pitot-tube reading into velocity and mass flow, and includes a typical data sheet used for stack-flow measurements. This figure shows the conversion between velocity and velocity pressure, including the necessary corrections for the properties of the flowing gas. Account is made of sampling position, the averaging of the square root of the velocity-pressure reading (not the velocity pressure itself), and the gas specific gravity.

Once a flow profile has been obtained, sampling strategy can be considered. Since sample collection can be simplified and greatly reduced, depending on flow characteristics, it is best to complete the flow-profile measurement before sampling or measuring pollutant concentrations. Often it seems convenient to determine flow and concentration simultaneously, but this can require an unnecessarily large number of samples and analyses.

Sampling Strategy

Types of sources have been characterized by Achinger and Shigehara.[10] They concluded that the source characteristics may be either cyclical or continuous as well as either variable or constant throughout the cross-section of the stack (uniform vs. nonuniform). Using these parameters, source characteristics may be placed in four different categories as shown in Table III and discussed in the following paragraphs.

Category 1 presents no time variation, and the emission is relatively uniform across the cross-section of the stack. In this case, only one concentration measurement is needed for accurate results.

Category 2 involves a steady generation of contaminant, but because of ducting configurations, etc., the flow is not uniform across the sampling location, so a traverse is necessary to measure the average concentration. Typically, this is done at the points selected for the velocity traverse. The time of sampling at each point should be the same so as to get a representative, composite sample.

Category 3 is characterized by cyclical operations in which the actual sampling location is ideal, and the variation across the stack is relatively uniform when the operation is running. Because the process involves time

Location of Traverse Points in Circular Stacks—Table II

Traverse Point Number On a Diameter	Number of Traverse Points on a Diameter									
	6	8	10	12	14	16	18	20	22	24
1	4.4	3.3	2.5	2.1	1.8	1.6	1.4	1.3	1.1	1.1
2	14.7	10.5	8.2	6.7	5.7	4.9	4.4	3.9	3.5	3.2
3	29.5	19.4	14.6	11.8	9.9	8.5	7.5	6.7	6.0	5.5
4	70.5	32.3	22.6	17.7	14.6	12.5	10.9	9.7	8.7	7.9
5	85.3	67.7	34.2	25.0	20.1	16.9	14.6	12.9	11.6	10.5
6	95.6	80.6	65.8	35.5	26.9	22.0	18.8	16.5	14.6	13.2
7		89.5	77.4	64.5	36.6	28.3	23.6	20.4	18.0	16.1
8		96.7	85.4	75.0	63.4	37.5	29.6	25.0	21.8	19.4
9			91.8	82.3	73.1	62.5	38.2	30.6	26.1	23.0
10			97.5	88.2	79.9	71.7	61.8	38.8	31.5	27.2
11				93.3	85.4	78.0	70.4	61.2	39.3	32.3
12				97.9	90.1	83.1	76.4	69.4	60.7	39.8
13					94.3	87.5	81.2	75.0	68.5	60.2
14					98.2	91.5	85.4	79.6	73.9	67.7
15						95.1	89.1	83.5	78.2	72.8
16						98.4	92.5	87.1	82.0	77.0
17							95.6	90.3	85.4	80.6
18							98.6	93.3	88.4	83.9
19								96.1	91.3	86.8
20								98.7	94.0	89.5
21									96.5	92.1
22									98.9	94.5
23										96.8
24										98.9

Note: Figures in body of table are percent of stack diameter
from inside wall to traverse point.

STACK VOLUME DATA

STACK NO._____ STATION_____ DATE_____ PAGE_____

NAME OF FIRM_____

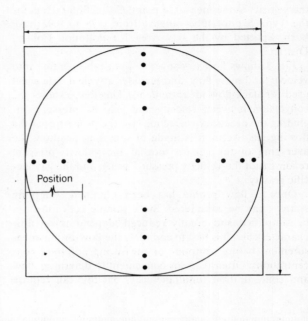

Position

Point	Position, in.	Reading, H, in. of H_2O	$\sqrt{H}$	Temp., t_s, °F.	Velocity, V_s, ft./sec.
1					
2					
3					
4					
5					
6					
7					
8					
9					
10					
11					
12					
13					
14					
15					
16					
	Totals				
	Average				
	Abs. temp., $T_s = t_s + 460 =$			°R.	

Dry bulb temp., $t_d =$ _____ °F. Barometer, $P_b =$ _____ in., Hg

Wet bulb temp., $t_w =$ _____ °F. Stack gage pressure = _____ in., H_2O

Absolute humidity, $W =$ ____ lb. H_2O/lb. dry gas Stack abs. pressure, $P_s =$ ____ $\dfrac{\text{in., } H_2O}{13.6} \pm P_b =$ _____ in., Hg

Stack area, $A_s =$ _____ sq. ft. Pitot correction factor, $F_s =$ _____

Component	Vol. fraction, dry basis	× mol. wgt.	=	wgt. fraction, dry basis
Carbon dioxide		44	=	
Carbon monoxide		28	=	
Oxygen		32	=	
Nitrogen		28	=	

Average dry gas molecular weight, $M =$ _____

Specific gravity of stack gas, $G_s = \dfrac{0.62\,M\,(W+1)}{18 + MW} = \dfrac{0.62 \times \_\_\_ \times \_\_\_}{18 + \_\_\_\_\_} =$ _____
(Ref. dry air at same conditions)

Velocity, $V_s = 2.9\,F_s \sqrt{\dfrac{29.92 \times T_s}{P_s \times G_s}}\ \sqrt{H} = 2.9 \times \_\_\_\_ \sqrt{\dfrac{29.92 \times \_\_\_}{\_\_\_ \times \_\_\_}}\ \sqrt{H} =$ _____ ft./sec.

Volume = _____ ft./sec. × _____ sq. ft. × 60 _____ = _____ cfm.

Standard volume = cfm. × $\dfrac{530}{T_s}$ × $\dfrac{P_s}{29.92}$ = _____ × $\dfrac{530}{\_\_}$ × $\dfrac{\_\_\_}{29.92}$ = _____ scfm.

PITOT TUBE calculation sheet—Fig. 3

Source Characteristics—Table III

Category	Time	Variation Condition Cross-Sectional Velocity
1	Steady	Uniform
2	Steady	Nonuniform
3	Unsteady	Uniform
4	Unsteady	Nonuniform

COMMERCIALLY available stack-sampling train.

variation only, the sampling is conducted at one point for extended periods, usually related to one or more operational cycles.

Category 4, where both the source and flow conditions are nonuniform, requires the most complicated procedure. If there is some measurable cycle related to the process, the sampling can be conducted over this period, using simultaneously collected samples. One sample is collected at a reference point and the other at selected traverse points. This is repeated until a complete traverse is made. Results are corrected by using the reference point data as a measure of the time variation.

Gas Sampling

When sampling for gases, it is necessary to study the temperature variation across the stack. This is done as part of the velocity traverse and indicates variation in gas distribution. If the temperatures are relatively constant, then a single sample point may be all that is necessary. If temperature variation is considered to be large (greater than 5%), traversing is necessary and is usually done at the same points that the velocity traverse is made. The sampling rate and time of sampling at each point are often kept the same to simplify calculations.

In gas sampling, a straight probe fitted with an integral filter (e.g., glass or Pyrex wool) is placed in the stack.

This filter removes particulates at stack temperature, thereby preventing downstream fouling as well as minimizing losses of gaseous pollutants due to reaction with the particulates on cooling. Suction on the nozzle draws the sample into a collection device (such as a bubbler filled with collecting solution) or into a freeze-out trap. The volume of gas remaining after the collected constituent has been removed is measured with a wet or dry test-meter downstream of the collection device; sampling is completed when either a cycle in the process has ended or when sufficient sample has been obtained for analysis.

Stainless steel is usually an acceptable probe material. Glass may also be used, but its fragile nature makes it less desirable. In some special applications (e.g., moderate- or ambient-temperature H_2S and mercaptan sampling), Teflon is the preferred material. Even when grab samples are being obtained materials must be carefully chosen. In general, glass bombs or Teflon, Tedlar or Mylar bags are best. Most plastics, except those cited, should be avoided in stack sampling.

In all cases, minimum probe-lengths should be used, and extended flushing with stack gas should precede sample collection. This flushing is very important if losses to the walls of the probe and sample collector are to be avoided. Often the entire sampling system must be heated to prevent condensation of water, heavy hydrocarbons, or sulfuric-acid mist.

Particulate Sampling

Sampling for particulates requires more-detailed concern about the sampling rate than does gas sampling. Depending on the reason for sampling, the variety and extent of components used in the sampling train will vary. For example, if the chemical and physical characteristics of the aerosols are to be measured, a multicomponent train, or even multiple sampling trains, may be required. On the other hand, if mass loading alone is being measured, a lesser number of components will be needed.

As can be seen from Fig. 4, representative sampling is obtained only if the velocity of the stack-gas stream entering the probe nozzle is the same as the velocity of the stream passing the nozzle. If the sampling velocity is too high (super-isokinetic sampling), there will be a smaller concentration of particles collected (because the inertia

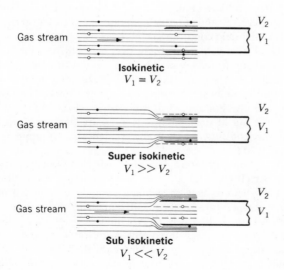

PARTICLE COLLECTION and sampling velocity—Fig. 4

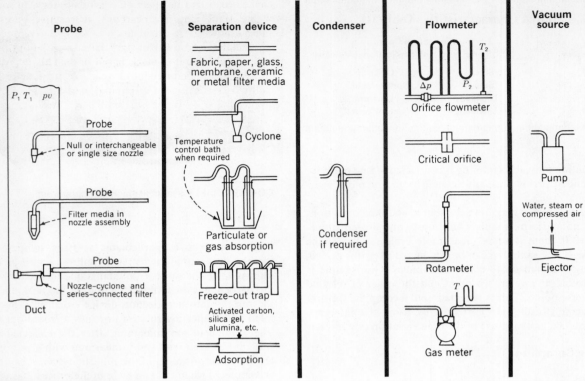

COMPONENTS of common sampling systems—Fig. 5

of the larger particles prevents them from following the stream lines into the nozzle). Alternatively, in subisokinetic sampling, where the sampling velocity is below that of the flowing gas stream, the gas samples would contain a higher-than-actual particulate concentration (because heavier aerosol particles will enter the nozzle, but light particles will be diverted).

It has been found[11] that inertia effects become more significant when particle size exceeds about 3 microns dia. Therefore, if a reasonable proportion of the particles exceed this size, isokinetic sampling is necessary.

The Environmental Protection Agency (EPA)[7] believes that samples that are more than about 20% from isokinetic—i.e., (nozzle-velocity)/(stack-velocity) is not between 0.82 and 1.2—should be rejected and sampling repeated. Even samples within this range, they say, should be corrected by means of a complicated expression. In simplified form, this expression can be stated as follows: (true-concentration)/(sampled-concentration) = ½(1 + nozzle-velocity/stack-velocity). Naturally, correction factors such as this one are based on the assumption of a "normal" particle-size distribution. If a source contains an unusual distribution, correction factors must be avoided. In many cases, isokinetic sampling (with or without correction factors) is used without particle-size data, since isokinetic conditions are needed to obtain valid samples for particle-size-distribution evaluations.

Sampling-Train Construction and Operation

The typical sequence of components in a sampling train is:

- Nozzle.
- Probe.
- Particulate collector.
- Cooling and/or gas collector.
- Flow-measurement devices.
- Vacuum source.

Flow measurements can be made preceding or following the vacuum source. However, vacuum pumps can leak, with the result that gas volumes measured downstream may be greater than those actually sampled. Some commonly used components are pictured in Fig. 5.[11] Ball-and-socket joints and compression fittings allow any desired arrangement of components to be set up rapidly under field conditions.

There is much controversy over particulate-sampling component arrangement. The major point of contention is whether to sample the particulates at stack temperature or at cooler temperatures outside the stack. The Los Angeles Air Pollution Control District (LAAPCD) prefers the out-of-stack sample to measure particulates including constituents that are condensable at approximately 70 F.

The (San Francisco) Bay Area Air Pollution Control District (BAAPCD) and the Environmental Protection Agency believe that particulates should be collected at stack temperature so they will remain in their original form. Though this was EPA's final decision, the controversy surrounding this point is indicated by the fact that the preliminary procedure published by EPA included condensable material in the mass measurement and the final train still collects condensables, though they are not now included in the particulate mass calculation.

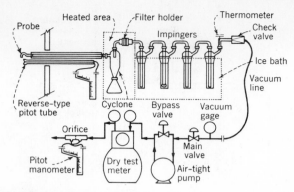

EPA particulate-sampling train—Fig. 6a

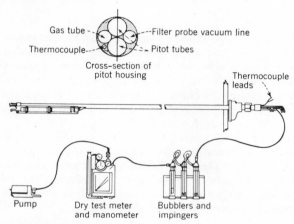

BAAPCD particulate-sampling train—Fig. 6b

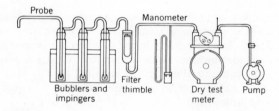

LAAPCD particulate-sampling train—Fig. 6c

BAAPCD uses a deep-bed Pyrex glass filter inserted in the stack with glass probes and connectors. Downstream and outside of the stack are cooling devices (impingers, a flow meter and the vacuum source). EPA's train is similar, but the filter is in an outside heated enclosure.

The LAAPCD train uses impingers rather than filters to collect the particles. Its experience indicates that impinger collection efficiency is usually sufficiently high that a downstream filter thimble rarely collects more than 5 to 10% of the total weight of particulates captured. This means that the LAAPCD sampling train runs essentially at a fixed pressure drop, hence corrections for changes in pressure are not often needed. However, other trains involving filters do require periodic adjustments in pump setting, to overcome increased resistance from collected particulate matter. Fig. 6 shows three commonly employed sampling trains.

This difference in official methods is of great signifi-

cance because of its effect on control strategy. In sources where significant hydrocarbons or reactive gases are present as well as particulates, it is possible to remove the particulates, and meet in-stack based regulations, while still exceeding out-of-stack based ones. This may mean serial use of different controls (e.g., electrostatic precipitation followed by wet scrubbing) or rejection of a technique that is optimum for particulates in favor of one adequate for both particulates and gases.

The probe nozzle is selected—after accounting for changes in temperature, pressure and moisture content (from condensation) in the train—so that the pump can maintain isokinetic velocity. For measurements at a single point this may not be difficult, but for multipoint sampling (which is most common) the mathematical and physical manipulations are often troublesome.

A simplification is to use the null probe, examples of which are shown in Fig. 7. Null-probe designs all involve measurement of static pressure perpendicular to flow both inside and outside of the nozzle. The static-pressure taps are connected to opposite sides of an inclined manometer. In operation, the flow through the sampling train is adjusted until there is no pressure differential across the manometer. This is the null condition, which

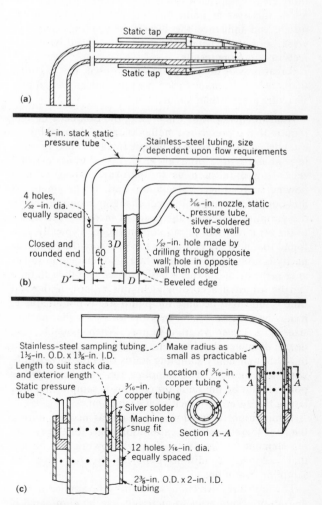

NULL PROBES of various configurations—Fig. 7

in theory presents a situation where isokinetic sampling occurs.

Null sampling probes, however, do not guarantee isokinesis because (even though the static pressures are equal) there may be differences in velocity between the inside and outside of the probe. The differences in turbulence for duct and probe flow; the nozzle shape, and its degree of surface roughness; and the location of the static holes—all may affect the relation between balanced static pressure and isokinetic flow. However, the error (estimated to be no more than 20% for a balanced-null probe) may still be acceptable in rapidly changing flow conditions, because there is also a high possibility of error using the standard nozzle method of sampling. To simplify source testing, some governmental agencies use the null method when checking compliance with air pollution regulations.

As a check on the stability of gas flow, a pitot tube can be placed at a reference point in the stack or located at the traverse point just prior to sampling. In a design adopted by EPA, an S-type pitot tube and a hook sampling nozzle are mounted together so as to continuously measure velocity while sampling. It was felt by EPA that adjustments in sample flow could be more rapidly applied to meet changes in stack flow conditions, and thus more closely approach isokinetic sampling overall. Critics of this system point out that the proximity of the nozzle can influence the pitot tube readings and vice versa, so that the benefits may be outweighed by the effects on sampling results.

Regardless of the sampling system used, it is essential that attempts be made to sample isokinetically when particles are greater than about 3 μm.

Adjustments in flow, when using the isokinetic design, require that a prescribed methodology of data-taking be established. A preferred technique involves making readings at fixed time intervals (e.g., 2 to 5 minutes) and recording data such as temperature, pressure and flowmeter reading. Calculation and readjustment of flowrate are then made to meet changing requirements in the sampling system as the result of increased resistance to the air flow when filtration is included. Typically, it is found that near the end of the run or where the pump capacity is no longer adequate, the rate of change in resistance is the greatest. This is easily sensed and the run time can be adjusted accordingly. In this regard, if the run time is too small because the pump size is not adequate, it may be necessary to rerun with either a smaller nozzle-size or a larger vacuum-source. When possible, the nozzle diameter should not be less than approximately ¼ in., although many sampling trains operate successfully with ⅛-in. nozzles. The sampling time at each traverse point should be the same for composite samples, regardless of the differing velocities at each point.

Temperature corrections are made as needed during the run; there must be a calibration chart for the metering element (viz., pressure drop vs. flowrate), to ensure that isokinetic conditions exist for the full range of sampling conditions.

The moisture correction, which must be applied if condensation occurs prior to metering, can be eliminated if the particulate collection and metering are conducted

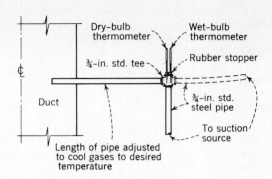

HOT-GAS psychrometry measurement—Fig. 8

above the dewpoint. Rate meters that can be used in this way include orifices and venturi meters. If condensation does occur, it is necessary to determine the condensate's vapor fraction in the total gas-volume samples. This can be done simply by drawing a sample of the hot stack-gas through a condenser and gas meter, or by use of hot-gas psychrometry, which is illustrated in Fig. 8.

At the completion of the run, the sampler must be taken to a clean area, and dust that has been collected in the nozzle, probe and collecting elements must be washed or brushed into the succeeding collector. The total catch in each stage is measured and all stages are summed to obtain total particulate-mass loading.

Fixed Gases

It is essential to know the average molecular weight of the flowing gases to determine the actual velocity or volumetric flow in a stack. In stacks where air is present or where combustion gases are emitted, fixed gas (N_2, CO_2, O_2) concentration data are required for this calculation. In addition, these gases are indicators of equipment operation; thus, their measurement is often desirable as a process-control tool.

Orsat Analysis—The above three gases are most easily measured by the Orsat technique. While Orsat analysis is not extremely accurate, it is sufficient for the purpose of determining average molecular weight and many unit operating parameters. Portable Orsat analyzers are available from many suppliers and have been in use for flue-gas analysis for decades. In these devices a measured gas sample is passed through several reagents, and the decrease in gas volume after passage through each reagent is determined. Reagents that are reasonably selective absorbers include 20%-40% aqueous KOH for CO_2 (and other acid gases), alkaline pyrogallol for O_2, and acidic cuprous chloride for CO. Nitrogen is determined by difference.

GC Analysis—In many places, gas chromatographic (GC) analyses are used to determine effluent composition. Generally a sample is caught in a glass bomb or plastic bag and returned to the laboratory for analysis. This technique has the advantage of measuring minor constituents accurately, as well as major ones.

O_2 and CO_2—Continuous monitoring of O_2 and CO_2 is possible using electrometric and nondispersive infrared (NDIR) analyzers, respectively. Normally, these continuous measurements are used for monitoring the operation of process equipment and are not justified for pollution monitoring alone.

Carbon Monoxide

Of the fixed gases, CO is the only pollutant, and so is the only one for which very accurate or continuous analysis may be necessary for pollution reasons alone. The major source of CO is automotive emissions, but significant emissions also occur from stationary-source fuel combustion and a myriad of industrial processes.

The classical procedure for measuring CO has been by passage of the gas sample over hot iodine pentoxide, followed by titration of the iodine generated. In general, gas chromatography (GC) has supplanted this method for source measurements. For very low CO levels (less than 50 ppm.), it is necessary to convert the CO to methane prior to GC analysis, since the very sensitive flame ionization detector does not respond to CO. Infrared spectrophotometry is also often used for spot CO analysis.

Continuous monitoring for CO is usually performed by nondispersive infrared (NDIR) analyzers. These instruments require that the gas sample be filtered and cooled, but no other pretreatment is necessary. Often a narrow-pass optical filter is included in the instrument to minimize CO_2 and H_2O interference. NDIR monitors have the advantages of rapid response and good sensitivity over a wide range of concentrations. Unfortunately, they are prone to drift, and fairly frequent zeroing and calibration may be necessary.

Sulfur Compounds

The sulfur compounds (SO_2, SO_3, H_2S, and mercaptans) comprise a major class of pollutants. They are generated during combustion, and also ore roasting, paper manufacture, and a wide range of other industrial operations. In many stacks, only one type of sulfur compound is present. In these situations, a total-sulfur analyzer may be preferable to a measurement specific for SO_2 or H_2S. Most of the wet-chemical systems are total-sulfur analyzers in any case, but instrumental monitors also exist. One that is finding increasing use is the flame photometric analyzer. In this device, the gas sample is burned in a hydrogen flame, and the intensity of a sulfur emission line is measured by a photomultiplier. This device is very sensitive, requires little sample pretreatment, and is not much different than a flame ionization detector (which many process instruments use). However, it does drift and is sensitive to small variations in gas pressures and flows.

Sulfur Oxides

In general, the measurement of sulfur dioxide involves two problems: obtaining a valid sample and eliminating interferences. Because of its reactivity, SO_2 is best captured by using bubblers. A heated or well-insulated, and flushed, sample line is best so as to prevent losses to the walls of the line. The probe system should contain a fil-

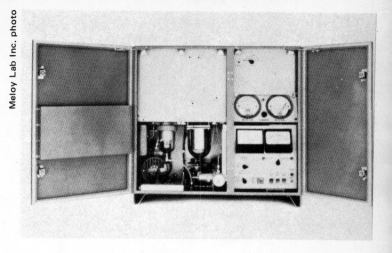

Meloy Lab Inc. photo

FLAME photometric analyzer for sulfur compounds.

ter, which could be packed with quartz or Pyrex wool, in order to remove particulates.

Perhaps the most widely used scrubber for SO_2 collection is hydrogen peroxide. This solution converts the SO_2 to sulfate, which can be determined by a number of standard analytical methods. Since this scrubber is not specific for sulfur dioxide, a sulfate-specific analytical procedure is best. A popular technique is titration with barium perchlorate, using thorin as an indicator. A pink coloration indicates the endpoint. Alternatively, the sulfate ion can be determined colorimetrically with barium chloroanalate reagent, which releases highly colored chloroanalate ion on reaction with sulfate. Color intensity is measured at 530 nanometers. In all these analyses, it must be remembered that SO_3 (but not sulfuric-acid mist, which is trapped on the filter) and H_2S will be measured along with SO_2.

In the sulfuric acid industry, and in some industrial processes using sulfuric acid, there are emissions of SO_2, SO_3, and/or sulfuric acid mist. A variety of sampling schemes, some of which are illustrated in Fig. 9,[12] have evolved for determining SO_3 and SO_2 separately. Sulfur trioxide and sulfuric acid mist are collected and separated from SO_2 by either filtration above the dewpoint of water or absorption in 80% isopropyl alcohol. In either case, the SO_2 passes through. The sulfate is analyzed colorimetrically or by titration as mentioned above for SO_2. Ref. 12 details sulfuric-acid-industry air-pollution problems, and many of the measurement methods used to characterize them.

Since sulfur dioxide was one of the first pollutants to be regulated, considerable effort has been devoted to instrumental monitors. Table IV lists some of the commercially available instruments. In general, all of these suffer from a lack of adequate sampling and sample pretreatment. Thus, they are frequently fouled by mist and particulates, interfered with by unremoved gases and water vapor, or overloaded by widely fluctuating sample concentrations, temperatures or pressures. If installation of a monitor is required, it must be individually selected for the specific stack it will monitor, and as much attention

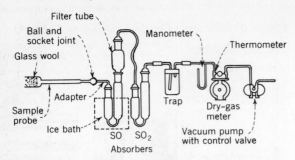

SAMPLING train for SO_2/SO_3 (Shell)—Fig. 9a

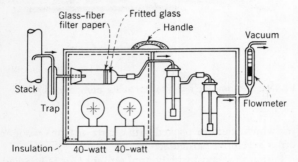

PORTABLE SO_2/SO_3 apparatus (Chemico)—Fig. 9b

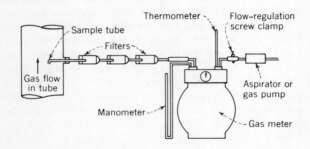

MIST sampling train for SO_3 (Monsanto)—Fig. 9c

must be paid to the sampling system and to ease of calibration and maintenance as to the analyzer itself.

Spectrometry—The spectrometric methods are all prone to fouling by particulate. In addition, those using the IR (infrared) portion of the spectrum require removal of water from the sample. Achieving this without also removing some SO_2 is difficult.

The correlation spectrometer (which uses the ultraviolet region of the SO_2 spectrum) is worth special mention because it is a true in-stack monitor. While the other types of monitors require withdrawal of a sample from the stack, this instrument uses the stack as its optical path, thereby providing cross-stack average measurements. This approach can be a major advantage, particularly in large-diameter stacks.

In this instrument, the fine structure of the SO_2 absorption spectrum is matched against a reference pattern in such a way that other materials do not interfere, even if they have some overlapping absorptions. Unfortunately, the engineering problems associated with using this spectrometer routinely, without fouling and with adequate

calibration, have not been completely resolved. These problems and its relatively high price have prevented a completely successful application of its potential.

Wet-Analysis Instruments—The many wet-analysis instruments available suffer from lack of specificity and the operational problems associated with flowing liquids. In some cases, the conductiometric analyzers require precise dilution of the sample. An advantage of these instruments, however, is the ability to calibrate the analyzer by using a liquid standard.

Electrochemical Sensors—The electrochemical sensors are a relatively new class of SO_2 analyzers. In these devices, an electrochemical reaction selectively measures the SO_2 in a gas sample, which has been extracted from the stack, cooled and had the particulates removed. These sensors are reasonably specific for SO_2, but suffer from drift problems.

Sulfides

Classically, sulfides have been determined by wet-chemical techniques. Specifically, hydrogen sulfide can be scrubbed from a gas sample and determined tritrimetrically. If, as in kraft-mill stacks, SO_2 is present, a more specific procedure, such as the colorimetric methylene blue method, can be used.

When process monitoring has been necessary, organic sulfides and hydrogen sulfide have been determined in kraft-mill stacks by first removing the sulfur oxides in a condenser and scrubber, and then oxidizing the sulfides in a quartz oven at 700 to 800 C. and analyzing them in an amperiometric titrator.

Paper-Tape Monitor—Hydrogen sulfide is frequently determined by using a paper-tape monitor. On exposure to H_2S, the lead-acetate-impregnated tape turns black. The density of the black spot is measured in a transmission photometer. This measurement gives the average H_2S concentration for the sample period (typically 1 to 4 hr.). Besides the problem of periodic tape replacement, this type of monitor requires frequent recalibration and careful tape handling. Sometimes, ambient H_2S per-

Types of SO₂ Monitors—Table IV

Approach	Manufacturers*
UV Absorption	DuPont
	Honeywell
Correlation spectrometry	CEA Barringer
NDIR	Beckman
	MSA
Titrimetric	CEC
	ITT Barton
Conductometric	Davis
	Wösthoff
Electrochemical	Dynasciences
	Theta Sensors
	EnviroMetrics
Flame photometric	Mel-Labs, Inc.

*Partial List

Theta Sensors Inc. photo

ELECTROCHEMICAL sensor for NO_x and SO_2

Types of Nitrogen Oxide Monitors—Table V

Approach	Oxides Measured	Manufacturers*
UV photometric	NO_2, NO	DuPont
	NO_2	Honeywell
Correlation spectrometry	NO	CEA Barringer
Electrochemical	NO_x, NO_2	Dynasciences
	NO, NO_2, NO_x	Environmetrics
	NO_x	Theta Sensors
Chemiluminescence	NO	Aerochem
NDIR	NO, NO_2	MSA
	NO	Beckman
	NO	Intertech
NDUV	NO	Beckman

*Partial List

meates the unexposed tape and darkens it beyond use.

Chromatography—More recently, a GC analyzer, which separates the individual sulfides, has been developed. It uses a Teflon column system and a polyphenyl-ether substrate. A sulfur-specific flame-photometric detector is employed to eliminate possible interferences from nonsulfur compounds. This technique has been successfully[13] used for process monitoring of sulfide effluents.

Nitrogen Oxides

The oxides of nitrogen play a major role in the formation of smog and accordingly are important air pollutants. The most important nitrogen oxide pollutants are nitric oxide (NO) and nitrogen dioxide (NO_2). These two gases are sometimes measured individually, but in the waste gases from stationary sources either NO or NO_2 is usually predominant and often a measurement of total nitrogen oxides (NO_x) is sufficient. Nitrogen oxides are present in the effluent gases from all combustion sources, and in the waste gases generated from the production of nitric acid, nitration processes, metal pickling, and the lead chamber process.

Laboratory Analyses—When the oxides can be determined without differentiation as NO_x, the phenoldisulfonic acid method of analysis is usually used. While this method is tedious to run, it is one of the few air pollution methods generally recognized to be accurate and reliable.

The gas sample is collected in an evacuated flask containing dilute sulfuric acid - hydrogen peroxide absorbing solution. The nitrogen oxides, except for nitrous oxide (N_2O), are oxidized to nitric acid by the hydrogen peroxide. After careful destruction of the peroxide with heat, the nitric acid produced is measured colorimetrically as nitrophenoldisulfonic acid. This method is suitable for NO_x concentrations between 15 and 1,500 ppm. by volume and has a sensitivity of about 1.5 ppm.

When no acidic gases (e.g., SO_2) other than nitrogen oxides are present, the nitrogen oxides can be deter-

mined acidimetrically. This method is much simpler and more rapid than the phenoldisulfonic-acid procedure. The gas is sampled into a gas buret and vigorously shaken with dilute peroxide and an antifoam agent. The nitrate formed is determined by titration with sodium hydroxide to the methyl-orange endpoint.

Table V lists some of the nitrogen oxide instrumentation now available. Many of these instruments were developed for automobile-emission monitoring and are untried for stack use.

Photometric Analyzer—A completely packaged split-beam photometric analyzer system, including sample handling, is available for continuous monitoring of stationary source emissions of NO_2 and NO_x. The gas sample is continuously drawn through a filtered probe. Sampling components, including the lines and the sampling cell, are kept at elevated temperature to avoid condensation. The analyzer provides automatic compensation for light decreases caused by changes in source intensity. Nitric oxide is essentially transparent in this region of the spectrum and cannot be detected directly. It is measured by quantitatively converting the NO to NO_2, using oxygen under pressure, and measuring the NO_2.

Spectroscopic Analyzer—A spectroscopic technique has been used to continuously monitor industrial emissions of NO. In this technique, the radiation source is mounted on one side of a tube that traverses the stack. This tube is slotted perpendicular to the gas stream to provide a fixed path-length of sample. This permits determination of the spatial average of the NO across the stack.

Electrochemical Method—A portable instrument is available that is based on an electrochemical transducer capable of selectively monitoring nitrogen oxides at concentrations up to 5,000 ppm. This transducer is a sealed faradic device in which the direct electro-oxidation or electro-reduction of absorbed gas molecules at a sensing electrode results in a current directly proportional to the partial pressure of the gas. In operation, the pollutant gas diffuses through a selective membrane and a thin electrolyte layer, to become absorbed at the sensing electrode

DuPont. photo

SPLIT-BEAM photometric analyzer for nitrogen oxides.

where it undergoes reaction. This is the same electro-chemical device described previously for sulfur oxide measurement, except that the sensing cell itself is different. SO_2 is not differentiated from NO_x by the NO_x sensor, but NO_x does not interfere with the SO_2 sensor. Common practice, therefore, is to use both types of sensors and to correct the NO_x measurement based on the SO_2 level. In fact, the difference is taken automatically in some models of this instrument.

Chemiluminescent Detectors—An optical detection device which is based on the chemiluminescent reaction of NO with ozone has recently become available. This device is selective for the measurement of NO for concentrations up to 1,000 ppm. Application of this device has been limited to automobile-emission measurement, but extension to stationary-source monitoring can be expected to occur quickly.

Particulates

Particulates are defined as all airborne solid and liquid matter. They include solid particles and liquid aerosols. Particle diameters range from a few hundred angstroms to larger than 50 microns (50,000Å).

Particulate determinations fall into two categories: opacity measurements and mass measurements. No generally acceptable correlation between these parameters exists except in very special cases.

In some particular situations (mostly research programs), detailed breakdowns of particulates into size ranges are required. Though there are many different approaches to size classification, the most generally accepted one is the use of inertial impaction devices. Since size segregation is a very specialized problem, it will not

be discussed further in this particular article.

Opacity

Ringlemann Numbers—The grossest and most widely used particulate measurement is opacity. Most pollution codes limit emissions of dark particulates by setting "Ringelmann Number" limits. In theory, such a number corresponds to the light transmittance of a plume. Zero Ringelmann indicates 100% transmittance, 1 indicates 80% transmittance, 2 indicates 60% transmittance, etc. In practice, Ringelmann "measurements" are made by "trained" observers. As a result, there is considerable variation in the measurement. Particulate size and color, background effects, time of day, distance, and other factors all affect the observer's judgment. However, these measurements have been upheld in court and are in use almost everywhere in the country.

Photometers—Recently, a number of photometers have been introduced to quantify the Ringelmann measurement. While these devices are new and untested, they are expected to improve the reproducibility of the measurement. They still suffer from the basic problems of this measurement—the requirement of daylight and black smoke.

Obviously, many stacks require virtually continuous monitoring to control smoke emissions, and so continuous in-stack opacity measurements are attempted. These measurements are not useful when there is a detached plume that is essentially invisible in the stack, but becomes visible some short distance after leaving the stack. Also, when the stack gases are opaque because of condensed water vapor, this method does not serve a useful purpose. Water is generally not considered to be an air pollutant. However, the in-stack measurement of opacity is valuable in those situations where suspended particulates are the major contributor to opacity. In one State, this type of device will be required in all suitable stacks with a flow greater than 10,000 cfm. The opacity limitation will be a minimum of 70% light transmittance across the stack, which corresponds typically to a 1-2 Ringelmann Number in the plume.

In order to provide continuous opacity measurements, a large number of stacks have been outfitted with transmission photometers. As shown in Fig. 10, these devices have a light source on one side of the stack and a detector on the other side. The reduction in light intensity on the beam's passing through the stack is measured.

The major problems with these devices are their propensity for becoming fouled and the difficulties associated with *in situ* calibration. Attempts to keep the optics clean and dustfree have taken two main forms. First, the components are recessed away from the stack; second, clean air is continually swept over the exposed areas. Often these actions are inadequate and measurement accuracy suffers.

After the instrument has been in use for a time, the optics can become dirty and scratched and the detector and light-source characteristics can change. Therefore, regular recalibration is necessary. But there is no really acceptable way of doing this recalibration, except for 0% transmission, unless the stack is taken out of use. Since

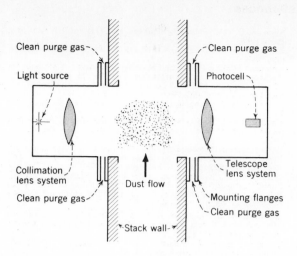

TRANSMISSION PHOTOMETER schematic section—Fig. 10

shutdowns are often impossible, some systems use a tubular sleeve to connect the light source and detector. A transmission of 100% can be checked by flushing the tube with clean gas.

Setting the 100% transmission by removing the detector from the stack and calibrating separately is the normal procedure (when recalibrations are done at all), but this will not really represent 100% in a real stack, primarily because of in-stack alignment variations and differing light-source characteristics. Furthermore, no test of intermediate transmission levels is possible except under laboratory conditions.

Mass Monitors

Opacity measurements are useful for controlling particularly bad (and nuisance-type) emissions, especially black particulates, such as found in incinerators and power-plant stacks. However, newer air-pollution codes are aimed at controlling all particulates, day and night, regardless of color. To achieve this, regulations controlling the total mass of emissions have been promulgated. In order to measure mass, two factors must be known: particulate concentration in terms of weight, and the gas flowrate. In the previous discussion of sampling, determination of gas flowrate was discussed, as was isokinetic sampling. If we assume that gas flowrate has been determined and that isokinetic sampling is used, it is only necessary to determine particulate concentration to know the mass emitted.

Before discussing particulate-concentration measurements it is important to reiterate the problem of defining particulates. If the particulates are collected at stack temperature (e.g., Bay Area approach), only solids are retained. If the sample is cooled prior to collection (e.g., Los Angeles approach), condensable liquids will be included. This definition is of great concern to the chemical process industries because in many processes the condensable mass equals or exceeds the mass of solids present.

Many different approaches to particulate-mass measurements are in use, including some continuous and semicontinuous ones; but the most common method by far is collection of the particulates by filtration, followed by drying and weighing. It is generally agreed that no completely acceptable filter medium is available. However, flash-fired glass fiber is usually used in hot service, and paper is commonly used in cool service. The glass-fiber paper itself is usable to temperatures as high as 900 F. Unfortunately, this temperature limit is misleading because most commercial filter holders have a 400 F. limit.

Once a filter sample has been collected, it is possible to perform detailed particulate analysis. Most metals, for instance, can be determined by standard methods such as extraction and spectrophotometry. Often microscopic examination yields valuable data on the types, and possibly sources, of the particles.

Considering the cumbersomeness of particulate collection trains and the sampling time needed to obtain a single mass measurement, it is readily apparent why so few good data exist. Obviously, one major improvement would be a continuous measurement. Isokinetic sampling would still be required, as would sampling traverses; but the time at each sampling point and the filter pretreating and weighing times would be slashed.

Piezoelectric Monitor—One currently marketed instrument that may achieve this result is the piezoelectric mass monitor. In this device, particles in the sample stream are electrostatically deposited onto a piezoelectric sensor. The added weight of particulate changes the oscillation frequency of the sensor in a known way. This instrument cannot handle the very high particulate loadings found in many stacks without dilution of the sample. This dilution step greatly complicates the sampling problems associated with use of the device since two isokinetic samples are required: one of the stack and one of the diluted stack sample (see Fig. 11[14]). These problems are being studied under an EPA contract, and a generally applicable system could well result.

Beta Attenuation Monitor—A monitoring system that is sometimes used on stacks is beta attenuation; commer-

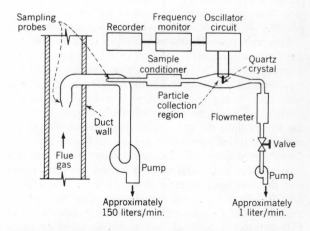

DUAL ISOKINETIC SAMPLING for particle-mass—Fig. 11

cial instrumentation is available that permits a particulate measurement every 15 to 30 min. In this type of device, the particulate sample is filtered using a continuous filter tape, and the mass of particulate filtered out is determined by measuring its attenuation of beta radiation. Since beta attenuation characteristics are not very different for a wide variety of stack particulate-matter compositions, a direct mass-measurement is possible. The major problem with this system is still sampling, though the difficulty is somewhat reduced because in a monitoring application the probe can be fixed and very accurate numbers are unnecessary.

Hydrocarbons

Hydrocarbons are emitted in the waste gases from numerous petroleum and chemical operations, from solvent cleaning systems, coke manufacture and many other sources. These compounds enter into atmospheric photochemical reaction processes that lead to the products and manifestations commonly associated with photochemical air-pollution. Hydrocarbon pollutants are usually determined either as a class, e.g. total hydrocarbons, or as individual chemical species. In the former case, the separate determination of methane is normally necessary, because most class regulations are based on non-methane hydrocarbons. These are determined by difference unless knowledge of the composition of the stream makes this unnecessary.

Flame Ionization Detector—The flame ionization detector (FID) is the most sensitive and most common technique for the continuous detection of total hydrocarbons. In this technique, a sensitive electrometer detects the increase in ion intensity resulting from the combustion of any organic compound in a hydrogen/air flame. Response is proportional to the number of carbon atoms combusted per unit time. As a result, FID data must be expressed with reference to the calibration gas used—e.g., "ppm. of carbon as propane."

The FID is very suitable for hydrocarbon measurement because it does not respond to other air contaminants such as CO, CO_2, H_2O, SO_2 and nitrogen oxides, but merely indicates compounds with C-H bonds. Many companies manufacture FID instruments, but their application to date has been primarily to ambient-air analysis. Thus, source-sampling systems and even explosion-proofing are only available on a special-order basis.

Spectroscopic Methods—Analysis for a specific hydrocarbon requires the isolation of the compound to be determined from a gas stream that normally contains many similar compounds. Except for a few hydrocarbons with very strong spectral adsorption bands, the interferences produced by other hydrocarbons present in the gas stream limit the application of all but very-high-dispersion, and thus very-expensive, spectral techniques.

Chromatography—Gas chromatography (GC) provides a convenient and tested method for the analysis of specific hydrocarbons. In most cases, the same type of process GC used for process monitoring can be used for measuring air emissions of a particular compound. The FID detector is generally preferable to other detectors because of its high sensitivity and good reliability.

References

1. Stern, A. C., "Air Pollution," 2nd ed., Vol. I, II and III, Academic Press, 1968.
2. Air Pollution, *Anal. Chem.*, **43**, 1R (1971); **41**, 1R (1969); **39**, 10R (1967).
3. American Soc. for Testing and Materials, Standards of Methods for the Sampling and Analysis of Atmospheres, Part 23, 1971.
4. "Source Sampling Manual," Los Angeles Air Pollution Control District, Los Angeles, Calif., 1963.
5. American Soc. Mechanical Engineers, Power Test Codes, N.Y., 1941.
6. Cooper, H.B.H., Jr., Rossano, A. T., Jr., "Source Testing for Air Pollution Control," Environmental Sciences Services Div., Wilton, Ct., 1970.
7. Environmental Protection Agency, "Standards of Performance for New Stationary Sources," Notice of Proposed Rule Making, *Fed. Reg.*, **36**, No. 159, pp. 15704-15722, Aug. 17, 1971.
8. Duprey, R. L., "Compilation of Air Pollutant Emission Factors," Public Health Service Publ. No. 999-AP-42, Washington, D.C., 1968.
9. Brief, R. S., State of the Art of Emission Testing, *Heating Piping Air Conditioning*, **43**, pp. 93-96, Sept. 1971.
10. Achinger, W. C., Shigehara, R. T., A Guide for Selected Sampling Methods for Different Source Conditions, *J. Air Pollution Control Assn.*, **18**, p. 605 (1968).
11. Bloomfield, B., in "*Air Pollution*," Vol. II, 2nd ed., A. C. Stern, ed., Chap. 28, Academic Press, New York, 1968.
12. "Atmospheric Emissions from Sulfuric Acid Manufacturing Processes," Public Health Service Publ. No. 999-AP-13, Raleigh, N.C., 1965.
13. Mulik, J. D., others, An Analytical System, Designed to Measure Multiple Malodorous Compounds Related to Kraft Mill Activities, Presented at TAPPI Water and Air Conference, Apr. 4-7, 1971, Boston, Mass.
14. Sem. G. J., Borgos, J. A., Olin, J. G., "Automatic Monitors of Particulate Mass Emissions from Stationary Fossil-Fuel Combustion Sources," Presented Am. Inst. Chem. Eng. 68th National Meeting, Houston, Tex. Mar. 3, 1971, Paper 68d.

Meet the Authors

Morrow   **Brief**   **Bertrand**

Norman L. Morrow is Head of the Special Problems Group in the Laboratory and Environmental Control Dept. of Enjay Chemical Co., P. O. Box 241, Baton Rouge, LA 70821, where he is concerned with all aspects of petrochemical pollution abatement. Previously, he spent three years at Esso Research and Engineering Co., specializing in air pollutant measurements. He has a B.S. from Stevens Institute of Technology and a Ph.D. in physical chemistry from the University of Connecticut. He is a member of Sigma Xi, ACS and Air Pollution Control Assn.

Richard S. Brief is a Senior Engineering Associate in the Medical Research Div. of Esso Research and Engineering Co., Linden, N.J. 07036. He has specialized in air pollution sampling and related activities. He graduated from The Cooper Union and has worked for the U.S. AEC and as a commissioned officer in the U.S. Public Health Service. He is a member of many technical societies and is a registered professional engineer in the State of New York.

Rene R. Bertrand is a Senior Research Chemist with the Government Research Laboratory at Esso Research and Engineering Co., Linden, NJ 07036. During the past two years, he was project leader for a team defining the R&D needs in air-pollution-measurement techniques and the market for air-pollution instrumentation, under an Environmental Protection Agency contract. He has a B.S. in chemistry from Worcester Polytechnic Institute and a Ph.D. in physical chemistry from MIT. He is a member of the Air Pollution Control Assn.

Air-pollution instrumentation

Instrumentation is available for monitoring ambient-air quality and stack emissions. And some devices may be required by law for your plant.

Herbert C. McKee, *Southwest Research Institute*

☐ Did you ever think of an air-pollution control engineer as a quality control engineer? You should. His job is to control the quality of the atmosphere around the plant, to the extent that the plant has an effect on that quality.

As applied to air quality, the basic control principles are much the same as in any other quality-control problem: find out which quality variations can be accepted, identify what must be measured to determine significant variations, find or develop a method of measurement adequate for the intended purpose, obtain representative samples of the material being measured, apply the method of measurement, and then use statistical analysis on the results to estimate the true quality of the atmosphere or of the emissions from the plant.

In considering different measurement methods, it is helpful to distinguish between ambient-air measurements and emission measurements. The term "ambient atmosphere" is used to mean, literally, the atmosphere around us. This usually means a general community atmosphere not influenced by any single nearby source of pollution. Emission measurements, as the name indicates, are measurements of what is going out of a stack or vent line to enter the atmosphere (and, later, influence ambient-air quality). Ambient-air measurements are usually expressed in units of concentration, i.e., parts per million. Emission measurements are also frequently expressed as concentration, such as grains per cubic foot, but are normally measured in such a way that rates can be calculated in terms such as pounds per hour. Governmental standards may be expressed as either emission rate or as concentration in the flue gas or other waste stream emitted to the atmosphere.

Ambient-air measurements

Some cities have been maintaining at least a minimum ambient-air monitoring network for several decades. During that time, many different means have been used to measure air quality, and frequently these are not comparable. As one example, the results obtained with a paper-tape sampler that measures the soiling properties of the atmosphere cannot be related directly to the total weight of dust and particulates in the atmosphere that will be measured by a high-volume sampler.

The most significant action so far that has resulted in at least some standardization of methods of measurement was the promulgation by the U.S. Environmental Protection Agency (EPA) of ambient-air-quality standards in 1971, shortly after the passage of the Federal Clean Air Act Amendments of 1970[1]. By this action, ambient-air-quality standards were established for the contaminants shown in Table I, and states were required to develop implementation plans to achieve these standards. To demonstrate compliance, monitoring networks were also required, and Table I also summarizes the methods of measurement specified.

Despite this standardization, techniques are still less than desirable in the field of ambient-air measurement. With some contaminants, the limitations are more serious than with others. For example, the methods specified for particulates and sulfur dioxide (and the original method for nitrogen dioxide) are manual methods in which samples are collected in either a bubbler or a mechanical sampler for later weighing or chemical analysis. Labor costs for these manual methods are relatively high, and results cannot be obtained until some time after collection of the sample. This is not too serious with 24-hour particulate samples by the high-volume method, but SO_2 data are needed on a short-term basis, especially if concentrations increase to the level requiring emergency control actions during an air-pollution episode.

To avoid the unreasonable cost inherent in using manual methods for a comprehensive community network, it has been necessary to permit the use of continuous instruments wherever possible. For sulfur dioxide, instruments are available that use the same chemical reaction system as does the manual method, so these devices can be made reliable and accurate. For particulates, no satisfactory method is now available for im-

Originally published June 21, 1976

Ambient-air quality standards and methods of measurement	Table I

Contaminant	Method of measurement
Sulfur dioxide	Manual; bubbler sample, analyzed by colorimetric *p*-rosaniline procedure.
Particulate matter	Manual; filter sample collected with high-volume sampler, deposit measured gravimetrically.
Carbon monoxide	Instrumental; infrared absorption, instrument calibrated with standard gas mixtures.
Photochemical oxidant	Instrumental; chemiluminescent reaction with ethylene detected by photocell, instrument calibrated with ozone from ozone generator, which in turn is calibrated by potassium iodide procedure.
Nitrogen dioxide	Originally, manual bubbler, analyzed by colorimetric procedure. Recently, an instrumental method has been proposed, based on chemiluminescent reaction with ozone*, because of lack of accuracy of the original method.
Hydrocarbons (referred to as non-methane hydrocarbons)†	Instrumental; flame ionization detector for total hydrocarbons, minus methane as determined by gas chromatography.

Reference: 40 CFR Part 50. Originally published in *Federal Register*, vol. 36, no. 84, Apr. 30, 1971, pp. 8186-8201.
Federal Register, vol. 41, no. 53, Mar. 17, 1976, pp. 11258-66.
†Not a mandatory standard, but "... for use as a guide in developing implementation plans to achieve oxidant standards."

mediate short-term results comparable to those obtained with a high-volume sampler.

In the case of oxidant measurements, questions have arisen because the chemiluminescent method specified has been shown to produce different results from the potassium iodide method that had been used previously[2].

To alleviate some of these problems, EPA has specified that any "equivalent" method can be used rather than the reference method specified, provided it is approved by the EPA's Administrator upon submission of evidence showing that the proposed method does indeed provide data equivalent to those obtained with the reference method[3].

The demonstration of this equivalence is difficult, especially because of the problem of interference. For example, the potassium iodide method, used for many years to measure oxidant, actually measures the net difference between all oxidizing and reducing substances in the atmosphere. This means that sulfur dioxide, if present, will decrease the total measurement, while halogens, if present, will increase it.

This raises questions about interpretation of the data with respect to ozone and other constituents of the photochemical oxidant being measured. The chemiluminescent method specified by EPA, however, is thought to be specific for ozone and is not affected by potential interferences. In the laboratory, it is easy to demonstrate that the two methods are equivalent. In a real-life atmosphere, however, this equivalence is more difficult to demonstrate. Even if the two instruments are operated side by side in a community atmosphere, and if the results agree, this only proves that they are equivalent at that particular place and time. At another location or at another time, the two cannot automatically be assumed to agree.

Emission measurements

Engineers in the CPI will probably be involved more with emission measurements than with ambient-air measurements. Many problems exist, although they are somewhat different from those encountered in ambient-air measurement. Here also, EPA regulations have been a dominant influence in establishing standard methods.

Probably the first generally accepted standard stack-sampling method was the Power Test Code developed by the American Soc. of Mechanical Engineers to measure particulate emissions from utility boilers [4]. With this one exception, measurements of other contaminants were performed by a variety of methods, and the plant engineer was left with a difficult choice in selecting the one to use. Under another provision of the 1970 Clean Air Act, EPA was required to set emission standards for stationary sources, to be applicable to all new construction commenced after the establishment of the regulations. These so-called "New Source Performance Standards" have now been developed for a number of different industrial processes, as shown in Table II. Engineers involved in designing or operating facilities in these categories are likely familiar with these standards.

To determine compliance, EPA has also specified reference methods for stack sampling or other emission measurements. For opacity, visual observations by a trained observer are prescribed. With only two exceptions (hydrocarbons and one of the alternatives for carbon monoxide), the standards require the use of manual methods; i.e., an operator climbs up on the stack or duct, measures flowrate with a pitot tube, and then withdraws a sample of the gas stream through a sampling device. This sample is then chemically analyzed or is weighed to determine the concentration, and the emission rate is calculated. The duration and frequency of sampling is specified for the various methods.

Such a procedure has two major disadvantages. First, the labor cost is extremely high if it is necessary to obtain measurements either very often or on a more or less continuous basis. Second, the results of the measurement are not available until after the analysis or weighing has been completed, which involves a time delay ranging from a few hours to a day or two. This makes it impossible to use the resulting data for plant operation. Hence, some other monitoring method is necessary for many operations if operating characteristics are to be kept within prescribed ranges and thus maintain adequate quality control over the waste gases emitted. A kiln operator cannot adjust feedrate or combustion conditions to correct a malfunction that increases atmospheric emissions if he does not know about that malfunction until the following day!

Because of these difficulties, EPA has also included continuous-monitoring requirements in the New Source Performance Standards. Fortunately, commercial instruments are available that continuously monitor either the contaminant being controlled or some characteristic of the gases in the system that is related to the contaminant. These instruments are automatic and can be equipped with warning systems to inform the operator if a malfunction occurs or is imminent, so that corrective action can be taken. Usually, a continuous recorder is also included to provide a permanent record of the measurements.

As with ambient-air quality, however, the question arises: Does the continuous instrument really give results equivalent to those obtained by the (manual) reference method? To help answer that question, EPA has published guidelines for selecting and operating monitoring instruments and for demonstrating that the results are equivalent to the reference method [5]. Usually, periodic tests using the reference method are required as a check on the continuous-monitoring system. While some questions may remain about this procedure, most industrial processes do not change significantly from day to day. Periodic checks with the reference method probably provide reasonable assurance of equivalency.

Optical measurement

Perhaps the most progress in stack monitoring has been made in instruments that measure the optical properties of the dust and particulate matter being emitted. These instruments have now augmented or replaced the visual observation specified in many air-pollution regulations to control visible emissions. The original reason for visual observation was very simple—no other method was available. The original Ringelmann method [6], using a printed chart for reference in making visual observations, was developed about the turn of the century. It has been widely employed ever since, with the more-recent addition of variations to measure visible emissions other than black smoke. Even when stack sampling to measure the weight of dust emitted became possible, visual observation still had the advantage of being inexpensive and simple to carry out.

Later, transmissometer instruments were developed, in which a light source was placed on one side of the stack, with a beam of light directed across to the opposite side where some type of light-detection device was mounted. Dust going up the stack obscured part of the light beam, and this effect was translated into a reading of percent opacity to indicate the optical properties of the material in the stack. Unless atmospheric reactions or absorption of water changed the optical properties after emission, the instrumental reading usually could be correlated reasonably well with the observed opacity of the plume to determine compliance with regulations that limited the allowable opacity to 20% or 30%.

These instruments have been further developed and now are reasonably automatic and dependable. One problem with the early instruments was that they had to be calibrated with the light off (to establish 0% transmittance), and with the light on but without any dust in the stack (to establish 100% transmittance). This is no problem for processes that run only intermittently. However, in some continuous operations, recalibration is not possible. For example, in a petro-

Standards of performance for new stationary sources (EPA)		Table II
Reference	Type of process	Emissions controlled
(a)	Fossil-fuel-fired steam generators	Particulates (weight and opacity limits), sulfur dioxide, nitrogen oxides
(a)	Incinerators	Particulates (weight limit)
(a)	Portland cement plants	Particulates (weight and opacity limits)
(a)	Nitric acid plants	Nitrogen oxides
(a)	Sulfuric acid plants	Sulfur dioxide, acid mist, opacity
(b)	Asphalt concrete plants	Particulates (weight and opacity limits)
(b)	Petroleum refineries	Particulates (weight and opacity limits), carbon monoxide, sulfur dioxide
(b)	Storage vessels for petroleum liquids	Hydrocarbons
(b)	Secondary lead smelters	Particulates (weight and opacity limits)
(b)	Secondary brass and bronze ingot production	Particulates (weight and opacity limits)
(b)	Iron and steel plants	Particulates (weight limit)
(b)	Sewage treatment plants (sludge incinerator)	Particulates (weight and opacity limits)
(b)	Phosphate fertilizer plants (5 categories)	Fluorides
(d)	Electric-arc furnaces (steel industry)	Particulates (weight and opacity limits)
(e)	Coal-preparation plants	Particulates (weight and opacity limits)
(e)	Primary copper, zinc, and lead smelters	Particulates (weight and opacity limits), sulfur dioxide

Reference: 40 CFR Part 60. Standards of Performance for New Stationary Sources. Originally published in the *Federal Register* on the following dates (note individual references in first column of table):

(a) Dec. 21, 1971
(b) Mar. 8, 1974
(c) Aug. 6, 1975
(d) Sept. 23, 1975
(e) Jan. 15, 1976

leum refinery, a large catalytic cracking unit can usually be operated continuously for three or four years following a maintenance turnaround, so there is never an opportunity to recalibrate the instrument with an empty stack. The newer instruments incorporate an automatic calibration device in which a mirror diverts the light beam out of the stack through a calibration chamber, so that the 100% transmittance point can be checked as often as desired. Other automatic features have been incorporated, and several manufacturers now provide instruments that are reported to be quite acceptable.

Transmissometer instruments are often used to determine compliance with the particulate standards that establish weight limits. For each process unit, however, one must find the relationship between the pounds per hour permitted by the regulation, and the transmissometer-reading on the stack corresponding to that emission rate. Some workers feel that this is unnecessary and that the optical properties of the plume alone can be specified in such a way that legal control is possible and perhaps preferable, irrespective of emission rates based on pounds per hour.

The primary reason for this is the effect of particle size on the weight of an emission. A few large particles weigh more than a large number of small particles, and therefore a dust emission that contains particles as large as 10 to 50 microns may easily exceed the applicable weight limits. However, if these large particles are removed, it might be possible to meet the weight limits and still allow the emission of a very large number of particles in the micron and submicron range. Since it is these particles that are thought to be primarily responsible for adverse effects on human health, and since optical instruments measure primarily particles in the micron and submicron range, such instruments measure a property more directly related to adverse effects of air pollution that need to be controlled [7, 8]. Undoubtedly, there is still a lot to learn about the best ways of measuring and controlling dust emissions in process plants of various kinds, but significant progress has been achieved in the past few years.

ASTM standard methods

Mention should also be made of the work of Committee D-22 of the American Soc. for Testing & Materials in developing standard methods for both ambient-air and stack measurement. Unfortunately, many of the present ASTM methods were developed and published after the establishment of legal regulations by EPA, and the adoption of reference methods of measurement required for compliance with the EPA standards. Therefore, the ASTM methods have been used less than would otherwise have been the case, but even so are still useful for research purposes where legal compliance need not be demonstrated, and for studies in which a comparison of different methods is valuable in indicating the nature of air quality or of some emission from an industrial plant.

Future trends

No radically new technology appears imminent for measuring and controlling air pollution. However, substantial improvements are being made in the existing technology, and this trend likely will continue for several more years. The cost of instruments and their operation is great enough to provide a significant incentive for greater automation and for increased reliability with less maintenance.

As in many areas, space age technology has been investigated for possible use in emission monitoring. One interesting device uses a laser-type beam that can be aimed at the top of a stack, and reflected energy is measured to estimate the amount of dust, sulfur dioxide, nitrogen dioxide, or other contaminants present. Such devices need not be mounted on the stack, but can be placed some distance away at any convenient location. They could also be used by a control agency without the necessity of working on the stack, or even going onto the plant property. At present, the very high cost (tens of thousands of dollars), size (a large truck trailer) and complexity of these experimental devices prevents their widespread use. There is also the disadvantage that they would not be accepted legally to demonstrate compliance with regulations, unless the regulations were changed or unless the new device could be shown to give equivalent results compared to the legally accepted method. For these reasons, development and application of radically new and different concepts for emission measurement will probably occur rather slowly.

In ambient-air measurement, basic research is needed to identify the specific contaminants responsible for various effects on human health and welfare, so that these contaminants can be monitored routinely. Progress in measurement technology will continue, in parallel with progress in basic research in atmospheric chemistry and in the effects of various pollutants.

This progress will be reflected in the emission measurements required. In addition to limiting total dust and particulates emitted, more-restrictive limits may be placed on heavy metals, polynuclear aromatic compounds, or other constituents. Rather than measuring total hydrocarbons, measurements may be needed for olefins or other constituents that are more reactive in the formation of photochemical smog.

By developing more-sophisticated monitoring methods, and by considering air pollution control as another quality-control problem to be solved by good engineering, the chemical process industries can continue to reduce their portion of the nation's air-quality problems.

References

1. Clean Air Act as Amended in 1970, Public Law 91-604, U. S. Congress, Dec. 31, 1970.
2. Severs, Richard K., Simultaneous Total Oxidant and Chemiluminescent Ozone Measurement in Ambient Air. *J. Air Poll. Cont. Assn.*, vol. 25, no. 4, pp. 394-96. Apr. 1975.
3. Environmental Protection Agency, Ambient-Air Monitoring Reference and Equivalent Methods, 40 CFR, Parts 50, 51 and 53, *Federal Register*, vol. 41, no. 53., Mar. 17, 1976, pp. 11252-57.
4. Determining Dust Concentration in a Gas Stream, Power Test Code No. 27, *American Soc. of Mechanical Engineers*, New York, 1957. Test Code for Dust Separating Apparatus, Power Test Code No. 21, 1941. (Earlier versions of these codes were first published in 1925.)
5. Environmental Protection Agency, Emission Monitoring of Stationary Sources, Emission Monitoring Requirements and Revisions to Performance Testing Methods CFR Parts 51 and 60, *Federal Register*, vol. 40, no. 194, Oct. 6, 1975, pp. 46240-271.
6. U. S. Bureau of Mines. "Ringelmann Smoke Chart," Information Circular 7718 (1955). (Method originally developed by Maximilian Ringelmann prior to 1900.)
7. McKee, Herbert C., Instrumental Method Substitutes for Visual Estimation of Equivalent Opacity, *J. Air Poll. Cont. Assn.* vol. 21, no. 8, pp. 488-90, Aug. 1971.
8. McKee, Herbert C., Texas Regulation Requires Control of Opacity Using Instrumental Measurements, *J. Air Poll. Assn.*, vol. 24, no. 6, pp. 601-604. June 1974.

The author

Herbert C. McKee is Director, Houston Laboratories, Southwest Research Institute—Houston, 3600 Yoakum Blvd., Houston, TX 77006. He has about 20 years of experience in air- and water-pollution control, and holds a B.S. in chemistry and mathematics from Muskingum College and an M.S. and Ph. D. in chemical engineering from Ohio State University. He was formerly a member (and chairman) of the Texas Air Control Board and is a member of the Advisory Committee to the Governor's Energy Advisory Council, as well as being a member of, and participant in, several scientific and engineering societies.

Online Instruments Expedite Emissions Test

Complicated chemical tests for analyzing SO_x and NO_x in stack gases may soon be replaced by instrumental techniques that are quicker, simpler and more reliable.

THOMAS T. SHEN, New York State Dept. of Environmental Conservation

Online instrumental techniques should replace classical chemical methods for compliance testing of SO_2 and NO_x contaminants in stack gases. For short-term emissions testing, instrumentation offers the advantages of simplicity, reliability, reproducibility, quick results, continuous data output and labor-saving operation.

The U. S. Environmental Protection Agency (EPA) currently recommends traditional wet-chemical techniques for measuring SO_2 and NO_x emissions—the barium-thorin titration method for SO_2 and the phenoldisulfonic (PDS) method for NO_x [1]. However, these methods are inadequate for field application because of their overly specific sampling techniques and their tedious analytical procedures.

Consequently, the New York State Dept. of Environmental Conservation set out to explore instrumental methods for emission testing and evaluate their analytical equivalence to current chemical methods.

Tests have shown (Fig. 1) that SO_2 and NO_x emissions can vary as much as 15% and 20%/day, respectively, under normal operating conditions without adjusting air-fuel ratio. Because of this large variation, representative examples may not always be obtained by the sampling techniques normally used in wet-chemical methods. The

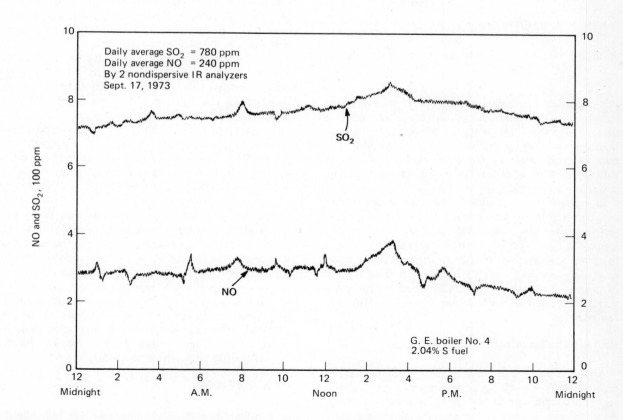

TYPICAL 24-h SO_2 and NO_x emission in stack gases from fired boiler plants—Fig. 1.

Originally published May 26, 1975

III

Type of SO$_2$ and NO$_x$
Emission-Testing Instruments — Table I

Instruments	Price Range, $	Manufacturer*
—— Absorption Spectrophotometry ——		
UV	5,000-9,000	Du Pont[†]
		Honeywell Inc.
		Mine Safety Appliances Co.
		Beckman Instruments
Nondispersive IR	5,000-9,000	Beckman Instruments
		Mine Safety Appliances Co.[†]
		Inter-Technical Group, Inc.
Dispersive IR	5,000-9,000	Wilks Scientific Corp.[†]
		Beckman Instruments
—— Emission Spectrography ——		
Chemiluminescence (for NO$_x$ only)	5,000-7,000	Aerochem[†]
		Thermo Electron Engineering Corp.[†]
		Scott Laboratories
		REM Co.[†]
		Monitor Labs, Inc.
		Bendix Corp.
Fluorescence (for SO$_2$ only)	5,000-7,000	Thermo Electron Engineering Corp.[†]
		Celesco Industries
—— Coulometry ——		
Electrochemical transducer	3,000-6,000	Dynasciences Corp.
		Theta Sensors
		Envirometrics[†]

*Partial list of manufacturers
[†] Instruments investigated

IIIIIIIIIIIIIIIIIIIIIIIIIIIIIIIIIIIIIIIIIIIIIIIIIIIIIIIIIIIIIIIIIIIIIIIIIIIIIIIII

PDS method, for instance, employs a grab-sampling technique, which involves an instantaneous sampling time; results from this method represent only the contaminant emission for the brief period of sampling. For more accurate and representative results, an average value of a few hours of continuous testing, as afforded by online instrumentation, is desirable.

In addition, wet-chemical methods are time consuming; one step in the PDS method takes 16 h for sample recovery. Even more important, multistep procedures increase the likelihood of human error, making it difficult to obtain reproducible results even under the most rigorous laboratory conditions [2]. For these reasons, interest in instrumental techniques as a replacement for traditional wet-chemical methods has waxed over the past few years.

Knowledge Lacking

Although interest in instrumental methods has increased lately, knowledge of instrumental performance for short-term emission testing has been lacking. Recently, two contract reports on the evaluation of commercial instruments for continuous monitoring (as op-

posed to short-term testing) of SO$_2$ and NO$_x$ in stationary combustion sources have been published by EPA [3,4]. Other relevant papers have been authored by Nader [5,6], Halstead [7], Paules [8], and Shen [29].

These reports, however, focus on the evaluation of instruments for continuous source-monitoring rather than short-term compliance testing, which requires instruments that are portable, simple to operate and maintain, and easy to calibrate. Data on instruments for stack-emission testing are fragmented, and conclusions are a matter of preference. Furthermore, much of the information needs to be updated as technology improves.

An important element in implementing an air-quality control program is the method of measurement, which should be simple, fast, accurate and low in cost. Thus, the development of an instrumental technique equivalent to the approved wet-chemical methods is a necessary prerequisite to any good air-quality program. Ideally, remote-sensing techniques such as ultraviolet correlation spectroscopy, or laser and Raman scattering, are the best for field-enforcement officers. But, unfortunately, these techniques are still primarily confined to the research stage and prototype development, and so have not been included in this study.

Selecting Instruments

After an extensive literature review of the most promising instrumental techniques currently available for SO$_2$ and NO$_x$ source testing, we selected spectrophotometry, coulometry and spectrography for investigation. Table I lists many commercially available instruments that employ these techniques. In these categories, we shall discuss six types of instruments (those indicated by the superscript † in the table were investigated).

First, let us outline the three instrumental techniques:

1. Absorption spectrophotometry in the IR and UV regions measures the relative amount of light absorbed by a sample, as a function of the wavelength of the light selected. Generally, the light absorption is not linearly proportional to the concentration of SO$_2$ and NO$_x$. Two types of infrared analyzers were tested. A Dispersive Infrared (DIR) analyzer measures light absorbed at different wavelengths across the infrared spectrum in one operation, while a Nondispersive Infrared (NDIR) analyzer measures only one wavelength per sample.

2. Coulometry measures the current induced by electrochemical oxidation at a sensing electrode. Sensitivity of the reaction cell, or transducer, is determined by the semipermeable membrane, the electrolyte, the electrode material and the retarding potential. The outputs are linear with respect to concentration.

3. Emission spectrography of chemiluminescence or fluorescence measures the energy or light emitted by atoms or molecules when excited by an electric charge or energized by a high-intensity light source. The emitted light is linearly proportional to the SO$_2$ and NO$_x$.

How We Tested

A schematic diagram of the sampling train and gas pretreatment system appears in Fig. 2. It consists of a

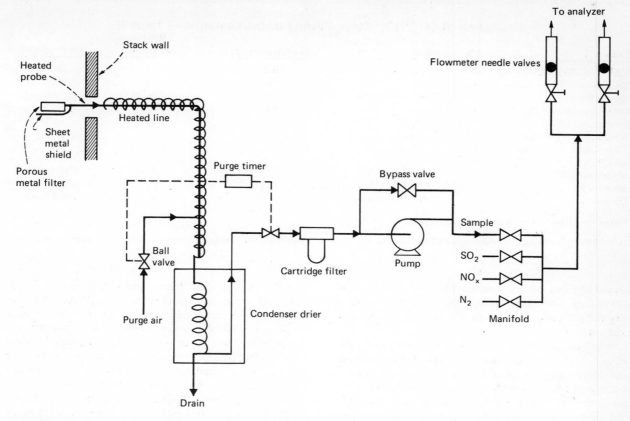

STACK-GAS sampling and conditioning system for compliance tests on emissions—Fig. 2.

standard stainless-steel probe with porous metal filter, heated line, condensate unit, cartridge filter, pump, valves and flowmeters. The sampling system and analytical instruments were used to monitor emissions from two oil-fired boiler stacks at the General Electric (G.E.) power plant in Schenectady, N.Y.

A continuous monitoring of SO_2 and NO_x, along with intermittent monitoring of CO_2 and O_2, established baseline emission studies of boiler operation for comparing separate instrument trials. Concurrently, researchers kept a record of fuel consumption, steam flow, load demand and operating conditions to aid data interpretations. During tests, fuel-oil samples were also collected and analyzed for elemental sulfur, and a sulfur material-balance calculation was performed. Typical 24-h emissions of SO_2 and NO_x appear in Fig. 1.

No attempt was made to calibrate instruments against the extant chemical methods, for reasons previously mentioned. For convenience, and to assure a constant concentration of SO_2 and NO_x, we performed calibration measurements of the instruments against commercial calibration gases.

Instruments were evaluated on the basis of accuracy, interfering substances, instrumental stability, simplicity, cost, size and weight, capability of multifunction, durability, and response time. Operating data from the G.E. plant determined performance characteristics. An acceptable emission-testing instrument must have:

■ Setup time of less than 1 h.

■ Zero drift of less than 1% of full scale in 8 h.

■ Calibration drift of less than 2% of the measured value in 8 h.

■ Response time of less than 1 min for 95% of the instrument's full scale.

■ Relative accuracy within 5% of the measured values.

We tested instruments by comparing their readings—two instruments at a time. This was done by simultaneously testing both instruments with the same commercially purchased calibration gas. If the results obtained on the two instruments under comparison were within 5% of each other over the period of the tests, the instruments were qualified as accurate for sampling stack gases, which were emitted from the oil-fired boilers chosen for study.

All source-testing instruments require the removal of particulates, water vapor and other interfering gases from the sample. The effect of interfering gases is generally compensated for electronically or optically within the instruments themselves, so no attempt was made to remove them in sampling. Particulates can be removed efficiently by two-stage filtration. But removal of water vapor without affecting SO_2 and NO_x measurements is relatively difficult. As shown in Fig. 2, a heated line prevents water condensation in the sample system. A refrigeration unit then removes water vapor by condensing it from the sample.

For compliance testing, typically less than 8 h daily, the sampling system can be simplified by replacing the

Evaluation of SO_2/NO_x Stack Testing Instrumentation — Table II

Criteria	UV	NDIR	DIR	Chemiluminescence	Fluorescence	Electrochemical
Cost range, $1,000	5-9	5-9	4-6	4-6	4-6	3-5
Setup time, <1h	Yes	No*	No	Yes	Yes	Yes
Zero drift, <1%	Yes	Yes	No	Yes	Yes	Yes
Calibration drift, <2%	Yes	Yes*	No	Yes*	Yes	No*
Interference effect	Some	Some	Yes	Little*	Little	Little*
Accuracy of analyzing stack gases, <5% error	Yes	Yes	No	Yes	Unknown	Yes*
Portability	No	No*	Yes	Yes*	Yes	Yes
Multifunction	Yes	No	Yes	NO_x, NO	SO_2	SO_2, NO_x
Simplicity	No	No	No	Yes*	Yes	Yes
Durability	Yes	Yes	Unknown	Yes*	Unknown	Unknown

*Depends on particular model and manufacturer.

heated line and refrigeration unit by an ice-bath condensation trap. This simplified system has proved satisfactory for electrochemical, chemiluminescent and fluorescent analyzers.

Most instruments listed in Table I operate at a sampling rate of 1 ft³/h or higher, as originally designed for ambient-air monitoring. If sampling rates of these instruments could be reduced to less than 0.1 ft³/h, water vapor in the gas stream could be removed more efficiently and economically by membrane techniques, such as permeation drying or permeation/distillation, without affecting SO_2 and NO_x measurements. Contrary to ambient-air monitoring instruments, stack-monitoring instruments should be designed to take as small a sample as possible because of high pollutant concentrations.

Which Instruments?

Table II summarizes our general finding. Spectrophotometric techniques (IR and UV regions) have proved the least advantageous for field application. These techniques require efficient small-particle filtration, water removal, and sulfuric acid mist removal. They also require a skilled operator and considerable time to be set up and calibrated. Moreover, the instruments are difficult to transport from one location to another because of size and weight.

Even though the electrochemical instrument has demonstrated less accuracy than the spectrophotometers due to calibration drift, it appears more suitable for short-term field applications. Since a compliance test is usually conducted for less than 8 h, and since the instrument is calibrated at the beginning and end of each test, long-term stability is less important in this application. If the calibrations at the beginning and end of the test show a significant difference, the final results can be easily corrected. This type of instrument offers the advantages of simplicity, portability, immediate results, lower costs, and the multifunction of measuring SO_2 and NO_x simultaneously.

In situations where more-accurate measurements are desired, emission spectrographic techniques should be employed. The chemiluminescence technique for NO_x and NO source measurement has proved accurate, specific, sensitive and highly reproducible. The fluorescence method for SO_2 source measurement shows promise, but its accuracy has not yet been established. As pointed out earlier, this type of instrument evaluation study should be periodically updated, particularly when new or improved techniques become available.

When selecting instruments for routine compliance tests of SO_2 and NO_x emissions, portability, setup time, and simplicity in operation and maintenance should be considered. The instrument should be usable by nonscientific personnel faced with day-to-day enforcement tasks. For control procedures in short-term source testing, some accuracy of measurement can be sacrificed for economy, immediate results and simplicity. But no matter what instrumental method one selects, an adequate sample pretreatment system must be used to ensure accurate results.

Acknowledgements

The author wishes to acknowledge his appreciation to Barry Siegal and Timothy Ross of the New York State Dept. of Environmental Conservation for their assistance in field work. The assistance of Larry Ayer, of General Electric Co., is also gratefully appreciated. Most of this work was done at General Electric's power plant at Schenectady, N.Y.

Meet the Author

Thomas T. Shen is a Senior Research Scientist for the New York State Dept. of Environmental Conservation. In this capacity, he is primarily responsible for research in stationary combustion sources and source emission measurement. He has been involved in civil and environmental engineering for the past 20 years and is a diplomat of the American Academy of Environmental Engineers. An author of many papers in the areas of fuel combustion, fuel additives, air-pollutant measurements, and air pollution control, he earned his Ph.D. in environmental engineering from Rensselaer Polytechnic Institute, Troy, N.Y.

Analyze stack gases via sampling or optically, in place

Stack flues can be monitored by extracting and conditioning samples, or by analyzing the gases spectroscopically. Both methods yield accurate, reliable results, but the second offers additional advantages.

H. A. Klasens, N. V. Philips' Gloeilampenfabrieken

☐ Stack monitoring for environmental protection has stimulated the development of new measuring systems that are more resistant to the corrosive action of flue gases at elevated temperatures.

The usual approach has been to continually extract gas from the stack, and to remove interfering components, such as water and particles, before the gas is admitted to the detector. Or, the water is prevented from condensing by the heating of transport lines and detectors.

In a more recently developed method, the gas is analyzed *in situ*, by optical methods in which interference from particles and water is minimized.

Sample conditioning and analysis

In the first approach, gas is extracted from the stack through a filtered probe that removes most of the dust (Fig. 1). (When the filter clogs, it is cleaned by back-flushing it with compressed air.) The gas then passes through a system, frequently a refrigerator dryer, that continually removes the condensate. The dried gas then enters the analyzer, usually after passing through a second filter. The system commonly contains some provisions for the introduction of zero- and span-gases. The gas may be analyzed by one of several methods, depending on its nature, such as spectroscopy, flame photometry, chemiluminescence and chromatography.

In this sampling method, many potential sources of error exist, such as: chemical adsorption on the dust, chemical absorption in the condensate, and leakage from sampling lines.

Condensate from flue gases containing sulfuric acid is very corrosive and may cause trouble in pumps, filters and transport lines. Many of these problems can be

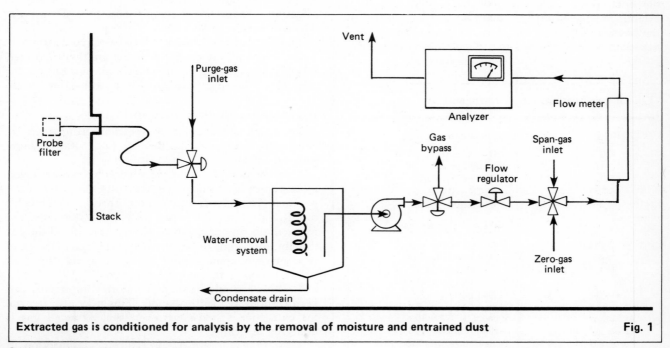

Extracted gas is conditioned for analysis by the removal of moisture and entrained dust   **Fig. 1**

Originally published November 21, 1977

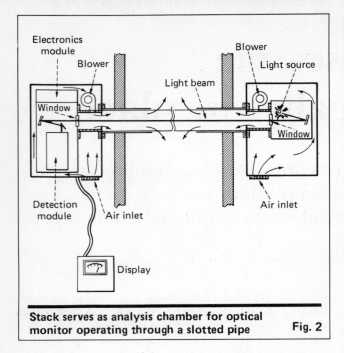

Stack serves as analysis chamber for optical monitor operating through a slotted pipe   Fig. 2

partially overcome by using reflux probes, diaphragm pumps coated internally, and membrane dryers, and by heating sampling lines. Not only the lines but also the analyzer may be heated.

Despite such measures, this system suffers from considerable downtime, although downtimes as small as 5 to 10% have been claimed.

Analyzing stack gases in place

In *in situ* monitoring, the stack itself serves as an analysis chamber (Fig. 2). The system illustrated consists of a light source and a receiver, at opposite ends of the duct. Both are connected to a slotted pipe that extends across the duct to maintain the alignment of the beam and to restrict, when necessary, the absorption-path length of the light. A collimated polychro-

matic light beam passes through the gas to the analyzer at the opposite side. The analyzer and source are protected from the gas by two windows, which are kept clean by air curtains maintained by blowers. Since the advent of dual-wavelength spectroscopy and correlation techniques, little or no interference results anymore from dust or other components such as water vapor.

In dual-wavelength spectroscopy, one beam of polychromatic light is sent across the cell. Two adjacent wavelengths are chosen so that only one is absorbed by the gas, but both are similarly attenuated by interfering components. This takes into account the effect of interference on signal strength (see the later discussion of spectroscopy principles).

In the correlation technique, the beam is split in half. One half is sent through a cell containing a high concentration of the gas to be analyzed and through the measuring cell. The other half is sent through the measuring cell, and through an adjustable attenuator to equalize the zero signals. The signal from the first beam now contains only information on the interfering components, because the light of all wavelengths that can be absorbed by the gas to be analyzed has been removed by the correlation cell. Because the other beam contains information on both the gas to be analyzed and the interfering components, the influence of the latter can easily be corrected.

The table presents some specifications for such an analyzer in typical stack applications.

Zero- and span-checks are made, using a light source mounted in the analyzer box. To set span, one passes the light through sealed cells containing precisely-known gas concentrations to the analyzer. A dynamic calibration can also be used to check the entire system, using the light beam across the duct. Standard cells of known concentrations are incrementally added to the unknown gas concentration in the duct, and readings are made to match the calibration curve. In some applications, optical filters, equivalent to a known gas concentration, can be used.

The dynamic calibration can be performed automatically. The calibrated cells or filters are then mounted on a timer-actuated solenoid arm. A pushbutton for overriding the timer can provide a calibration check.

Advantages of *in situ* monitoring

The chief advantage of *in situ* monitoring is that it eliminates corrosion problems. Other advantages claimed include:

1. *In situ* monitoring provides better averages because measurements are made over the entire diameter of the gas duct. This can also be accomplished in principle by multi-point sampling. However, averaging over the whole diameter can produce erroneous results if the gas distribution is not centrosymmetric. If such is the case, trial measurements should be made across the duct before choosing the sampling location.

2. With a single instrument, more components can be measured at identical spots. This may also be true for extractive systems, as when the extracted gas is analyzed optically.

3. Components are not lost through absorption in dust filters or in condensates. This is a valid claim.

Typical specifications for an *in situ* analyzer	
Gas	**Standard ranges, ppm**
NO	0–100 to 0–5,000
CO	0–500 to 0–3,000
CO_2	0–15% (typical)
SO_2	0–25 to 0–10,000
C_xH_y	0–25 to 0–6,000
	Special-order ranges, ppm
H_2S	0–100 to 0–2%
H_2O	0–1,000 to 0–80%
NO_2	0–1,000 (typical)
NH_3	0–1,000 (typical)
HCl	0–1,000 (typical)

Accuracy	±2%	Span drift	±2%/90 d
Linearity	±1%	System response	1 s (1/e)
Zero drift	±2%/90 d	Temperature	−40 to +65°C

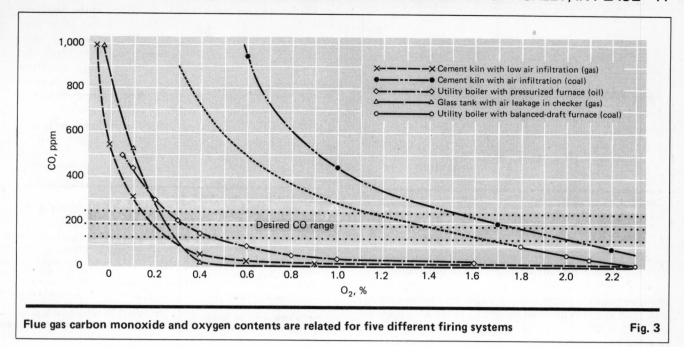

Flue gas carbon monoxide and oxygen contents are related for five different firing systems Fig. 3

4. *In situ* monitoring presents fewer maintenance problems. This also is a valid claim.

5. Such monitoring is better suited for process control because of the small response time. This also is true, although there remains the transport time of the gas. In fuel-fired power stations, this is the transport time of the gas between burners and lightbeam.

Disadvantages of *in situ* monitoring

Some disadvantages cited include:

1. The windows get dirty, and the interfering components, dust and moisture, cannot be removed. True to a certain extent, but with dispersive double-wavelength spectroscopy or correlation techniques (see later discussion), interference and dirty-window effects are virtually eliminated.

2. Oxygen cannot be measured optically because it does not suitably absorb light. Its measurement was once required for combustion control because of EPA regulations. However, CO monitoring is more suitable than O_2 monitoring for combustion control, and EPA has changed its regulations to make it possible to use optical *in situ* monitoring.*

3. Calibration cannot be done with calibration gas mixtures. Although this is partly true, calibration can be done easily by inserting closed calibration cells or filters in the optical light path. This method is allowed by EPA for emission control.

4. It is more expensive because time sharing between several stacks is not possible. This is true, because every

stack must have its own monitor. However, because isolated and heated sampling lines, and such equipment as manifolds, are expensive, the price difference between having analyzers at each chimney or time sharing between several chimneys is small for extractive systems.

5. At high opacities, the optical signals become too weak. Although true, signal strengths in practice can be lowered considerably before the signals become drowned in noise. At opacities higher than 95%, the optical path length must be shortened.

6. Initial cost of an *in situ* monitor is higher. This may be true when monitoring single components, although higher costs for upgraded sampling equipment have lowered the price difference. When more components must be measured and when maintenance costs are taken into consideration, cost comparisons may favor *in situ* monitoring.

Although it is difficult to say which of the two methods is more accurate and reliable, generally the method that manipulates the sample the least without sacrificing precision is the better one.

One application: combustion control

Maximum combustion efficiency occurs when the air supply to burners is just sufficient to completely burn all carbon to carbon dioxide and all hydrogen to water. Too little air leads to the production of less heat; too much causes heat to be carried away by the excess air. Therefore, the air-to-fuel ratio must be exactly balanced, and this can be controlled by means of the chemical composition of the flue gas.

Of course, combustion processes are never perfect and there are always traces of C, CO and O_2, due to incomplete combustion. Most plants have measured the excess O_2 in the flue gas to gauge the excess air in the fuel-burning zone, and have regulated the air-to-fuel ratio so that excess oxygen is kept at 1% or lower.

However, CO is more useful for determining fuel-

*EPA regulations require that sulfur dioxide, nitrogen oxides, opacity, and oxygen or carbon dioxide be continuously measured and registered when existing coal- and oil-fired steam generator capacity exceeds 250-million-Btu/h heat input. The measure of one of the last two gases is necessary for reference and to convert monitored data into the required pollution data (g/million cal). Carbon dioxide has been included as an alternative to oxygen, as a revision because many plants had already installed, or were planning to install, optical *in situ* monitors in stacks [1]. The revision was necessary because such monitors cannot measure oxygen, since it does not absorb light in the accessible spectral region. Carbon dioxide does show characteristic absorption bands in the infrared region.

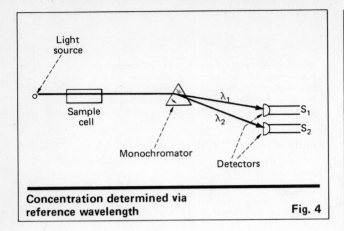

Concentration determined via reference wavelength   **Fig. 4**

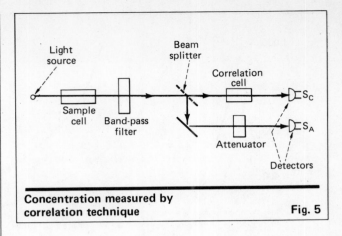

Concentration measured by correlation technique   **Fig. 5**

burning efficiency, because oxygen is present in flue gas not only from excess air supplied to the burner but also from air entering through idle burners, observation ports, expansion joints, casing leaks, soot-blower openings, and such. When the burners are set at 1% excess air, a 1% infiltration of air into the flue gas before the point of measurement can cause an oxygen analysis to read 20% higher. On the other hand, carbon monoxide, not being present in air, will only be diluted 1% by a 1% infiltration of air.

In flue gas, CO can easily be measured *in situ* by light absorption in the infrared region of the spectrum. Fig. 3 illustrates the relationship between the content of CO and O_2 in flue gas [2].

In a typical oil-fired boiler, the CO content should be maintained at 100 to 200 ppm to keep the fuel-to-air ratio close to the stoichiometric ratio.

A further advantage of CO analysis is that individual burners can easily be checked by turning them off one by one and noticing the changes in CO content. Burners that cause the largest changes in CO content are likely to be dirty, to have oversized tips, to pass less air flow and to cause poor air and fuel mixing. Those that cause the smallest change are likely to be partially plugged, to have undersized tips, or to pass larger-than-average air flow.

Another use: basic-oxygen steelmaking

The basic-oxygen process enlarges and refines the old Bessemer process.* In the latter, compressed air is forced through blowholes in the base of the vessel into the molten iron. This elevates the temperature of the molten mass, as the oxygen of the air oxidizes unwanted elements in the iron (sulfur, phosphorus, excess carbon, etc.), which is removed either as fumes or in the slag. Unfortunately, some of the nitrogen of the air remains in the steel and makes it brittle.

In the basic-oxygen process, air is replaced by pure oxygen, which is blown in from the top. The vessel is loaded with scrap, molten iron from the blast furnaces and certain additives. A fume head is lowered over the open neck of the vessel. A heat-resistant oxygen lance is lowered through the hood until the nozzle is a few feet above the molten mass, and pure oxygen is then blown

*The word "basic" refers to the non-acid refractory lining of the vessel. In Europe, the process is better known as the LD process, LD standing for Linz-Donawitz, an Austrian plant where the process was first carried out.

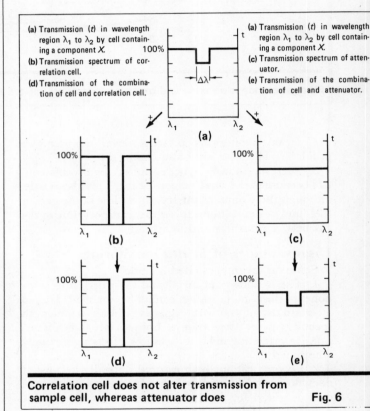

(a) Transmission (*t*) in wavelength region λ_1 to λ_2 by cell containing a component *X*.
(b) Transmission spectrum of correlation cell.
(d) Transmission of the combination of cell and correlation cell.

(a) Transmission (*t*) in wavelength region λ_1 to λ_2 by cell containing a component *X*.
(c) Transmission spectrum of attenuator.
(e) Transmission of the combination of cell and attenuator.

Correlation cell does not alter transmission from sample cell, whereas attenuator does   **Fig. 6**

into it with great force to cause turbulence. The temperature of the mass rises and any remaining scrap is melted. The waste gas, which also contains the fumes of the oxides of the unwanted impurities, is collected by the hood.

The blow lasts about 20 minutes. A sample is then taken and transported to the laboratory for an X-ray analysis, mainly of carbon. On the basis of the analysis, which is cumbersome and time-consuming, it is decided whether, and how long, the oxygen blowing has to be continued. Now the end-point can be determined by analyzing the waste gas.

Principles of spectroscopic analysis

The absorption of monochromatic radiation in a cell having a path length, *l*, and containing a gas, *X*, of concentration c_X is given by Beer's law as follows:

$$I = I_0 e^{-k_X c_X l} \qquad (1)$$

In Eq. (1), I_0 is the intensity of radiation entering the cell, I is the intensity of the radiation after traversing the cell, and k_X is the absorptivity of gas X.

The transmittance of the cell is defined as:

$$t = I/I_0 \qquad (2)$$

For values of $k_X c_X l \ll 1$, $t = 1 - k_X c_X l$.

When another gas is present that also absorbs the incoming radiation, or when the gas contains particles or water droplets that scatter the radiation, or when the windows at the entrance or exit of the cell are darkened, the radiation is attenuated by an additional factor, $e^{-\alpha}$, or:

$$I = I_0 e^{-(\alpha + k_X c_X l)} \qquad (3)$$

Methods are now discussed that allow c_X to be determined without knowledge of α.

1. Using a reference wavelength. The principle of the method is shown schematically in Fig. 4. The light transmitted by the cell is dispersed by a monochromator. Two wavelengths are isolated and measured separately by a detector. The two wavelengths λ_1 and λ_2 are so close that I_0 and α can be considered the same for both wavelengths, whereas only light with wavelength λ_1 is absorbed by gas X. Therefore:

$$s_1 \propto I_0 e^{-(\alpha + k_X x_X l)}$$
$$s_2 \propto I_0 e^{-\alpha}$$

or

$$s_1/s_2 = e^{-k_X c_X l} \qquad (4)$$

By measuring the ratio of the signals s_1 and s_2, not only are additional attenuations accounted for but also other changes in I_0 (such as from aging) that do not affect the results. Of course, if the disturbing factors get too large, the signals become drowned in noise, and accurate measurements are not possible.

2. Correlation spectroscopy. Fig. 5 shows this method schematically. The band-pass filter transmits a narrow region, $\lambda_1 - \lambda_2$, which is chosen so that it embraces one or more characteristic absorption lines of gas X. These lines are almost completely absorbed by the correlation cell, which is filled with a high concentration of the gas X.

The attenuator has, in general, a constant absorption over the wavelength region transmitted by the band-pass filter. This absorption is adjusted for the two signals s_C and s_A to be equal when no X is present in the sample cell.

The transmissions resulting from the combination of this cell and the correlation cell and attenuator, respectively, are illustrated in Fig. 6.

For simplicity's sake, gas X is assumed to have only one absorption line with a constant absorption in the wavelength region $\Delta\lambda$. The band-pass filter transmits all radiation between λ_1 and λ_2, and nothing outside that region.

The total transmission of the correlation cell is:

$$T_C^o = \int_{\lambda_1}^{\lambda_2} t_C d\lambda = (\lambda_2 - \lambda_1) - \Delta\lambda \qquad (5)$$

The total transmission of the attenuator between λ_1 and λ_2 is:

$$T_A^o = \int_{\lambda_1}^{\lambda_2} t_A d\lambda = t_A(\lambda_2 - \lambda_1) \qquad (6)$$

The value of t_A is chosen so that $T_C^o = T_A^o$.

When X is present in the sample cell, the total transmission of this cell and the correlation cell is hardly affected, as all radiation in the region $\Delta\lambda$ is almost totally absorbed by the correlation cell, or $T_C = T_C^o$. Here T_C is the total transmission of the combination of sample cell and correlation cell, assuming zero attenuation outside $\Delta\lambda$. The combined transmission of sample cell and attenuator is:

$$T_A = t_A(\lambda_2 - \lambda_1 - \Delta\lambda) + t_A t_X \Delta\lambda \qquad (7)$$
$$T_A = t_A(\lambda_2 - \lambda_1) - t_A \Delta\lambda(1 - t_X) \qquad (8)$$
$$T_A = T_C^o - t_A(1 - t_X)\Delta\lambda \qquad (9)$$

The signals s_C and s_A are proportional to T_C and T_A, with the proportionality constant the same.

The concentration, c_X, then follows from:

$$s_C - s_A \propto t_A(1 - t_X)\Delta\lambda \qquad (10)$$
$$s_C - s_A = \text{constant } (1 - t_X)$$

For small concentrations of X:

$$t_X = 1 - c_X k_X l \qquad (11)$$

And $s_X - s_A$ is proportional to c_X.

Now assume that the radiation is also attenuated in the sample cell, for reasons already discussed, by a factor $e^{-\alpha}$ over the wavelength region between λ_1 and λ_2. In such a case, both signals, s_C and s_A, are reduced by the same factor, $e^{-\alpha}$. The same holds for the difference $s_C - s_A$, and this would result in an error in the determination of c_X. To avoid this, one does not measure this difference but rather the ratio of it and another signal, for instance s_C, which is also reduced by the same amount and which does not depend on c_X. The ratio $s_C - s_A/s_C$ is a monotonic function of c_X and directly proportional to c_X at low concentrations.

References

1. *Federal Register*, Vol. 39, 1974, p. 32852.
2. Gilbert, L. F., Precise combustion-control saves fuel and power, *Chem. Eng.*, June 21, 1976, p. 145.

The author

H. A. Klasens is a consultant to the Scientific and Industrial Equipment Div. of N. V. Philips' Gloeilampenfabrieken (Bldg. TQ III-1, Eindhoven, The Netherlands). He was appointed a consultant upon his retirement from the position of deputy-director of N. V. Philips' Central Research Laboratories, in which he had previously served as a senior chemist and head of the department of luminescence, and as a research chemist.

Holder of a doctoral degree in organic chemistry from the University of Groningen, he has authored and coauthored a number of papers in the field of luminescence of the solid state.

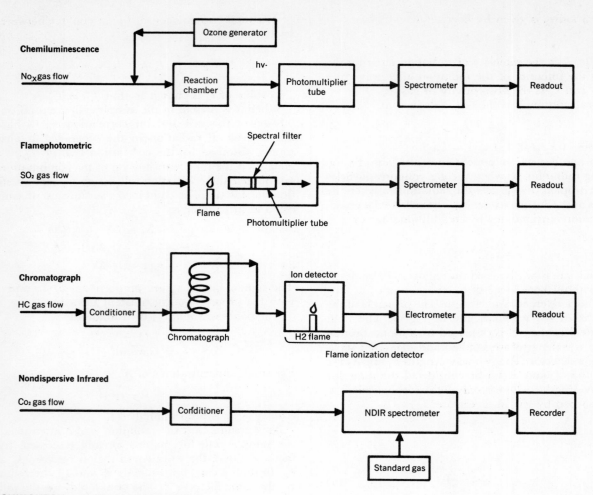

Chemiluminescence

Ozone generator

No$_x$ gas flow → Reaction chamber → hv → Photomultiplier tube → Spectrometer → Readout

Flamephotometric

Spectral filter

SO$_2$ gas flow → Flame / Photomultiplier tube → Spectrometer → Readout

Chromatograph

HC gas flow → Conditioner → Chromatograph → Ion detector / H2 flame → Electrometer → Readout

Flame ionization detector

Nondispersive Infrared

Co$_2$ gas flow → Conditioner → NDIR spectrometer → Recorder

Standard gas

TECHNIQUES for monitoring gases—Fig. 1

Continuous Source Monitoring

The need for continuous monitoring of air pollutants from stacks is relatively new. Here is a rundown on the latest techniques and equipment.

FRANK L. CROSS, JR., Roy F. Weston, Inc., and HOWARD F. SCHIFF, Gilbert Associates, Inc.

Air-pollution-source monitoring may be divided into four categories: gaseous pollutants, particulates, visible emissions and particle-size determination. Of these, particulates are the most difficult stack-emission parameter to monitor. Some of the problems are:
- Wide concentration ranges.
- Large particle-size variation.
- Gas-stream stratification.
- Stack-pressure differences and variations in moisture content and temperature.

Basic components of all monitoring systems are easily listed: a probe or inlet device; a sample conditioning unit; a detector-analyzer; and a data recording unit. Still,

there is much confusion over differences in monitoring methods. The Environmental Protection Agency is now comparing various techniques and equipment to develop a service called, "Performance Specifications for Continuous Monitoring Units."

Particulate Sampling

Most stack monitoring for particulates is still manual. Sampling trains using various procedures (WP-50, ASTM, ASME, EPA, etc.) only collect a sample for a limited, integrated time period (e.g., over a minimum of 1-to-2 hr.).

Originally published June 18, 1973

Criteria for Analytical Methods—Table I

Specificity	Response should be only to trace material(s) of interest.
Sensitivity, range	Method must be sensitive over the concentration range of interest.
Stability	Sample must be stable in the analyzer.
Precision, accuracy	Results must be reproducible, and must represent the actual TQ concentration.
Sample average time	Method must fit into the required sample averaging time for control.
Reliability, feasibility	Instrument investment and maintenance costs, analysis time, and manpower must be consistent with needs and resources.
Calibration	Instrument should not drift; calibration and other corrections should be automatic.
Response	Instrument must function rapidly enough to record significant process changes as they occur.
Effect of ambient conditions	Changes in temperature and humidity must not affect the accuracy of the observed results.
Data output	For some applications, output of the analyzer should be in a machine-readable format.

Gaseous Detector & Analyzer Techniques—Table II

Gas(es)	Technique
Nitrogen oxides	Chemiluminescence
Sulfur dioxide	Flame photometric
Carbon dioxide	Nondispersive infrared (NDIR)
Carbon monoxide	Nondispersive infrared
Oxygen	Paramagnetic
Hydrocarbons	Gas chromatographic —flame ionization detection

Cost of Continuous Monitors—Table III

Pollutant	Method	Approximate Cost, $
Sulfur dioxide	Spectrophotometric	3,000–5,000
	Coulometric	3,000–7,500
	Conductometric	2,000–4,000
	Flame photometric	3,000–10,000
Carbon monoxide	Nondispersive IR	3,000–7,000
Nitrogen oxides	Colorimetric	2,500–5,500
	Coulometric	3,000–7,000
	Chemiluminescence	5,000–7,000
Hydrocarbons	Flame ionization	3,000–5,000
Oxygen	Paramagnetic analyzer	600–800

Continuous-particulate-monitoring equipment have been under development for several years. A promising commercial unit uses a filtration, beta-radiation-attenuation technique that includes dilution, filtration, detection of attenuated beta radiation, and recorders.

The main advantage of particulate monitoring using beta-radiation attenuation is that the analysis depends only on the weight of collected material, and not on its chemical or physical properties.

One available instrument is the piezoelectric mass monitor. In this device, particles in the sample stream are electrostatically deposited on a piezoelectric sensor. The added weight changes the oscillation frequency of the sensor in a known manner.

This instrument, however, cannot handle the very high particulate loadings found in many stacks without sample dilution. The dilution step greatly complicates sampling since it requires two isokinetic samples—one of the stack and one of the diluted sample. This problem is being studied under an EPA contract, with the expectation that a more widely-applicable system will be available in the future.

Other available methods using optical density or light scattering are excellent for measuring visible emissions, but are not good for determining weight rate of emissions.

Gaseous Emissions

The state-of-the-art for continuous monitoring of stack emissions for gases is in a constant state of flux. Many instruments do not necessarily give the most accurate analysis of the concentration of the particular gas in question. Most units are subject to interference from other pollutants as well as electronic instability.

EPA's Office of Air Programs, together with the Scientific Apparatus Manufacturers Association, is promulgating performance specifications for continuous monitoring units. These will cover analytical procedures but will not mention any manufacturer by name. Specifications will include specificity, sensitivity, range, interferences and other criteria as listed in Table I. While EPA will not indicate preference for one procedure over another, the agency is now promulgating equivalency guidelines.

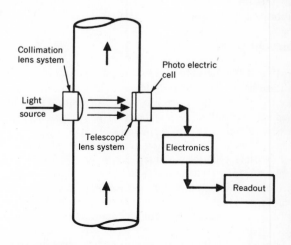

VISIBLE-emission detector—Fig. 2

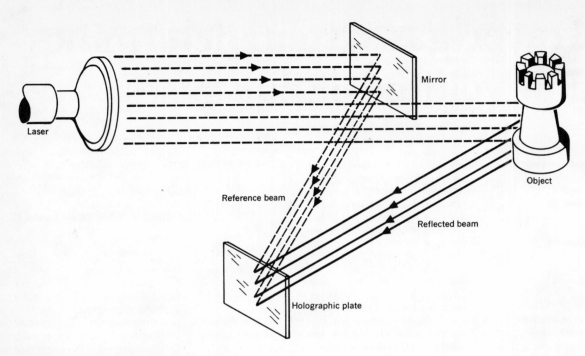

HOLOGRAPHY is promising for particle sizing—Fig. 3

Preferred techniques for continuous monitoring of primary gaseous pollutants are listed in Table II. Schematics of some of the sampling techniques are shown in Fig. 1. Price ranges for a number of continuous monitors are presented in Table III.

Visible Emissions

Many states now require visible-emission monitors on old and new point-source installations. All regulations now include Ringelmann or equal-opacity testing for visible emissions.

Most continuous monitors use photoelectric techniques such as illustrated in Fig. 2. The major problem with this approach is the likely possibility of coating the lens or detector with particulates. Some instruments compensate for this decrease in light transmission with a self-calibrating technique.

Particle Sizing

Most of the standard methods for determining particle size are based on static samples. However, laser holography has recently developed to the point where it is being used commercially. Laser systems are available for particle sizes from 0.1 to 1,000 microns.

In the most common method of making holograms, a film plate records the interference pattern made by interaction of two coherent light beams. One is a reference beam and the other is bounced off the object (Fig. 3). Real and virtual images appear when the original wave front is reconstructed by shining coherent light through the hologram.

The velocities that can be examined by laser holography range from slow-moving particles to those with velocities of 10,000 ft./sec., requiring an exposure time of 20 nanoseconds. New techniques like holography may enable engineers to be more accurate in sizing equipment to meet the latest air pollution control codes.

Bibliography

1. Fair, James, R., et. al., Sampling and Analyzing, *Chem. Eng.,* Sept. 18, 1972.
2. McShane, Wm. P., Bulba, E., Automatic Stack Monitoring of a Basic Oxygen Furnace, Paper 67-120, APCA, June 11–16, 1967, Cleveland, Ohio.
3. Morrow, N. L., et. al., Sampling and Analyzing Air Pollution Sources, *Chem. Eng.* Jan. 24, 1972.

Meet the Authors

Frank L. Cross, Jr., is principal air pollution control consultant and manager of environmental training for Roy F. Weston, Inc., Lewis Lane, West Chester, PA 19380. He has over 18 years sanitary engineering experience with the National Air Pollution Control Administration and various state agencies. Mr. Cross has B.S. degrees in chemical engineering (Northeastern University, 1953) and civil engineering (University of Florida, 1959), as well as an M.E. in air pollution from the University of Florida. He is a licensed professional engineer in five states.

Howard F. Schiff is field coordinator, air quality dept., Gilbert Associates, P.O. Box 1498, Reading, PA 19603. Previously he has worked at Roy F. Weston, Boyce Thompson Institute and Albert Einstein College of Medicine. Mr. Schiff received his B.S. degree in chemistry and M.A. in biochemistry from the City College of the City University of New York. He is a member of the American Chemical Society and ASTM Committee D-22.

Tracer-gas system determines flow volume of flue gases

Injecting a tracer gas into a flue at a known rate, and then analyzing the mixed gases for the concentration of tracer gas, enables ready calculation of the volumetric flowrate.

John Knoepke, *Metallurgical Engineer*

☐ The measurement of flue-gas flowrates has traditionally been performed using the pitot-tube flow-measurement system. This system has several disadvantages that limit its effectiveness. These include the geometric requirements of the flue system, and the relatively long time needed to traverse the cross-sectional area of the flue. Therefore, a tracer-gas flow-measurement system has been developed and tested at Bunker Hill Co.'s lead smelter in Kellogg, Ida.

In the tracer-gas system, a known flowrate of a tracer gas—in this case, sulfur hexafluoride (SF_6)—is injected into the flue. The flue gas downstream of the injection point is then sampled and analyzed for the tracer gas. From the known flowrate of the injected tracer gas and its concentration in the flue gas, the flowrate of the flue gas can be easily calculated.

The tracer-gas system in the test program has given results similar to the pitot-tube system over a wide range of flue conditions, including temperatures up to 870°C, high dust loadings, SO_2 up to 2%, wet gas, and very turbulent gas flows.

A tracer gas has been previously used to study dispersion of stack effluents in air [1] and in mine-ventilation analysis [2].

The tracer-gas system is shown schematically in Fig. 1. Its use involves:

1. Injecting tracer gas into the flue system.
2. Taking a flue-gas sample downstream of the injection point.
3. Analyzing the flue-gas sample for the tracer gas.

To determine the flue-gas flowrate accurately, the flowrate of the injected tracer gas must be both constant and accurately measured. The apparatus used to inject the tracer gas is shown schematically in Fig. 2. At the heart of the injection system is the mass flowmeter used to measure the tracer-gas flowrate.

The flue gas sample taken must be a representative one if there is to be an accurate measurement of the

Originally published January 31, 1977

total flowrate. A schematic diagram of the sampling apparatus is shown in Fig. 3. The vapor pressure of water in the flue gas is measured to correct for any moisture condensing in the sampling apparatus.

Analyzing for the tracer gas

The flue gas is analyzed for SF_6, using a 5700 series Hewlett-Packard gas chromatograph equipped with an electron detector and a gas-sampling valve. A stainless-steel column packed with Chromasorb 102, held at 50°C, is used to separate the SF_6 from the other components in the flue gas (O_2, N_2, SO_2, H_2O, CO_2, and CO). This system has been used to measure, quantitatively, SF_6 concentrations from 4 parts per billion (ppb) to 1 part per million (ppm). The gas chromatography can

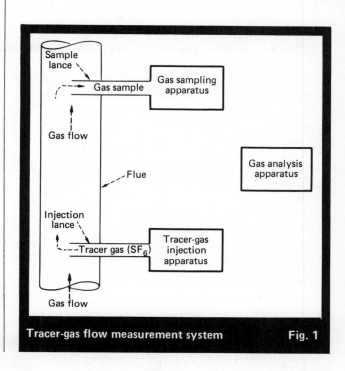

Tracer-gas flow measurement system   **Fig. 1**

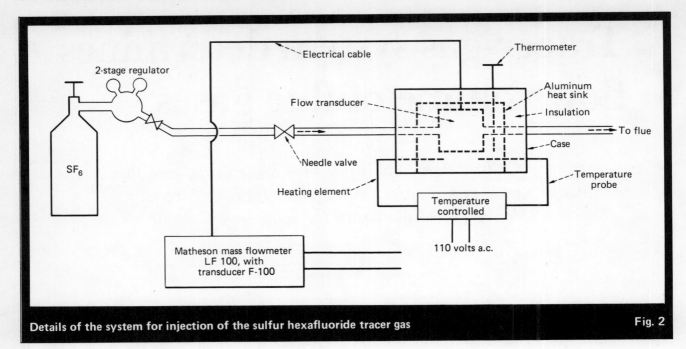

Details of the system for injection of the sulfur hexafluoride tracer gas   **Fig. 2**

measure higher concentrations of SF_6, but for this test program it was not necessary to go higher than 1 ppm.

The formula used to calculate the volume of the flue gas—knowing both volume of tracer gas injected and its concentration in the flue-gas sample—is:

$$F_{stp} = \frac{F_{TG}}{C_{TG} \ (T/T_o)(P_o/P)} \ (3.53 \times 10^{-5})(H_r)$$

(where 3.53×10^{-5} converts cm^3 to ft^3)
The correction for relative humidity (H_r) is made as follows:

If the vapor pressure of water in the flue gas is less than or equal to the vapor pressure of liquid water at temperature T, $H_r = 1$.

If the vapor pressure of the water in the flue gas is greater than the vapor pressure of water at temperature T, $H_r = (p_v)_{FG}/(p_v)_w$.

Sulfur hexafluoride was selected as the tracer gas because it:

■ Can be accurately measured from a few ppb to many ppm by using a gas chromatograph equipped with an electron-capture detector.

■ Is chemically stable at red heat.

■ Is not present in the atmosphere.

■ Is nontoxic and odorless.

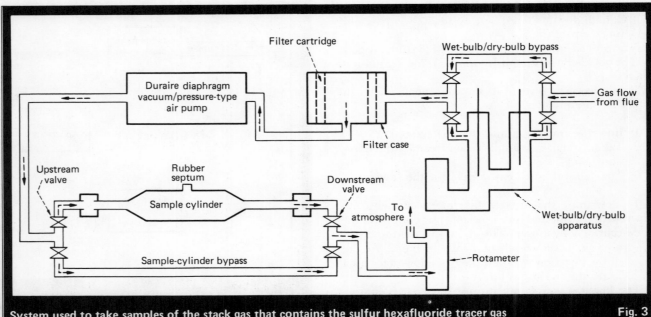

System used to take samples of the stack gas that contains the sulfur hexafluoride tracer gas   **Fig. 3**

Nomenclature

C_{TG} Volume concentration ratio of the tracer gas in the flue-gas sample, e.g., ppb

F_{stp} Volumetric flowrate of the flue gas at any standard temperature and pressure, ft³/min

F_{TG} Volumetric flowrate of the tracer gas at standard temperature and pressure, cm³/min

H_r Relative-humidity correction factor, used to correct for the condensation of water from the gas sample, dimensionless

P Pressure of the gas sample in the gas chromatograph's sampling loop at the time of analysis, e.g., mm Hg

P_o Standard pressure, e.g., 760 mm Hg

$(P_v)_{FG}$ Vapor pressure of water in the flue gas (as determined by the wet- and dry-bulb method, e.g., mm Hg)

$(P_v)_w$ The vapor pressure of liquid water at temperature T, e.g., mm Hg

T Absolute temperature of the gas sample in the gas chromatograph's sampling loop at the time of analysis

T_o Standard temperature, absolute, e.g., K

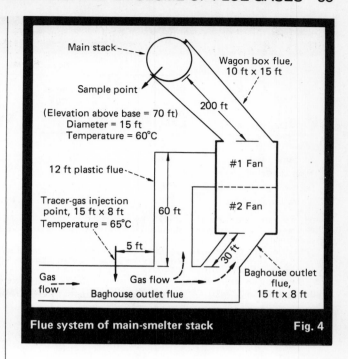

Flue system of main-smelter stack   **Fig. 4**

The SF_6 gave no indication of either sorption or chemical decomposition in any of the tests—the most severe test being one where the temperature ranged from 870°C at the injection point to 122°C at the sample point. Examples of other possible tracer gases are the Freon series of fluorocarbons.

Comparing the two systems

A comparison of the tracer-gas and pitot-tube flow measurements is given in Table I. As can be seen, the two systems agree very well, with the largest deviation being 7%, and average deviation being 4%.

Since the two agree so well, the advantages of the tracer-gas system can be fully exploited. A primary advantage is that truly instantaneous flow measurements can be made. Instantaneous flow measurements are impossible when using pitot tubes, because of the time required to traverse the cross-sectional area of the flue. By using two or more sampling apparatuses at the same time at different points along a flue, simultaneous flow measurements can be made. Such measurements allow greater accuracy in the calculation of leakage of air into or out of the flue system. The greater accuracy is obtained because the pitot-tube system must allow for a time-dependent flow fluctuation of the flue gas.

Another major advantage is that since the tracer-gas method takes less time to determine a single gas flow, and since several gas flows can be determined simultaneously, a flue survey can be completed in less time with a resulting saving in manpower.

This method also gives an accurate flowrate where the pitot-tube system is inaccurate. An irregular flow pattern due to the flue geometry, or the presence of a fan, causes inaccuracies in the pitot-tube system but increases the accuracy of the tracer-gas one (because an irregular flow pattern in the flue would increase the mixing of the gases). This independence of the flue geometry and gas flow patterns allows flow measure-

ments not possible using the pitot-tube system, such as determining the gas flow through a lead blast-furnace.

Possible errors in the tracer-gas method

Even though the tracer-gas method has many advantages, several factors can produce errors:

1. Inaccuracies in the level, or variation in the flow, of tracer gas injected into the flue.
2. Errors in the analysis of the gas sample when using the gas chromatograph.
3. Leakage of the tracer gas out of the flue, or sorption of the tracer gas on the flue walls or on flue dust.
4. Chemical decomposition of the flue gas.
5. Imperfect mixing of the flue and tracer gases.

Of the five types of errors possible, only errors 1 and 2

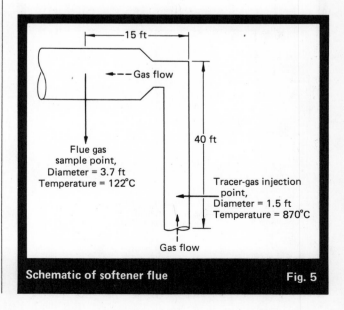

Schematic of softener flue   **Fig. 5**

Tracer gas flow measurement and pitot tube flow measurement comparison   **Table I**

	Tracer-gas flow measurement					Pitot-tube flow measurement, std ft³/min*	Difference between the tracer-gas and pitot-tube flow measurements, std ft³/min*	%
Description	Volume of SF_6 injected, m³/min*	Concentration of SF_6 in the flue-gas sample, ppb	$\dfrac{T}{T_o} \times \dfrac{P_o}{P}$	Relative-humidity correction	Calculated volume of the flue gas, std ft³/min*			
Main smelter stack†‡	28.8	4.15	1.187	1.0	291,000	313,000	−22,000	−7
Softener flue†	32.1	156	1.172	1.0	8,600	8,500	+100	+1
First Lurgi tail gas†¶	33.1	60.9	1.180	1.0	22,600	22,900	−300	−1
Second Lurgi tail gas†¶	26.0	46.9	1.180	1.0	23,000	21,700	+1,300	+6
Third Lurgi tail gas†¶	28.6	62.7	1.185	1.0	19,200	18,600	+600	+3

*At standard condition of 0°C (273 K) and 760 mm Hg.
†Schematic diagrams of the flue systems for the main smelter stack, softener flue, and Lurgi tail-gas flue are given in Fig. 4, 5 and 6 respectively.
‡ The flue-gas flowrate varied from 237,000 std ft³/min for the NE-SW pitot-tube tranverse to 313,000 std ft³/min for the SE-NW pitot-tube traverse. The tracer-gas system was compared to the NE-SW traverse because the flue-gas sample was taken at the same time as the NE-SW pitot-tube traverse.
¶ The first, second and third Lurgi tail-gas flow measurements were taken on different days at the same sample point.

could have affected the accuracy of the system installed at Bunker Hill Co. The maximum magnitude of Error 1 is set by the mass-flowmeter used to measure the tracer-gas flowrate. According to the factory specifications for the mass flow meter used in this study, the maximum error is 2%. The magnitude of Error 2 is set by both the gas chromatograph and the preparation of analysis standards. The maximum analysis error is estimated to be 7%.

Errors 3 and 4, if present, will always result in a calculated flowrate higher than the actual one. These errors are systematic and strongly dependent on the flue system. Their magnitude is not constant and will vary with the flue system.

Leakage of the tracer gas out of the flue has always been much less than 1%, and there have been absolutely no signs of either sorption or decomposition of the tracer gas in all of the tests performed to date at Bunker Hill Co.

Error 5, if present, will result in either a high or low calculated flowrate. This error results when the flue-gas sample is taken from a flue that has pockets of both high and low tracer-gas concentration. In the ideal case, the concentration of the tracer gas would be completely homogeneous. This error is also systematic and is strongly dependent on the flue system.

Flue-gas samples were taken at different sample-lance insertion positions at the same sample point in the different flue systems to determine whether mixing showed less flow variation than would be expected from normal flow fluctuations and analysis deviations.

Because of possible errors, a comparison of the tracer-gas and the pitot-tube systems should be made occasionally to ensure that the two methods agree reasonably well. Two or more samples of flue gas should be taken occasionally from the same sample point, with the sample lance in a different position, to determine the degree of mixing of the gases in the flue.

References

1. Turk, Amos, others, Sulfur Hexafluoride as a Gas-Air Tracer, *Environ. Sci. Tech.*, Vol. 2, No. 1, 1968, pp. 44–48.
2. Thimons, E. D., and Kissell, F. N., Tracer Gas as an Aid in Mine Ventilation Studies, U.S. Bureau of Mines, RI 7917.

The author

John Knoepke, 235 Nagogami Terrace, Rolla, MO 65401, is presently on a one year leave of absence from Bunker Hill Co. (a subsidiary of Gulf Resources and Chemical Corp.) to pursue a Ph.D. in metallurgical engineering at the University of Missouri, Rolla, Mo. He presently holds a B.S. and M.S. in that subject from Montana College of Mineral Science and Technology, Butte, Mont. He is a member of the Metallurgical Society of the American Institute of Mining, Metallurgical and Petroleum Engineers (AIME).

50 ft — Gas flow

30 ft

Flue-gas sample point,
Diameter = 4.1 ft
Temperature = 210°C

Fan

100 ft

Tracer-gas injection point
Diameter = 4.1 ft
Temperature = 250°C

Gas flow

Lurgi tail-gas flue system   **Fig. 6**

CO$_2$ measurements can correct for stack-gas dilution

To calculate permissible pollutant emissions, it is necessary to know the amount of excess air fed to the furnace plus the amount infiltrated from other sources. Here is a simple way of determining this dilution.

Harry C. Lord, *Environmental Data Corp.*

☐ Instantaneous gas concentrations out of the stack are dependent not only upon the process but also upon dilution of the gas stream. This dilution occurs from the excess air in the boiler, infiltration of air into the ducting and air preheaters, etc., water injection (on a wet analysis), as well as from other sources.

However, it is necessary to report emissions data on a standardized basis. The Federal Standards of Performance for New Stationary Sources as applied to fossil-fuel-fired steam generators sets gaseous-emission limits in terms of mass per unit of heat input (lb/10^6 Btu or g/10^6 cal). Another accepted reporting base is a volumetric analysis at a fixed dilution, such as "ppm's dry, at 3% O$_2$."

The CO$_2$ content of the flue-gas stream in a fossil-fuel-fired steam-generating plant has a unique source, the combustion of the carbon in the fuel. Since air contains only trace amounts of CO$_2$ (typically, 0.03%), leakage or excess air will not add more to the flue-gas CO$_2$ concentration. In fact, any dilution variation will equally affect the CO$_2$ content as well as the regulated parameters, such as SO$_2$ and NO. Therefore, a comparison of the actual flue-gas CO$_2$ concentration with the calculated theoretical content for complete combustion of that particular fuel provides an exact dilution factor.

Since fossil fuels (gas, oil, coal) consist primarily of carbon and hydrogen (with secondary amounts of sulfur, oxygen, etc.), the primary combustion products are the oxidized forms of carbon and hydrogen, CO$_2$ and H$_2$O. When there is exactly enough air present to completely oxidize the carbon and hydrogen with no oxygen left over, this is called stoichiometric combustion. Air, however, is a mixture of gases, consisting primarily of approximately 78% nitrogen (N$_2$), 21% oxygen (O$_2$), and 1% trace species. Only the oxygen takes part in the combustion process (although a minor reaction of oxygen and nitrogen forms some nitrogen oxides).

In the combustion of CH$_4$, one has (approximately):

$$CH_4 + 2O_2 + 8N_2 \Rightarrow CO_2 + 2H_2O + 8N_2$$

Originally published January 31, 1977

Because of incomplete mixing of the fuel and the air, extra air will be added to the fire to ensure complete combustion. This extra unreacted air is called "excess air." (It is not uncommon to have 25 to 30% excess air in a boiler.)

Calculating dilution-free concentrations

The flue-gas emissions, are primarily CO$_2$, H$_2$O, N$_2$, excess O$_2$, as well as trace emissions (such as NO, CO, SO$_2$, etc.). These result from some incomplete combustion, as well as oxidation of fuel impurities, such as sulfur. If you let the total emissions equal 100% by volume, you can then calculate the exact percentage of each product for a given fuel (of known H and C content) and a given excess oxygen content. This is normally done by assuming that the trace emissions (NO + SO$_2$ + CO, etc.) are minimal ($\leqslant$10,000 ppm or 1%).

Fuel analyses for composition and heat content are standard procedures, and the information is available from the plant engineer. Fuel composition numbers are in weight percent. For the combustion calculation, we need percent by numbers of atoms. Therefore, one must divide the weight percent value by the atomic weight.

Let us assume the following fuel analysis:

$$
\begin{aligned}
C &= 86.500 \\
H &= 11.750 \\
N_2 &= 0.390 \\
S &= 0.430 \\
Ash &= 0.013 \\
O_2 &= 0.960 \\
CO_2 &= 0.0
\end{aligned}
$$

Then we have:

$C = 86.5/12 = 7.21$ (since atomic weight of $C = 12$)

$H = 11.75/1 = 11.75$ (since atomic weight of $H = 1$)

giving:

$$H/C = 11.75/7.21 = 1.63$$

CO₂ theoretical values at 3% O₂ dry   Table I

H/C ratio	CO₂	H/C ratio	CO₂
[Carbon] 0	0.18	2.1	0.12723
0.1	0.17651	2.2	0.12548
0.2	0.17816	2.3	0.12378
0.3	0.16993	2.4	0.12212
0.4	0.16682	2.5	0.12050
0.5	0.16382	2.6	0.11893
0.6	0.16093	2.7	0.11740
0.7	0.15814	2.8	0.11590
0.8	0.15544	2.9	0.11445
0.9	0.15283	3.0	0.11303
1.0	0.15031	3.1	0.11165
1.1	0.14787	3.2	0.11029
1.2	0.14551	3.3	0.10898
1.3	0.14323	3.4	0.10768
1.4	0.14101	3.5	0.10643
1.5	0.13886	3.6	0.10520
1.6	0.13678	3.7	0.10400
1.7	0.13476	3.8	0.10283
1.8	9,13279	3.9	0.10168
1.9	0.13089	[Methane] 4.0	0.10056
2.0	0.12903		

From Table I, which shows the calculated CO_2 generation in a fossil-fuel-fired steam generator with fuel of a certain H/C, we find:

Theoretical CO_2, dry, (for H/C = 1.63) @ 3% $O_2 \simeq 13.6\%$

For an *in situ* measurement, wherein the gas analysis is made on a wet gas stream (wet from water of combustion, etc.), we have:

Nomenclature

C	Pollutant concentration, dry basis, lb/std. ft³ (or g/std. m³)
E	Pollutant emissions, lb/10⁶ Btu (or g/10⁶ cal)
F_c	V_c/H_{HV} (values are tabulated in Table II)
H_{HV}	Higher heating value of a fuel, 10⁶ Btu/lb (or 10⁶ cal/g)
K	Defined in Eq. (6)
$(\%CO_2)_m$	Measured CO_2 content, dry basis, %
$(\%CO_2)_t$	Theoretical CO_2 produced by complete combustion of fuel, dry basis, %
Q_h	Heat input rate, 10⁶ Btu/h (or 10⁶ cal/g)
Q_s	Volumetric flowrate of effluent, dry, std. ft³/h (or std. m³/h)
V_c	Theoretical volume of CO_2 produced for a given mass of fuel burned, dry, std. ft³/lb, (or std. m³/g)
V_t	Theoretical combustion products, dry basis, for a given mass of fuel, std. ft³/lb (or std. m³/g)

$$SO_2 \text{ [dry @ 3\% } O_2] = SO_2 \text{ [wet (stack)]} \times \frac{\text{theoretical } CO_2 \text{ [dry @ 3\% } O_2]}{CO_2 \text{ [wet @ stack } O_2]}$$

Let us assume that the stack measurement shows:

$$SO_2 \text{ (wet) at stack } O_2 = 210 \text{ ppm}$$
$$CO_2 \text{ (wet) at stack } O_2 = 10.2\%$$

then,

$$SO_2 \text{ dry @ 3\% } O_2 = 210 \text{ ppm} \times \frac{13.6\%}{10.2\%}$$
$$= 280 \text{ ppm}$$

This calculation adjusts the readings for all dilution effects, including the air preheater leakage, breeching leakage, and water dilution (including water injection).

Calculation of mass emissions

The volumetric concentrations of emissions from a variety of emission points is of little benefit in assessing their impact upon a given air-quality region. Because there are significant variations in stack diameters, exit-gas velocities (and hence total volume being released from each point), a very low concentration from a large-gas-volume emission point may dominate the emissions in that area. In order to derive data directly proportional to the source's impact on the air-quality region, most regulatory agencies are requiring that the emissions be calculated in terms of the mass being released. The two most common units are lb/h (g/h), or lb/10⁶ Btu (g/10⁶ cal).

EPA (Environmental Protection Agency) has promulgated in the "Federal Register"* the following calculation of mass emission rates for fossil-fuel-fired steam generators:

$$E = C \frac{Q_s}{Q_h} \qquad (1)$$

In addition, the EPA scientists have developed a formulation for mass-emission-data calculations based upon volumetric gas concentrations, including the CO_2-based dilution analysis. This is founded upon the observation that within any group of fossil fuels (gas, oil, coals), the ratio of the volume of CO_2 released to the quantity of heat released is essentially a constant.

The higher heating value, HHV, of the fuel (the total heat released in the combustion process, including the heat released from the condensation of the water vapor produced from the combustion of the fuel hydrogen) is a standard component of a fuel analysis. The ratio of the theoretical total combustion products per unit of mass of the fuel burned to this higher heating value is:

$$\frac{V_t}{H_{HV}} = \frac{Q_s}{Q_h} \frac{(\%CO_2)_m}{(\%CO_2)_t} \qquad (2)$$

The ratio of $(\%CO_2)_m$ to $(\%CO_2)_t$ is a measure of the total dilution, as described in the preceding section. Furthermore, note that if we let V_c equal the theoretical volume of CO_2 produced for a given mass of fuel burned, then:

*Federal Register, Standards of Performance for New Stationary Sources, Vol. 36, 247 Part II, Dec. 23, 1971.

Average F_c factors		Table II
Fuel type	Std. m³ of CO₂/10⁶ cal*	Std. ft³ of CO₂/10⁶ Btu*
Coal		
Anthracite	0.222	1,980
Bituminous and lignite	0.203	1,810
Oil	0.161	1,430
Gas		
Natural gas	0.117	1,040
Propane	0.135	1,200
Butane	0.142	1,260

*Standard conditions are 70°F, 29.92 in. Hg, and 0% excess air

$$V_c = (\%CO_2)_t/100 V_t \qquad (3)$$

Substituting the last two equations into the basic mass-rate equation, we have:

$$E = C \frac{100}{(\%CO_2)_m} \frac{V}{H_{HV}} \qquad (4)$$

and if we let $F_c = V_c/H_{HV}$ (the ratio of the theoretical CO₂ generated to the higher heating value of the fuel being burned, std ft³/10⁶ Btu, or std. m³/10⁶ cal), we have:

$$E = C \frac{100}{(\%CO_2)_m} F_c \qquad (5)$$

However, the concentration of the pollutant that is measured is in terms of parts per million by volume. In a standard cubic foot there is a known number of molecules. The emissions, measured in ppm, indicate the proportion of the pollutant of interest to the total number of molecules. Multiplying the mass of each molecule by the number of molecules gives the mass emission per dry standard cubic foot.

C [in g/dry std. m³] =
$$(ppm\ emission)(4.15 \times 10^{-5})(M)$$

C [in lb/dry std. ft³] =
$$(ppm\ emission)(2.59 \times 10^{-9})(M)$$

where M = molecular weight of the gas

$$= 64.07\ for\ SO_2$$
$$= 46.01\ for\ NO_x$$

and where 4.15×10^{-5} (or 2.59×10^{-9}) converts the volume of any ideal gas to the number of mols per m³ (or per ft³) of flue gas at the standard conditions of 70°F and 29.92 in. Hg.

For the NO$_x$ emissions, the mass emission rate is calculated by assuming that the emissions are 100% NO₂. This calculation is used even if the emissions are exclusively or predominantly NO (and all that is measured is NO). The reason for this is that the emitted NO molecules have a relatively short life in the atmosphere, being oxidized to NO₂ by the oxygen content of the ambient air.

Therefore, for a mass-emission analysis of SO₂ in terms of lb/10⁶ Btu, we have:

E [in lb/10⁶ Btu] =
$$SO_2\ [in\ ppm,\ dry] \times \frac{100}{(\%CO_2)_m} \times K \qquad (6)$$

where

$$K = F_c\ (2.59 \times 10^{-9})(64.07)$$

Furthermore, for a wet measurement (either by means of extractive sampling with heated lines, or an *in situ* measurement), the same equation applies, since the factor to convert from wet to dry, $100/[100 - (\%H_2O)_v]$, is in the numerator of the equation, affecting ppm SO₂, and also in the denominator, affecting $(\%CO_2)_m$, and hence cancelling.

$$E = (ppm)_{wet} \frac{100}{(100-\%H_2O)} \times$$
$$\frac{100}{(\%CO_2)_{wet}\ 100/(100-\%H_2O)} K \qquad (7)$$
$$= (ppm)_{wet} \times \frac{100}{(\%CO_2)_{wet}} K$$

The F_c values of a variety of fossil fuels have been calculated and found for each specific type to be a constant having a maximum deviation of ±5.9%. The average F_c values for each fuel type are given in Table II.

Also observe that the emissions measured by this procedure are independent of the gas stream temperature. For a monitoring application where the gas temperature can vary, one normally uses a temperature compensation circuit to correct for the variations. As the gas stream temperature varies, the gas density varies, and thus the concentration of the measured parameter. Since $PV = nRT$, and if P, V, and R are constant, then $n_1 T_1 = n_2 T_2$, or $n_1/n_2 = T_2/T_1$. Again, however, the correction factor for the temperature variation affects the pollutant concentration (in the numerator) as well as the CO₂ concentration (in the denominator) and temperature cancels from the equation.

Summary

To report emissions data on the standardized basis required by recent EPA regulations, it is necessary to be able to determine such data on the basis of either the amount of fuel fired, or by volumetric analysis at a fixed dilution. The techniques reported in this article will enable the engineer to make the needed calculations easily and correctly.

The author

Harry C. Lord, III, is founder and vice-president of Environmental Data Corp., 608 Fig Ave, Monrovia, CA 91016, a company that designs, manufactures and services instrumentation for the continuous monitoring of gaseous emissions. He holds a B.S. in chemistry from Tufts University and a Ph.D. in physical chemistry from the University of California, San Diego, Calif. He is a member of ACS, Air Pollution Control Assn., Instrument Soc. of America, The Combustion Institute, Sigma Xi and the New York Academy of Sciences.

Particulate Collectors

a. Introduction
Gas/solid separations
Removing particulates from gases
Controlling fine particles
Predicting efficiency of fine particle collectors

b. Particulate Scrubbers
How to choose a particulate scrubber
Practical process design of particulate scrubbers
Halt corrosion in particulate scrubbers
Maintaining venturi-tray scrubbers
Get better performance from particulate scrubbers:
    Upgrading existing particulate scrubbers
    Troubleshooting wet scrubbers

c. Dust Collectors
How to choose a cyclone dust collector
New design approach boosts cyclone efficiency
Selecting, installing and maintaining cyclone dust collectors
Calculator program solves cyclone efficiency equations
How to specify pulse-jet filters
Cooling hot gases before baghouse filtration
Baghouses: separating and collecting industrial dusts
Baghouses: selecting, specifying and testing industrial dust collectors

d. Electrostatic Precipitators
Selecting and specifying electrostatic precipitators
Electrostatic precipitators in industry
Electrostatic precipitators: How they are used in the CPI
Improving electrostatic precipitator performance
    Tuning electrostatic precipitators
    Specifying mechanical design of electrostatic precipitators

Gas/Solid Separations

Equipment for separating solid particles from a gas stream may be based on dry, wet or electrostatic collection. Selection of the optimum system is a complex procedure, requiring careful analysis of many variables.

GORDON D. SARGENT, Nopco Div.,
Diamond Shamrock Chemical Co.

The engineer faced with a dust collection problem must make a thorough evaluation of the situation. If he tries to shortcut with an equipment specification limited to a "horseback" estimate of gas flow and efficiency, he risks wasted pilot trials or costly, inadequate installations.

What data are needed? What type of equipment might be suitable? When choosing a pump or heat exchanger there are well-established routes to answering the usual questions. By contrast, the heterogeneous nature of particulates in gas streams has led to a wide variety of separation devices, for which engineering principles are either lacking or not readily available. Equipment manufacturers must be relied upon for proprietary designs and performance guarantees.

This article presents available facts and sources that the engineer needs to avoid false starts and quickly develop a complete definition of the problem. When checklists of data, equipment selections and process features have been satisfied, then vendors can be chosen for help in completing the solution.

Gas-cleaning equipment discussed here will handle dust particle sizes between 0.1 and 100 microns and concentrations from 0.1 to 100 grains/cu.ft. The micron (μ) is the commonly used unit of particle-size measurement and is defined as 1/1,000 mm. or 1/25,400 in.

Dust concentrations are usually given in terms of grains/cu.ft. of gas (7,000 grains = 1 lb.).

Air cleaners for fumes with much smaller particle size and loadings are not covered here nor are combustion or catalytic incinerators, gas or odor absorbers or adsorbers, ventilation-air cleaners and mist eliminators.

The need for gas cleaning may be for process operation, protection or profit. A collector may be an integral part of a process such as spray-drying or an auxiliary to recover valuable byproduct. The main requirement may be safety, as in reducing toxic or combustible dust. People and property in the plant or neighborhood may need to be protected by a good dust collection system, or the requirement may be to meet air pollution laws and to clean up an unsightly stack plume.

The many different dust collectors available today are summarized in Table I, which is a simplified review of the whole collector field. Ranges and limits tabulated are typical values but naturally may vary widely for unusual applications. Suppliers can be found listed in the directory at the back of this issue and should be consulted especially for integrated systems, packaged units and air cleaners.

DRY INERTIAL COLLECTORS

A dry collector has certain advantages compared to a wet collector. If the dust is a useful product, dry collection saves the cost of reprocessing. Handling the collected material can give rise to additional dust problems. Dry, dusty material has the disadvantage of requiring ventilation and if hygroscopic, caking can be a problem. The cleaned gas will not be cooled or completely free of fines. Without cooling,

Originally published February 15, 1971

Equipment Application—Table I

Types of Dust Collecting Equipment	Particle Size Microns	Loading Grains/ Cu. Ft.	Collection Efficiency Weight %	Pressure Loss		Utilities Per 1,000 Cfm.	Gas Velocity, Fpm.	Size Range Limits, 1,000 Cfm.	Space Required, (Relative)
				Gas, In. W.G.	Liquid Psi.				
Dry inertial collectors									
Settling chamber	>50	>5	<50	<0.2	———	———————	300–600	None	Large
Baffle chamber	>50	>5	<50	0.1–0.5	———	———————	1,000–2,000	None	Medium
Skimming chamber	>20	>1	<70	<1	———	———————	2,000–4,000	50	Small
Louver	>20	>1	<80	0.5–2	———	———————	2,000–4,000	30	Medium
Cyclone	>10	>1	<85	0.5–3	———	———————	2,000–4,000	50	Medium
Multiple cyclone	>5	>1	<95	2–6	———	———————	2,000–4,000	200	Small
Impingement	>10	>1	<90	1–2	———	———————	3,000–6,000	None	Small
Dynamic	>10	>1	<90	Provides head	———	1–2hp.	———————	50	
Wet scrubbers									
Gravity spray	>10	>1	<70	<1	20–100	0.5–2 gpm.	100–200	100	Medium
Centrifugal	>5	>1	<90	2–6	20–100	1–10 gpm.	2,000–4,000	100	Medium
Impingement	>5	>1	<95	2–8	20–100	1–5 gpm.	3,000–6,000	100	Medium
Packed bed	>5	>0.1	<90	1–10	5–30	5–15 gpm.	100–300	50	Medium
Dynamic	>1	>1	<95	Provides head	5–30	1–5 gpm., 3–20 hp.	3,000–4,000	50	Small
Submerged nozzle	>2	>0.1	<90	2–6	None	No pumping	3,000	50	Medium
Jet	0.5–5	>0.1	<90	Provides head	50–100	50–100gpm.	2,000–20,000	100	Small
Venturi	>0.5	>0.1	<99	10–30	5–30	3–10 gpm.	12,000–42,000	100	Small
Fabric filters	>0.2	>0.1	<99	2–6	———	———————	1–20	200	Large
Electrostatic precipitators	<2	>0.1	<99	0.2–1	———	0.1–0.6 kw.	100–600	10–2,000	Large

Note: The terms expressing concentration, or loading, can be defined as light = ½–2, moderate = 2–5, and heavy = 5+ grains/cu. ft. Particle size: fine, 50% in ½–7 micron size range:, medium, 50% in 7–15 micron size range; coarse 50% over 15 microns.

the temperature limits of equipment will have to be considered. Corrosion will be minimum unless the fumes contain corrosive mists. Equipment generally is bulky.

Inertial or mechanical collectors are best suited for medium or coarse particulates. High dust-loadings can be handled at moderate pressure drops and power consumption. Simple construction of this type of collector results in lower cost and maintenance than other types. Efficiency is not very high; hence, for a really clean effluent, some other type of collecting device must be used in combination with or in place of the inertial collector.

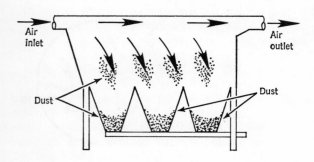

GRAVITY CHAMBER has simple construction but low collection efficiency—Fig. 1

Gravity Settling Chamber

Principle—Dirty gas is directed through an oversized duct where velocity drops low enough to let large particles settle out by gravity. (Fig. 1.)

Comments—Flow may be horizontal or vertical. Dust separation suffers from reentrainment from eddy currents. In the Howard dust chamber, horizontal shelves or trays have been added to shorten the settling path of the particle, improving collection efficiency, but making cleaning more difficult.

The gravity settling chamber is seldom used today, but can be designed for a specific application—ventilation contractors are familiar with them. Space requirement is large and efficiency low, which limits this type to precleaning gas to be fed to a more efficient collector. A combination of settling chambers and radiant-cooling connecting ducts is used in the metal refining industry.[1]

Baffle Chamber

Principle—Settling is aided by using momentum from a direction change. Gas flow is directed downward through a chamber containing a baffle around which the gas is deflected; meanwhile the larger dust particles tend to continue moving downward to be collected in a hopper for later use or removal. (Fig. 2.)

Comments—This collector takes less space than the straight-through settling chamber and has similar efficiency. One chip-trap design[1] (intended to protect downstream fans from very coarse materials) has dimensions given in terms of the inlet duct diameter.

Skimming Chamber

Principle—The dirty gas stream enters a scroll tangentially; the dust is carried to the periphery by inertia. Concentrated dusty gas is skimmed by slots and led to a dust hopper or secondary collector. The cleaned gas stream from the hopper is combined with that leaving axially from the skimming chamber.

Comments—Dry collectors of medium efficiency such as this one have an exit-gas stream that will probably not satisfy most requirements of dust collecting. A secondary collector may well be required, the skimming chamber being used to reduce the load of coarser particles that are carried into the secondary collector.

Louver-Type Collector

Principle—Gas passes into the wide end of a wedge or cone and must take a sharp bend in order to escape through slots or louvers in the walls. The larger particles are carried by inertia to the narrow end of the chamber where they are purged with a small fraction of the gas stream. (Fig. 3.)

Comments—This collector must be followed by a second collector, such as a high-efficiency cyclone, to separate the dust from the gas. One author (Strauss)[2]

shows cone-shaped louvered collectors followed by a baffle chamber for coarse dust, followed in turn by a cyclone for fine dust. The gas purged, usually less than 10%, is returned to the inlet of the louvered collector and recycled.

Cyclones

Principle—In the most common arrangement, the gas enters the cyclone tangentially at the top of the cylindrical section and spirals downward into the bottom section, which is usually conical in shape. Dust particles, which have a greater applied centrifugal force than the gas molecules, accumulate at the wall and are carried down, held against the wall by the gas velocity. At the bottom of the cyclone the gas separates from the dust, flows back up in a smaller spiral and exits at the top. Solids are collected in a hopper and removed by a rotary valve, screw conveyor, or other such means.

Comments—Cyclones are one of the most widely used collectors. The unit is low cost, has no moving parts and can be constructed with refractory linings for high temperature, up to 1,800 F. Units can be designed for high dust capacity at medium efficiencies and medium pressure drop. High efficiencies are obtained with smaller diameters and higher velocities, which in turn take higher pressure drops. Units may be installed in parallel for large gas flows and in series for higher efficiencies (or for both advantages, in combined series-parallel).

Cyclones may have various configurations and still operate on the same basic principle of centrifugal separation. Gas entry may be tangential or axial. Gas exit may be axial or axial combined with a pressure-recovery device. Dust exit may be with gas purge or for solids only, with the configuration being axial

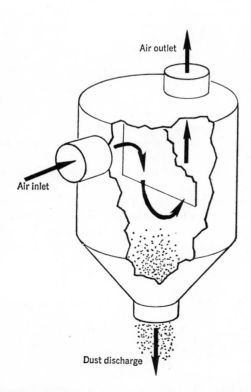

BAFFLE COLLECTOR is similar to gravity settler but occupies less space—Fig. 2

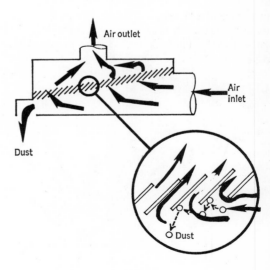

LOUVER SEPARATOR should be followed by a more effective collector—Fig. 3

or peripheral. The skimming chamber, multiple cyclone tubes, and the Uniflow cyclone are variations in the standard design.

Multiple Cyclone

Principle—Since small-diameter cyclones are more efficient than large ones (centrifugal force for a given tangential velocity varies inversely as the radius of the cyclone), banks of small (10 in. or less) cyclones are arranged in parallel with feed gas from a plenum chamber. (Fig. 4.)

Comments—The major advantage is high efficiency —the disadvantage is plugging of the small tubes. The individual small cyclone does not operate as efficiently in multiple installation as it would by itself. This difference arises from unequal gas or dust distribution to inlets and recirculation of gas from dust hopper back through dust outlets. Consequently, multiple cyclones are best obtained from a manufacturer as a complete unit in order to ensure good design.

Impingement Collectors

Principle—Gas velocity is increased in a venturi and particle momentum carries the particles through slots to a flat plate where they drop to a collector.

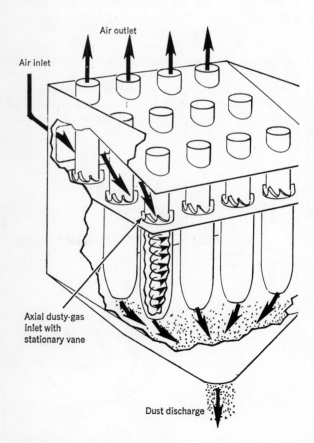

Air outlet

Air inlet

Axial dusty-gas
inlet with
stationary vane

Dust discharge

CYCLONE efficiency can be raised by placing several small units in parallel—Fig. 4

Dust particles are collected on a surface while the gas stream is diverted around the plate.

Comments—Collection of mist is simplified with this collector, since liquid merely runs down the baffle plate. These devices may require rappers to free dust that builds on surfaces. If the solids are sticky, the surfaces may be continually washed by circulating water; this film, besides cleaning the surface, prevents re-entrainment of particles.

Dynamic Collector

Principle—This is a fan with a specially designed impeller and casing that employs centrifugal force to collect particles at the periphery, where they are drawn off in a concentrated stream.

Comments—The unique features of this collector are small space requirement and low pressure drop. The packaged unit by American Air Filter combines exhauster, dust separator and storage hopper. The dynamic precipitator acts as a true fan although at somewhat lower efficiency, 40 to 50% as against 60 to 65% for a fan designed simply for gas service.

WET SCRUBBERS

The wet scrubber recovers solids as a slurry or solution that requires further processing either to obtain product or to dispose of as waste.

In a wet scrubber:
- The gas is both cooled and washed.
- Gases may be removed as well as particulates.
- Corrosive gases may be neutralized by proper choice of scrubbing medium.
- The stack effluent usually will be well cleaned but will contain some unwetted fines, mists, and a steam plume.
- The temperature and moisture content of the inlet gas is essentially unlimited.
- Freezing conditions must be considered.
- Hazards of explosive dust-air mixtures are reduced.
- Equipment occupies only a moderate amount of space.

Efficiencies vary with power input and can extend over a wide range depending on the design. Equipment size and initial cost are reasonable, but operating cost is high, especially for high efficiency, which requires large power consumption.

Gravity Spray Scrubber

Principle—Liquid is sprayed into the top of the tower and coarse droplets fall by gravity through a countercurrent flow of the gas being scrubbed. Dust particles are collected mainly by inertial impaction and interception. (Fig. 5.)

Comments—Efficiencies and pressure drops are low. This device is useful for a heavy loading of coarse particulates, or for absorption accompanied by solids removal. The wet cap used on top of foundry cupolas is a spray and baffle arrangement in which water both cools the baffle and conveys the collected dust. Boiler and process stacks have been scrubbed with sprays installed in the stack, thereby avoiding an additional fan and separate scrubber. Entrainment is controlled by using low gas velocities when designing spray towers, but this results in large equipment. Sprays are selected to give large droplets that must be heavy enough to fall counter to the gas flow, even though they may be further reduced in size by evaporation.

Centrifugal Scrubber

Principle—Liquid is sprayed into the unit and mixed with the rising vortex of gas. By impaction and interception the gas and liquid particles combine and are accelerated to the vessel wall by centrifugal force. There they are collected. The wetted wall also aids collection. (Fig. 6.)

Comments—A variety of types have been developed based on differing ways of droplet forming and of promoting cyclonic gas action. Vaned baffles direct gas flow and convert velocity pressure to droplet formation energy. Sprays may be installed axially for radially-directed droplets, or circumferentially for tangentially introduced sprays. For higher pressures of 400 psig. or so, droplet size is 50 μ or less, instead of, say, 500 μ.

Impingement Scrubber

Principle—The gas stream, carrying both dust particles and water droplets from preconditioning sprays, is directed through perforated plates to impinge on baffle plates. Gas velocity acts to atomize water on the perforated plate. Enlarged particles are collected on vaned mist eliminators and are withdrawn along with the solids collected in the liquid overflow from the impingement plate.

Comments—The scrubber is similar to a sieve-plate column and usually has from one to three plates, although there may be more. Extra stages can be added later. Each hole in the "sieve" plate has a baffle or "target" above. Flow is countercurrent. The gas rate in the perforations is high, 75 ft./sec. or more, and is used to provide atomization of the liquid on the plate. Plugging of holes, which may be ¼ in. or less, is not as much of a problem as might be expected, owing to the agitation as well as to the gas preconditioning sprays that wet the underside of the plate. Soluble gases can be effectively removed along with the dust.

Packed-Bed Scrubber

Principle—Wetted packing provides an impingement surface that prevents reentrainment. The liquor provides a means of washing off dust and conveying it in a slurry or solution. (Fig. 7.)

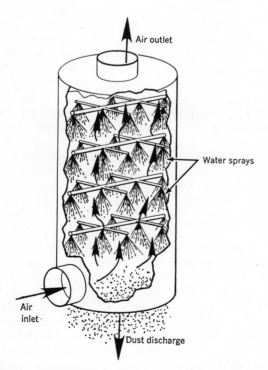

GRAVITY SCRUBBER is useful for collecting heavy loadings of coarse particles—Fig. 5

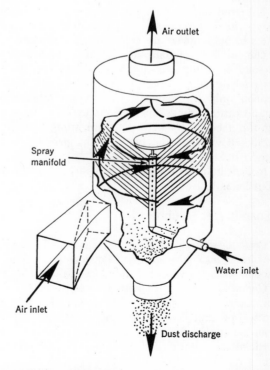

CENTRIFUGAL SCRUBBERS are available in many different designs—Fig. 6

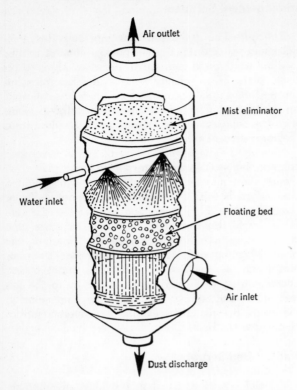

Air outlet

Mist eliminator

Water inlet

Floating bed

Air inlet

Dust discharge

PACKED-BED scrubber can be applied for low dust loadings and gas absorption—Fig. 7

Comments—Packing may be fixed or it may be a floating bed of low-density spheres. The advantages are: low cost and simplicity, corrosion resistance, and no moving parts. Dust collecting may be secondary to direct-contact cooling and gas absorption. Gases to be removed are below 1% by volume.

The usual countercurrent packed tower has almost no solids-handling capacity, since solids tend to plug the packing and support-plates, which can then be cleaned only by removal. Crossflow scrubbers can handle dust loadings up to 5 grains/cu.ft. by washing the face of the packing with spray nozzles in parallel flow while the body of the packing is irrigated from the top in crossflow.

In a floating bed of plastic spheres, packing movement helps to free the solids.

Dynamic Wet Scrubber

Principle—Liquid is sheared mechanically to break the liquid into droplets for collection by inertial impaction between droplets and dust particles.

Comments—In the simplest form of dynamic scrubber, water is sprayed into the suction of a fan, and the wetted impeller and housing holds dust particles from reentrainment. The efficiency is high for fine particles and utilities are 3 to 5 gpm./1,000 cfm. and 2 to 4 hp./1,000 cfm.

Another unit, the disintegrator, has rotating and stationary bars to break up the water feed stream into fine droplets. This machine seems to be displaced by the venturi today.

Power consumption is very high, 10 to 20 hp./1,000 cfm., and efficiency is correspondingly high. Rotor speed is 350 to 750 rpm., and buildup must be avoided to prevent rotor unbalance. Consequently, a precleaner, such as a cyclone or centrifugal scrubber, is needed to hold the inlet dust loading on the disintegrator below 0.5 grains/cu.ft., and the temperature below 125 F. (Other collectors with mechanically driven elements and a pool of scrubbing liquid are included in the following paragraphs on submerged-nozzle scrubbers.)

Submerged-Nozzle Scrubber

Principle—Gas passing through a nozzle or orifice is scrubbed by the liquid and also atomizes the liquid (assisted in some cases by mechanical means) for further collection of dust particles on the droplets. These droplets are removed in a disengagement chamber, aided by baffles.

Comments—This type of collector includes scrubbers that atomize liquid entirely by gas kinetic energy, as well as scrubbers in which the gas merely passes through a mechanically formed spray. High dust-loadings can be handled, especially if the units are designed for continuous sludge removal (by conveyor, screw, etc.). Plugging is prevented by designing to avoid close clearances in the area where dry dust meets the spray.

Jet Scrubber

Principle—Water flow is used in a jet ejector, both to aspirate dusty air and to provide droplets for collecting particulates. The conditioned-dust and the water droplets are separated from the gas in a settling chamber, which may be baffled. (Fig. 8.)

Comments—An induced draft of a few inches of water is usual because higher values require very high water rates as well as excessive power. The jet scrubber can be used where it is not economical to add a fan for a dust collection system and where either mist or easily absorbed gas is to be removed from the gas stream.

Venturi Scrubber

Principle—Water is introduced into the throat section and atomized by the high-velocity gas stream. The high relative velocity between the accelerating solid particle and the liquid droplet makes for high efficiency by impingement. Collection is aided by condensation if the gas is saturated in the reduced-pressure section of the venturi, since the solid particles serve as nuclei for condensing in the pressure-regain section. Agglomerated particles built on droplets of 50 μ or more can be collected with high efficiency in a subsequent centrifugal collector.

Comment—High efficiencies require high power input; the venturi scrubber can be designed for large pressure-drops, as much as 80 in. water gage, to collect submicron dusts. Water can be introduced by spraying or by weir overflow. Efficiency approaches 100% at 2 μ and 22 in. water gage. The equipment is simple and can be fabricated in a variety of materials for corrosion resistance.

FABRIC FILTERS

Dry collectors have already been compared to wet scrubbers in the previous section on dry inertial collectors. Some functional characteristics of fabric filters are similar to those of inertial collectors: product is collected in usable condition, cleaned gas is uncooled, and secondary dust problems can be created in handling recovered dust. However, the fabric filter can handle much smaller particles at high efficiencies. Temperature is limited by the fabrics. Moisture content is limited because cold spot condensation can cake solids and also cause corrosion.

Principle—Filters for industrial gas cleaning are both bag and envelope type, with woven or felted fabric, and made from natural or synthetic fibers. Dirty gas flows through a porous medium and deposits particles in the voids. As the voids fill and a cake builds on the fabric surface, the pressure drop increases to a point where the solids must be removed.

Comments—Filters are used for high efficiency, 99+%, on small particles in the submicron range. Filters will continue to function effectively even when gas properties and process conditions vary. Costs are moderate.

The principal limitation is temperature, with a maximum of 550 F. Cooling a very hot dirty gas may be worthwhile in order to permit using a fabric filter. If the gas is below the dewpoint, heating some 50 to 75 F. above the dewpoint will allow a dry unit. Space required is large but may be acceptable because the recovered material is dry and ready for either use or disposal (unlike wet-scrubber recoveries).

Intermittently Cleaned Filters

One type of fabric filter is cleaned intermittently by shutting down the process and using shakers or reverse-air-jets to remove the dust cake. Another type operates continuously by sequentially cleaning one isolated compartment after another. As the compartments are cleaned they are put online again. These filters operate at air-volume/cloth-area ratios of 1.5 to 3 cfm./sq.ft. of filter area, or in terms of velocity, 1.5 to 3 ft./min. Filter media are woven fabrics chosen for thermal, chemical and mechanical endurance.[3] (Fig. 9.)

Continuously Cleaned Filters

Continuous cleaning of filter media without isolating any part of the equipment is accomplished by a travelling blow-ring or by reverse-air-jets. These cleaning methods are so thorough and leave so little filter cake that woven fabrics cannot be used without loss of efficiency. Hence felted fabrics are employed. Air flowrates of 15 ft./min. are usual and result in more compact baghouses. This kind of filter permits higher dust-loads, but is more complex, hence has higher first cost and maintenance.

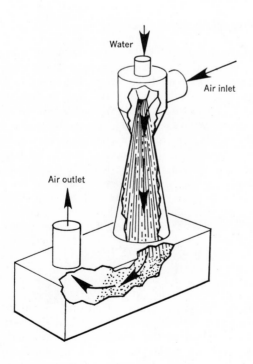

JET SCRUBBER does not require fan in the dust collection system—Fig. 8

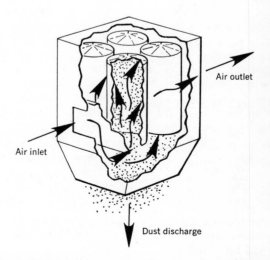

FABRIC FILTERS offer high efficiencies when separating small particles—Fig. 9

ELECTROSTATIC PRECIPITATORS

The electrostatic precipitator has the advantage of permitting dry collection, and is highly efficient on small particles. A fabric filter might be a first choice, however, unless the process stream is hot or corrosive. If the particulate is coarse, a wet scrubber is lower in cost. But the pressure drop for the filter or scrubber will be higher than for the precipitator, and the scrubber's operating costs are likewise higher. Capital cost of electrostatic precipitators is usually the highest of all collectors, but a complete economic comparison must be made for a true picture.

Principle—The gas between a high-voltage electrode and a grounded (or oppositely charged) electrode, is ionized. Dust particles are charged by the gas ions and migrate to the grounded collecting electrode, where they adhere. Mists run off the collecting surface, often aided by addition of irrigation streams. Solids are most commonly removed by rapping with hammers or vibrators, although removal can be by washing or scraping. Pipe-type precipitators are used for mists or water flushing, and plate-types are used for dry collection and large gas flows.

Comments—Although there is widespread use of electrostatic precipitators and fundamentals are well-developed, piloting a precipitator application is often worthwhile, since particle properties may vary between installations and theoretical efficiency is never attained. Some common applications are the collection of fly ash from pulverized-coal-fired boilers, cement-kiln dust, sulfuric acid mist, catalyst dust in oil refineries, and steel blast-furnace dust. Operating temperatures range from 0 to 700 F. although units have been designed for temperatures of −70 or +1,000 F. Pressure drop is small due to low velocities, but this makes for a large installation.

In considering electrostatic precipitation, the most important characteristic of a dust is its electrical conductivity. The dust conductivity must lie between that of a good conductor such as a solid metal and that of a good electrical insulator—and not too near either extreme. The reciprocal of conductivity, electrical resistivity, is used to define this property. For most-effective operation of a precipitator the dust resistivity should be between 10^4 and 10^{10} ohm-cm. Particles with a very low electrical resistivity readily lose their charge on the collecting electrode, only to be reentrained. On the other hand, materials with high electrical resistivity coat and insulate the collecting electrode, thereby reducing potential across the gas stream; this may lead to a spark discharge that reverses ionization and causes reentrainment. Conditioning agents such as moisture, acid mist, and ammonia can be added to the gas to reduce resistivity.

PROCESS APPLICATIONS

Whenever a granular solid is mixed with air or other gas for conveying, drying, classifying, etc., a collector of some type is necessary to separate the product at a later stage. Often this collection will be best accomplished in two or more stages. The optimum system design will require comparison of several alternates. Some common situations are discussed below.

Spray Drying

Air from a spray dryer or flash dryer carries a heavy loading of solids, although in some cases the spray tower serves as a primary separator. Solids should be recovered dry in a cyclone followed by a secondary collector. When the temperature is not too high, a fabric filter is often used with the spray dryer, collecting essentially 100% of the product in dry form. Sticky materials that build up on the filter cloth rule out bag collectors. Condensation on the filters and housing would be a problem for a soluble product and would require insulation or heating of the unit. Mechanical collectors offer easy cleaning, which is needed to avoid cross-contamination from product changes. If the fabric filter does not have too many bags they may be removed for washing or replacement.

A small-tube multiple cyclone may give the right combination of high efficiency, dry recovery, and cleanability. Two or three collectors in series present the opportunity to segregate fines for reworking or discarding. The composition of a spray-dried mixture may vary with particle size and should be checked. It may be necessary to redissolve the fines and blend them with the dryer feed liquor. A wet scrubber is often the secondary collector in a spray drying system. However, if the fines are to be discarded the advantages of handling dry versus wet scrap is important.

Drying with Solvent Recovery

Where a product is dissolved in a solvent other than water, the solvent as well as the solids will have to be recovered. A wet scrubber utilizing the solvent as a scrubbing medium will permit direct-contact absorption at high efficiency, and recycle of the solvent with some solids, suspended or dissolved. If the solids are objectionable in the scrubber or recycle stream, a dry collector would be required before the solvent recovery step. The collector may be heated or insulated if necessary, and the dry solids collected in a receiver. Then solvent vapor can be recovered in a condenser without solids fouling.

In a vacuum dryer system, the vacuum pump can be followed by a condenser for additional recovery. If the solvent forms an explosive mixture with air, the system may use an inert gas to control or eliminate oxygen content. The amount of solvent vapor going uncondensed will suggest recycling the inert atmosphere, with solids collecting equipment after the dryer.

Pneumatic Conveying

Frequently after a dry product has been separated from a process gas stream, it will have to be elevated to permit gravity flow through operations such as storage, blending and packaging. A pneumatic conveying system offers economy, simplicity and flexibility, and if purchased as a package system will include a separating device, most commonly a cyclone or fabric collector.

The specification for the collector develops in conjunction with conveyor tests or specifications. Careful collector selection is important because poor efficiency can give problems in any pneumatic conveying system. It would seem that in a closed-loop system where dust is recycled, the collector would not be a problem. But low separating efficiency builds up a heavy solids conveying load, resulting in plugging or other problems.

The simplest separator for pneumatic conveying uses the receiving bin as a collector with a bag tied over the vent. Bin capacity is sacrificed in this instance, and the bin cannot be maintained under pressure or vacuum. The bag filter may exhaust some dust, but a good ventilation system will take care of any hazard or nuisance.

The subject of ventilation suggests one possible abuse of collectors. A dust separation system is a handy vacuum cleaner around the plant. "Temporary" ventilation can be provided with flexible ducts tied into the blower suction duct. This increased load can create problems if the collector does not have extra capacity.

ESTIMATING COLLECTOR SPECIFICATIONS

To make a preliminary selection of a suitable gas-cleaning plant, only four basic items are required: dust loading, particle size, gas flow and allowable emission rate. Other factors that must be considered before final selection are listed in Table II.

Requirements can be estimated for an existing process except for particle size. However, for a process in the design stage, both dust size and loading may be difficult to estimate, especially if process experience is not available within the organization or from suppliers or the literature. A broad list of processes and usual collector applications has been compiled by Kane.[4]

Once a tentative choice has been made, pilot plant trials may be the simplest approach. Particle size distribution of dust can be ignored and the dust loading will be checked in the normal course of efficiency measurement, permitting direct evaluation for simple scaleup. Many suppliers have pilot-model collectors available for rental, often with some refund on purchase of a full-scale unit. Installation and operation of a 1,000 cfm. pilot device may cost from $500 to $5,000 or more, depending on the complexity of installation and the auxiliaries that may be required.

Going a step further before making a choice, some simple lab work may prove economical in the long run. Laboratory data can be obtained from an existing process by collecting dust samples for particle size determination, either in your plant's lab or by suppliers or consultants.

Lab data on crudely collected samples can be obtained quickly and cheaply but will probably have to be supplemented by isokinetic sampling (described later), depending on the size of the project and the economic risk that can be tolerated.

It may be feasible to proceed with installation and plan to modify, supplement, or add stages to the collector. For example, a venturi with adjustable pressure drop can be installed, a wet collector can be added after a cyclone, or more stages may be added to some types of centrifugal wet scrubbers.

Determining the desired emission level depends on one or more factors: air pollution codes, toxicity limits, and economics. Legal requirements of air pollution control in the U.S. are very complex, depending on federal, state and local governments. Laws are concerned with the type and amount of particulate and gaseous emissions and also the appearance of smoke. A dust-control system is subject to future, more limiting legal requirements and should be selected with some thought toward future improvement.

Hazard control may be an added requirement. Data are available for many dusts and vapors, although new materials may require some lab work. Dust hazards to consider are explosive concentrations with air, flammability, and toxicity to plant or animal life, including radioactive hazards. Vapors in the gas stream must be checked for explosive limits

Data Required for Equipment Selection—Table II

Particulate characteristics
* *1. Particle size distribution
* *2. Concentration—average and extreme values
* 3. Particle density (and viscosity)
* 4. Bulk density
* 5. Moisture content
* 6. Electrical resistivity and sonic properties
* 7. Handling characteristics—erosion, abrasion, frangible, flocculent, adhesive, sticky, linty, bridging
* 8. Composition
* 9. Recovery value
* 10. Flammability or explosive limits
* 11. Toxicity limits
* 12. Solubility

Gas characteristics
* *1. Flowrate—average and extreme values
* 2. Pressure
* 3. Temperature
* 4. Moisture content, condensable vapors
* 5. Composition and reactivity
* 6. Corrosive properties

Effluent
* *1. Desired emission of contaminant in clean gas
* 2. Method of disposal or recovery of collected contaminant

---

* Required for preliminary equipment selection.

and toxicity. A summary of properties is given in one handbook.[5] Concentrations involved in gas cleaning are usually below the minimum explosive concentration. Recommendations for safeguards are contained in the safety codes of the National Fire Protection Association.[6]

SAMPLING AND ANALYSIS

Sizing a proposed collector or checking the performance of installed equipment will require sampling and analysis to determine dust loading at inlet and outlet. Or a sample may be needed for particle size determination. There is a variety of techniques that can be used, permitting a choice depending on convenience, cost or accuracy. A summary is presented in "Aerosol Sampling and Analyzing Instruments."[7] As will be seen, test equipment choice depends on the process characteristics, so the investigator must have some idea of the magnitude of the answers he is setting out to find; for example, what rate and time of sampling will yield a significant weight of solids.

The gas flowrate must first be measured. For most process gas streams the pitot tube and inclined manometer are suitable. For obtaining a rough measure of velocity below 10 ft./sec., a van anemometer may be used. High velocities in small ducts may be measured with an orifice or a venturi meter. Methods are discussed in Perry.[3]

There are many means of measuring gas velocity but the standard pitot tube and differential pressure gage is relatively simple and reliable. The operator must be careful that the section of duct is straight for 6 to 10 duct diameters upstream and 2 to 4 diameters downstream, to ensure uniform flow; also that the dust is easily accessible, and that the pitot tube is pointing directly upstream. The combined pitot-static tube consists of two concentric tubes, one for impact pressure and the outer one for static pressure. The pressure difference, measured by an inclined manometer, is the velocity pressure. The standard type "L" pitot tube is suitable unless dust loading or gas humidity is high enough to cause plugging, in which case, the special type "S" pitot tube is used. Wet- and dry-bulb gas temperatures are taken for use in calculating both gas density and humidity.

To determine the dust loading, isokinetic sampling is used to withdraw and collect all dust in a measured volume of gas, matching the sampler velocity with that in the main duct and ensuring a representative dust sample. There are two methods of doing this. Duct velocity may be measured and the sampler velocity adjusted to the isokinetic velocity. Or a null-balance probe can be used to match static pressures, thereby matching velocities without measuring duct velocity. The former method is described in several references,[8, 9] and the latter is described by Lapple.[10] The sampling train is assembled from a choice of components. Thanks to the growing in-terest in air pollution, components that once had to be made by the user are now commercially available.

Sampling Techniques

Equipment is connected as shown in the schematic arrangement, Fig. 10, with a sharp-edged nozzle, usually ¼ in. or more in diameter of opening, pointing upstream. Dusty gas is withdrawn at a rate fixed in such a manner that the velocity at the face of the probe is within 10% of the velocity in the duct near the probe.

The gas is drawn through the sampling train by a vacuum pump—or an ejector operating on air, steam or water—and capable of maintaining a flow of ½ to 2 cfm. as resistance increases in the filter owing to dust buildup.

The gas sample stream must be metered by some suitable instrument, such as a rotameter, orifice and manometer, or gas meter. The rate of sampling is controlled by a valve for variations in temperature, vacuum or density.

The dusty gas picked up by the sample probe is filtered in a preweighed paper or sintered-metal thimble. The filters or separating devices used are usually at least 99% efficient. The probe is traversed like the pitot tube and held at the same locations with the sampling flow set to equal duct velocity at that location.

A simplified method may be used on the exit of a wet scrubber if it is felt that all particles are less than a 5μ and sampling need not be isokinetic. Temperature of the gas up to the filter must be kept above the dewpoint by heating with an electric resistance-heater. Placing the filter in the probe assembly (where it is heated by stackgas) avoids condensation, loss of sample, or plugging in the tube between probe and remote filter.

The sample collector is not limited to paper filters and in fact other types may be more suitable. A description of a variety of such devices is given in Stern II p. 497[11] and Western Precipitation bulletin WP-50.[9] The filter medium can be chosen from papers, fabrics, porous metals, or cellulose ester

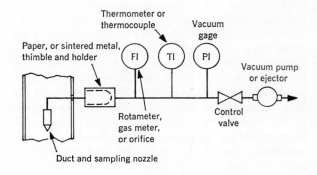

TYPICAL EQUIPMENT arrangement for collecting dust samples prior to analysis—Fig. 10

membranes, thereby permitting for special conditions —high temperatures, extremely small particles, low ash content media, or transparency of media with oil or water treatment after sample collection.

Small-diameter cyclones can be used to collect samples if the dust is coarse, or if total dust measurement is not important. One manufacturer (Dustex) uses the cyclone efficiency for simple scaleup (assuming no efficiency is lost in gas distribution) to the multiple cyclone needed for the commercial-sized collector.

Another common sampling device, the impinger, consists of a flat surface with sample stream directed through a jet to impinge on the plate, usually submerged. Solid particles down to 1μ are determined after evaporation.

Particle Size Analysis

A particle-size analysis should be run on the sample collected since some idea of particle size is a prerequisite to selecting a collector. There are many approaches depending on the situation. Adequate facilities may already be available within the organization, or it may be worthwhile to obtain analytical equipment for the current study, if future gas-cleaning problems will justify the effort and equipment. On the other hand, the process being studied may be a duplication of an existing plant with much background experience to draw upon.

If answers are needed with a minimum of cost and time, it may be advantageous to call in outside analytical services that can provide the analysis for less than $100 per sample. Many of the dust collection equipment manufacturers—if one can be chosen at this point—can provide the service.

A summary of size analysis techniques and devices appeared in CHEMICAL ENGINEERING recently.[12] Methods of particle-size analysis are varied to suit the nature of the particulate, the needs of the analyst or the method of sample collection; a complete picture of test methods and devices is beyond the scope of this article. However, it might be well to give some specific examples of the commonly used methods that are applied in dust collector work.

The Bahco Micro-Particle Classifier is a combination air centrifuge-elutriator, and is an ASME standard for particle-sizing.[13] A weighted sample of 10 to 20 g. is charged into the hopper, and introduced through a feed mechanism into a spiral of air having suitable tangential and radial velocities. A portion of the sample is carried by centrifugal force against the air flow and toward the periphery of the spiral. The remainder of the sample—the smaller, lighter particles—is carried with air flow toward the center. The air is pumped through the unit, by means of an integral fan impeller, at a rate controlled by adjusting the air inlet orifice. The residue is weighed after removal of a light fraction, and the light-cut weight is determined by difference. The process is repeated to obtain nine fractions. Each

orifice setting must be calibrated on a dust of known size distribution and the terminal-velocity values assigned. The development of standardization techniques are discussed by Crandall.[14]

Other useful methods include electronic counting by the Coulter counter in which a suspension is put through an aperture, essentially one particle at a time, with electronic recording of each particle by size.

The cascade impactor projects particles through a jet onto a plate where large particles are collected, the fines going on to the next stage.

Elutriation is applied in the Roller analyzer on the principle of the settling chamber described under collectors, using an airstream to remove a fines fraction, depending on velocity. These methods and others are reviewed in Perry as well as others.[11, 15, 16]

A sieve shaker can provide several particle size ranges down to 50μ particles (43μ equals 325 mesh). About 100 g. of sample are needed. There are some limitations: thin, flat flakes may give artificially large weight fractions of the larger particles, needle-like crystals may conversely appear in the analysis as smaller sizes, some materials finer than 150μ may flocculate, and other materials may be degraded.

The sieve analysis should be supplemented by a microscopic examination to check particle shape. In the particle sizes between 0.2 and 100μ the optical microscope can be used without sieves. The method is laborious but can be simplified by using photomicrographs or by projecting the image of the particles on a screen.

An important property needed for selecting an electrostatic precipitator is the electrical resistivity of the dust to be handled. The measurement can be omitted if a precipitator is commonly used for the dust or if the equipment designer is otherwise familiar with it.

Several methods for particle resistivity measurement are discussed in White,[17] and resistivities of some representative dusts and fumes are given. A standard method along with apparatus construction details is given in the ASME Power Test Code 28,[13] and in the API manual Electrostatic Precipitators.[18]

Choosing a Collector

Let us look at the selection of a collector from meager preliminary information. Take as an example a process in which organic chemicals are spray-dried and collected by a cyclone, the dusty gas then going to a spray tower. It is desirable to replace the existing inefficient wet collector to meet air pollution abatement requirements and to recover product.

Material-balance calculations of product recovery efficiency can be used to determine dust-loading and water evaporated in the dryer. Microscopic counts of dust samples collected downstream from the cyclone give particle size. A manometer can be used to measure pressures and, with the fan curves, the gas flow is determined.

State air pollution laws fix the emission based on stack height and distance from the property line.

A summary of data looks like this:

Dust loading to collector	50 to 200 lb./hr.
Particle size—Product A	50% < 30 μ
Product B	50% < 20 μ
Product C	50 % < 5 μ
Gas flow at inlet conditions	20,000 cfm.
Gas temperature	250 F.
Gas humidity	3,000 lb./hr.
Dust emission allowable from collector	20 lb./hr.

Converting the dust loading to grains/cu. ft.:

$$200 \; \frac{lb.}{hr.} \times 7{,}000 \; \frac{gr.}{lb.} \times \frac{1}{20{,}000} \; \frac{min.}{cu.ft.} \times \frac{1}{60} \; \frac{hr.}{min.}$$
$$= 1.17 \; gr./cu.ft.$$

The requirement is to emit no more than 20 lb./hr. from a maximum of 200 lb./hr., or 10% emission. Collection efficiency is then 90% for a dust loading of a little more than 1 grain/cu.ft. of particulate with a mean particle diameter of 5 μ.

From the note on Table I, we find that the loading is "light" and the particle size "fine." From the table in Ref. 4, Industrial Collector Applications, Lines 4 and 39 favor high-efficiency centrifugals, wet collectors and fabric filters.

Cyclones are unsuitable, which is evident since we are collecting dust that already has passed a cyclone. Electrostatic precipitators are not likely since the gas-handling capacity is somewhat small for good operating efficiencies, and also the electrical resistivity of an organic dust will probably be too high. Multiple high-efficiency cyclones are small in diameter and subject to plugging with sticky organics. Fabric filters are excellent collectors for fine materials, especially when we would like to recover usable product, but we should keep the sticking problem in mind. Wet scrubbers will give a clean stack but with a wet plume; the recovered product will be degraded in value, since it will have to be dried again. Product value is 2¢/lb. dry and 1¢/lb. wet.

Referring to Table I, we can choose centrifugal, impingement and venturi scrubbers. Other equipment types are excluded because, let us assume, they are unsuited for scrubbing with dust slurries. Recovered dust will have to be recycled to build up concentration for economical recovery.

References

1. "Air Pollution Manual," Part I, Evaluation (1960), Part II, Control Equipment (1968), American Industrial Hygiene Assn., Detroit, Mich.
2. Strauss, W., "Industrial Gas Cleaning," Pergamon Press, New York, 1966.
3. Perry, J. H., Ed., "Chemical Engineers' Handbook," 4th ed., McGraw-Hill, New York, 1963.
4. Kane, J. M., *Heating and Ventilating*, 51, No. 10, pp. 77-82, Oct. 1954.
5. Marks, L. S., Ed., "Mechanical Engineers' Handbook," 5th ed., McGraw-Hill, New York, 1951.
6. National Fire Codes, Vol. 3, "Combustible Solids, Dusts and Explosives," National Fire Protection Assn., Boston, 1967.
7. Whitby, K. T. Liu, B. Y. H., Dust (Engineering), "Kirk-Othmer Encyclopedia of Chemical Technology," 2nd ed., Vol. 7, Wiley, New York, 1965.
8. Industrial Gas Cleaning Inst., "Test Procedure for Gas Scrubbers," Pub. No. 1, Rye, N. Y., 1964.
9. Western Precipitation Corp., Bull. WP-50, "Methods for Determination of Velocity, Volume, Dust and Mist Content of Gases," 7th ed., Los Angeles, 1968.
10. Lapple, C. E., *Heating, Piping and Air Conditioning*, 16, pp. 410, 464, 578, 635 (1944); 18, p. 208 (1946).
11. Stern, A. C. Ed., "Air Pollution," 2nd ed., 3 vol. Academic Press, New York, 1968.
12. Lapple, C. E. Particle-Size Analysis and Analyzers, *Chem. Eng.*, 75, No. 11, pp. 149-156, May 20, 1968.
13. ASME Power Test Code 28, "Determining the Properties of Fine Particulate Matter," American Soc. of Mechanical Engineers, 1965.
14. Crandall, W. A., "Development of Standards for Determining Properties of Fine Particulate Matter," American Soc. of Mechanical Engineers, 1964.
15. Orr, C., Jr., Dallavale, J. M., "Fine Particle Measurement," Macmillan, New York, 1959.
16. Cadle, R. D., "Particle Size Determination," Interscience, New York, 1955.
17. White, H. J., "Industrial Electrostatic Precipitation," Addison-Wesley Pub. Co., Reading, Mass. 1963.
18. "Removal of Particulate Matter From Gaseous Wastes: Electrostatic Precipitators," American Petroleum Inst., New York, 1955.

Meet the Author

Gordon D. Sargent is chief project engineer for the Nopco Div., Diamond Shamrock Chemical Co., in Harrison, N.J. The design of pollution-control installations has been part of his project engineering experience at Nopco, and previous to joining Nopco, with Celanese Fibers Co. and Hooker Chemical Corp. He is a graduate of M.I.T. with a B.S. in chemical engineering and is a licensed engineer and a member of AIChE.

Removing particulates from gases

Year by year, requirements for controlling particulates in gases entering the atmosphere tend to become more stringent. Engineers need to know what equipment is available for designing new installations, or for upgrading old ones.

Ted M. Rymarz, Flex-Kleen Corp. *and* *David H. Klipstein,* Research-Cottrell Inc.

☐ There is a wide variety of devices for separating particulates from gas streams before the gases are discharged into the atmosphere. These many devices have evolved because industrial processes vary greatly in the nature of the particulate matter that they generate, as well as in the conditions under which particulate collection must be accomplished.

Particulate collectors may be classified as:
- Mechanical collectors.
- Wet scrubbers.
- Fabric filters.
- Electrostatic precipitators.

Some collectors may combine the operating principles of two such classes; fabric filters incorporating settling chambers are an example. Some, principally scrubbers, may also remove some gaseous pollutants. Additionally, a particulate collector may include a secondary function such as gas cooling.

Application of a given collector to a specific particulate source is subject to change, because of evolution both in collection technology and collection requirements. A change in process may mean that a different collector will be needed. And, as the requirements for "successful" collection become steadily more demanding, a level of particulate collection from a process that was hitherto satisfactory may now be unacceptable, or will eventually become so.

The term "particulates" includes both solid particles and liquid droplets light enough to be swept along in a flowing gas stream. All such particles are small, but size difference is so significant that particle size is measured in microns (25,400 microns, or μm, = 1 in). The rate at which particulates enter a collector is usually expressed as "inlet grain loading," in grains per actual ft^3/min of gas (7,000 grains = 1 lb). The capability of a collector is expressed as "collection efficiency," which means the percentage by weight of particulate load at the inlet that the collector will remove from the gas stream.

Many things influence collection efficiency. However, the related factors of particle size, shape, density, and tendency to agglomerate are of principal importance—it is practical for comparison purposes to characterize capability simply in terms of particle size.

If a gas stream contained only particles of uniform size, a single collection-efficiency figure would suffice to express pollution control capability. But this is rarely the case. Thus, precise evaluation of a collector for a specific application requires knowing both the distribution by weight of particle sizes in the gas stream and the collector's efficiency for each size fraction. Nonetheless, it is common for rough comparison purposes to cite single particle-size and efficiency figures for a collector: The particle size cited is that below which the collector is essentially ineffective; to the extent that particles smaller than this figure represent a significant fraction of the load, collection efficiency will suffer. The efficiency figure cited usually represents a collector's performance in its typical applications.

Mechanical collectors

The simplest type of particulate collector, one of two types described as "inertial," is the mechanical collector. In this type the gas stream is made to flow in a path that particles, because of their inertia, cannot easily follow. Most mechanical devices are dry collectors only—solid particles wetted by liquid particulates would plug them up.

As a class, mechanical collectors are low in initial cost, with their operating and maintenance costs being low to moderate. Most involve no moving parts. In their usual applications, gas pressure drop varies from less than 1 in. water gage to, in the most efficient types, 4 to 6 in.

An important advantage of mechanical collectors is their relatively simple configuration. This makes the use of abrasion-, corrosion-, or temperature-resistant materials more practical than in some other types of particulate collectors. Another advantage of many mechanical collectors is their ability to handle a high inlet-grain-loading. A disadvantage is the tendency to plug up if particles to be collected are wet or otherwise sticky.

Originally published October 6, 1975

Hence, the gas temperature must be above its dewpoint throughout the collector, which usually must be carefully insulated; also, gas reheat may be required ahead of the collector.

However, the chief limitation of inertial collectors is their inability to collect tiny particles. Inertia is a function of mass, and mass is the product of density and volume. In a given application the particles generally have substantially the same density, so it is volume, or particle size, that governs inertial collectibility.

Energy input to most mechanical collectors is provided by induced-draft or forced-draft fans, and appears as gas pressure drop across the collector. In one type of design, however, energy is supplied by a motor-driven impeller. Either way, the smaller the particle size that a collector is expected to separate from a gas stream, the greater the required power input. The measures needed to improve mechanical collection of small particles ultimately become self-defeating by increasing problems. Particles may tend to bounce off baffles and walls and be reentrained in the passing gas stream. Or, equipment may plug up owing to the particles' inability to drain out of complicated flow paths as fast as they are separated from the gas stream. Therefore, mechanical collectors are limited in power input and, hence, cannot achieve high operating efficiency in collection of particles much smaller than 5–10 microns.

The simplest form of mechanical collector is the subsidence or settling chamber, a chamber larger in cross-section than the gas duct leading to it. Gas velocity through the chamber is low, allowing heavy particles to settle out. Little pressure drop is involved, and separation capability is limited to particles larger than about 50 microns; efficiency seldom exceeds 50%. Baffles are often added, which do not materially improve collection efficiency but achieve the required efficiency with a reduced chamber length.

Higher collection efficiency can be had if gas velocity is not reduced. In the louver collector and similar devices involving multiple, closely spaced provisions needing abrupt changes in flow direction, efficiencies can reach 80% and particles down to 20 microns can be collected, at pressure drops up to 2 in. water gage.

In another type of collector, the path of the gas undergoes a continuous direction change that is difficult for particles to follow. Gas flows around a scroll-shaped chamber while particles tend to move straight out to the periphery where a concentrated particulate stream is bled off, through slots, in a small portion of the gas stream. Such a skimming collector is effective down to 20 microns and is very compact, but the bleed stream of gas requires a secondary collection step if it is to be vented to the atmosphere.

Skimming action is enhanced in a dynamic collector by employing a motor-driven impeller. With this effective power input, efficiencies as high as 90% can be achieved with particles down to 10 microns. Abrasion problems are the chief drawback of this approach. The advantage is that no gas pressure drop need occur and, in fact, pressure can increase across the collector, which can be useful in retrofit situations (where providing power input to the impeller may be more convenient than repowering or replacing an existing fan).

In the impingement collector, an accelerated gas stream is directed against a flat surface. Particles strike this surface, lose momentum, and drop out of the gas. This equipment can collect liquid droplets larger than 10 microns. Solid-particle collection efficiency is low because of bounce-back into the gas stream. Wetting the impingement surface can improve solid collection efficiency but results in a wet slurry or sludge usually more difficult to reuse or dispose of than dry material.

The most familiar and widely used mechanical collector is the cyclone. Gas enters tangentially at the top of a cylindrical shell and is forced downward in a spiral of decreasing diameter in a conical section. Particles are centrifugally thrust outward and forced to spiral downward to the bottom, which is closed by an air lock. Since gas cannot escape at the bottom it is forced to turn and travel, still whirling, back up the center of the vortex and out the top. Particles are discharged from the bottom through the air lock.

The tighter the spiral in which gas must flow, the greater the centrifugal force acting on a particle of given mass, and thus the more efficient a cyclone can be. Top diameters of cyclones range from more than 120 in. to as little as 24 in., and capabilities reach 85% efficiency with particles as small as 10 microns, at pressure drops from 1/2 to 3 in. water gage.

Cyclones alone are seldom adequate for pollution control, except where the load consists almost entirely of coarse particles, as in woodworking shops, or where particle density is unusually high. Cyclones are often

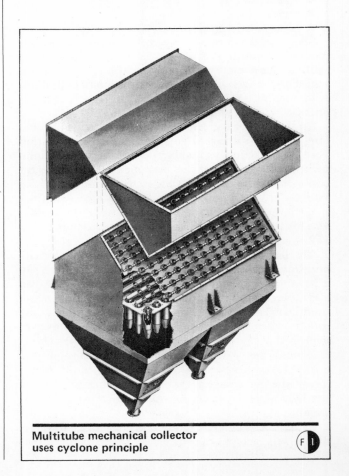

Multitube mechanical collector uses cyclone principle

used when there is a special reason to separately collect reusable coarse particles from useless fines (in fluid-bed catalyst regenerators, for example, where the fines pass through for subsequent separation by more-efficient means). Cyclones are sometimes applied where gas cooling is necessary, as where the cyclone is followed by a fabric filter; the cyclone scalps the coarse fractions of the dust load and at the same time provides cooling.

Cyclones are either applied singly, or manifolded so as to divide large gas flows between several individual units. They are also applied as two or more units of decreasing diameter in series, for large loadings with a substantial percentage of small particles. The simple cyclone design, however, is not practical in very small diameters, for capture of very small particles. Instead, the cyclone principle is employed in a different way in what is called a multitube mechanical collector.

As the name implies, the multitube collector involves a large number of individual cyclone tubes in parallel within a single collector housing or shell. Dividing the gas stream among many tubes permits small tube diameters without requiring excessive gas velocity or pressure-drop through the individual cyclone tube. Tube diameters range from 4 in to 10 or 12 in, with the larger sizes more common. Particles as small as 5 microns can be collected; efficiencies range up to 95%.

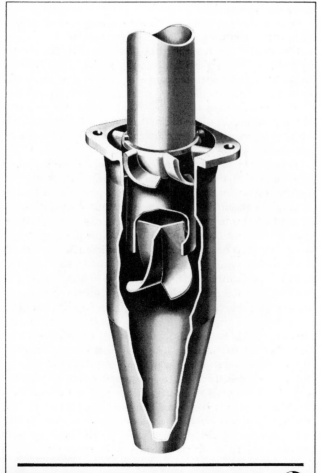

Single tube of multitube mechanical collector

In a tube of a typical multitube collector, gas flows from the inlet section of the housing directly downward through an annular space between the outer tube wall and a smaller, concentric, exit tube. Fixed vanes in the annulus impart helical motion to the gas stream, thus applying centrifugal force for particle separation. As the gas stream slows down, reverses, and flows up to the exit tube, exit vanes may straighten out the cleaned-gas flow (which accomplishes a measure of gas pressure recovery to minimize overall pressure drop across the collector).

The smaller the tube diameter, the greater the tendency for large particles to bounce back off walls into the rising exit-gas stream. Tube diameter, inlet velocity to the collector, and the number of tubes in parallel (which determines the velocity through the individual tube) must be carefully selected for each individual application. It is vital that careful design ensure equal distribution of inlet gas among all tubes to avoid excessive velocity in some and inadequate velocity in others, a special concern where angled or vertical gas flues are used to or from a multitube collector. In applications where particles have a tendency to be sticky, the smallest tube diameters cannot be used and it may be necessary to eliminate exit recovery vanes.

Multitube mechanical collectors can be used alone for pollution control where the load fraction represented by particles less than 5 microns is small and the strictest emission codes do not apply. Increasingly, however, multitube collectors are used to scalp coarse particles from a gas stream ahead of a more efficient type of collector. Application to boilers that have been converted from oil or gas to pulverized coal is an example. Here, multitube collectors separate, for return to the boiler, coarse particles with their content of unburned carbon, while ash fines pass through to a more efficient particulate collector.

Scrubbers

Scrubbers are compact inertial collectors capable of separating solid or liquid particulates (or both simultaneously) from a gas stream.* The many types are generally divided into low-energy, medium-energy, and high-energy designs. As the terminology implies, scrubbers can be designed to accept enough energy input for high-efficiency collection of very fine particulates. If pressure drop as great as 10–30 in. water gage can be justified, some scrubbers can achieve 99+% efficiency and collect particles of submicron size.

Particulate scrubbing is a two-step process. In the contact step, particles are wetted or, more commonly, captured by drops of scrubbing liquid. This, often aided by agglomeration, solves the problem of collecting tiny low-mass particulates by creating particulates of greater mass. These are separated from the gas stream by simple inertial means in the second, mist-elimination step. In low-energy scrubbers for collection of relatively large particles, the two operations proceed simultaneously in

---

*A single scrubbing stage cannot achieve high efficiency in removal of both gas and fine particulate matter simultaneously. To achieve high gas-removal efficiency, long contact time and high gas-liquid interface area are necessary, whereas high gas velocity is necessary for efficient scrubbing of fine particulates. Both tasks can be accomplished by two-stage scrubbing with particulate removal first, followed by gas removal.

the same area of the scrubber, and gravity may be enough for separation. In high-energy scrubbers, mist elimination is a distinctly separate stage following the contact stage. It is accomplished, typically, with a baffled gas flow path, or spiral gas flow to create centrifugal force, for separation.

Scrubber capability for collection of fine particulates is directly proportional to power input and is essentially independent of scrubber configuration. The energy derived from pressure drop in the gas is supplemented in some designs by supplying scrubbing liquid at high pressure. No design tricks can reduce the energy input needed for a given particulate collection requirement, although, conversely, poor design can reduce the effectiveness with which the energy input is utilized.

As a class, scrubbers have the advantage of small size, which is especially useful when it is necessary to retrofit a particulate collector to an existing pollution source. Also important is the indifference of scrubbers to the temperature and moisture content of a gas. This not only allows cleaning of a gas having an inherently high moisture content, but it allows cooling of hot gas with water sprays. Scrubbers can be economically built to accept high-temperature gas since only the inlet requires temperature-resistant material; however, providing upstream gas cooling does make heat-resistant ductwork unnecessary and in addition, reduces the gas volume for which the scrubber must be sized.

While scrubbing is advantageous when inlet gas conditions are considered, the accompanying discharge of cool, moist, exit gas and wet collected particulates can be disadvantageous. Exit gas reheat may be necessary to eliminate objections to a visible water-vapor emission or to ensure proper chimney action. Also, scrubbers can present corrosion problems if a neutralizing scrubbing liquid cannot be used. A scrubber can readily be fabricated of, or lined with, corrosion-resistant material, but gas reheat may be necessary for protection of such downstream equipment as induced-draft fans and chimney linings.

The fact that collected solid particulates leave a scrubber as a slurry or sludge is usually cited as a disadvantage, but, depending on the nature of the material and of subsequent reuse or disposal methods, it can be an advantage instead. It is often said that a scrubber for particulates that cannot be recycled simply converts a problem of air pollution into one of water pollution. However, in many situations it may be easier to handle, transport and dump a slurry or sludge than dry particulates (which have the potential of recreating an air pollution problem).

Water can be employed as the scrubbing medium even where water supply is limited, since some scrubbers can recirculate scrubbing liquid that contains a substantial amount of particulate matter. In such applications, makeup water is required only to replace evaporation losses and liquid that exits in the sludge of collected particulates.

Types of scrubbers

In the simplest application of scrubbing, liquid is sprayed in at the top of a column and collision for particulate wetting or capture occurs as drops fall through a rising gas stream. Pressure drop is low, but application is limited to situations where 50% efficiency or less is acceptable and the percentage of particles smaller than 10 microns is low, or to scalping coarse particles ahead of a precipitator or a more efficient scrubber. A spray column is often used primarily for quenching hot gas, with coarse particulate removal a useful but incidental effect.

In a centrifugal scrubber, column design and directed sprays cause drops and gas to mix in a rising vortex so that centrifugal force increases the momentum of collisions between particles and drops. Thus, smaller particles can be captured; and efficiencies as high as 90% can be achieved with particles as small as 5 microns, at pressure drops from 2 to 6 in. water gage.

A column may be fitted with impingement plates, wetted mesh, or fibrous packing, or packed with saddles, rings or other solid shapes. In such scrubbers, at typically 1 to 10 in. water-gage pressure drop, efficiencies can range up to 95% with particles as small as 5 microns. Packed beds designed for gas absorption, however, are subject to fouling if the gas stream contains a significant fraction of solid particulates. In designs that use sprays to wash the packing, or that are packed with small spheres agitated by the gas flow, the fouling problem is reduced. A recent development employs a moist chemical-foam packing, which drains slowly from the scrubber with captured particulates and is replaced with fresh material.

A venturi operating on the eductor or ejector principle, with scrubbing liquid as the motive fluid, can collect particles down to submicron size with efficiencies as high as 90%, if grain loading is low. Gas pressure drop is not depended on for power input and there can even be a gain in gas pressure across such a scrubber. The disadvantages are: (1) a requirement for substantial scrubbing-liquid flow at high pressure and (2) inability to remove large particles because the induced gas-stream velocity is low.

A scrubber sometimes referred to as a submerged-jet type actually discharges gas through an orifice at, rather than under, the surface of a pool of scrubbing liquid. As it skims the surface the gas atomizes some of the liquid into droplets for particle capture. A baffled flow path causes the heavier particle-droplet agglomerations to drop back into the pool. Efficiencies can reach 90%, with particles down to 2 microns in size, at pressure drops from 2 to 6 in. or more water gage. The special characteristic of this type of scrubber is its almost negligible power requirement for liquid pumping.

In what is usually called the venturi scrubber, the drop in gas pressure across a venturi or orifice accelerates the gas stream. Scrubbing liquid is sprayed in at the throat or orifice (or is flowed in to be atomized by the passing gas stream), and mixes turbulently with the gas. Turbulence is ordinarily undesirable in fluid flow because it represents a useless consumption of power. But in these scrubbers, turbulence is the cause of collison between liquid droplets and particulates at high relative velocity, which permits capture of very small particles.

If the venturi throat, or orifice, area of a venturi scrubber is fixed, pressure drop (and, therefore particulate capture capability) will vary with gas flow rate. If gas flow drops significantly below the design value, pressure drop will be inadequate to achieve design efficiency. Therefore, venturi-type scrubbers have been designed to permit variation of the throat or orifice area, manually or under control of pressure-differential sensors, to maintain optimum pressure drop for stable operation and constant collection-efficiency.

When a venturi is employed, mechanical means can be provided for adjustment of throat area while maintaining the proper throat configuration. In the Flooded-Disc scrubber the orifice through which gas is accelerated is the annular space between a horizontal disk positioned in a tapered duct, and the duct wall. Pressure drop is maintained, despite flow variations, by raising or lowering the disk in the tapered duct to increase or decrease the orifice area.

Scrubbing liquid is supplied at the center of the upper face of the disk, where gas pressure forces it outward. At the disk periphery, liquid is sheared off and atomized by the passing high-velocity gas stream. Since this type of unit involves no spray nozzles or other constricted liquid passages it is particularly suitable for recirculating liquid with a high solids content; high liquid pressure is not required. And because the orifice is a narrow annulus its area can be large, where a high rate of gas flow is involved, without requiring distribution of scrubbing liquid over an excessive area.

Fabric filters

Fabric filters collect solid particulates by passing gas through cloth that most particles cannot penetrate. As the layer of collected material builds, the pressure differential required for continued gas flow increases; consequently, at frequent intervals the accumulated dust must be removed. Cotton, wool, glass, and various synthetic fibers are used in filter fabrics, which are shaped in the form of cylindrical bags or envelopes of roughly elliptical cross section.

The term "air-to-cloth ratio," used in characterizing fabric filters, defines the cleaning capability, for a specific inlet loading, of the type of fabric and fabric-cleaning method employed in a proposed filter. The ratio specifies the number of cubic feet per minute of the given contaminated gas that can be handled by 1 ft^2 of fabric area. The ratio is used to determine how many bags of given size the installation must comprise. If flow is to be continuous in the types of fabric filters that require shutting-off of contaminated gas flow from bags being cleaned, there will be a gross air-to-cloth ratio larger than the net ratio, to allow for the bags that will be out of service while being cleaned.

Fabric filters are capable of 99 + % collection efficiency with particles down to submicron size. High efficiency is attained with moderate pressure drops, typically in the range 2 to 4 in. water gage. Power input is thus comparable to that of multitube mechanical collectors, while capability in collection of fine particle sizes is much greater. Operating cost, including maintenance, is somewhat higher, however, because some moving parts are involved and bags must be periodically replaced. Unlike wet scrubbers, fabric-filter performance is relatively unaffected by variations in gas

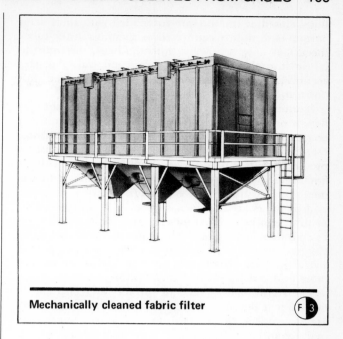

Mechanically cleaned fabric filter   F 3

flowrate. Fabric filters are sometimes preceded by mechanical dust collectors, or settling chambers in the bag house, when excessive grain loading or abrasive, coarse particles, or both, are involved.

The advantages of fabric filters are offset for many potential applications by limitations of the fabric itself. Continuous gas temperature may not exceed 550°F where glass fiber is employed, and the limit is lower for other fibers. Thus, gas cooling is often required ahead of fabric filters. Water sprays may be used if precise sensing and control is employed to avoid condensation. Moisture need not be a problem if it can be assured that the gas temperature in the fabric filter will be above its dewpoint, otherwise blinding—plugging of the fabric with wetted particulates—will occur.

There are two basic classes of fabric filter. Surface filters use woven fabrics, which cannot prevent passage of submicron particles. However, the accumulation of coarser particles on the fabric surface serves to filter out smaller particles. Cleaning methods are tailored so that complete removal of collected particles is not achieved (since thorough cleaning would exaggerate the slight drop in efficiency that occurs after cleaning, before a fully effective layer of collected particles is restored).

Depth filters use a felt-like fabric of fibers that are attached, by a needling process, to a coarse-woven backing. In felted fabrics, particles are collected within the mazelike gas paths through the felt, and by surface accumulation. Thorough cleaning methods can be used while keeping very low emission levels; the cleaning methods employed with woven fabrics are not adequate for dislodging particulates trapped in felt.

Types of fabric filters

The simplest type of fabric filter is the mechanical—or more properly, mechanically cleaned—type. Near the bottom of the filter housing is a horizontal tubesheet separating the inlet dusty-air plenum from the clean-air plenum into which the bags discharge.

Bags, closed at the top, are suspended with their open bottom ends tightly connected to openings in the tube-sheet. Gas flow is upward, into the bags, through the fabric into the clean-air plenum, and out. Particles accumulate inside the bags and, when dislodged, fall down to hoppers for disposal.

If intermittent gas cleaning is permissible, the gas flow is periodically shut off so that the entire unit can be cleaned. If gas flow must continue for long periods, then more bags—greater total fabric surface—must be provided than is necessary to handle the specified gas flow. The housing is divided into two or more compartments, each containing a set of bags. Inlet-gas ductwork and dampers are provided so that each compartment, in turn, can be isolated for cleaning, while bags in the remaining compartments continue to accept and clean the gas stream. The more elaborate construction and the extra bags make such an installation larger and more costly than one for interruptible gas flow.

The traditional method of surface filter cleaning is mechanical shaking of the structure from which bags are suspended. Design and operation are simple, but maintenance requires access to the shaking mechanism within the housing. The same gas flow and bag arrangement is employed in the so-called shakerless, or reverse-air, fabric filter. A fan, ducting, and dampers are provided to supply a reverse flow of air from the clean side of bags through to their interiors, to dislodge accumulated particles. Bags are usually stiffened by wire rings to prevent their complete collapse when the air flow reverses.

The mechanical-shake action is violent enough to accelerate bag wear if the fabric is not resistant to abrasion or if unusually abrasive particles are being collected. In fact the action can be positive enough for cleaning felted fabrics in some applications, although most mechanical fabric filters are surface filters. The gentler reverse-air-flow cleaning action is preferred where abrasion will be a problem, and is usually required when high gas-temperature dictates use of glass-fiber bags; glass fabrics cannot be expected to withstand the mechanical shaking action employed in most applications even when the fibers have been lubricated with graphite or silicone.

If air jets are used for cleaning, instead of low-velocity reverse air flow, felted fabrics can be used and bags can be cleaned while contaminated gas continues to flow to them. A reverse-jet type of fabric filter employs a ring of nozzles around each bag and a mechanism to traverse the rings up and down the bag length. It does not require the elaborate inlet duct-and-damper arrangement necessary for continuous duty with mechanical fabric filters. But this advantage is substantially offset by the cost of providing and maintaining the complex cleaning mechanism, all of which is inside the filter housing, and by the possibility of misalignment that will allow abrasion of bags by moving rings.

It is more common for depth-fabric filters to use both a gas flow and bag arrangement that is opposite that of mechanical filters. The tubesheet is placed near the top of the housing and bags with closed bottoms are suspended beneath it. A cage-like framework in each bag prevents collapse as gas flows from the outside of the bag in through the fabric up to the clean air plenum, and out. The bags hang in the contaminated gas compartment. The system for providing bag cleaning air is not in the dusty compartment, but in the clean air plenum. And in some designs none of the system's moving parts are in the housing at all. But if a bag change must be made from inside the housing, it must be done from the dusty, rather than the clean, compartment.

One fabric filter using this arrangement is a surface filter with reverse-air-flow bag cleaning. The housing is compartmented, and a compartment must be shut off from the contaminated gas flow while being cleaned. The clean-air plenums of each of the compartments are pressurized, one after the other, to clean all the compartment's bags at once.

Another type of reverse-air fabric filter uses a rotating distributor in the clean-air plenum above bags that are arranged in radial rows like the spokes of a wheel. As the distributor rotates, it supplies air in turn to the interiors of each row of bags.

Entry into the plenum is required for mechanism maintenance. It is not necessary to cut off contaminated gas flow to bags while they are being cleaned. This design can utilize the gentle cleaning action of air flow-

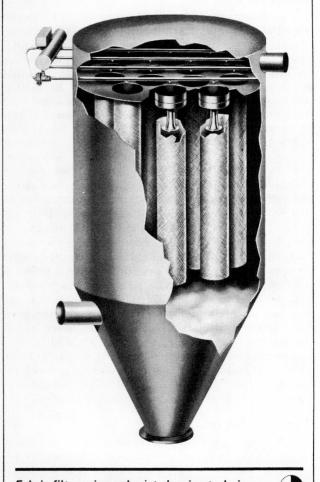

Fabric filter using pulse-jet cleaning technique

ing through bags at low velocity, or bursts of high-pressure air can be used to flex the bag fabric, or both. Felted fabrics can be used.

In the pulse-jet type of depth filter, a compressed-air system in the clean-air plenum provides an opening in a distribution pipe above each felted-fabric bag opening, usually with a venturi that accelerates air flow to enhance the cleaning action. Automatically controlled solenoid valves, all located, accessibly, outside the housing, admit momentary bursts of air, in turn, to pipes above one or more rows of bags at a time. The pulse of air from the opening above each bag jetting down the length of the bag, momentarily interrupts contaminated gas flow through it, flexes the fabric, and dislodges accumulated particles.

Depth filters can usually be designed with air-to-cloth ratios 2–3 times greater than that of a surface filter for the same application. Offsetting the advantage is the cost of providing and maintaining a compressed air supply and its distribution system, and the higher cost of felted fabric. And where gas temperatures above about 440°F are involved, felted fabric cannot be used. Development work is being done on felted fabrics employing only glass fibers, but generally the felts include other fibers that fix the fabric's temperature limit.

In sizing a fabric filter, the difference between net and gross air-to-cloth ratios will depend on not only how much surface will be cleaned at once, but how long it will take (when allowance is made for the bag surface out of service while being cleaned). Damper diversion of gas flow requires appreciable time. In contrast, the pulse-jet action needs only a fraction of a second, and restoration of flow of gas through the fabric is instantaneous. In most installations the area out of service being cleaned is negligible and sizing can be based on a net air-to-cloth ratio; even in the largest installations the amount of extra bag surface provided is much less than that required by filters in which bags must be shut off from inlet gas flow for cleaning.

The fabric filter arrangement in which bags are suspended below the tube sheet permits designs that allow top removal and replacement of bags, through the clean-air plenum. In these designs it is not necessary to enter the dusty area of the filter housing for bag change. A further advantage of top bag change is the accessability of any bag without removal of others; bag change from inside often requires removal of bags that do not require replacement, in order to reach an individual bag or bags in which seams have burst or abrasion has worn holes. When such breaks occur it is not easy to determine from inside the housing where the faulty bags are located and considerable trial-and-error time is usually involved. In a top-removal installation it is often possible to immediately identify bags that require replacement, by telltale traces of dust where they discharge through the tube sheet.

Electrostatic precipitators

The electrostatic precipitator, an extremely efficient air-pollution-control device, can remove in excess of 99% of the undesirable particulates in a gas stream. This high efficiency is possible because, unlike other pollution-control devices, the precipitator applies the

collecting force only to the particles to be collected, not to the entire gas stream. Thus, an extremely low, energy-saving power input—about 200 watts per 1,000 cfm (0.5 m³/s)—is required.

How precipitators work

In operation, a voltage source creates a negatively charged area, usually by means of wires suspended in the gas flow path. On either side of this charged area are grounded collecting plates. The high potential difference between these plates and the discharge wires creates a powerful electric field. As the polluted gas passes through this field, particles suspended in the gas become electrically charged and are drawn out of the gas flow by the collecting plates. They adhere to these plates until removed for storage or disposal. Removal is accomplished mechanically by periodic vibration, rapping or rinsing.

The gas stream, now free of particulate pollution, continues on for release to the atmosphere.

Precipitators may be designated as either plate or pipe varieties. The plate type, described above, is widely used for dry dust. The pipe type, for removal of liquid or sludge particles and particulate fumes, uses the same principle, but with a different mechanical setup; the discharge electrodes are suspended within a series of pipes (collecting plates) contained in a cylindrical shell, under a header plate.

As noted earlier, the precipitator applies the separating force directly to the particles, regardless of gas

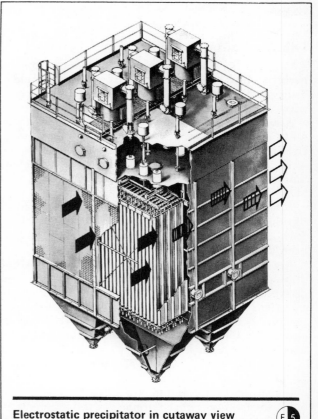

Electrostatic precipitator in cutaway view

stream velocity, thereby requiring less power than other control devices. And, because they can operate completely dry, they are ideal for use in the following situations:

■ Where water availability and disposal are problems.

■ Where very high efficiency on fine materials is required. For example, electrostatic precipitators are removing particulate matter as low as 0.05 microns from zinc oxide fumes with 97 to 98% efficiency.

■ Where large volumes of gas must be treated. In nonferrous metals production, gas flows of 600,000 cfm (250 m³/s) at temperatures up to 800°F (425°C) are treated with up to 99.6% efficiency.

■ Where valuable dry materials must be recovered, as in rotary-kiln or spray-drying operations, and in calcining.

Advantages of electrostatic precipitators

Initial installation cost is higher for precipitators than for most other collection systems (fob. costs for a 1,000,000 acfm, 470 am³/s, unit average $2.00 acfm, or $4,250 per m³/s, with erected costs about 1.7 times that). Electrostatic precipitators, aside from being the only feasible approach for specific situations, offer other advantages. The operating cost, for example, because of the low power requirements, is only about $0.03 yearly per acfm (or $65 per am³/s); maintenance costs run from $0.02 to $0.03 acfm ($40–65 per am³/s).

The precipitators are not limited by stream temperatures (units are processing gas streams from −70°F to 1,000°F (−55°C to 540°C), by dust structure or by corrosive gas problems, and can be designed to overcome a variety of other problems.

Other considerations

Consider first the collection, or pollution-control, problem. If a low cleaning efficiency is acceptable or if the particulate matter is large, other systems may be adequate.

Also, unlike cyclone filters, for instance, which require high pressures to force the flow stream to be cleaned through the filters, electrostatic precipitators work efficiently at a very low pressure drop. This eliminates the need for extra fans or similar pressure-increasing devices.

Electrostatic precipitators are not usually recommended for explosive or nonconducting materials, or for odor control, although successful installations have been designed for such applications.

Also bear in mind that although the particle size is a minor consideration, the electrical conductivity and density of these particles are critical factors. Assuming all other factors equal, the higher the gas density, then the higher the field strength, and the higher the efficiency. Thus, the more efficient precipitators operate at about 15,000 V/in.

Particulate conductivity is also a critical operating factor. Conductivity must be somewhere between that of a good conductor and that of a good insulator without being too near either extreme. If conductivity is too high, there will be a loss of charge for particles on the collecting plates and subsequent reentry into the prod-

uct stream. On the other hand, too little conductivity may coat and insulate the discharge electrode, creating a spark discharge that also causes reentry.

Particulate conductivity can usually be controlled by regulating temperature, or by using a conditioning agent such as water.

Recent developments

Two recent developments, automatic control and the "hot" precipitator, add additional degrees of efficiency. The all-solid-state control system automatically senses and reacts to the three factors—sparking, high current demand, and high voltage demand—that are involved in maintaining the highest possible field charge under all load conditions.

The "hot" electrostatic precipitator is based on the principle that most dielectric materials show decreasing resistivity with increasing temperature. Thus, by operating in the general range of 500°F to 700°F (260°C to 370°C), the "hot" precipitator ensures both particle mobility in hoppers and disposal systems and improved corrosion resistance. Air heater surfaces contact only "clean" gas; thus, wear and maintenance are held to a minimum.

System selection

Though many operational factors are involved in system selection, a preliminary selection may be made on just five—dust loading, particle size, gas flow, gas temperature and allowable emission rate.

An efficiency requirement in the 99% range, for example, with particulate sizes below 10 microns, could not be met with a cyclone. A gas temperature of 600°F (315°C) would rule out the use of fabric filters unless the gas were precooled.

Preliminary screening will define the solutions available. Final selection will require a detailed analysis of additional technical and economic factors.

The authors

Ted M. Rymarz is manager of applications research for Flex-Kleen Corp. (subsidiary of Research-Cottrell Inc.), 222 S. Riverside Plaza, Chicago, IL 60606. He is responsible for application of pulse-jet filters and related research and development activities. He joined the company in 1970 as assistant chief engineer after working in fine-particle research at the IIT Research Institute. He received a B.S. in chemical engineering from the University of Illinois in 1952.

David H. Klipstein is Vice President, Particulate Operations, Utility Gas Cleaning Div., Research-Cottrell Corp., Box 750, Bound Brook, NJ 08805, where he is in charge of electrostatic precipitator operations. He received his B.S. in Ch.E. from Princeton University, and his S.M. and Sc.D. degrees from M.I.T. (where he served as an Assistant Professor of Chemical Engineering). He is a member of AIChE, ACS, Soc. of Chemical Industry, the Commercial Development Assn. and the Environmental Advisory Council for the Federal Energy Administration.

Controlling Fine Particles

Particles smaller than two microns are a major factor in air pollution, affecting both human health and physical properties of the atmosphere. Moreover, they are extremely difficult to measure and control. Here is a report on collection equipment efficiency and the latest work on new ways to cope with these pollutants.

A. E. VANDEGRIFT, L. J. SHANNON and P. G. GORMAN, Midwest Research Institute

When air pollution became a major concern in the late 1950's and early 1960's, particulates was the first class of air pollutants to receive attention. We are succeeding in controlling a significant portion of the smoke and dust emitted from industrial operations. In fact, it is a fairly rare sight to see black clouds boiling out of a smoke stack these days.

However, to anyone who travels by air, it is obvious that today there is an almost permanent brown or yellowish-brown haze on the horizon. If we have made strides in controlling particulate emissions, then why has the visibility of our atmosphere deteriorated instead of improved? The causes may be twofold: (1) gaseous pollutants, which are generally not as well controlled as particulates, can react in the atmosphere to form particulates such as sulfates and photochemical smog; and (2) while we have succeeded in removing the black clouds from the smoke stack and with them the major fraction of the mass emitted from industrial processes, we have not been as successful in eliminating the particulates that can cause deterioration in the visibility of the atmosphere—the particulates below about 2 μm. in size.

Fine particulates, then, create most of the adverse effects that can be attributed to particulate pollutants. It is extremely important to document all the adverse effects of these fine particulates, to define the extent of the fine particulate problem by accurate measuring techniques, and to develop methods for controlling these particulates as they are emitted.

The Fine Particle Problem

The effects of particulate matter on human health are mainly related to injuring the surfaces of the respiratory system. Such injury may be permanent or temporary and may extend beyond the surface, sometimes producing functional or other alterations. Particle size, specific gravity and chemical nature largely determine: (1) where particles are deposited in the respiratory system; (2) the fate of the particles after deposition; and (3) to a considerable extent their physiological action. Particulate material in the respiratory tract may produce injury itself, or it may act in conjunction with gases, producing synergistic effects.

Originally published June 18, 1973

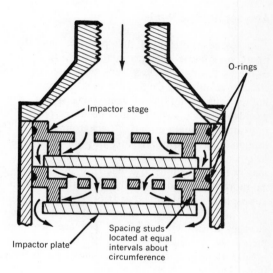

CASCADE IMPACTOR for source testing—Fig. 1

The analyses of numerous epidemiological studies clearly indicate an association between air pollution, as measured by particulate matter accompanied by sulfur dioxide, and health effects of varying severity.[1] This association is most firm for the short-term air pollution episodes. The connection between long-term residence in a polluted area and chronic disease morbidity and mortality is somewhat more conjectural. However, in the absence of other explanations, the findings of increased morbidity and increased death rates must still be considered consequential.

Fine particulate pollutants also affect the physical properties of the atmosphere—its electrical properties; its ability to transmit radiant energy; its ability to convert water vapor to fog, cloud, rain, and snow; and its ability to damage and soil surfaces. While concern with atmospheric transmission of radiant energy encompasses the entire electromagnetic spectrum, of particular importance is the infrared region, as it affects the terrestrial heat balance; the ultraviolet region, as it affects both biological processes and photochemical reactions in the atmosphere; and the visible spectrum, as it affects both our ability to see things and our need for artificial illumination.

Recent advances in our knowledge of the size distributions and optical properties of well-aged atmospheric aerosols (fine particles) have resulted in a clearer description of visibility reduction by such aerosols. For instance:

• Aerosols in the lowest region of the atmosphere (troposphere), whether over urban areas or not, tend to have similar size distributions.

• A narrow range of particle sizes, usually from 0.1 to 1.0 µm. radius, controls the extinction coefficient and thereby visibility.

• The mass concentration of a well-aged aerosol is approximately proportional to the light-scattering coefficients for atmospheric aerosols originating naturally or as the result of man's activity.

Atmospheric Effects

Increased emission of fine particles into the atmosphere may cause changes in the delicate heat balance of the earth-atmosphere system, thus altering worldwide climatic conditions. Comparisons at different sites covering periods of up to 50 years suggest that a general worldwide rise in turbidity may be taking place. This may indicate a gradual buildup of background levels. The net influence of atmospheric turbidity on surface temperature is uncertain, but the emission of long-lived particles may well be leading to a decrease in air temperature. As more is learned about the general circulation of the atmosphere and the balance between incoming and outgoing radiation, it seems increasingly probable that small changes in particle loads may produce very long-term meteorological effects.

Suspended particulate matter may alter weather patterns. There is evidence that some of the particles introduced into the atmosphere by man's activities can act as nuclei in processes that affect formation of clouds and precipitation. Investigations of snow and rainstorm patterns suggest that submicroscopic particulates from man-made pollution may be initiating and controlling precipitation in a primary manner, rather than being involved in the secondary process in which precipitation elements from natural sources serve to remove the particles by diffusion, collision, and similar scavenging processes.

MEASUREMENT OF FINE PARTICULATES

Before discussing sampling and sizing, we should define what we mean by particle and particle size. For the purposes of this discussion, a particle is a solid or liquid drop between 0.01 µm. and 2.0 µm. in size. This definition is rather arbitrary but most adverse effects are confined to this size range. In addition, the efficiency of particulate collection devices drops off markedly below 2 µm.

Particle size or diameter usually depends on the type of measuring instrument or the purpose for which the size measurement is to be used. In general, size is some consistent measure of the spatial extent of the particle, such as diameter, area or volume. It may also be a measure of the behavior of the particle as a function of its spatial extent. If particles are spherical or cubical, size corresponds to a single, measureable geometric dimension. Other regular geometric shapes may be defined in terms of dimension ratios. If particles are irregularly shaped, diameter is defined by consistent, arbitrary measures.

The variability in particulate properties has led to the development of numerous methods for the measurement of specific properties such as particle size. Selecting a method for determining particle size or particle-size distribution is generally based on (1) the particle-size range, (2) the form in which results are desired (i.e., number or weight-size curves), (3) application of the results, and (4) accuracy required. Different methods of determining particle size may give quite different results, and the method used to prepare the samples can also be a deciding factor in what the analysis actually measures.

Accurate collection of a representative particulate sample from a gas stream is by no means a simple matter. Usually, a representative sample can only be withdrawn if the probe does not disturb the stream pattern. Therefore, an ideal probe samples isokinetically and has infinitely thin walls. Both of these conditions are impossible to achieve in practice, and the problem is one of determining how far from the ideals one can depart and still obtain representative samples. Ref. 2 presents a more detailed discussion of sampling criteria.

Sample Collectors

Standard techniques are inadequate for collecting particles for accurate size determinations, particularly in the micron or less size-range. Cyclone separators tend to break up aggregates and friable particles; stainless steel or alundum thimbles probably do not capture all of the submicron particles; and small particles may penetrate deep into the body of a fiber filter, where they cannot be removed or observed microscopically. Another source of sampling error for particle-size analysis is leading the aerosol through a long and/or tortuous path before the particles are removed. In this procedure, particles are caught on the walls of the tubing, particularly near bends. Preferential trapping of certain sizes can produce serious errors in the particle-size determination.

For accurate determination of particle-size distribution, a collection mechanism that collects submicron particles and causes neither formation nor breakup of aggregates is necessary. Small electrostatic precipitators, thermal precipitators and cascade impactors come close to meeting these requirements.

Impaction devices:

Cascade impactors, constructed of stainless steel, are beginning to be used for in-stack source testing. Cascade impactors, such as shown in Fig. 1, consist of several stages containing jets and collector plates. The jets in each stage are progressively smaller. Particles are classified by inertial impaction according to their mass: The larger (heavier) ones are collected on the plate opposite the first stage and the smallest (lightest) on the plate opposite the last stage.

Classification of particles with impactors has certain advantages for microscopic determination of size distribution when there is a large variation in particle size. Furthermore, this method provides a way to obtain mass distribution curves, normally in the size range of 0.3 to 60 μm. The impactor is calibrated to determine the smallest particles collected at each stage, or, preferably, the size of particles sufficiently small so that they are collected with only 50% efficiency at each stage. Size calibration depends on the gas velocity through the instrument. Following calibration, the dust whose particle-size distribution is to be determined is collected with the impactor. The weight of material collected at each stage is measured and the results are plotted as cumulative weight-distribution curves.[3]

Impactors have very high collection efficiency under rigid experimental conditions. An efficiency of 100% for particles as small as 0.6 μm. has been obtained.[3] Commercial impactors for stack sampling are now available,[4,5] and are reported to be capable of sizing particles down to 0.1 μm.[4]

Particle-size distributions made with cascade impactors may be distorted by particle bounce, surface overloading (i.e., particle buildup and blowoff) and wall losses. Wall losses can be minimized by good impactor design and proper operation. Particle bounce may be reduced by using a viscous coating on the collector plates.

Thermal precipitators:

These units have been used for many years in England and South Africa for ambient sampling and have also been used to a limited extent for stack sampling in this country.[6,8] The thermal precipitator uses a radiometer force produced by a thermal gradient between two different surfaces. Direct measurements of collection efficiency show that thermal precipitators deposit virtually all particles of most materials from 5 down to 0.01 μm., and possibly smaller.

By depositing particles on an electron-microscope grid, the thermal precipitator can be used with an electron microscope for sizing particles. If the deposit is not so dense that overlapping occurs and attention is paid to statistical errors, the thermal precipitator is one of the most accurate methods for sampling nonvolatile solid particles. The upper limit of number concentration that can be determined depends upon the measurement accuracy and the sample volume drawn through the instrument.

Small electrostatic precipitators:

Small-scale electrostatic precipitators in various configurations have been used with some success for sampling and subsequent particle sizing by optical methods. The essential features of many of these devices are a tubular electrode through which the particles pass and a central ionizing electrode maintained at a high potential. The electric field gives the particles a high charge so they travel to the outer electrode, where they are deposited.

A drawback of electrostatic precipitators is that the deposit varies in density and size distribution. This means that a full assessment entails considerable counting and sizing. Electron microscope screens have also been used as deposit sites.[9]

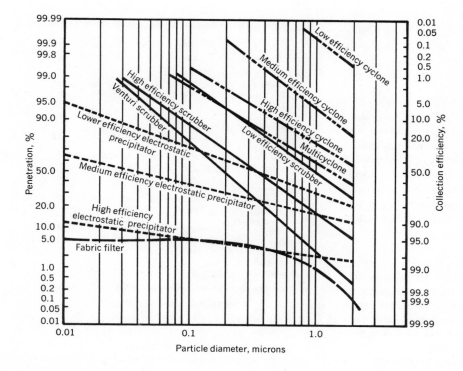

FRACTIONAL efficiency in the fine-particle range for various control devices—Fig. 2

Particle-Size Measurement

Following collection of a particulate sample, a method for size determination must be selected. With the standard sampling techniques utilizing filters, redispersion of the particles is a potential problem. Cascade impactors, thermal precipitators, or electrical precipitators minimize this troublesome factor because optical techniques are generally used with these devices.

A major portion of the data currently available on the particle size of particulates emitted from industrial sources has been obtained by using the Bahco Micro-Particle Classifier, which is a combination centrifuge-elutriator. The Coulter Counter, Whitby Centrifuge/MSA Sedimentation, and microscopic techniques have also been used. Cascade impactors, such as the Andersen and those developed by Pilat[10,11] are now being used in routine stack sampling.

Coulter Counter:

The Coulter Counter determines particle size by a change in electrolytic resistivity. The instrument measures a number-volume particle-size distribution of particles suspended in an electrically conductive liquid. The suspension flows through a small aperture with an electrode immersed on either side. The concentration is low enough so that the particles move through the aperture one at a time in most cases.[12] The size range of this instrument is approximately 0.3 to 100 μm.

Differential sedimentation:

The MSA-Whitby device is based on the differential-layer-sedimentation technique. A basic part of the method is specially designed centrifuges that have constant speeds to within 1% and repeatable speed-versus-time curves during starting and stopping so that corrections do not vary more than ±0.5 sec.

At the beginning of a size analysis, the clean tube is filled to the line near its top with a suitable sedimentation liquid. Next, a suspension of particles is made up in a liquid that is miscible with the sedimentation liquid but has a slightly lower density.

Sedimentation height is assumed to be proportional to sediment weight. With complete dispersion of particulate matter, this is a good assumption, because same-sized particles are essentially settling at one time and the void space is independent of size for monodisperse systems. In some cases, however, 2 μm. particles may be aggregates of 0.01 to 0.03 μm. solids. In these situations, compression of the sediment column with increasing centrifuge speed takes place and bulk-density-correction factors must be used.[12]

Optical techniques:

These are the most direct methods for particle-size-distribution measurements. Theoretically, the applicability range of optical methods is unlimited; but practical considerations and availability of more expedient techniques make microscopy a less-desirable tool in certain size ranges.

Microscopic examination, however, is generally considered most reliable for particle-size analysis—provided samples are properly prepared, the particles are all or nearly all of one geometric shape, and enough of them are measured to give statistically-reliable data. Inevitably, these methods are difficult and time-consuming. The sample must be magnified to such an extent that individual particles can be examined and their dimensions measured. (Of course, the images may be photographed and the analysis made from the photographs.) Optical techniques are generally used in conjunction with impactors, thermal precipitators and electrostatic precipitators.

Equipment Under Development

Work is continuing on various types of cascade impactors to determine their range of applicability as well as their limitations. This includes comparative tests of the Brink, Andersen, and University of Washington Impactors. Another inertial device that is presently undergoing testing is a parallel cyclone, developed by McCrone Associates. This consists of three parallel-cyclones and a bypass filter in series with a larger cyclone. The larger cyclone has a 50% cutoff point at 6 μm. and the smaller cyclones have cutoff points at 4, 1, and 0.5 μm. respectively. This looks promising because it is able to handle very high dust loadings and gas volumes. Other devices undergoing testing and development include the diffusion battery, condensation nuclei and single particle counters. The diffusion battery has been around since 1900. Present work is in the 0.01 to 1.0 μm. range.

Whitby's electron mobility device is still another size-measurement technique. This consists of an aerosol-charging section and an electric-mobility analyzer. Particles smaller than a certain size, with high electric

Average Collection Efficiencies—Table I

Type Collector	Overall	Efficiency, %, in Micron Range				
		0–5	5–10	10–20	20–44	>44
Baffled settling chamber	58.6	7.5	22	43	80	90
Simple cyclone	65.3	12	33	57	82	91
Long-cone cyclone	84.2	40	79	92	95	97
Multiple cyclone (12-in. dia.)	74.2	25	54	74	95	98
Multiple cyclone (6-in. dia.)	93.8	63	95	98	99.5	100
Irrigated long-cone cyclone	91.0	63	93	96	98.5	100
Electrostatic precipitator	97.0	72	94.5	97	99.5	100
Irrigated electrostatic precipitator	99.0	97	99	99.5	100	100
Spray tower	94.5	90	96	98	100	100
Self-induced spray scrubber	93.6	85	96	98	100	100
Disintegrator scrubber	98.5	93	98	99	100	100
Venturi scrubber	99.5	99	99.5	100	100	100
Wet impingement scrubber	97.9	96	98.5	99	100	100
Baghouse	99.7	99.5	100	100	100	100

Data based on standard silica dust with the following particle distribution:

Particle Size Range	Weight %
0–5	20
5–10	10
10–20	15
20–44	20
>44	35

mobility, are drawn to a collecting rod, while the larger particles escape the collection rod but are trapped by a filter where their charge drains off through an electrometer. Changes in rod voltage produce changes in electrometer current that can be related directly to the number of particles in the aerosol between two discrete particle sizes. This device was intended for ambient sampling but may be adaptable to stack sampling.[13, 14]

CONTROL OF FINE PARTICLES

Equipment available for removing fine particles is generally the same as that usually considered for controlling total particulate emissions, that is, electrostatic precipitators, fabric filters, scrubbers and afterburners. As yet, there has been little incentive for manufacturers to develop new devices for removing fine particles. Instead, they are usually asked to meet limitations on total quantity (mass) emitted. However, the manufacturers may also be required to meet opacity regulations, which may in effect correspond to a limitation of fine particles because it is these particles that contribute most to plume opacity.

Even the new-source-performance standards currently being established by the Federal Environmental Protection Agency (EPA) are based primarily on total mass emissions. However, increasing evidence and awareness of the importance of fine particles, and hazardous pollutants that often are predominantly in the fine-particle size range, has prompted consideration of emission regulations based on particle size.[15,16] As a first step, efforts are being made to examine the capability of conventional control devices to remove fine particles.

Fractional Efficiencies

Currently, control equipment is rated mainly by one parameter—overall mass efficiency. Specification of equipment efficiency by overall performance, however, is inadequate with respect to fine particle emission. Penetration in specific size ranges is a more revealing term for rating equipment performance. Correlations of penetration vs. particle size are called "fractional efficiency curves." Table I presents typical data on collection efficiencies for various particle sizes and control equipment.[17]

Midwest Research Institute, under contract from EPA, has been conducting studies on control of fine-particulate emissions from stationary sources. A part of this effort has involved evaluation of information that is available on the fractional efficiency of different control devices.[14]

Our analysis indicated that over 95% of the data currently available on the particle size of materials emitted from industrial sources has been obtained by sampling and sizing techniques that are not suitable for the size range of less than 2 μm. As a result, only a meager quantity of data are available on particle sizes in the less than 2 μm. range and the ability of control devices to collect these particles is ill-defined.

We evaluated the available fractional efficiency data, using log-probability coordinates to magnify the efficiency relationship for smaller particles. This included all available information on specific types of equipment, overall efficiencies, operating conditions (pressure drop, water rate, etc.), and sampling and analysis techniques. The data for each type of device varied over a wide range, but this is not surprising considering the variations in equipment design, testing procedures, etc.

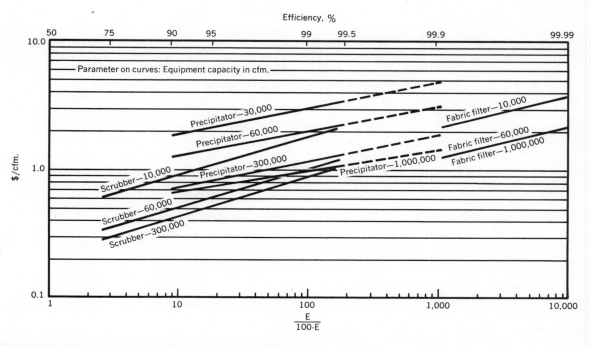

INSTALLED COSTS for control equipment in several capacity ranges—Fig. 3

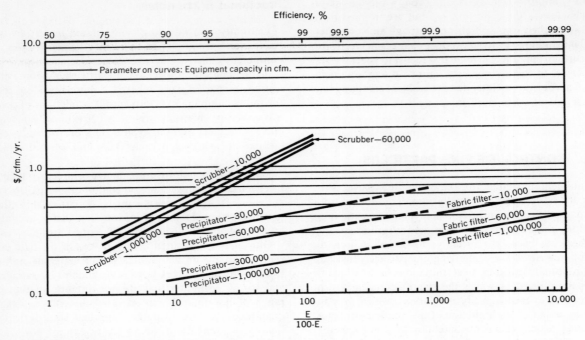

ANNUALIZED COSTS for collection equipment according to capacity—Fig. 4

Since the data for each collection device did vary over a wide range, we examined the individual figures with the objective of drawing general curves that would represent low, medium and high overall efficiencies for each type of equipment. The data for four kinds of control devices were carefully assessed to determine the fractional efficiency curves shown in Fig. 2.

Cost is another aspect of control devices that is especially important in connection with fine-particle removal. In many cases, the efficiency required to meet standards based on particle size approaches the limits of equipment capability. This can mean large cost increases for seemingly small gains.

Attempts have been made to correlate cost of control devices as a function of volume and efficiency using the data in Ref. 18. The basic flaw, however, is that these cost curves reflect only general efficiency ranges—low, medium and high. Specific values of efficiency can be assigned to the extremes of these ranges. The relationships depicted in Fig. 3 and 4 were developed from the resulting data.[14] The relationships shown in the figures give a very rough approximation and their accuracy is, admittedly, highly questionable.

Other Control Approaches

The preceeding discussion of equipment efficiencies and costs for fine-particle control has been limited to conventional control methods. However, novel devices as well as basic collection mechanisms must be considered for the specialized area of fine-particulate control.

There are several avenues that might lead to improved control of fine particulates. The more promising approaches can be grouped into two broad categories: de-

velopment of new or novel control devices; and augmentation of commonly used collection mechanisms or agglomeration-promoting techniques.

A variety of devices touted as having high collection efficiencies in the fine-particle range have been reported in the literature.[19] For most of these the only supporting data are often meager, unavailable or inconclusive. Extensive testing programs will be required to determine the full potential of these devices.

Theoretically, improvements in control of fine particles might result from better exploitation of collection mechanisms (e.g., diffusiophoresis,* thermophoresis*) or agglomeration methods (e.g., condensation, acoustic agglomeration). Such phenomena might be used advantageously if devices can be designed without having unreasonably high energy requirements.

A brief presentation of the theoretical aspects and experimental results for these mechanisms is given here. A more in-depth analysis of these mechanisms, with emphasis on wet scrubber technology, has been carried out by Calvert, et al,[20] among others. Equations developed by Calvert plus the many important references will be of interest to those who may want to continue investigations in this area.

Thermophoretic Forces:

Exploitation of thermal forces to improve collection of fine particles might be accomplished in one of two ways: design and construction of equipment based entirely on particle deposition by thermal forces; and use

---

*Diffusiophoresis: Movement of aerosols due to a concentration gradient of some gaseous component, e.g., movement to a surface where water is condensing.

Thermophoresis: Movement of aerosols due to a temperature gradient, e.g., away from a hot surface toward a cold surface.

of thermal forces as a contributing mechanism in existing dust collectors.

From the standpoint of collection efficiency, thermal precipitators appear quite promising. However, consideration of energy requirements presents a different picture. Comparison of high-efficiency wet scrubbers with an ideal thermal precipitator illustrates the disadvantage of the thermal precipitator. For a collection efficiency of 99.3%, the power required for wet scrubbers is estimated at about 10 hp./1,000 cfm. The comparable figures for the ideal thermal precipitator are an efficiency of 100% for particles smaller than 5 μm. with a power requirement of about 100 hp./1,000 cfm.

On the basis of this comparison, the power required for thermal precipitation would be at least an order of magnitude greater than that required for most other air cleaners. However, if waste thermal energy is available from the gas stream that is being cleaned, then a thermal precipitator becomes more attractive from an energy-consumption standpoint.

Thermal forces might be exploited to enhance the performance of certain existing control devices. Wet scrubbers and packed columns are two cases in point. Research on the relative contribution of thermophoresis to the collection of 0.1–2 μm. particles in wet scrubbers indicates that for air in the temperature range 20–100 C., the predicted effect of thermophoresis in bubbles and drops is small.

Deposition of particles from a hot gas in a cooled, packed bed is another instance where thermal forces may improve collection efficiency. Experiments have shown that when a packed bed is initially cold, particle collection is more nearly complete. Theoretical calculations indicate that, since the passages in a bed are narrow, a temperature difference of 50 C. might give rise to a temperature gradient of 1,000 C./cm. in the passages. Calculations show that this could result in deposition of 98.8% of the 0.1 μm. particles in a 9-in. deep bed from a gas at 500 C.

Diffusiophoretic Forces:

Diffusiophoresis might be a useful mechanism to exploit together with other methods for removing small particulates from gas streams. Diffusiophoresis has two advantages:

1. The fundamental mechanism is independent of particle size and becomes more important compared to other removal techniques for particles below 2 μm.

2. The expected particulate removal efficiency depends on operating conditions and equipment design, but can theoretically reach 100%. In a more practical situation, theory indicates that diffusiophoresis might account for more than 30% of the total collection efficiency.

To exploit the mechanism of diffusiophoresis, the collecting device must be designed so that one component in the gas phase is diffusing toward a collecting surface. The most practical case would be the diffusion of water vapor toward a surface. Theoretical calculations indicate that in a spray tower a collection efficiency of 30% for 2 μm. particles can be expected from diffusiophoresis with an energy expenditure of about 1 hp./1,000 cfm. of gas.

Condensation Phenomena:

Condensation (of steam) to enhance particle collection has attracted interest for several years. Several collection mechanisms may contribute to this phenomena. Some experimental devices have demonstrated very high collection efficiency but they required excessive quantities of steam. Calculations by Davis[21] confirmed that large quantities of steam would be required to produce uniform growth of particles, assuming heterogeneous nucleation, up to a size that could be efficiently removed in conventional devices.

Some interesting experimental work in condensation has been done by Lancaster.[22] He concluded that the dominant mechanisms influencing collection in the scrubber under test were particle buildup and subsequent impaction. Correlation of scrubber efficiency with steam-jet parameters showed a relationship between efficiency and the mixing characteristics of the system at steam injection. The equation developed indicated that steam-injection rate was the most significant variable influencing scrubber performance. It also suggested that scrubber performance relates to quantity of steam injected and not to the amount condensed. Lancaster found that a large percentage of the total injected steam could condense prior to complete mixing of steam and air.

The inefficient steam use appears to be due primarily to inadequate steam distribution between potential condensation nuclei. If efficiency can be increased by special condensation scrubbers that adequately distribute the condensed steam among aerosol particles, the system could be competitive with conventional scrubbers in normal applications. For the removal of ultrafine particles, the condensation scrubber may prove to be superior. In addition, the large surface area of condensation nuclei could be expected to efficiently absorb soluble gaseous pollutants such as SO_2.

Agglomeration of Particles

Agglomeration or coagulation of particulates could be used as one step in a sequence of operations aimed at controlling emissions. If sufficiently large particles could be produced, it may be possible to use conventional, low-cost techniques as the final collection step. At this time, only sonic agglomeration and the agglomeration of charged particles appear attractive.

Many theoretical examinations of agglomeration have assumed that all particles cohere if collisions occur. However, this may be untrue, as evidenced by the stereo-ultra microscope study described in Ref. 23.

Sonic Agglomeration:

The principal advantage of coagulation by a sonic field is its applicability to any aerosol, including those with submicron particles. The main disadvantage of this method is relatively high energy requirements. A second problem is the low efficiency of acoustic coagulators and their inability to handle highly dispersed suspensions.

Water augmentation to sonic agglomeration appears necessary to remove a large fraction of the particulates.

If a pound of water per 1,000 cfm. is used, this would increase power requirements up to 1 hp./1,000 cfm., depending on the device used to introduce the water and the mean droplet size. The minimum energy requirements would then be 5.5 to 10 hp./1,000 cfm. Whether this is competitive with other devices capable of removing fine particles remains to be seen.

Agglomeration of charged particles:

One method of increasing the agglomeration rate is to add a bipolar charge, either with or without an externally imposed field. To a limited degree, this occurs in a standard electrostatic precipitator, but not sufficiently to obtain efficient collection of fine particulates. With proper conditions, however, electrostatic forces between particulates can produce a large increase in agglomeration of submicron particulates.

Space-charge precipitation is a method of removing particulates from gas streams. This is based on migration of charged particles in their own space-charge fields. However, particulate-gas systems are usually not sufficiently concentrated to provide adequate fields. It is, therefore, necessary to add a cloud of charged drops to produce fields that will remove the particles.

Basically, a space-charge precipitator consists of two parts: a short section where both the particles and the added drops are charged by high voltage coronas; and a section of grounded tubes or plates on which the particles and drops are collected. In the collecting section, the drops and particles migrate to the surfaces, where they coalesce and flow from the precipitator as a slurry.

Theoretical and limited experimental studies on space-charge precipitation have been conducted at the University of California.[24] While these experiments may not accurately model a practical system, the work done to date indicates that space-charge precipitation may be a viable approach to the collection of particulates. Energy requirements are relatively uncertain. Calculations indicate a total requirement of about 0.5 hp./1,000 cfm. of gas. Based on a preliminary cost estimate, space-charge precipitation appears to be economically competitive with conventional electrostatic precipitation. Additional testing with a pilot-scale model on an actual industrial source will be needed to allow precise analysis of potential of full-scale units.

References

1. Speizer, F. E., An Epidemiological Appraisal of the Effects of Ambient Air on Health: Particulates and Oxides of Sulfur, *Journal of the Air Pollution Control Association,* **19,** 647–656, 1969.
2. Parkes, G. J., Some Factors Governing the Design of Probes for Sampling in Particle- and Drop-Laden Streams, *Atmospheric Environment,* **2,** 477–490, 1968.
3. Cadle, R. D., "Particle Size Determination," Interscience Publishers, New York, 1955.
4. Sales Brochure, Andersen Air Samplers, 2000, Inc., South State Street, Salt Lake City, Utah.
5. Brink, J. A., Jr., Cascade Impactor for Adiabatic Measurements, *Ind. Eng. Chem.,* **50** (4), 645 (1958).
6. Bush, A. F., and E. S. C. Bowler, Electron Microscopy Studies of a Coal Fired Steam Electric Generating Stack Discharge, Report No. S.I. 395, Dept. of Engineering, University of California, Los Angeles, Cal., Dec. 1965.
7. Bush, A. F., and E. S. C. Bowler, Electron Microscopy Studies of a Bag Filterhouse Pilot Installation on a Coal Fired Boiler, Report No. S.I. 405, Dept. of Engineering, University of California, Los Angeles, Cal., Sept. 1966.
8. Bush, A. F., and E. S. C. Bowler, Electron Microscopy Studies of Steam Electric Generating Stack Discharges, Report No. C 60–79, S.I. 8007, Dept. of Engineering, University of California, Los Angeles, Calif., 1960.
9. Green, H. L., and W. R. Lane, "Particulate Clouds: Dusts, Smokes, and Mists," Second Ed., D. Van Nostrand Company, Inc., Princeton, N.J., 1964.
10. Private communication, Prof. M. J. Pilat, University of Washington, Seattle, Wash., Nov., 1970.
11. Bosch, J. C., Size Distribution of Aerosols Emitted from a Kraft Mill Recovery Furnace, M.S. Thesis, University of Washington, Seattle, Wash., 1969.
12. Irani, R. R., and C. F. Callis, "Particle Size: Measurement, Interpretation, and Application," John Wiley, New York, 1963.
13. Flesch, J. P., Calibration Studies of a New Submicron Aerosol Size Classifier, *Journal of Colloid and Interface Science,* **29** (3), 502–509, 1969.
14. Particulate Pollutant System Study, Volume II—Fine Particulate Emissions, EPA Contract No. CPA 22-69-104, Midwest Research Institute, 1 August 1971.
15. Corn, M., Particle Size: Relationship to Collector Performance, Emission Standards and Ambient Air Quality, paper presented at the Second International Air Pollution Conf. of the International Union of Air Pollution Prevention Associations, Washington, D.C., 10 December 1970.
16. Air Quality Control Regulation Number 504 (as amended), New Mexico Environmental Improvement Board, Santa Fe, N.M., 1972.
17. Ottmers, D. M., Jr., and Ben R. Breen, Technical Basis for Texas Air Control Board Particulate Regulations, Radian Corporation, Technical Note No. 100-007-01, 20 August 1971.
18. Control Techniques for Particulate Air Pollutants, U.S. Dept. of Health, Education, and Welfare, National Air Pollution Control Administration, Publication No. AP-51, Washington, D.C., 1969.
19. Iammartino, N. R., Technology Gears Up to Control Fine Particulates, *Chemical Engineering,* 21 August 1972.
20. Calvert, S., *et. al.,* "Scrubber Handbook," study prepared for the Environmental Protection Agency by A.P.T., Inc., Riverside, Cal., July 1972.
21. Davis, R. J., *et. al.,* The Function of Condensing Steam in Aerosol Scrubbers, Oak Ridge National Laboratory, Oak Ridge, Tennessee, ORNL-4654, Mar. 1971.
22. Lancaster, B. W., and W. Strauss, A Study of Steam Injection into Wet Scrubbers, *Industrial Engineering Chemistry Fundamentals,* **10**(3), 1971.
23. Wilson, L. G., and P. Cavanagh, A Stereo-Ultramicroscope for Studying Sub-micron Aerosols, *Atmospheric Environment,* Pergamon Press, **3,** 47–53, 1969.
24. Kostow, Lloyd Pl., Design and Testing of Space-Charge Precipitators, M.S. Thesis, University of California, Berkeley, Cal., 6 March 1972.

Meet the Authors

A. Eugene Vandegrift is head of the environmental sciences section at Midwest Research Institute, 425 Volker Blvd., Kansas City, MO 64110. He has managerial and technical responsibility for environmental science programs including air and water pollution, and solid waste disposal. He has a B.S. in chemical engineering from the University of Kansas (1959) and a Ph.D. in chemical engineering from the University of California at Berkeley (1963).

Larry J. Shannon, principal chemical engineer at Midwest Research Institute, has particular expertise in air pollution control technology and solid waste management. Before joining MRI in 1969, Dr. Shannon spent six years as a senior chemical engineer with United Technology Center, Div. of United Aircraft. He received his B.S. in chemical engineering in 1959 from Seattle University and a Ph.D. in chemical engineering from the University of California at Berkeley in 1963.

Paul G. Gorman, associate chemical engineer at Midwest Research Institute, has extensive experience in developing air pollution control equipment, sulfur oxide removal processes, plant operations and process design. He worked as a process engineer with the J. F. Pritchard Co. and Cooperative Farm Chemicals Assoc. before joining MRI in 1969. Mr. Gorman has a B.S. in chemical engineering from the University of Kansas (1958) and a M.S. in chemical engineering from the University of Missouri (1972).

Predicting Efficiency of Fine-Particle Collectors

A good first approximation for selecting and modifying fine-particle collection systems relies on graphical integration for interrelating the particle-size distribution with operating parameters of the collector to get its efficiency.

JAMES M. PETERS, El Paso Natural Gas Co.

The state-of-the-art of particulate control is well developed. Although requirements are stiffer now than ever before, the technology can be extended, at considerable cost, to meet the regulations. Nevertheless, little attention has been paid to very fine particles until recently, so that it is difficult to assess how well a particular control system can capture them.

Here is a simple graphical method for predicting small-particle capture, along with specific examples to illustrate the procedure for a number of systems.

Predicting Small-Particle Capture

Any method for estimating the efficacy of fine-particulate control depends upon a knowledge of the particle-size distribution and other characteristics, along with the operating parameters of the specific control device used.

For example, uncontrolled fly ash from pulverized-coal-fired boilers typically has a mass median diameter of 15 to 20 μm. and a logarithmic normal-size distribution similar to that shown in Fig. 1.[2-4] The dashed ends of the curve indicate rough extrapolation—a dangerous practice in the absence of more-detailed information. Diminution of particles due to collisions with surfaces and with each other could change the particle-size distribution somewhat. However, very fine particles in the submicron range may tend to coagulate as a function of Brownian motion, particle concentration and other factors. As these particles approach the size of the mean-free-path of air (about 0.07 μm.), they act more like true gases rather than like finite entities.[5] In reality, then, the lower end of the distribution curve may swing away from linearity either horizontally or vertically, depending on whether very small particles are continuously created or

Originally published April 16, 1973

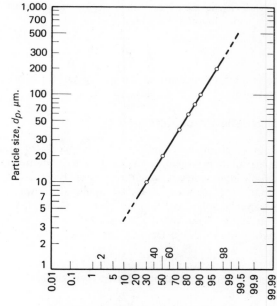

PROBABILITY plot of fly-ash size distribution—Fig. 1

Amount less than stated size, % by weight

Particle size, d_p, μm.

removed. The same phenomena may be observed with other kinds of industrial dusts.

Assuming that the extrapolation in Fig. 1 is reasonably accurate, it is possible to predict the overall efficiency of small-particle collection by almost any of the modern control equipment available. If the operating parameters are known, along with appropriate design equations, the

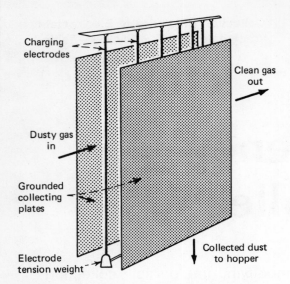

CONFIGURATION for electrostatic precipitator—Fig. 2

graphical method to be described here is straightforward. The basic theory is derived from:

$$E_t = 1 - \int_0^\infty Kf(d_p)dd_p \qquad (1)$$

where K is a constant.

For the particle-size range of interest ($0 < d_p < 2 \mu$m.), the total collection efficiency becomes:

$$E_{0\text{-}2} = 1 - \int_0^2 Kf(d_p)dd_p \qquad (2)$$

The graphical technique of integrating Eq. (2) is best illustrated by way of examples.

Electrostatic Precipitator

It may be estimated from Fig. 1 that about 4% of the total particulate mass will be less than 2 μm. in diameter. Let us consider an electrostatic precipitator (ESP) having an electrode plate area of 10,000 sq.m., as shown in Fig. 2, that handles 20,000 cu.m./min. of fly-ash-laden gases at 150 C. The overall efficiency of the ESP is expressed by the modified Deutsch equation:

$$E = 1 - \exp(-Aw/100Q) \qquad (3)$$

where the particle migration velocity, w, is given by:

$$w = CE_oE_pd_p/4\pi\mu \qquad (4)$$

In Eq. (4), d_p has dimensions of cm., and Cunningham's correction factor for particle slip is:

$$C \sim 1 + (1.72\lambda/d_p) \qquad (5)$$

Next, we assume likely values of E_o and E_p at 5 and 3 kv./cm., respectively,[5,7] and note that $\mu = 2.38 \times 10^{-4}$ poise for air at 150 C.

By substituting this information into Eq. (3) to (5), we may obtain Table I for a range of particle sizes. The percent weight less than the stated particle size is obtained from Fig. 1.

Fig. 3 is a plot of the last two columns in Table I. The area under the curve between 0 and 4% by weight is graphically integrated by determining the point at which the areas above and below the line will be equal. The result is that theoretically some 85% of the particles less than 2 μm. can be collected.

If the regulation required 90% removal, for example, the ESP as designed probably would fall short. Accordingly, it might be necessary to consider parametric adjustments or add-on devices to polish the remaining 5% in bringing the plant into compliance.

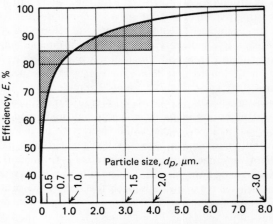

COLLECTION efficiency for particles ≤ 2 μm. dia.—Fig. 3

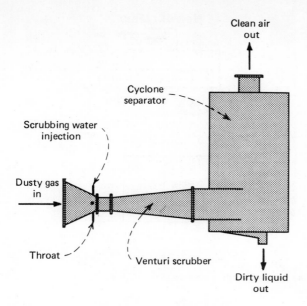

VENTURI scrubber followed by cyclone separator—Fig. 4

Collection Efficiency of Electrostatic Precipitator for Fine Fly Ash—Table I

Particle Size, d_p μm.	Migration Velocity, w cm./sec.	Amount $< d_p$ % by weight	Efficiency %
5	28.5	14.0	99.98
4	23.0	11.0	99.90
3	17.4	8.0	99.46
2	11.8	4.0	97.10
1.5	9.0	3.0	93.28
1.0	6.2	1.0	84.43
0.8	5.1	0.7	78.35
0.6	4.0	0.35	69.88
0.5	3.4	0.22	63.94
0.4	2.9	0.13	58.10
0.3	2.3	0.06	49.84
0.2	1.75	0.02	40.84
0.1	1.2	<0.01	30.23

Venturi Scrubber

The same graphical technique may be applied to fine-particle collection with a venturi scrubber, like that in Fig. 4. Again, Eq. (2) is applicable. As with the ESP, particulate-size distribution and other characteristics and scrubber operating parameters are important. For instance, favorable particle wettability is important for promoting high collection efficiency, and the fly ash from some coals may not favor the use of a scrubber at all.

Assuming that it is feasible to use a wet scrubber for a given particle emission, the efficiency of collection of a given particle size is:

$$E = 1 - \exp(-kL\sqrt{P}) \qquad (6)$$

where k is an experimental factor determined by throat geometry and other parameters. The value of k may be on the order of 0.1 to 0.2.[8] P in Eq. (6) is an inertial impaction parameter evaluated from:

$$P = C\rho_d u_t (d_p)^2 / 18\mu d_o \ 10^4 \qquad (7)$$

The expected droplet diameter created by high-velocity gases impacting against injected scrubbing water is:

$$d_o = 1.45L^{1.5} + (16,400/u_t) \qquad (8)$$

where u_t is expressed in ft./sec. Research has shown that droplets 30 to 200 μm. in size have the best target efficiencies for fine particles.[8]

If a certain venturi scrubber has a throat velocity of 500 ft./sec., and L is 10 gpm./1,000 cu.ft./min., d_o is about 78 μm. Assuming a value of $k = 0.2$, fly ash with a density of 0.7 g./cu.cm. can be collected according to:

$$E = 1 - \exp(-4.11d_p) \qquad (9)$$

Now a table similar to Table I may be constructed, and an efficiency curve like that in Fig. 3 may be drawn for the venturi scrubber. The same graphical technique will provide the theoretical system efficiency for the size range of interest.

Fabric Filtration

Filtration of dusty air in a baghouse filter (Fig. 5) is the most efficient means of removing fines, especially if the filter fabric is felted or matted.[9] In filtration, three collection mechanisms are dominant: interception, inertial impaction, and diffusion.[10] The equations devised for these are quite distinct and are complicated; and the

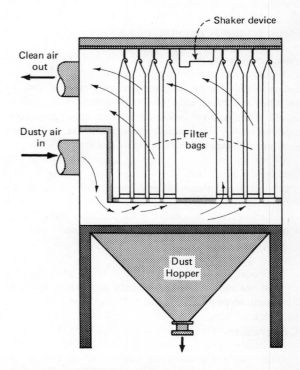

SINGLE-COMPARTMENT baghouse filter with shaker—Fig. 5

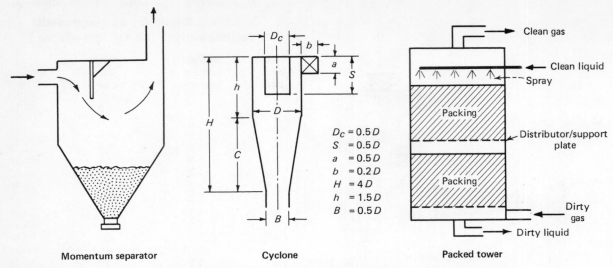

$D_c = 0.5 D$
$S = 0.5 D$
$a = 0.5 D$
$b = 0.2 D$
$H = 4 D$
$h = 1.5 D$
$B = 0.5 D$

Momentum separator   Cyclone   Packed tower

SOME of the other common types of collectors for trapping and separating particulate materials—Fig. 6

overall efficiency in the size range of interest would involve a composite of the three mechanisms:

$$E = 1 - (1 - E_i)(1 - E_m)(1 - E_d) \qquad (10)$$

See Strauss[11] for additional information on Eq. (10).

Research has shown that:

$$E_m = f(\sqrt{P}) \qquad (11)$$

$$E_i = 0.0076 C_f N_{Re} \, (d_p/d_o)^{1.5} \qquad (12)$$

$$E_d = 0.375 (C_f N_{Re})^{0.04}/(v d_o)^{0.6} D \qquad (13)$$

Eq. (11) through (13) have been obtained from Ref. 10.

While these equations are somewhat difficult to evaluate, the end-results are similar to the foregoing. The same graphical technique may be used to estimate the overall efficiency of small-particle collection.

Other Collectors

Fig. 6 shows some of the other common wet and dry collectors. Of these, only the wet-type devices are capable of removing particles less than 2 μm. to any appreciable extent. Dry devices such as the momentum separator and cyclone will generally collect 0 to 25% or so.

Collection efficiencies of wet collectors such as the packed tower are similar to those given for the venturi scrubber. A more concise relationship for all wet scrubbers incorporates the concept of contacting power, which is a function of the number of transfer units (NTU) required to effect collection of the particles:[10]

$$E = 1 - \exp(-NTU) \qquad (14)$$

The concept represented by Eq. (14) is commonly used in many facets of chemical engineering and will not be discussed further.

Conclusion

While the graphical technique is simple, it can only be regarded as an approximate method of predicting the collection of fines. Accurate estimates of particle-size distribution, collector-operating parameters, and other information must be available to engender confidence in the results; but it is advantageous to have a simple predictive technique for design purposes. Graphical integration can provide a good first approximation for use in selecting and modifying process equipment to control fine particles.

References

1. "Particulate Emissions From Coal-Burning Equipment," Air Quality Control Regulation 504, New Mexico Environmental Improvement Board, Santa Fe, N.M., Mar. 25, 1972.
2. Engdahl, R. B., Stationary Combustion Sources, in "Air Pollution," Vol. III, A. C. Stern, Ed., Academic Press, New York, 1968.
3. "Air Quality Criteria for Particulate Matter," Publication AP-49, National Air Pollution Control Admin., Washington, D.C., 1969.
4. Cuffe, S. T. and Gerstle, R. W., "Emissions From Coal-Fired Power Plants: A Comprehensive Summary," PHS Publication 999-AP-35, U.S. Public Health Service, Durham, N.C., 1967.
5. White, H. J., "Industrial Electrostatic Precipitation", Addison-Wesley, Reading, Mass., 1963.
6. Gottschlich, C. E., Source Control by Electrostatic Precipitation, in "Air Pollution," Vol. III, A. C. Stern, Ed., Academic Press, New York, 1968.
7. "Air Pollution Engineering Manual," USPHS Manual 999-AP-40, U.S. Public Health Service, Cincinnati, 1967.
8. "Removal of Particulate Matter From Gaseous Wastes: Wet Collectors," American Petroleum Institute, New York, 1961.
9. "Removal of Particulate Matter From Gaseous Wastes: Filtration," American Petroleum Institute, New York, 1961.
10. "Air Pollution Manual, Part II—Control Equipment," American Industrial Hygiene Assn., Detroit, 1969.
11. Strauss, W., "Industrial Gas Cleaning," Pergamon Press, Oxford, 1966.

Meet the Author

James M. Peters is an environmental engineer in the engineering department of El Paso Natural Gas Co., El Paso, TX 79978. He is involved in air-quality and air-pollution modeling, design of air-pollution-control equipment and processes, and studies on noise, odors and visibility. He has an M.S. in environmental engineering from the University of Texas, and a B.S. from the U.S. Military Academy. Formerly he served as an editor for the Pennsylvania State University Press. He is a member of the Texas Soc. of Professional Engineers, Air Pollution Control Assn., Water Pollution Control Federation, and AIChE.

How to choose a particulate scrubber

In this feature report, the author groups the myriad of wet scrubbers used for product recovery and pollution control according to the physical mechanisms by which they operate. Also covered is the cut/power method of analyzing scrubber performance—a technique developed by the author in various EPA work.

Seymour Calvert, *Air Pollution Technology (APT), Inc.*

☐ Wet scrubbers use a liquid stream to recover small particles from a gaseous flow. Whether the particles are caught first by the liquid, or first on the scrubber structure and then washed off by the liquid, we will call any device that uses a liquid to clean up a gas a "scrubber." Its duty might be to purify a process stream or to remove a pollutant from a stream being emitted to the outside atmosphere.

To select the best scrubber for an installation, it is essential to know precisely what the device is supposed to do—often easier said than done. For example, you might need to clean up an effluent gas stream so that its opacity will be less than 20%. But if you do not know the particle size distribution, particle concentration, particle physical properties, stack diameter, and a few other things (such as the influence of these variables on plume opacity), you cannot specify the scrubber.

For process cleanup, the allowable concentration and/or particle size range will be set by process criteria. Air-pollution control regulations generally specify a maximum mass rate of emissions, and often set a concentration limit as well. Translating an opacity limit into a particle concentration limit takes some work [1].

It is possible to predict the efficiency and power requirement for most scrubber types with reasonable accuracy. Performance depends on the particle diameter, collection mechanism(s), collecting-element dimensions, gas flowrate, and liquid flowrate. Particle size must be known if any quantitative prediction of collection efficiency is desired.

Fine-particle diameters are usually expressed in microns and there is generally a distribution of particle diameters present in the process stream. For example, a typical dust emission could have 95% of its mass distributed over a 100-to-1 particle diameter range. If the mass median diameter were 5 microns (i.e., 50% of the

Originally published August 29, 1977

particle mass were smaller than 5 microns), about 7% of the mass would be smaller than 1 micron diameter, and 1% smaller than 0.4 micron diameter. Thus, if one needed 99% collection efficiency for the dust, it would be necessary to collect all particles larger than about 0.4 micron diameter.

Particle concentration and size distribution can be measured in several ways, as described in the literature on particle technology [2,3,4]. The U.S. Environmental Protection Agency (EPA) has issued reports presenting recommended methods for particle sampling and size analysis [5,6]. The best advice to follow is to be sure that particle sampling and size analysis are done by someone who has had experience and is technically competent to understand the principles involved.

The scrubber system

The overall scrubber system includes a number of components essential to its operation. These are the hooding and ductwork conveying the gas to the scrubber, the scrubber supporting structure and foundations, the entrainment separator, ductwork to the fan (assuming an I.D. fan is used), the fan itself, and the smokestack and its breeching. Fig. 1 offers a schematic diagram of items to be considered in a scrubber system.

A hot gas must be cooled (or quenched) before it contacts scrubber or duct components that cannot withstand high temperature. Sprays are generally used for this purpose, and it is often advisable to use clean water at high pressure (>100 psi) in order to get finely atomized drops and rapid heat and mass transfer. Duct walls can become coated with deposited solids in the quencher and connecting ducting, so provisions should be made for keeping the walls washed.

The world's greatest scrubber

Scrubber manufacturers offer a bewildering array of products. The devices are available in a wide range of designs, sizes, advertised performance capabilities, and capital and operating costs. Choosing the right scrubber for a particular job requires an understanding of the alternatives that various units may present.

For the past several years, the author has been involved in an EPA-funded program for making detailed performance tests on "novel devices" for fine-particle scrubbing. The program has reached the point where there are not many more things to test, and in retrospect, very few new principles have been discovered.

The lesson from years of particulate scrubber study, analysis and development has been that there are relatively few fundamental mechanisms for particle collection. After one has studied a scrubber in detail, he generally finds that one or more of these same basic mechanisms are responsible for the scrubber perform-

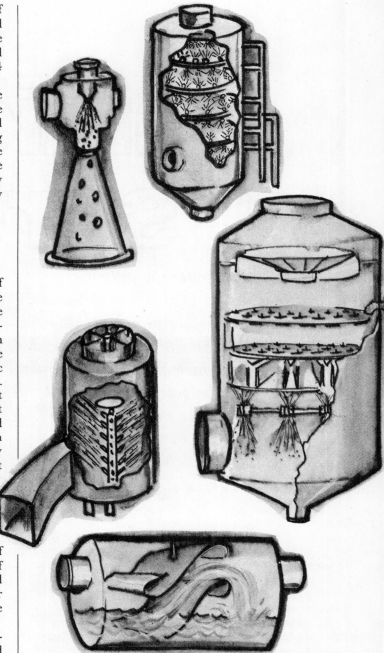

ance. This is analogous to the menu in a good Chinese restaurant. There may be a bewildering array of dozens of dishes offered but these are made from a much smaller number of basic ingredients.

All of this is written to convince you that it is worthwhile to spend a little time in understanding how small particles can be separated from a gas stream. This will prove invaluable to you if you are trying to get the best out of an existing scrubber system, or if you are trying

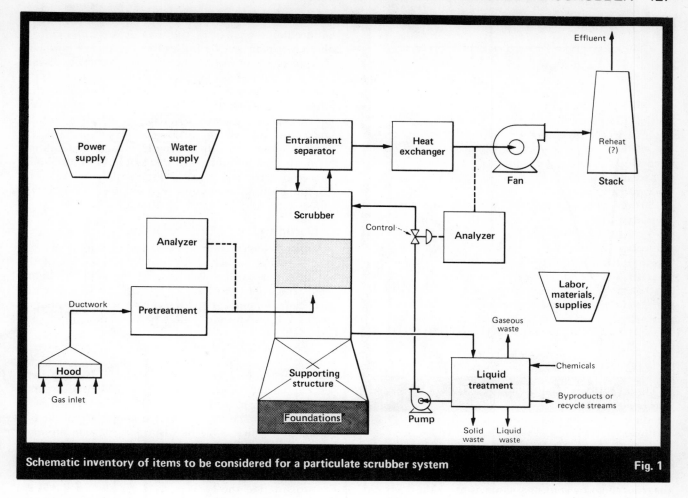

Schematic inventory of items to be considered for a particulate scrubber system   **Fig. 1**

to select a new system from among the dozens offered by vendors.

Collection mechanisms

Basic mechanisms for particle collection from gas streams include:

1. *Gravitational sedimentation.* This mechanism is usually of little consequence for any particles small enough to require consideration of a scrubber.

2. *Centrifugal deposition.* Particles may be "spun out" of a gas stream by centrifugal force induced by a change in gas flow direction. Large-scale changes in flow direction, as would be encountered in a cyclone separator, are not very effective on particles smaller than about 5.0 microns dia.

3. *Inertial impaction and interception.* When a gas stream flows around a small object, the inertia of the particles causes them to continue to move toward the object, and some of them will be collected. (This is really the same thing as centrifugal deposition, and the distinction is made here on the basis of common usage.) Inertial impaction customarily describes the effects of small-scale changes in flow direction.

Because inertial impaction is effective on particles as small as a few tenths micron dia., it is the most important collection mechanism for the usual particle scrubber. Since this mechanism hinges on the inertia of the particles, both their size and density are important in determining the ease with which they may be collected. All important particle properties may be lumped into one parameter, the *aerodynamic diameter*, defined as:

$$d_{pa} = d_p \, (\rho_p C')^{1/2} \qquad (1)$$

where d_{pa} = particle aerodynamic diameter, μmA; d_p = particle physical diameter, μm; and C' = Cunningham's correction factor, dimensionless. By a fortunate

Nomenclature

c	Mass concentration, g/cm^3
C'	Cunningham correction factor, dimensionless
d_p	Particle physical diameter, μm
d_{pa}	Aerodynamic particle diameter, μmA = $d_p \, (C'\rho_p)^{1/2}$, μm (g/cm^3)$^{1/2}$
d_{pg}	Geometric-mean particle diameter, μmA
d_{PC}	Performance cut diameter, μmA
d_{RC}	Required cut diameter, μmA
Pt	Penetration, fraction or percent
$\overline{Pt}$	Overall penetration, fraction or percent
T	Temperature, °F
ρ_p	Particle density, g/cm^3

Subscripts

i	Inlet
o	Outlet

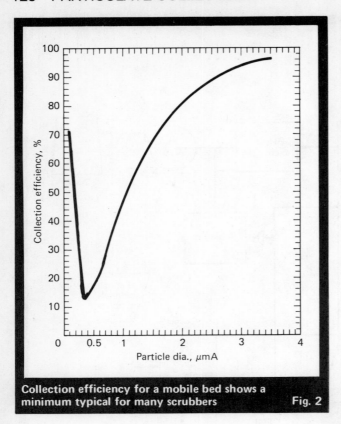

Collection efficiency for a mobile bed shows a minimum typical for many scrubbers   Fig. 2

7. Electrostatic precipitation. If an electrostatic charge is induced on the particles, they can be precipitated from the gas stream by the influence of a charge gradient. This mechanism can be effective on all particle diameters and can provide high collection efficiency.

8. Particle growth. While it is not a collection mechanism in itself, the enlarging of particle mass by such means as having water condense in a film around it makes the particles more susceptible to collection by inertial impaction. This phenomenon, in combination with diffusiophoresis and thermophoresis, can take place in scrubbers where condensation occurs. The combination of mechanisms is referred to as "flux force/condensation" (F/C) scrubbing.

Cut diameter

A very convenient parameter for describing the capability of a particle scrubber is the diameter of the particle that it will collect at 50% efficiency. This diameter is referred to as the *cut diameter*, generally given in aerodynamic units. Thus, a scrubber with a cut diameter of 1.0 μmA would collect particles of 1 μm size at 50% efficiency.

The reason cut diameter is so useful a parameter is that a curve of collection efficiency vs. particle diameter for collection by inertial impaction is fairly steep. Several important types of scrubbers have performance characteristics such that a particle whose aerodynamic diameter is half the cut diameter would be collected at about 10% efficiency, whereas a particle with an aerodynamic diameter twice the cut diameter would be collected at about 90% efficiency.

Because the cut is fairly sharp, one can use as a rough approximation the concept that the scrubber collects everything larger than the cut diameter and passes everything smaller.

Types of scrubbers

This section will organize scrubbers currently available into generic groups, according to the basic particle-collection mechanism, and then summarize the characteristics of each group. The organizational system was first devised and used in the "Scrubber Handbook" [1]. Tradenames and specific designs will not be discussed. To identify the group in which a specific scrubber belongs, you will have to look at a drawing of its internals and read the manufacturer's description of the way it operates. Then you can decide which group(s) utilize(s) the same mechanism(s) of particle collection.

Some scrubbers are designed primarily for the collection of particles, and others for mass transfer. Since good liquid-gas contact is needed for both operations, all scrubbers can collect both particles and gases to some extent. The degree to which particle-collection and mass-transfer characteristics of a scrubber can be exploited, as well as the associated cost, will determine how suitable the scrubber is for a job with specific purification requirements.

The approach here will be to group scrubbers into a number of categories: plate, massive packing, fibrous packing, preformed spray, gas-atomized spray, centrif-

coincidence, most methods for measuring particle size (such as the cascade impactor) determine the aerodynamic diameter. Since this is the most important parameter where inertial impaction is at work, one need not know the actual physical diameter or particle density (ρ_p).

4. Brownian diffusion. When particles are small enough (i.e., less than about 0.1 micron dia.), they are buffeted around quite a bit by gas molecules, and they begin to act like gas molecules. That is, they diffuse randomly through the gas because of their Brownian motion. In general, inertial impaction and Brownian diffusion are the two principal mechanisms operating in particulate scrubbers. As a consequence, there is generally a minimum point when collection efficiency is plotted against particle diameter (Fig. 2). Above about 0.3 micron dia., inertial impaction becomes important and efficiency rises with particle diameter. Below 0.3 micron dia., diffusion begins to prevail and efficiency rises as particle diameter falls below that size.

5. Thermophoresis. If there is heat transfer from the gas to the liquid, there will be a corresponding temperature gradient, and fine particles will be driven toward the cold surface by differential molecular bombardment arising from the gradient. This effect will rarely be of much significance in a scrubber.

6. Diffusiophoresis. Mass transfer within the scrubber —as might be caused by condensation of water vapor from the gas onto a cold liquid surface—will exert a force upon particles that causes them to deposit on the surface. Diffusiophoretic deposition can be significant; the fraction of particles removed will roughly equal the fraction of the gas stream condensed out.

Photo: Koch Engineering Co.

Venturi cuts dust from iron-ore sintering   **Fig. 3**

ugal, baffle, impingment-and-entrainment, mechanically aided, moving-bed, and combination.

Plate scrubbers

A plate scrubber consists of a vertical tower with one or more plates (trays) mounted transversely inside. Gas comes in at the bottom of the tower and must pass through perforations, valves, slots, or other openings in each plate before leaving through the top. Usually, liquid is introduced to the top plate, and flows successively across each plate as it moves downward to the liquid exit at the bottom. Gas passing through the openings in each plate mixes with the liquid flowing over it. Gas-liquid contacting causes the mass transfer or particle removal for which the scrubber was designed.

Fig. 4 shows several types of plates and a tower. Plate scrubbers are generally named for the type of plates they contain; for example, a tower containing sieve plates is called a sieve-plate tower.

In some designs, impingement baffles are placed a short distance above each perforation on a sieve plate, forming an impingement plate (Fig. 5). The impingement baffles are below the level of liquid on the perforated plates, and for this reason are continuously washed clean of collected particles. The chief mechanism of particle collection is inertial impaction from gas jets impinging on the liquid or on solid members. Particle collection may be aided by atomization of liquid flowing past openings in the irrigated perforated plate.

Collection efficiency increases as the perforation diameter decreases and can enable a cut diameter of about 1.0 μmA for 1/8-in.-dia. holes in a sieve plate.

Engineers are accustomed to the notion that plate columns become more efficient for mass transfer as the number of plates increases. This generally does *not* hold for particle collection whenever a range of particle diameters are present. *A plate does not have the same efficiency for all particle sizes, but rather shows a sharp efficiency change around the cut diameter. Once particles larger than this size are removed from the gas, additional plates can do little good.* This kind of behavior is characteristic of most types of scrubbers and should be kept in mind whenever one is tempted to try two scrubbers in series.

Plate-scrubber capacity, entrainment, pressure drop, and stability properties are well known to chemical engineers through experience with gas absorption and distillation. Care must be taken in selecting plates for systems that have a tendency toward scaling—adherence of solids to the plates, resulting in plugging of the perforations.

Massive packing

Packed-bed or tower scrubbers are familiar as gas absorbers or fractionators and can also be used as particle scrubbers. They may be packed with a range of manufactured elements, such as various ring- and saddle-shaped packings, or with commonly available materials like crushed rock. The gas-liquid contacting may be cocurrent, countercurrent (Fig. 6), or crossflow. Mist collection in packed beds with subsequent drainage can be accomplished without additional liquid flow.

Collection in packing works mainly by centrifugal deposition due to curved gas-flow through the pore spaces, and by inertial impaction due to gas-jet impingement within the bed. The good mass-transfer characteristics of packings can also make for efficient collection of particles by diffusion if the particles are small enough.

Collection efficiency for particles in the inertial size range (larger than 0.3 μmA) rises as packing size falls. A cut diameter around 1.5 μmA can be reached using columns packed with 1-in. Berl saddles or Raschig rings. Smaller packing gives higher efficiency: a $\frac{1}{2}$ in. packing can achieve 0.7-μmA cut diameter at 30-ft/s gas velocity. Packing shape does not appear to be very important so far as collection efficiency is concerned.

Packings are subject to plugging, but can be removed for cleaning. Temperature limitations are of special importance when plastics are used. Likewise, corrosion can have a severe effect on metallic packings.

Fibrous packing

Beds of fiber can be employed in various configurations for the collection of particles (Fig. 7). The fibers are made from materials such as plastic, spun glass, fiberglass and steel. Fibrous packing usually has a very large void fraction ranging around 97–99%. Fibers should be small in diameter for efficient operation, but strong enough to support collected particles or droplets without matting together. A liquid flow flushes away collected material from the fibers in cocurrent, counter-

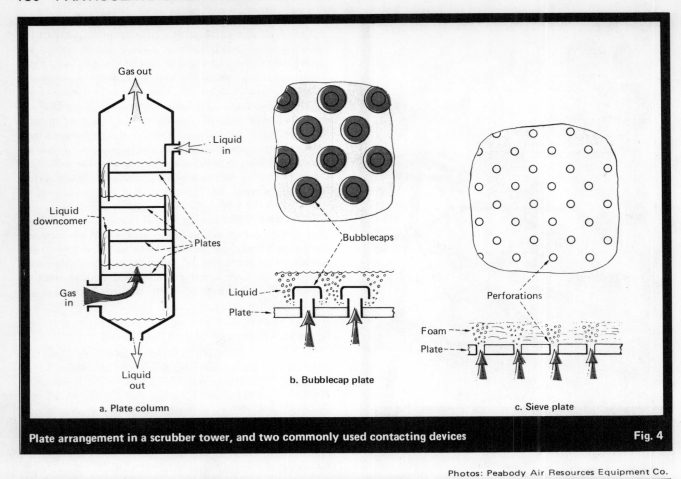

Gas out

Liquid in

Liquid downcomer

Plates

Gas in

Liquid out

Bubblecaps

Liquid

Plate

b. Bubblecap plate

Perforations

Foam

Plate

c. Sieve plate

a. Plate column

Plate arrangement in a scrubber tower, and two commonly used contacting devices

Fig. 4

Photos: Peabody Air Resources Equipment Co.

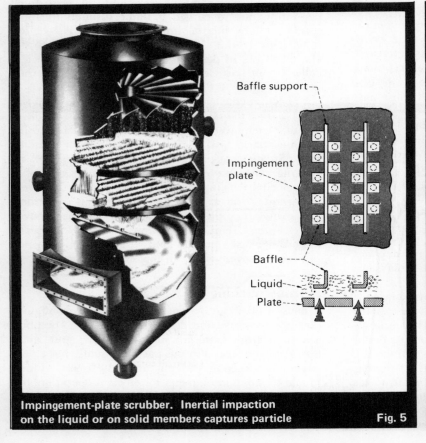

Baffle support

Impingement plate

Baffle

Liquid

Plate

Impingement-plate scrubber. Inertial impaction on the liquid or on solid members captures particle

Fig. 5

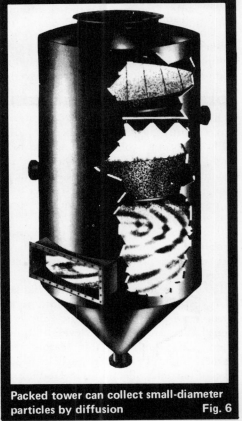

Packed tower can collect small-diameter particles by diffusion

Fig. 6

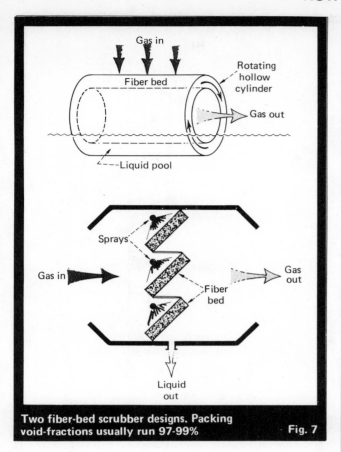

Two fiber-bed scrubber designs. Packing void-fractions usually run 97-99%   **Fig. 7**

particles, and the efficiency of this mechanism will improve as gas velocity diminishes through a given scrubber. Cut diameters can run as low as 1.0 or 2.0 μmA for knotted wire mesh with 0.011-in.-dia. wire, and to around 0.5 μmA for very fine wires and/or higher gas velocities.

Fibrous beds are very susceptible to plugging and can be impractical where scaling persists and conditions favor deposition of suspended solids. Obviously, they will also be especially sensitive to chemical, mechanical and thermal attack. Fibrous-bed scrubbers are widely used—so the previous comments should be viewed as words of caution, not rejection.

Preformed spray

A preformed-spray scrubber collects particles or gases on liquid droplets that have been atomized by spray nozzles. The properties of the droplets are determined by the configuration of the nozzle, the liquid to be atomized and the pressure to the nozzle. Sprays leaving the nozzle are directed into a chamber that has been shaped so as to conduct the gas through the atomized droplets. Horizontal and vertical gas flowpaths have been used, as well as spray-entry flowing cocurrent, countercurrent or crossflow to the gas (Fig. 8). If the tower is vertical, the relative velocity between the droplets and the gas is ultimately the terminal settling velocity of the droplets.

Ejector venturis are preformed spray devices in which a high-pressure spray is used both to collect particles and move the gas. High relative velocity between the droplets and the gas aids particle separation. Preformed sprays have also been installed in venturi scrubbers that use a fan to provide high gas-phase pressure drop.

Particle collection in these units results from inertial impaction on the droplets. Efficiency is a complex function of droplet size, gas velocity, liquid-gas ratio, and droplet trajectories. There is often an optimum droplet

current or crossflow arrangements similar to those for massive packings.

Collection is by inertial impaction accompanying the gas flow around the fibers. Efficiency rises as fiber diameter decreases and as the gas velocity increases. Diffusional collection can be important for very small

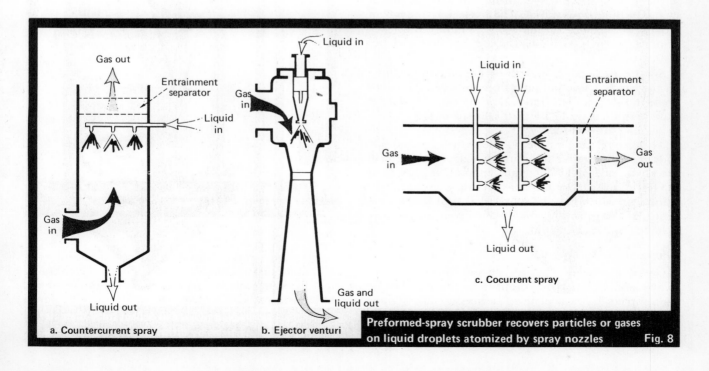

Preformed-spray scrubber recovers particles or gases on liquid droplets atomized by spray nozzles   Fig. 8

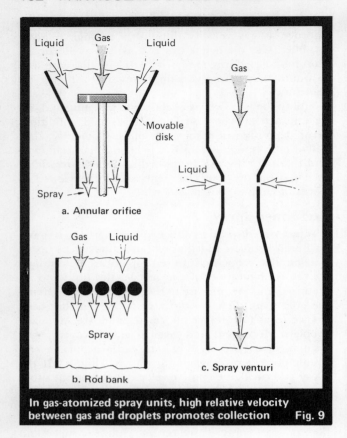

Liquid Gas Liquid

Movable disk

Spray

a. Annular orifice

Gas Liquid

Spray

b. Rod bank

Gas

Liquid

c. Spray venturi

In gas-atomized spray units, high relative velocity between gas and droplets promotes collection Fig. 9

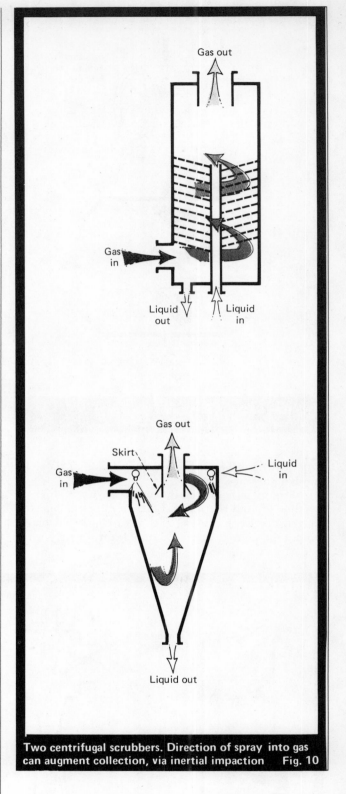

Gas out

Gas in

Liquid out Liquid in

Gas out

Skirt

Gas in Liquid in

Liquid out

Two centrifugal scrubbers. Direction of spray into gas can augment collection, via inertial impaction Fig. 10

diameter that varies with fluid-flow parameters. For droplets falling at their terminal settling velocity, the optimum droplet diameter for fine-particle collection is around 100 to 500 μm; for droplets moving at high velocity within a few feet of the spray nozzle, the optimum is smaller.

Spray scrubbers that take advantage of gravitational settling can achieve cut diameters around 2.0 μm at moderate liquid-gas ratios. High-velocity sprays can reduce cut diameters down to about 0.7 μmA. Efficiency improves with higher spray-nozzle pressures and liquid-gas ratios.

Spray scrubbers are practically immune to plugging on the gas-flow side but are subject to severe problems on the liquid side. The liquid-gas ratio required is high, usually running 30 to 100 gal/1,000 ft^3 of gas treated, depending on efficiency.

The recirculating scrubber liquor can erode and corrode nozzles, pumps and piping. Nozzles can plug with pieces of scale or agglomerates of particles. By their nature, sprays generate a heavy loading of liquid entrainment, which must be collected. Gas-phase pressure drop (ΔP) is generally low or may even be positive, enhancing the flow of gas.

Gas-atomized spray

Gas-atomized spray devices use a moving gas stream to first atomize liquid into droplets, and then accelerate the droplets. Typical of these devices are the venturi scrubber (Fig. 3) and the various orifice-type scrubbers. High gas velocities of 60–120 m/s (200–400 ft/s) raise the relative velocity between the gas and the liquid

drops, and promote particle collection. Many gas-atomized spray scrubbers incorporate the converging and diverging sections typical of the venturi scrubber, but this modification does not appear to yield much benefit. Various geometries have been used successfully, as illustrated in Fig. 9.

Liquid may be introduced in various places and in

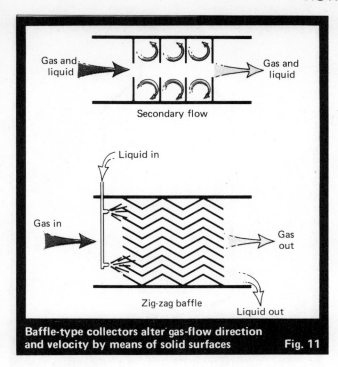

Baffle-type collectors alter gas-flow direction and velocity by means of solid surfaces **Fig. 11**

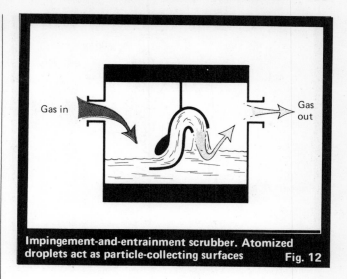

Impingement-and-entrainment scrubber. Atomized droplets act as particle-collecting surfaces **Fig. 12**

different ways without having much effect on collection efficiency. Usually, it is introduced at the entrance to the throat through several straight-pipe nozzles directed radially inward. Other gas-atomized-spray designs distribute a liquid film over the scrubber walls upstream from the throat.

Particle collection results from inertial impaction due to gas flowing around droplets. Velocity is so high (and droplet residence time so short) that diffusional collection and deposition by other forces, such as electrostatic, are not very effective. Efficiency increases with throat velocity and with liquid-gas ratio. Because there must be enough liquid to effectively sweep the gas stream, it is good practice to use a high liquid-gas ratio rather than a high gas velocity to get a given cut diameter. At least 8 gal/1,000 ft^3 should be specified. Cut diameters down to about 0.2 μmA have been achieved with venturi scrubbers.

Gas-atomized scrubbers have about the simplest and smallest configurations of all scrubbers. While fairly difficult to plug up, they are susceptible to erosion because of their high throat velocity. They can be built with adjustable throat openings to permit variation of pressure drop and collection efficiency. Liquid-gas ratios ranging from 5 to 20 gal/1,000 ft^3 have been used. All of this liquid is entrained and must be removed from the gas. In general, the entrainment separator is much larger than the gas-atomized scrubber.

Centrifugal scrubbers

Centrifugal scrubbers, usually cylindrical in shape, impart a spinning motion to the gas passing through them. The spin may come from tangential introduction of gases into the scrubber, or from direction of the gas stream against stationary swirl vanes. In a dry centrifugal collector (cyclone), the walls can be wetted down to hinder reentrainment of particles that collect there, and

to wash off deposits. Often, sprays are directed through the rotating gas stream to catch particles by impaction on spray droplets. Sprays can be directed outward from a central manifold, or inward from the collector wall, as shown in Fig. 10. Spray connections directed inward from the wall are more easily serviced, since they can be made accessible from the outside of the scrubber.

Particle collection operates by centrifugal deposition caused by the rotating gas stream. In the absence of spray, the efficiency will be the same as for a dry collector. If much spray is present, the scrubber performance compares more with that of a spray scrubber. In the latter case, the collection efficiency of the droplets overshadows that arising from centrifugal deposition, and the major effect of the swirling flow is to separate entrained droplets.

A particle cut-diameter of 4.0 or 5.0 μmA can be obtained with a centrifugal scrubber in the absence of spray. As more spray is introduced or generated inside, the unit's performance nears that of a preformed-spray scrubber.

Centrifugal scrubbers are fairly simple in form and have no small passages. Thus, they are not very susceptible to plugging, although solids can deposit on sections of the wall that are not adequately washed. If designed properly, centrifugals have the advantage of a built-in potential for entrainment separation. Tangential gas velocity should not exceed about 100 ft/s, and the internal configuration must prevent flow of the liquid along the wall into the gas exit.

Baffle and secondary-flow scrubbers

Baffle-type collection devices force changes in gas flow-direction and velocity by means of solid surfaces. Either the major direction of flow may be altered, or secondary flow patterns may be set up, as shown in Fig. 11. Louvers, zig-zag baffles, and disk-and-donut baffles are examples of target surfaces that produce changes in main flow direction. If the material to be collected is liquid, it runs down the target into a collection sump. If solid, collected particles may be washed intermittently from the target plates.

Particle collection follows centrifugal deposition

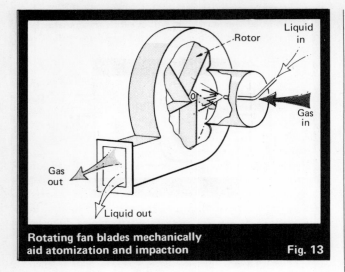

Rotating fan blades mechanically aid atomization and impaction　　　　　Fig. 13

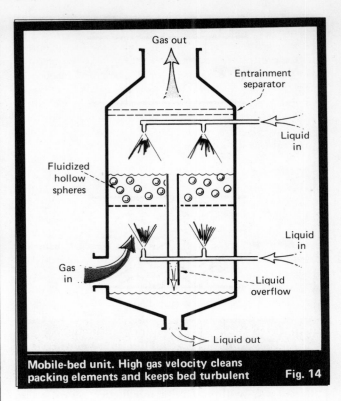

Mobile-bed unit. High gas velocity cleans packing elements and keeps bed turbulent　　　Fig. 14

caused by change in the main flow direction, or by rotating secondary flows. The performance potential of baffles causing only flow-direction change depends on the radius of curvature of the flow pattern and on the expanse of baffle surface. Thus, the cut diameter can go as low as 5.0 or 10.0 μmA for continuous and discontinuous zig-zags and similar arrangements.

The fine-particle collection capability of baffle scrubbers is so low that they are not suitable for this application. They are useful as precleaners and as entrainment separators. Heavy particle- or slurry-loadings can cause solids deposition, which can lead to plugging and corrosion.

Impingement-and-entrainment scrubbers

Impingement-and-entrainment (self-induced-spray) scrubbers feature a shell that retains liquid, so that gas introduced to the scrubber impinges on and skims over the liquid surface to reach a gas exit-duct (Fig. 12). This contact atomizes some of the liquid into droplets that are entrained by the gas and act as particle-collecting and mass-transfer surfaces. The gas exit-duct is usually designed so as to change the direction of the gas-liquid mixture flowing through it, reducing droplet entrainment.

Particle collection is attributed to inertial impaction caused by the gas jet's impinging on the liquid, and by the gas flowing around the atomized drops. Drop size and the liquid-gas flow ratio inside the scrubber depend upon scrubber geometry and gas flowrate, but are not controllable or measurable.

Generally, the performance of an impingement-and-entrainment scrubber seems to be comparable to a gas-atomized scrubber operating at the same gas-phase pressure drop. Cut diameter ranges from several microns for low-velocity impingement to around 0.5 μmA for high-velocity impingement.

Liquid flow in this type scrubber is boosted by the gas, so that liquid pumping requirements are mostly for makeup and purge streams. Solids deposition can pose a problem on the bottom and on portions of the wall that are not well washed. Good entrainment separation is required because of the amount of spray generated.

Mechanically aided scrubbers

Mechanically aided scrubbers incorporate a motor-driven device between the inlet and the outlet of the scrubber body. Often, the motor-driven devices are fan blades, used to move air through the scrubber. Particles are collected by impaction upon the fan blades as the gas moves through the device. Usually, liquid is introduced at the hub of the rotating fan blades. Some liquid atomizes upon impact with the fan, and some runs over the blades, washing them of collected particles; the latter portion atomizes as it leaves the fan wheel. The liquid is recaptured by the fan housing, which drains into a sump.

Fig. 13 shows one configuration that has been used to mechanically aid fluid flow. Liquid from mechanically aided scrubbers can usually be recirculated.

Disintegrator scrubbers draw on a submerged, motor-driven impeller to atomize liquid into small drops. The drops spin off the impeller across the gas stream, collecting particles on the way. Mechanically aided scrubbers are used almost exclusively for particle collection. Their mass-transfer capabilities are generally low, due to the relatively low amount of liquid available for contacting.

Particle collection mechanisms, in probable order of importance, are: inertial impaction on the atomized liquid, inertial impaction on the rotor elements, and centrifugal deposition on the housing. Cut diameters (aerodynamic) down to about 2.0 μmA have been achieved with devices having fine sprays and low-rpm fans. As low as 1.0 μmA can be reached with a disintegrator-type scrubber.

Mechanically aided scrubbers can provide the necessary power in a way that may or may not be superior to using a fan and/or liquid pumps in other types of scrubbers.

There seems to be no power advantage for mechanically aided units over other types. Disintegrators require more power than a gas-atomized scrubber with comparable efficiency. Further, high-speed impaction of liquid and slurry on the scrubber parts promotes severe abrasion and corrosion conditions. Rotating parts are also subject to vibration-induced fatigue caused by solids deposition, or wear leading to unbalancing.

Moving-bed scrubbers

Moving-bed scrubbers provide a zone of mobile packing—usually plastic or glass spheres—where gas and liquid can mix intimately. The vessel shell holds a support grid on which the movable packing is placed. Gas passes upward through the packing, while liquid is sprayed up from the bottom and/or down over the top of the moving bed, as shown in Fig. 14. Gas velocities are sufficient to move the packing material around when the scrubber operates. This movement aids in making the bed turbulent and keeps the packing elements clean. When hollow or low-density spheres are used, the bed fluidizes and bed depth extends to about double that of the quiescent bed.

Particle collection stems from inertial impaction on atomized liquid and on the packing elements. Cut diameters down to about 1.0 μmA are attained in the fluidized "ping-pong-ball" type of bed having three stages in the scrubber column. Performance of the less violently agitated "marble"-type bed resembles that of massive-packing beds unless the gas velocity rises so high as to cause significant liquid atomization and entrainment from the bed.

Moving-bed scrubbers prove beneficial where good mass-transfer characteristics are needed, as well as particle collection. The agitation cleans the packing and reduces problems with solids deposition. Ball wear can be severe, and the scrubber's hydrodynamic stability is limited by fluidization ranges and surging difficulties.

Combination scrubber designs

Numerous combinations of the scrubber types already discussed have been contrived and used. No attempt will be made here to describe these essentially geometric variants. Hybrids that combine particle deposition forces have also been devised, some of which are noteworthy.

Condensation of water vapor caused by steam addition to a saturated gas, or by scrubbing of a saturated gas with cold liquid, gives rise to F/C effects. Particles may grow to several microns diameter with steam addition, improving overall scrubbing efficiency. However, the cost of purchased steam will generally be prohibitive unless special conditions prevail.

Scrubbing a hot, humid gas with relatively cold liquid can be economically attractive in cases where the gas is hot enough. Such a scrubber operating at moderate pressure drop can provide the same efficiency as a high-energy venturi scrubber operating at a pressure drop over 100 in. H$_2$O, depending on how much condensation occurs. The scrubber design and economics have to be worked out for each case.

Electrostatically augmented scrubbers can be very efficient, depending on design and operating parameters. The variations available include: wet electrostatic precipitators, charged-dust/grounded-liquid scrubbers, charged-drop scrubbers and charged-dust/charged-liquid types. Performance forecasting methods are still in the elementary stages for any departures from traditional electrostatic precipitator geometry; hence, pilot tests are generally stipulated. Corrosion and voltage isolation problems can be severe with this class of scrubbers.

Survey of scrubber applications in a variety of installations										Table I
Process	Scrubber type									
	Plate[1]	Massive packing	Fiber bed	Preformed spray	Gas-atomized spray	Centrifugal	Baffle	Impingement	Mech. aided	Moving bed
Calcining	6 (1)[2]	2 (1)	— (0)	13 (5)	21 (23)	— (0)	— (0)	43 (3)	— (0)	— (0)
Combustion	17 (3)	— (0)	— (0)	5 (2)	2 (2)	2 (1)	— (0)	29 (2)	— (0)	9 (2)
Crushing	6 (1)	— (0)	— (0)	— (0)	— (0)	26 (11)	— (0)	14 (1)	— (0)	5 (1)
Drying	39 (7)	— (0)	— (0)	10 (4)	18 (19)	70 (30)	100 (1)	— (0)	25 (1)	64 (14)
Gas removal	17 (3)	72 (33)	40 (2)	45 (18)	9 (10)	2 (1)	— (0)	14 (1)	50 (2)	5 (1)
Liquid-mist recovery	0 (0)	24 (11)	60 (3)	7 (3)	— (0)	— (0)	— (0)	— (0)	— (0)	— (0)
Smelting	17 (3)	2 (1)	— (0)	20 (8)	50 (54)	— (0)	— (0)	— (0)	25 (1)	18 (4)

Notes:
1. Read vertically. Example: 39% of all plate-type scrubbers are used to control discharges from drying processes.
2. Numbers in parentheses refer to number of operators reporting information to the survey.

Scrubber applications

Patterns of scrubber usage for control of various air pollution sources were determined in a survey carried out during 1970 and 1971 as part of work reported in the "Scrubber Handbook" [1]. The number of responses was relatively small, so the survey results should be viewed as showing where scrubbers have been used, but not necessarily where they have not.

The processes to which each type of scrubber was applied are listed in Table I. In the following summary, there are notes [in brackets] on applications uncovered since the survey.

Plate-type scrubbers are most frequently applied to dry-ing processes, although they are also chosen for combustion and smelting and for pollutant-gas removal (Fig. 15).

Packed-bed scrubbers are prescribed almost exclusively for treating pollutant gases, and for collecting liquid mists. [They are also used on soluble particles and nonadhering, insoluble particles.]

Fiber-bed scrubbers, like packed towers, are chosen to capture pollutant gases and liquid mists. [They are also specified for soluble particles and occasionally for nonadhering insoluble particles.]

Preformed sprays are used mainly to capture gases. They are also employed for smelting units ("wet caps") and occasionally for calcining, combustion and drying processes, as well as for collection of liquid mists. [They are used for removal of SO_2 and particles in large power plants, and for general particle collection.]

Gas-atomized spray scrubbers appear most frequently in smelting operations. They are also used in calcining and drying processes, and occasionally for pollutant-gas collection and control of combustion processes. [They are used in large power plants for removing SO_2 and particles.]

Centrifugal scrubbers are most often coupled to dryers. Some are applied to crushers. Centrifugal scrubbers are seldom selected to control emissions from units other than these. [Internally baffled centrifugal units are used in metallurgical applications, and for moderate-efficiency collection of particles in rock processing.]

Baffle scrubbers are less used. The only application uncovered in the user-survey was for a drying process.

Impingement-and-entrainment scrubbers are given the nod for calcining, combustion and crushing processes, and for pollutant gas collection. [Ejector venturis find use in petroleum refining for recovery of entrained catalyst dust.]

Mechanically aided scrubbers appeared in a few applications in the user-survey. Services noted include drying and smelting processes, and pollutant-gas removal.

Moving-bed scrubbers crop up most frequently in drying processes. They also aid cleanup of emissions from smelting and combustion processes, and control of crusher dusts and polluting gases. [They are used for large power-plant scrubbing of SO_2 and particles.]

Describing particle size

Particle size would be easy to describe if all the particles in a gas stream had the same diameter, but this almost never happens. Unless very special conditions exist, there will be a distribution of particle size over a broad range. Since most industrial particulates follow a log-normal size distribution (or close to it), this type of curve offers a concise way of characterizing them.

In a log-normal distribution, the ratio of particle size to the mean size is "normally" distributed. This means that the probability of finding one diameter is the same as that of finding one twice the mean diameter. The distribution can be plotted on log-probability graph paper, which has a logarithmic scale on one axis and a probability scale on the other.

Fig. 16 shows a log-probability plot for a particle-size distribution having a mass median diameter of 10.0 μmA and a geometric standard deviation (σ_g) of 3.0.

Photo: Koch Engineering Co.

Self-contained plate-scrubber unit   **Fig. 15**

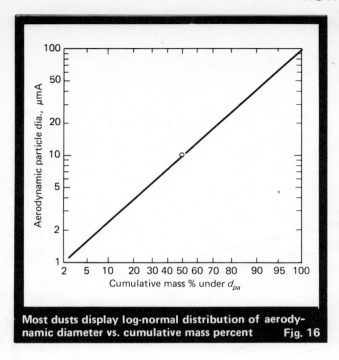

Most dusts display log-normal distribution of aerodynamic diameter vs. cumulative mass percent   Fig. 16

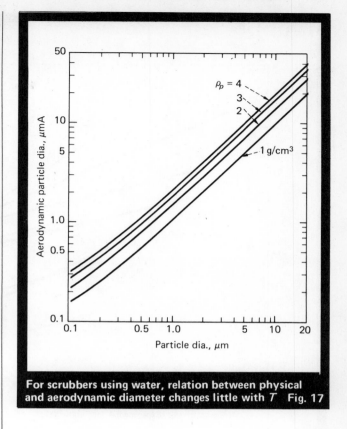

For scrubbers using water, relation between physical and aerodynamic diameter changes little with T Fig. 17

The aerodynamic diameter is plotted against the cumulative mass percent of particles smaller than that size. Fifty percent of the particle mass is smaller than (or larger than) the mass median diameter.

The geometric standard deviation of this dust is the ratio of the particle diameter at about 84.1% undersize to the mass-median diameter. If the size distribution is log-normal, the data points will follow a straight line on log-probability paper, and we can define the distribution in terms of d_{pg} and σ_q alone.

The size shown in Fig. 16 is aerodynamic, which has

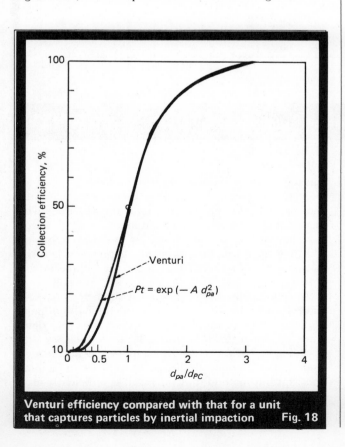

Venturi efficiency compared with that for a unit that captures particles by inertial impaction   Fig. 18

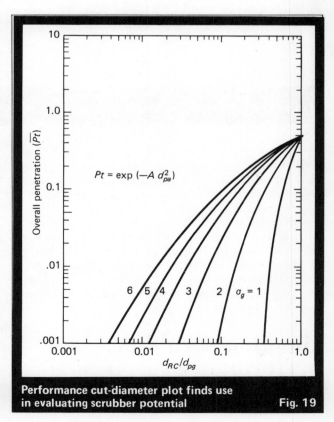

Performance cut-diameter plot finds use in evaluating scrubber potential   Fig. 19

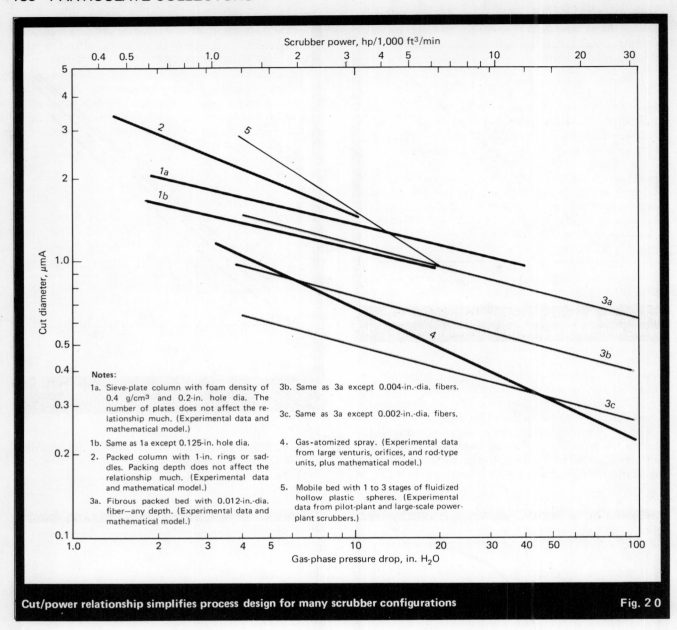

Notes:

1a. Sieve-plate column with foam density of 0.4 g/cm³ and 0.2-in. hole dia. The number of plates does not affect the relationship much. (Experimental data and mathematical model.)

1b. Same as 1a except 0.125-in. hole dia.

2. Packed column with 1-in. rings or saddles. Packing depth does not affect the relationship much. (Experimental data and mathematical model.)

3a. Fibrous packed bed with 0.012-in.-dia. fiber—any depth. (Experimental data and mathematical model.)

3b. Same as 3a except 0.004-in.-dia. fibers.

3c. Same as 3a except 0.002-in.-dia. fibers.

4. Gas-atomized spray. (Experimental data from large venturis, orifices, and rod-type units, plus mathematical model.)

5. Mobile bed with 1 to 3 stages of fluidized hollow plastic spheres. (Experimental data from pilot-plant and large-scale power-plant scrubbers.)

Cut/power relationship simplifies process design for many scrubber configurations　　Fig. 20

been defined previously in Eq. (1). In order to relate physical properties to aerodynamic diameter, it takes a considerable effort to compute the Cunningham correction factor and solve Eq. (1). To avoid this, a plot of aerodynamic diameter versus physical diameter, with particle density (ρ_p) as a parameter, can be substituted (Fig. 17). This plot is based on Eq. (1), with the Cunningham factor evaluated for room-temperature air. The relationship will not change much with temperature or gas composition within the operating range of scrubbers that use water.

Gauging scrubber performance

We have already noted that scrubber performance can be defined in terms of cut diameter (d_{PC}), which

has the virtue of being a simple parameter. It shows one side of the picture; it gives a measure of what the scrubber can do. The other side of the picture is what the scrubber needs to do. This is analogous to knowing how many transfer units can be provided by a gas absorber operating in a specified way, but not knowing how many are required to achieve the desired separation.

As it happens, the required cut diameter (d_{RC}) is a parameter that defines quite well the difficulty of particle separation. A simple, graphical method can determine d_{RC} for a given overall collection efficiency on a specified particle-size distribution. Some assumptions and approximations are necessary in order to facilitate this shortcut method, but the result is sufficiently accurate for preliminary selection of scrubbers.

Because particle collection efficiency changes with particle size for a given operating condition in a scrubber, we need a relationship between efficiency and particle size. Most scrubbers that collect particles by inertial impaction perform in accordance with the following relationship:

$$Pt = exp\left(-A\ d_{pa}{}^B\right) = \frac{c_o}{c_i} \qquad (2)$$

where Pt = particle penetration, fraction; A = empirical constant, dimensionless; B = empirical constant, dimensionless; c_i = inlet particle concentration, g/cm^3; and c_o = outlet particle concentration, g/cm^3.

Packed-bed and plate-type scrubber performance are described by a value of $B = 2.0$, whereas for centrifugal scrubbers of the cyclone type, $B = 0.7$. Gas-atomizing scrubber performance fits a value of $B = 2.0$ over a large portion of the usual operating range. Therefore, we can use a value of $B = 2.0$ as representative of most scrubbers operating in the inertial impaction regime. Fig. 18 plots collection efficiency against the ratio of aerodynamic particle diameter to performance cut-diameter, showing one line based on Eq. (2) and another for a venturi scrubber under typical operating conditions.

One can integrate Eq. (2) over the particle size distribution and, if the distribution is log-normal, arrive at the graphical relationship shown in Fig. 19. The overall penetration $\overline{(Pt)}$ for the entire size distribution is plotted against the ratio of required cut diameter to mass-median diameter, with geometric standard deviation as the parameter.

Fig. 19 can be used to determine what d_{RC} must be in order to get a specific $\overline{Pt}$ for a given size distribution. For example, suppose the size distribution has $d_{pg} = 10$ μmA and $\sigma_g = 3.0$, as in Fig. 16, and we need 99% collection efficiency. The penetration is 100% minus the percent collection efficiency, or 1%, which corresponds to $\overline{Pt} = 0.01$ in fractional units.

The diameter ratio corresponding to $\overline{Pt} = 0.01$ and $\sigma_g = 3.0$ is $d_{RC}/d_{pg} = 0.063$. Since $d_{pg} = 10$ μmA, $d_{RC} = 0.63$ μmA. This means that we will need a scrubber with a cut diameter of 0.63 μmA or less to achieve 99% collection of the particles in question.

Cut diameter vs. power input

For detailed design work, relationships between scrubber performance, cut diameter, scrubber geometry, and operating conditions can be worked out. A simpler relationship that describes the performance of many scrubbers on a single plot will be outlined here.

Mathematical models for scrubber performance and the cut-diameter approach developed in the "Scrubber Handbook" [1] led to the concept that performance cut diameter could be related to gas-phase pressure drop, or power input to the scrubber. Subsequent performance tests on a variety of scrubbers in industrial installations, combined with mathematical modeling, has led to the refinement of the cut/power relationship shown in Fig. 20. The curves give the cut diameter (μmA) as a function of either power input (hp/1,000 ft^3/min) or gas-phase pressure drop (in. H$_2$O) for a number of typical installations—sieve-plate column, packed column, fibrous packed bed, gas-atomized spray, and mobile fluidized bed.

The APT cut/power relationship has been devised and tested on the basis of all of the published data available to the author. It appears to be an accurate and reliable criterion for scrubber selection.

We can see from Fig. 20 that the only "unaided" scrubbers capable of giving a 0.6-μmA cut diameter are the gas-atomized and fibrous-packed-bed types. The separation would require a gas-phase pressure drop of about 13 in. H$_2$O for the gas-atomized scrubber. The fibrous packing would need 22 in. H$_2$O for 0.004-in. fiber diameter, and about 6 in. H$_2$O for 0.002-in.

It would take about 0.003-in. fiber diameter to achieve a $d_{PC} = 0.6$ μmA at slightly less pressure drop than for the gas-atomized scrubber. This is pretty fine fiber or wire, and serious questions would arise regarding its structural stability, and susceptibility to corrosion and plugging. The safe bet would be to choose the gas-atomized scrubber unless extensive pilot tests could be done with fine-fiber beds.

Other types of scrubbers could achieve the required performance if augmented by F/C effects or by electrostatic charging. Each system would have to be examined to determine whether it would be economically attractive.

Power and cost

The equivalent power axis plotted on the top of the cut/power plot is based on 50% efficiency for a fan and motor combination. The theoretical power requirement is approximately 0.16 hp/1,000 actual ft^3/min for each inch of water pressure-drop. Power costs can be based on twice the theoretical power, corresponding to 50% efficiency.

Equipment costs are best estimated from vendors' quotations. As usual, one must be sure that all prices for competing units are on the same basis. Materials, ducting, electrical work, foundations, supporting structure, etc., must be specified as included or not.

References

1. Calvert, S., Goldshmid, J., Leith, D., and Mehta, D., "Scrubber Handbook," Natl. Tech. Info. Service, Springfield, Va., #PB 213-016, 1972.
2. Calvert, S., *J. APCA*, Vol. 24, 1974, p. 929.
3. Harris, D. B., "Procedures for Cascade Impactor Calibration and Operation in Process Streams," EPA 600/2-77-004, 1977.
4. Mercer, T. T., "Aerosol Technology in Hazard Evaluation," Academic Press, New York, 1973.
5. Stern, A., "Air Pollution," Third Ed., Vol. III, Academic Press, New York, 1977.
6. Weir, A., Jones, D. G., Papay, L. P., Calvert, S., and Yung, S. C., *Environmental Sci. and Technol.* Vol. 10, 1976, p. 539.

The author

Dr. Seymour Calvert is President of Air Pollution Technology, Inc., 4901 Morena Blvd., Suite 402, San Diego, CA 92117, an air-pollution-control engineering firm that he founded in 1970. He earned his Ph.D. in chemical engineering at the University of Michigan. Formerly Air Pollution Control Officer for Brook Park, Ohio, and Chairman of the Cleveland Advisory Board on Air and Water Pollution, he has also served as Dean of Engineering at the University of California (Riverside) and Director of the Air Pollution Control Assn.

Practical process design of particulate scrubbers

Energy consumed by a scrubber reveals much about its performance—regardless of its geometry or internals. Here the author tells how to predict collection efficiency, using the "contacting power" technique.

Konrad T. Semrau, SRI International

☐ Only 30 years ago, it was widely held that scrubbers were ineffective for capturing fine particles of less than 2 to 3 micrometers. The assertion was not true even then, for units having such capabilities were known at least as far back as the 1890s, although most could not meet such standards.

Today, many scrubbers suitable for fine-particle collection are available. However, progress has resulted not from the introduction of new devices but from the use of a new design method that relates power dissipation to scrubbing efficiency. The method provides a systematic procedure for setting performance.

Lapple and Kamack [1] first demonstrated that, for a given particulate, there is a functional relationship between the collection efficiency of a scrubber and the energy expended in the gas-liquid contacting process. Their work was confirmed and extended in a series of three papers [2-4] by the present author and coworkers. It was concluded not only that efficiency is determined by power dissipation, but also that it is relatively independent of scrubber geometry, and of the way that the power is applied to the gas-liquid contacting. The validity of this conclusion is still not fully explored, but it is at least a good approximation.

With the importance of specific design configurations discounted, manufacturers and users of scrubbers have in recent years turned mostly to a few relatively simple designs. Venturi and orifice scrubbers having no essential design differences are now produced by most of the major manufacturers. One feature common to many of these units is a variable-area throat with provisions for automatically controlling the gas-pressure drop—and hence, the collection efficiency—at a constant level.

Grouping scrubbers by power input

The number and variety of scrubber designs that are in use or that have appeared in the past 100 years are so great as to defy any scheme of classification that is at once self-consistent and practical. However, power consumption is a salient characteristic of scrubbers. Hence,

Informal grouping of scrubbers according to mechanism of power input	Table I
Mechanism	**Scrubber[1]**
Gas-phase contacting power[2]	Plate (sieve, valve, bubblecap, etc.)
	Gas-atomized spray[3]
	Centrifugal
	Baffle
	Impingement-and-entrainment (self-induced spray)
	Moving bed[7]
Liquid-phase contacting power[4]	Preformed spray[5]
	Ejector venturi[6]
Mechanical contacting power	Mechanically aided devices (eg., disintegrator scrubber)[8]
Wet-film collection (no spray formation)	Massive and fibrous packing[9]

Notes:

1. See Aug. 29, 1977 issue, p. 54, for further description of the devices listed here.

2. This category claims the bulk of particulate scrubbers in use. Many designs get appreciable contributions of liquid-phase contacting power from spray nozzles.

3. Includes venturi and orifice scrubbers, and rod-bank devices.

4. All of these units have some gas-phase friction losses, and hence incorporate some gas-phase contacting power. Most are low or medium-energy. To obtain high-energy scrubbing, one must specify high liquid-gas ratios and/or high liquid-feed pressures.

5. Includes hydraulic (or pressure) spray nozzles, rotating nozzles, spinning-disk atomizers, as well as the simple spray chamber. Cyclonic spray devices (centrifugal) are also included if liquid-phase contacting power dominates.

6. Major subgroup of preformed spray scrubbers (also known as water-jet scrubbers). Two-fluid nozzles are sometimes used.

7. Other names: fluidized packing, mobile packing, floating or "ping-pong" type devices, and turbulent contacting absorber (TCA).

8. Includes all devices that use a motor-driven rotor to assist gas-liquid contacting.

9. Operated below floodpoint. Above flooding, spray formation contributes to contacting power mechanisms.

Originally published September 26, 1977

Formulas for predicting effective friction loss and contacting power   Table II

	Symbol	English units	Metric units	Dimensional formula* English units	Metric units
Effective friction loss	F_E	in. H_2O	cm H_2O	$\Delta p^\dagger$	$\Delta p^\dagger$
Gas-phase contacting power	P_G	hp/1,000 ft³/min**	kWh/1,000 m³**	$0.1575\,F_E$	$0.02724\,F_E$
Liquid-phase contacting power	P_L	hp/1,000 ft³/min	kWh/1,000 m³	$0.583\,p_f\,(L/G)^{\dagger\dagger}$	$0.02815\,p_f\,(L/G)^{\dagger\dagger}$
Mechanical contacting power	P_M	hp/1,000 ft³/min	kWh/1,000 m³	$1,000\,(W_s/G)^{\dagger\dagger}$	$16.67\,(W_s/G)^{\dagger\dagger}$
Total contacting power	P_T	hp/1,000 ft³/min	kWh/1,000 m³	$P_G + P_L + P_M$	$P_G + P_L + P_M$

*See nomenclature.
†Effective friction loss is usually approximately equal to the scrubber pressure drop, Δp.
**1.0 kWh/1,000 m³ = 2.278 hp/1,000 ft³/min.
††This quantity is actually power input and represents an estimate of contacting power.

it is convenient to characterize (if not to classify) scrubbers by the way, or ways, that power is introduced.

That scrubbers of widely different designs have about the same relationship of collection efficiency to contacting power implies that the underlying mechanisms are essentially the same. "High energy" scrubbers used to collect fine particles are not fundamentally different from other scrubbers, but merely incorporate mechanical arrangements that aid power input.

Energy for gas-liquid contacting can be supplied in three ways (Table I):
■ From the energy of the gas stream.
■ From the energy of the liquid stream.
■ From a mechanically driven rotor.

There are a number of variations on these. Liquid stream energy may be supplied from the pressure head, as in hydraulic nozzles; from a second, compressible fluid (air or steam) as in two-fluid nozzles; or from a mechanically driven atomizer, such as a spinning disk. In at least one scrubber, the high-pressure water is also superheated, so that the water-jet emerging from the nozzle is given kinetic energy by the flashing of part of the liquid into steam.

In ejector units, the liquid supplies energy directly, and also supplies, via the draft, energy to the gas stream that is partly used in scrubbing.

If a scrubber is supplied with power in a single way, one can correlate collection efficiency as a function of gross power input. But for better understanding, one should identify the component of power consumption that relates to collection efficiency. This is a practical necessity if power is being supplied via two or more paths, and the separate components are to be combined on a consistent basis to give the total power. The gross power input includes losses in motors, drive shafts, fans and pumps that obviously should be unrelated to scrubber performance.

The basic quantity that does correlate with scrubber efficiency is "contacting power," power per unit of volumetric gas flowrate that is dissipated in contacting and is ultimately converted into heat [4]. The correlation clearly implies that collection mechanisms are associated with fluid turbulence.

Calculation of contacting power

Although contacting power is easily defined in abstract terms, it is less easily specified, and still less easily measured. In the simplest case, when all the energy is obtained from the gas stream as pressure drop, contacting power is equivalent to friction loss across the wetted equipment (rather than to pressure drop), an observation which may reflect kinetic energy changes rather than energy dissipation.

Although friction loss across a scrubber is often nearly equal to the pressure drop, the distinction between the two is important. Pressure drops due solely to kinetic energy changes in the gas stream will not correlate with performance. Furthermore, any friction losses taking place across equipment that is operating dry obviously do not contribute to gas-liquid contacting and likewise do not correlate with performance [7].

The author has adopted "effective friction loss" to denote friction loss across the wetted system; the loss is measured in units of gas pressure-drop, in. or cm of H_2O. Conversion to contacting power requires only a change of units to power per unit of volumetric gas flowrate, hp/1,000 ft³/min, or kWh/1,000 m³.

Nomenclature

F_E Effective friction loss, in. or cm water
G Gas flowrate, ft³/min or m³/min
L Liquid flowrate, gal/min or L/min
N_t Number of transfer units, dimensionless
p_f Liquid feed pressure, lb/in² or atm, gage
Δp Gas-pressure drop, in. or cm water
P_G Gas-phase contacting power, hp/1,000 ft³/min, or kWh/1,000 m³
P_L Liquid-phase contacting power, hp/1,000 ft³/min, or kWh/1,000 m³
P_M Mechanical contacting power, hp/1,000 ft³/min, or kWh/1,000 m³
P_T Total contacting power, hp/1,000 ft³/min, or kWh/1,000 m³
W_s Net mechanical power input, hp or kW
α Coefficient of P_T in Eq. (6), [hp/1,000 ft³/min]$^{-\gamma}$ or [kWh/1,000 m³]$^{-\gamma}$
γ Exponent of P_T in Eq. (6), dimensionless
η Fractional collection efficiency, dimensionless
$1 - \eta$ Fractional penetration, dimensionless

Comparing gas- and liquid-phase inputs

This "gas-phase contacting power" is easily determined, since pressure drops across dry equipment can be measured and subtracted from the total, and those resulting from kinetic energy changes can be estimated or measured. However, "liquid-phase contacting power" supplied from the stream of scrubbing liquid, and the "mechanical contacting power" supplied from a mechanically driven rotor, are not readily measured. One is generally obliged to compute a theoretical contacting power supplied to the gas-liquid contacting process rather than measure actual contacting power consumed.

For example, if hydraulic spray nozzles are used, theoretical liquid-phase contacting power can be taken as the theoretical kinetic energy of the liquid jet emerging from the nozzle, which is equal to the product of the volumetric liquid flowrate and the gage pressure of the liquid upstream of the nozzle. However, some of this energy may be lost by friction in the nozzle itself, and after the liquid jet emerges from the nozzle, some of its remaining kinetic energy may be lost in ways that do not contribute to contacting. If a rotor is used, theoretical mechanical-contacting power delivered to the gas stream is equal to the input to the motor, minus losses in the motor, couplings and drive shaft.

If a two-fluid nozzle is used, analogy suggests that the corresponding theoretical contacting power be taken as the sum of the separate kinetic energies of the liquid stream and the compressible-fluid stream as they emerge from the nozzle. An approximate value for the kinetic energy of the compressible fluid can be calculated by assuming an isentropic expansion of the fluid through the nozzle. Unfortunately, almost no data are available to show the validity of this procedure.

In ejector-type scrubbers, and usually in scrubbers with mechanically driven rotors, part of the energy supplied is not dissipated but produces a rise in the pressure of the gas stream, and therefore is by definition not contacting power. To predict contacting power, power equivalent to the gas-pressure rise must be subtracted from theoretical power input to the scrubber.

A summary of formulas for computing actual or estimated contacting power in English and metric units is presented in Table II.

Correlation of efficiency data

Collection efficiency of a scrubber is an exponential function of contacting power:

$$\eta = 1 - \exp[-f(P_T)] \qquad (1)$$

where η is fractional collection efficiency, and $f(P_T)$ is a function of total contacting power, P_T. Collection efficiency is thus an insensitive function for purposes of data correlation, particularly in its high range. The penetration

$$1 - \eta = \exp[-f(P_T)] \qquad (2)$$

is better, but the number of transfer units, defined in this case by

$$N_t = \ln\,[1/(1 - \eta)] \qquad (3)$$

or $\qquad \eta = 1 - \exp(-N_t) \qquad (4)$

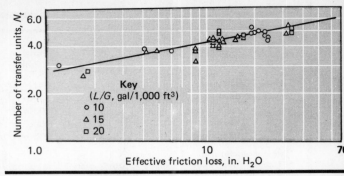

Performance curve for a venturi scrubber collecting fly ash   Fig. 1

is still more convenient for correlating performance data.

From these equations, it is seen that for a scrubber:

$$N_t = f(P_T) \qquad (5)$$

Data from many different sources [2–4,8,9] have shown that

$$N_t = \alpha\,P_T{}^{\gamma} \qquad (6)$$

where α and γ are empirical constants that depend on the aerosol (dust or mist) collected, and are little affected by the scrubber size or geometry, or the manner of applying the contacting power. This is the "contacting power rule."

Recent data

In a log-log plot of N_t vs P_T, γ is the slope of the curve, and α is the intercept at $P_T = 1$. The relationship is illustrated for an actual case in Fig. 1, which presents data for a pilot-plant venturi scrubber operating on fly ash from a power boiler fired with pulverized coal. The curve is fairly typical, although the slope is flatter than with most particulates. The data also show that the liquid-to-gas ratio in itself had essentially no influence on performance.

The author has adopted the term "scrubber performance curve" for plots such as Fig. 1. One must appreciate that scatter in the data does not result solely from experimental errors, but also from actual variations in the particle-size characteristics of the dust itself. In fact, it is difficult to maintain consistency even in test aerosols generated under controlled laboratory conditions. The curve for an industrial particulate necessarily represents an average material. Thus, for best definition, it should cover as wide a range of contacting power as is practical. Obviously, such data can be obtained only with pilot-plant equipment that has the flexibility to operate over a wide range of conditions.

Aerosol efficiencies

The efficiency of a test scrubber at a fixed contacting power—or, better, the performance curve—is the most sensitive indicator of variations in an aerosol, although it does not provide quantitative measurement of changes. For practical design, the performance curve is more useful than particle-size distribution. Especially for finding *in situ* aerosol particle size in waste gases, measurements of particle size are tedious and still subject to serious errors, some of which are not even fully defined yet.

Performance curves for an industrial particulate are straightforward, provided a small pilot-plant scrubber is used. A common error is to carry out tests with a large and inflexible unit. If a scrubber is to be used for the first time in a large, critical application, operational experience with a semiworks unit of reasonable size is appropriate. However, such a unit is *not* appropriate for gathering basic efficiency data.

Past experience [2-4] shows that scrubber size has little or no effect on the performance curve, at least down to 75 to 100 ft³/min capacity. The relationship may break down at some lower scrubber size, but if so, the critical size has not been established. At least three companies have employed microscale scrubbers for such test work [11-13].

Experimental caveats

Often, workers will measure only outlet dust concentration from a scrubber and correlate this as a function of scrubber pressure drop, without measuring inlet concentration. This is equivalent to correlating penetration through the scrubber on the basis of a constant inlet concentration [4].

This method is justifiable only with certain dusts. Those from iron blast furnaces, for example, may contain two dissimilar fractions: a fine-fume fraction with moderate concentration variation, and a second, larger fraction of coarse material that varies widely in concentration. Outlet concentration will be determined by inlet concentration of the fine fume, whereas the coarse fraction will be collected with virtually 100% efficiency at all scrubber operating conditions.

In such a circumstance, overall collection efficiency based on total input dust-concentration will be virtually meaningless for assessing performance.

Unresolved are the conditions at which gas flowrate should be evaluated in contacting-power correlations. The problem arises in calculating liquid-phase and mechanical contacting power, where gas flowrate appears explicitly. Where there are large changes in gas pressure, temperature, and water-vapor content across the scrubber, there can be large changes in volume (and considerable changes in mass) during passage of the gas.

In the case where all contacting power is gas-phase and is measured by gas pressure drop or effective friction loss, gas volume is implicit only, but represents an average value across the scrubber. It also happens to be the case for which the contacting-power correlation is best verified.

Hence, by inference, the average of inlet and outlet volumetric flowrates seems appropriate when the gas rate appears explicitly—or is at least consistent with the above computation of gas-phase contacting power.

What these data mean

The basis of the contacting power rule is almost entirely empirical, although there are fundamental reasons for expecting relationships between power consumption and collection efficiency for various scrubbing

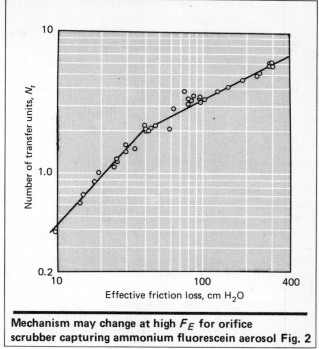

Mechanism may change at high F_E for orifice scrubber capturing ammonium fluorescein aerosol Fig. 2

devices; for example, particle collection by both inertial and diffusional mechanisms should increase with turbulence. However, there is no obvious reason to expect the relationship to be so nearly the same for radically different devices—as has been found by experiment. An interesting parallel occurs in the case of cocurrent gas absorbers, where investigators have correlated mass-transfer coefficients with energy dissipation [14,15].

The dominant collection mechanism in particulate scrubbers is inertial. On theoretical grounds, diffusion would be expected to become operative only at less than 0.5 μm, and to become a major factor only for particles under perhaps 0.1 μm, for which diffusivities are relatively high. There is little evidence that industrial aerosols contain appreciable mass fractions of particles under 0.1 μm; such aerosols coagulate rapidly except at very low concentrations. There has been no evidence that particulate scrubber performance has been appreciably affected by gas residence time, as would be expected if diffusion were important.

Recent investigations [8,9] have provided new insights into the efficiency/contacting power relationship, but have also produced new questions. Interpretation of the slope of performance curves has always presented difficulties [4]. A value of $\gamma < 1$ indicates that the aerosol becomes more difficult to collect at higher efficiencies. In keeping with this fact, scrubbers preferentially collect larger particles in a polydisperse aerosol (one containing particles of varying sizes), and higher contacting power is needed to collect finer particles. If diffusion were the dominant mechanism and also increased with contacting power, finer particles would collect preferentially, and collection would become harder with increasing efficiency and contacting power. This would again lead to a value of $\gamma < 1$.

For a monodisperse aerosol (all particles the same

size), one would logically expect $\gamma = 1$. It is difficult to imagine a physical process that would give a value of $\gamma > 1$, which indicates that collection becomes easier as collection efficiency increases. Yet correlations of data from scrubbing of some aerosols have produced such performance curves [3,4].

In recent investigations with small pilot-plant orifice-scrubbers, performance curves with two branches (Fig. 2) were obtained for fine polydisperse synthetic aerosols of ammonium fluorescein [8,9]. The lower branch had a value of $\gamma > 1$, and the upper branch a value of $\gamma < 1$. If investigations previously reported [3,4] had been extended to high-enough contacting-power levels, the performance curves with slopes greater than 1 might have been followed by additional branches with slopes less than 1.

Among aerosols that yield double-branched performance curves, or curves with slopes greater than 1, the only common feature is that all have large weight fractions composed of submicrometer particles. The steep lower branch of the curve suggests a region in which both inertial and diffusional collection increase with contacting power; the upper branch suggests a region in which inertial mechanisms dominate.

Gas- vs. liquid-phase contacting power

The precise degree to which contacting power supplied by each of the three basic methods is equivalent has still not been fixed. Previous correlations [2–4] indicate that liquid-phase contacting power supplied by hydraulic spray nozzles was generally equivalent to gas-phase contacting power, but the data were relatively few and were not generally obtained under closely controlled conditions. Performances of a number of widely different pressure-spray nozzles [9] were compared with that of the standard orifice contactor, using fine synthetic, ammonium-fluorescein aerosols under lab conditions. Spray-nozzle contactors always consumed some gas-phase contacting power, and the gas-phase and liquid-phase contacting powers varied widely, both absolutely and relative to each other. Some deviations, mostly moderate, appeared among spray nozzles, but all performed more or less inferior to the orifice contactor.

Deviation of spray-nozzle performance from that of the orifice scrubber was found to be correlated by a simple parameter, *the ratio of liquid-phase contacting power to total contacting power for the scrubber with spray contactor*. In addition, relative deviation of spray-contactor performance from that of the orifice contactor rose with a drop in the particle size of the test aerosol. It is still unclear how much of the apparent difference between gas-phase and liquid-phase contacting power is real and how much stems from the difficulty of estimating true liquid-phase contacting power.

Gas-phase power vs. other mechanisms

No similar comparison has yet been made between gas-phase and mechanical contacting power. Such data as have been available for mechanical scrubbers [3,4] show that these perform about as well as units using gas-phase contacting power. The mechanically driven rotor in such a device is commonly designed to promote draft as well as gas scrubbing, but in its latter role it is a turbulence promoter comparable to a venturi or orifice contactor.

To date, it appears that a venturi or orifice scrubber will give as good performance as can be expected from "normal" scrubbing, and that most other units using gas-phase contacting power will give essentially the same performance. Occasionally, a device has performed poorly, but none has done better [7]. Where superior performance has occurred (sometimes in conventional devices), it has followed some additional collection mechanism that is not ordinarily operative in scrubbing. A gas-liquid contactor composed of a packed bed of metal fibers gave superior performance, apparently because filtration was coupled to scrubbing [7]. Condensation of water vapor boosts efficiency in conventional venturi and orifice scrubbers [2,4,8].

It must be recognized, however, that condensation scrubbing requires transfer of heat energy, with its attendant costs, and may in some cases also demand heat energy that could be used in another way [4,8]. Hence, economics of condensation scrubbing will generally be decided by the needs of the associated process system. Condensation scrubbing may be practical when gases must be dehumidified as well as cooled and cleaned (e.g., blast-furnace gas, feed gas for contact sulfuric-acid plants), or when there may be a use for low-level heat, as in some kraft pulp mills.

References

1. Lapple, C. E., and Kamack, H. J., *Chem. Eng. Progr.*, Vol. 51, No. 3, 1955, p. 110.
2. Semrau, K. T., Marynowski, C. W., Lunde, K. E., and Lapple, C. E., *Ind. Eng. Chem.*, Vol. 50, 1958, p. 1,915.
3. Semrau, K. T., *J. Air Poll. Control Assn.*, Vol. 10, 1960, p. 200.
4. Semrau, K. T., *J. Air Poll. Control Assn.*, Vol. 13, 1963, p. 587.
5. Walker, A. B., *J. Air Poll. Control Assn.*, Vol. 13, 1963, p. 622.
6. Walker, A. B., and Hall, R. M., *J. Air Poll. Control Assn.*, Vol. 18, 1968, p. 319.
7. Semrau, K. T., Unpublished data, SRI International, Menlo Park, Calif., 1952–53.
8. Semrau, K. T., and Witham, C. L., "Condensation and Evaporation Effects in Particulate Scrubbing," Paper 75-30.1, Air Pollution Control Assn., Boston, Mass., June 1975.
9. Semrau, K. T., Witham, C. L., and Kerlin, W. W., "Relationships of Collection Efficiency and Energy Dissipation in Particulate Scrubbers," Second Fine Particle Scrubber Symposium, New Orleans, La., May 1977.
10. Raben, I. A., "Use of Scrubbers for Control of Emissions from Power Boilers—U.S.," Paper 13, U.S.-U.S.S.R. Symposium on Control of Fine-Particulate Emissions from Industrial Sources, San Francisco, Calif., Jan. 1974.
11. Hardison, L. C., *Amer. Ceram. Soc. Bull.*, Vol. 49, 1970, p. 978.
12. Riley Environeering Inc., Bull. 112, Skokie, Ill., Nov. 1976.
13. McIlvaine Co., "The McIlvaine Mini-Scrubber," Northbrook, Ill. undated.
14. Jepson, J. C., *AIChE Journal*, Vol. 16, 1970, p. 705.
15. Reiss, L. P., *I & EC Proc. Des. Develop.*, Vol. 6, 1967 p. 486.

The author

Konrad T. Semrau is a Senior Chemical Engineer at SRI International, 333 Ravenswood Ave., Menlo Park, CA 94025, where he is chiefly concerned with control of air pollution from stationary sources. His long interest in particulate scrubbers encompasses a major pilot-plant program. A graduate of the University of California at Berkeley, he holds an M.S. in chemical engineering. He is active in the Air Pollution Control Assn., ACS and AIChE.

Halt corrosion in particulate scrubbers

Careful design, proper materials, and close control of operating conditions can offset attack from hot inlet gases and contaminated scrubbing liquors.

Thomas G. Gleason, Swemco, Inc.

☐ Corrosion takes place in scrubbers chiefly because of acid contaminants and solids trapped in the scrubbing liquor. Inlet gases are usually hot (often 350° to 2000°F), and depending on the system, they may contain SO_2, weak sulfuric acid, plus chlorides or fluorides from either the dust or the makeup water. The potential for high concentrations of acids and halogens is enhanced by evaporative losses and attempts to minimize purge with tight recycle circuits.

The variables

The degree of attack varies with the type of scrubber used and with the different zones in the scrubber. Generally, attack is most serious in the inlet section, less so above the separator, and least in the main body of the scrubber.

Stainless steel alloys conventionally used for scrubbers are, in order of increasing cost: Types 304 and 316, the Alloy-20 series (Carpenter 20 Cb3, Hastelloy G, Incoloy 825, Uddeholm 904L), Inconel 625 and Hastelloy C. If the corrosion resistance is satisfactory, stainless-steel alloys may be chosen over rubber-lined steel or plastic construction for their mechanical strength and temperature resistance.

The basic mechanism by which a stainless steel resists corrosive attack is through the formation of a passive interface that resists penetration by corrosive media. This film provides a barrier consisting of absorbed gas and metal oxides formed by the alloy, and has been described as a thin, transparent, tenacious, molecular layer. The formation of the protective film and its breakdown in wet scrubber systems is caused by electrochemical or chemical reactions.

Very little is known about the basic nature of this film or the mechanisms by which it is destroyed. Corrosion data show only the end result of film breakdown. As reported in the general literature, they usually repre-

Originally published October 24, 1977

Erosion-corrosion in spray nozzles

sent uniform corrosion situations where the metal becomes thinner over its entire exposed surface.

These corrosion data do not necessarily represent conditions that exist in scrubber systems, especially at the gas inlet, in the high velocity zones, and above the entrainment-elimination section, exit duct, and exhaust fan. Serious failures can take place in these critical areas even though the main scrubber body is free from attack.

Selection of materials of construction for a scrubber and its auxiliary components can in many cases be difficult. It is helpful to have good understanding of corrosion-resistant materials and scrubber design, familiarity with the process variables of the given system, and experience in setting up scrubber process flow sheets and instrumentation.

Tailoring a system for corrosion resistance

These disciplines must be brought to bear on the many variables of a particular gas-cleaning application. It is not possible to construct a simple formula for avoiding corrosion in scrubbers. However, the following example should illustrate some general design and operating guidelines.

Fig. 1 shows a flow diagram for a scrubber system where acid gases are being scrubbed from a calciner.

For the given case, the solids-removal efficiency is fixed at 99% by weight. 394 lb/min of water is evaporated in cooling the hot inlet gas to its saturation temperature of 166°F.

For a purge concentration of 1% by weight, total makeup water is 2494 lb/min, which includes makeup for evaporation losses. Chloride content is 1.0% by weight in the entering solids, 50 ppm in the makeup water. Hence, chloride concentration in the recycle circuit and purge stream works out to be 160 ppm.

Since approximately 80% of the SO_3 is caught in the scrubber, an H_2SO_4 concentration of 0.13% by weight is the equilibrium content in the purge. If 316 ELC alloy is selected for the scrubber, good service life can be expected if the concentrations are held to the levels indicated by this flow scheme.

At the low pH that prevails in this particular system, chloride concentrations in excess of 200 ppm could readily produce corrosive attack in the alloy. If the purge stream were inadvertently reduced by blockage of the valve controlling the purge, so that a concentration of 2% solids resulted, then the chloride concentration would reach 408 ppm, and rapid attack would ensue.

Instrumentation on this system should provide for a constant minimum flow of makeup water with operating checks against high chloride levels. For upset conditions such as reduced flow of water makeup, alternative sources must be provided or the system should be shut down. Otherwise, even with the proper alloy, serious failure of the scrubber and its auxiliaries could result.

Let us look briefly at scrubber design so that its impact on corrosion resistance can be appreciated. Commonly used towers contain trays or packing, spray devices or venturi units. Design differences in these generic types influence resistance to corrosion in various parts of the equipment.

Towers with trays, packing and baffles

Scrubbers having internals such as trays, packing, or baffles that promote low-velocity inertial impact perform well in the 1 to 2μ size range and above. Their efficiency on sub-micron particles is usually low. However, they do have the advantage of low energy consumption: 3 to 8 in. H_2O pressure loss.

The areas of most serious attack in this type of scrubber are at the gas inlet and above the entrainment separator. The scrubbing internals, if designed well, usually produce gas-liquid contact without very high-velocity zones. Corrosion rates in the main contact area of the unit reflect uniform profiles of solids and acid buildup. This does not hold for the inlet zone; hot gases entering the unit impinge on partially wetted surfaces and create wet/dry interfaces. This is the primary area of corrosion in scrubbers that have internal contacting elements.

After the gas passes through the scrubbing zone, entrained scrubbing liquor is removed in a separation stage. The separation technique may use inertial im-

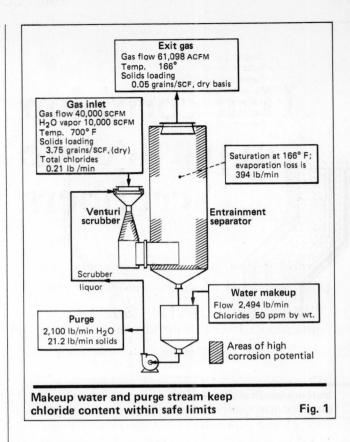

Gas inlet
Gas flow 40,000 SCFM
H_2O vapor 10,000 SCFM
Temp. 700° F
Solids loading
3.75 grains/SCF, (dry)
Total chlorides
0.21 lb/min

Exit gas
Gas flow 61,098 ACFM
Temp. 166°
Solids loading
0.05 grains/SCF, dry basis

Saturation at 166° F;
evaporation loss is
394 lb/min

Venturi scrubber

Entrainment separator

Scrubber liquor

Water makeup
Flow 2,494 lb/min
Chlorides 50 ppm by wt.

Purge
2,100 lb/min H_2O
21.2 lb/min solids

Areas of high corrosion potential

Makeup water and purge stream keep chloride content within safe limits **Fig. 1**

pact, a tortuous path, or cyclonic or low-velocity disengaging, but the result is the same: a zone above the separator where there is no excess water (beyond that due to saturation) in the gas stream. In this area residual particulates can gradually accumulate, as can unscrubbed acids. Dangerously high concentrations of contaminants can readily build up in this often-inaccessible, stagnant area.

Spray towers

Spray towers are widely used; they have good efficiency in the 3 to 5μ size range and above, and use a simple, open design. They are particularly well-suited for primary scrubbing and quenching upstream of high-performance scrubbers. They are used for absorption of SO_2, fluorides, and other acid gases.

The zones of attack in spray towers lie at the inlet area, above the entrainment separator, and in the spray nozzle itself. The spray pressure used for atomization of scrubbing liquor usually runs 15 psig and above. When weak-acid slurries are recycled through stainless-steel spray nozzles, serious breakdown of the corrosion-resistant film can result. Refractory-type spray nozzles such as those made of Refrax (silicon carbide) are often used to control this type of corrosion.

The separator section downstream of a venturi throat-section is generally comparable to a spray tower in that a mixed spray of agglomerated solids and scrubbing liquor spins tangentially into the unit. Although the inlet section does not suffer from wet/dry problems downstream of the venturi, the tangential spin imparted at the inlet can cause localized breakdown of the corrosion-resisting film of the alloy. Corrosion levels can

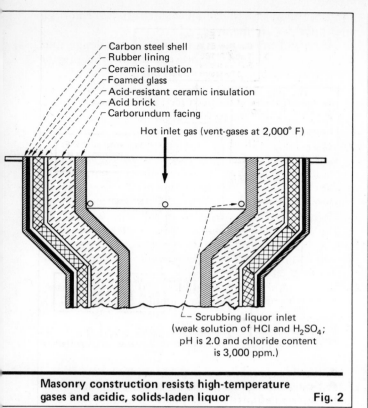

Carbon steel shell
Rubber lining
Ceramic insulation
Foamed glass
Acid-resistant ceramic insulation
Acid brick
Carborundum facing

Hot inlet gas (vent-gases at 2,000° F)

Scrubbing liquor inlet
(weak solution of HCl and H_2SO_4;
pH is 2.0 and chloride content
is 3,000 ppm.)

Masonry construction resists high-temperature gases and acidic, solids-laden liquor Fig. 2

likewise be high in the quiescent zone above the entrainment separator (or at the top of the tangential separator).

Venturi scrubbers

Venturi scrubbers are best suited for high performance on sub-micron fumes. This is achieved by accelerating the gas stream to very high speeds, on the order of 150 to 250 ft/s. The high-speed action generated in the venturi throat atomizes the feed-liquor and impacts the fine dust particles against the slower-moving liquor.

The venturi scrubber is gaining popularity due to its inherent ability to remove sub-micron particulates at high efficiency. Higher performance levels required in air-pollution control systems have dictated its use, so that it is replacing lower-energy scrubbers in many applications.

Under high-velocity scrubbing conditions, with solids present in the venturi scrubbing-liquor, severe localized attack can take place on the alloy's protective film.

This film is very sensitive to velocity effects, and solids in the gas stream or recycle scrubbing-liquor can wear it away. The alloy may easily handle corrosive acids, but high gas speeds in the scrubbing zone and the presence of solids may cause rapid breakdown.

The most serious failure is usually below the throat (in the diverging section and elbow) where the gas velocity is only on the order of 60 to 90 ft/s. Where alloy failure is due to high-velocity effects (i.e., abrasion), little or no improvement is seen when a higher grade alloy is substituted. If 316 ELC fails in four weeks, one can reasonably expect that Alloy 20, Inconel 625, or Hastelloy C would have the same life in a similar abra-

sion environment. As a practical matter, the corrosion-resistant film formed on many of the various stainless steel alloys appears to wear away at the same rate under abrasive conditions.

The corrosion problems that exist at the gas inlet are present also at the top of the scrubber, above the separator section. This includes the exit duct and fan.

The venturi has certain features that make it less susceptible to attack. It has large clearances for gas and scrubbing-liquor flow; it resists solids buildup extremely well; and its vertical gas inlet and shielded liquor-entry help to avoid wet/dry interface problems.

There is a close tie between solids buildup and corrosion resistance. Scrubbers designed to avert solids buildup provide complete washing of all scrubber surfaces in the hot inlet zone and in the particulate-separation section. Conditions leading to solids buildup also encourage localized acid formation. Solids buildup is serious enough in itself, but if associated with acid concentration, it can be devastating: crevice corrosion and concentration cells combine to produce a double-barreled attack on the alloy.

Good design of venturi inlet openings can greatly minimize corrosion in this critical zone. Continuous flushing of all inlet surfaces averts particle buildup. Uniformly wetted surfaces reduce damage from hot gases and also prevent localized hot-spots in which acid concentrations cycle rapidly. Proper design forestalls crevice corrosion due to solids buildup, and avoids localized high acid concentrations. Similar wetted-wall gas inlets using low pressure-drops can be incorporated upstream of tray, spray and packed towers. The substantial benefits of good inlet design in reducing corrosion are not often appreciated.

The inlet section of a scrubber often uses a much higher grade of alloy to combat the higher corrosion potential. For example, Inconel 625 may be used in the inlet section, whereas 316 ELC alloy is satisfactory for the main scrubber body. The zone above the separator may also be of a higher-grade alloy. Ironically, a superior entrainment separator can aggravate corrosion problems at the top of the scrubber vessel by retaining corrodents. Outlet corrosion can often be controlled by intermittent washing to eliminate acid concentration and solids buildup.

Lined construction using nonmetals

High-velocity corrosion can be controlled by using masonry lining backed by a corrosion-resistant membrane, usually rubber or lead, supported by a carbon steel shell.

Masonry linings are also used when high acid concentrations and very high inlet temperatures are involved. Where acid chlorides persist in combination with low-pH scrubbing liquor, even the higher-grade alloys are vulnerable. Masonry can take the severe temperature shock that may occur in the inlet quench zone and is also well suited for high-velocity scrubbing in venturi units. It is the only sure-fire construction wherever abrasion wear of an alloy is of concern.

Since acid brick and mortars are permeable to scrubbing liquors, the metal shell of the scrubber, usually carbon steel, is protected by an acid-resistant membrane. The membrane is usually rubber, fiberglass-reinforced plastic (FRP), or lead. Since these materials are vulnerable to high temperatures, the brick lining serves two main functions:

■ It provides the necessary temperature drop from the hot face of the brick to the membrane, for protection against failure. A reasonable limit for rubber is 180°F; for lead, 250°F; for FRP, 200°F.

■ It resists the abrasive effect of high-speed gas and scrubbing liquor.

The selection of brick and mortar linings covers a wide range of materials and designs, and only a brief discussion of this multi-layered construction will be given here.

Fig. 2 depicts a vent-gas stream at 2,000°F. The scrubbing liquor is a weak solution of HCl and H_2SO_4 (pH = 2), with chloride concentration at 3,000. The high temperature calls for acid-proof refractory that will take the severe temperature stress, which usually limits selection of hot-face brick to silicon carbide. Acid brick backs up the silicon carbide and provides the initial temperature drop.

Acid-proof insulation—Kaowool or foamed glass—provides the additional temperature-drop necessary to protect the membrane lining. Due to the high installation cost of brick linings, their heavy weight, and restrictions on the internal design of a scrubber, the use of this type of construction is limited to known critical areas. Masonry construction is also used when the potential seriousness of abrasion or corrosive attack is unknown, and the reliability of the equipment must be high.

The degree of temperature protection needed for this type of construction varies with individual client preference and the nature of the particular system. Normally masonry construction stops at the tee-elbow at the base of a venturi unit. Temperature-protection devices are then used to protect the balance of the scrubber from upset conditions. Emergency water is injected into the quench section, or into the venturi if a temperature rise should occur.

The need to ensure the stability of masonry linings limits their flexibility. They cannot be pinned or anchored through the walls of the carbon steel vessel, because the membrane lining would then be pierced in many places, a possibly serious defect. The difficulty of sealing the anchors or pins where they pass through the membrane rules out this construction. The use of acid-resistant, impermeable plastic refractory material can eliminate the membrane and provide substantial cost savings. Unfortunately, there is no refractory that can be reliably used in place of waterproof, acid-proof membranes, although a number have been tried.

The brick or masonry structure must provide its own structural integrity. Cylindrical and conical shapes are therefore required. Attachments for scrubber internals that control velocity through a venturi throat create further design and corrosion difficulties.

The corrosive environment that makes masonry construction advantageous often dictates the use of rubber-lined or FRP equipment downstream. Rubber lining or solid FRP, as well as FRP-flaked glass on steel, can take a much wider range of acid concentrations than many of the stainless-steel alloys—particularly when chlorides and fluorides are involved.

Of critical importance is basic scrubber design and backup systems that will prevent the overheating of these temperature-sensitive materials.

The venturi scrubber has a special advantage: the good gas-liquid contact produced in the throat rapidly quenches the hot gas to saturation. When gas enters a high-energy venturi at 2,000°F, the gas is cooled over a few feet to the saturation temperature.

It is therefore often convenient to follow a venturi with a lower-cost FRP or rubber-lined separator. To ensure that temperature upsets downstream are averted, emergency water is sent to the throat along with makeup water required by the scrubbing process.

Prudent scrubber design with good inlet-gas quenching and careful alloy selection avoids concentration cells or solids buildup. Alloy corrosion rates will then come closer to published data.

Where high-velocity scrubbing is done with a slurry in venturi scrubbers, and the potential for severe wear is unknown, a number of options are available:

■ Build critical zones for easy access and repair.

■ Install supports for refractory linings, and oversize the venturi section for later acid-brick construction.

■ Provide metal liners that are replaceable.

■ Design for abrasion-resistance throughout in masonry construction that uses membrane linings.

Exhaust fans

Material for exhaust fans, particularly on venturi units, is often the highest grade of alloy. In a venturi system there is high pressure-drop (20 in. H_2O and higher) and with system losses included, an exhaust fan generating 50 in. H_2O is not unusual. The temperature rise in such a fan is substantial; the heat of compression is approximately 0.4°F per inch of water column generated—a temperature rise of 20°F. Acid droplets carried through the separator or condensed from the ductwork leading to the fan can cause severe corrosion.

When superheat exists in the fan, droplet carryover is evaporated and strong acid conditions can prevail. The work done through compression of the inlet gas stream partially agglomerates residual solids entering the fan. These solids build up on the fan's high-speed surfaces, along with partially evaporated weak-acid carryover.

Good exhaust-fan design, particularly when tip-speeds of fan blades are high, provides for wheel washing to lessen acid and solids buildup. In addition, the chemical analysis of the fan drip should be known.

The author

Thomas G. Gleason is president of Swemco, Inc., 470 Park Ave. South, New York, NY 10016. He has more than twenty-five years experience in design and development of gas-cleaning and absorption devices. A former vice-president for Combustion Equipment Associates, Inc., he holds several patents related to particulate scrubbers. He received his B.S. in chemical engineering from Columbia University.

Maintaining venturi-tray scrubbers

Here are ways to design and maintain these scrubbers so as to prevent most problems. The author provides a detailed list of things to look for and do should problems occur.

William J. Kelly, Swemco Inc.

☐ Venturi-tray-scrubber maintenance starts with equipment design, continues with preventive maintenance, and ends with troubleshooting. This article gives design tips that will make maintenance easier, suggests a preventive-maintenance program, and tells what to do should trouble develop.

Design considerations

The following will determine service life and maintenance requirements of a combination tray-venturi scrubber:

- Corrosion protection.
- Temperature protection.
- Erosion protection.
- Access to equipment.
- Solids accumulation.

Corrosion protection

Corrosion considerations are complex and should be undertaken according to specific application. Corrosion has been given a great deal of attention in the technical literature. The protective measures discussed here are common-sense guidelines for those involved in purchasing scrubbing equipment. The purchaser should:

- Require mill certification for all alloy materials used in scrubber, fan and duct fabrication.
- Deal only with reputable suppliers and check their past performance.
- Request multi-source recommendations on selection of alloys, resins or refractories from materials manufacturers and users.
- Use test coupons (whenever possible) on existing equipment that is being replaced, to determine corrosion rates.

Temperature protection

When handling high-temperature inlet gas (above 700°F), it is imperative that proper scrubber-liquor distribution be maintained. This will ensure continuous liquor-film protection on surfaces exposed to the high temperature. Fig. 1 shows a typical venturi scrubber with

Originally published December 4, 1978

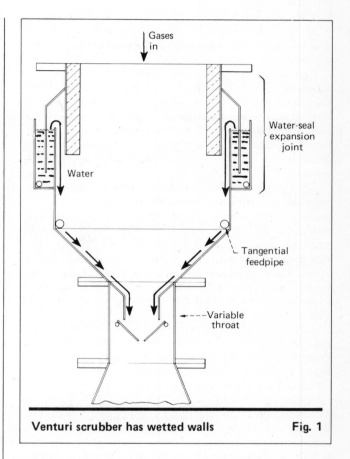

Venturi scrubber has wetted walls   **Fig. 1**

"wetted wall" protection. A water-seal expansion joint compensates for thermal expansion and contraction and prevents stress from developing in the high-temperature zones. This seal also floods the upper venturi section to keep all surfaces wet and at relatively low temperature. The continuous washing also prevents "wet-dry interfaces," where solids build up on the scrubber wall and where corrosion and stress cracking can occur.

Tangential feed pipes, although used primarily to inject scrubbing liquor to the throat, also provide a wall-washing action and assure that surfaces are protected from high temperature. Proper distribution of liquor on very-high-temperature applications (above 1,200°F) greatly reduces the possibility of stress cracking, carbide precipitation and sigma-phase contamination of alloy. In addition, the continuous flush prevents local high-acid concentration and corrosion. Another zone where corrosion is related to temperature is the area above the mist eliminator, where dewpoint corrosion is common. Proper material selection in this zone is critical.

Preventive maintenance   Table I

Equipment check	Frequency
Make internal inspection to remove construction debris.	Before startup.
Test level, temperature and pressure controls and alarms.	Before startup; 6 to 12-month intervals thereafter.
For automatic throats, use manual override to check free motion and the correct position of end-points. Refer to manufacturers' instructions for lubrication requirements.	Before startup; 6 to 12-month intervals thereafter.
Be sure that trays are bolted tight and level.	Before startup; 6 to 24-month intervals thereafter.
Check direction of sprays, and clearances of pipes and nozzles.	Before startup; 1 to 6-month intervals thereafter.
Make sure vessel is vertical. True vertical installation is important, due to gravity flow.	Before startup.
If water seal is provided, check vertical and horizontal clearances. Ensure correct openings for liquor flow, and proper tolerances for expansion.	Before startup; 3 to 12-month intervals thereafter.
Observe operation and pattern of quench sprays. If any problem is suspected, contact manufacturer.	Before startup; 3 to 6-month intervals thereafter.
Temporarily apply full water flow to the unit and check the liquor drain for fast draindown or pumpdown. If liquor holds up in bottom of unit for several seconds a constriction should be suspected.	Before startup; 12 to 24-month intervals thereafter.
Initiate gas flow and take pressure-drop readings across trays, throat and overall unit. Readings should be made at full gas flow. Refer to operating instructions for design pressure-drops.	Initial startup; 1-week to 1-month intervals thereafter.
When system reaches steady state, check gas and liquor temperatures in and out of the scrubber.	Initial startup; 1-week to 1-month intervals thereafter.
Observe stack appearance.	Initial startup; 1-day to 1-week intervals thereafter.
On units with hot inlet-gas, check outside shell temperature at the gas-inlet zone. A high relative temperature difference indicates local hot spots and possible stress-cracking zones.	Initial startup; 1 to 12-month intervals thereafter.
Analyze recycle liquor for unexpected components. Determine the effect of these constituents since they may greatly foreshorten service life (for example; high-chloride liquor in stainless steel service).	Initial startup; 1-day to 1-week intervals thereafter.
Note venturi-throat response to changes in pressure drop on units with automatically variable throats. (Typically, changes in gas flow occur at relatively slow rates so that hunting and lagging are almost never a problem.)	Initial startup; 1-week to 6-month intervals thereafter.
Check recycle and drain piping and pump for surging and cavitation. Causes may be scrubber induced—liquor foaming, high scrubber vacuum, loss of sump level, etc.	Initial startup; 1 to 6-month intervals thereafter.
After an initial period of operation (1-4 weeks) shut down and make a thorough internal inpection, specifically looking for: (a) Plugging or blocking of trays, nozzles or weirs. (b) High-velocity-zone wear (throat, sprays, turns, etc.) (c) High-temperature spots in gas-inlet zone (d) Signs of corrosion (e) General mechanical condition of internals. Detecting a progressive type of problem at this stage provides the time margin necessary to modify the equipment before serious damage can occur. If indications of a problem are so slight as to be questionable, schedule an additional shutdown at a similar interval for an additional checkout.	1 to 6-month intervals following initial shutdown.

Correcting scrubber problems

Table II

Major causes	Action
Poor cleaning performance	
a. Low scrubbing-liquor flowrate.	a. Check pump output. Look for plugged piping and nozzles, incorrectly opened valves, overthrottled pump-discharge valve.
b. Low pressure-drop across venturi.	b. Check for low scrubbing-liquor flowrate; low gas flowrate; inoperative or uncalibrated variable-throat controller; damaged variable-throat blade/disk.
c. Inlet dust loading or size distribution beyond scrubber design capability.	c. If operating modifications fail to correct the problem, analyze particle size and quantity.
d. Excessive gas flow.	d. Check fan-damper setting, venturi-throat setting, system fan operation vs. fan curve.
e. Partially blocked entrainment separator.	e. Check washdown sprays if installed. Check composition of spray liquor. If scaling occurs, investigate use of low-pH flushing liquor.
High exit-gas temperature	
a. Very low venturi-liquid flowrate.	a. Check pump output; look for plugged piping, nozzles, etc., incorrectly opened valves, overthrottled pump-discharge valves.
b. Low tray-water flowrate.	b. Check pump output, look for plugged piping, nozzles, etc., incorrectly opened valves, overthrottled pump-discharge valves.
c. Tray(s) (partially) plugged with solids.	c. Check condition of tray flushing sprays if installed. If scaling is observed, use a low-pH wash periodically to dissolve scale. Check percent solids in recycle liquor.
d. High cooling-water inlet temperature.	d. Check heat-exchanger operation and adjust cooling-water flowrate and temperature.
e. High scrubber-inlet temperature or excess gas-flow.	e. Check upstream equipment operation.
Exhaust-gas liquor entrainment	
a. Moisture eliminator drain plugged (tangential vane eliminators only).	a. Shut down and snake out the eliminator drain. If problem recurs, add flushing water to continuously irrigate drain pipe.
b. Excessive tray-liquor flow (tower flooding).	b. Reduce flow; calibrate flow-control mechanism if installed.
c. Excessive water frothing (possibly due to foaming agent in liquor).	c. Sparge a liquor sample. If liquor froth does not disappear quickly, foaming may be choking downcomers and drains. Analyze liquor for foaming agents.
d. Plugging of chevron-type eliminator.	d. Check flushing-spray conditions and pattern. Use more flushing periods per hour. Poor gas-cleaning performance will accelerate buildup. Check liquor chemistry for scaling agents.
e. Excessive gas flow.	e. Check fan damper position. Check variable venturi-throat opening.
Fan motor overload	
a. Low scrubber-pressure-drop due to excessive throat opening.	a. Normally, the fan damper will provide enough "choke" to prevent overload, so fan damper must be checked with variable throat opening. Check throat operator and liquor flow.
Plugged spray nozzles	
a. Nozzle openings too small.	a. Modify strainer/nozzle-opening ratio so that nozzle holes are at least twice the diameter of strainer openings.
b. Solids concentration too high in spray liquor.	b. Check separation equipment. Check for excessive dust load in gas stream. Check for purge-line malfunctioning.

Correcting scrubber problems	Table II (cont'd)

Major causes	Action
Plugged spray nozzles (cont'd)	
c. Pipe scale or debris entering liquid stream beyond strainer.	c. Remove spray heads and flush spray piping and nozzles. Replace piping if corrosion is apparent.
Excessive nozzle wear	
a. Solids concentration too high in spray liquor.	a. Check separation equipment. Check for excessive dust load in gas stream and for purge-line malfunctioning.
b. Abrasives in spray liquor.	b. Remove abrasives from liquor stream or install abrasion-resistant linings in wear zones.
c. Low pH in combination with abrasives, causing erosion/corrosion.	c. Check separation equipment. Check for excessive dust load in gas stream and for purge-line malfunctioning. Remove abrasives from liquor stream or install abrasion-resistant linings in wear zones. Add alkali for pH modification.
Plugged trays	
a. Hot gases entering equipment before liquid flow is initiated.	a. This condition causes "bakeout" of solids. Shut down. Try "washing" trays in place by recycling strong detergent. Otherwise, remove trays and scrape clean.
b. Inefficient venturi scrubbing, allowing high solids levels in gas stream to contact trays.	b. Check throat pressure-drop and venturi liquor rate.
c. High solids in tray liquor, formation of insoluble salts in scrubber.	c. Check solids-separation equipment. Analyze liquor and determine and eliminate cause of high solids.
d. No water flow or very low water flow to trays.	d. Check pump output; look for plugged piping, nozzles, incorrectly opened valves, overthrottled pump discharge valves.
Excessive throat wear	
a. High solids recirculation.	a. Check solids-separation equipment. Check for excessive dust load in gas stream and for purge-line malfunctioning.
b. Corrosion/erosion	b. Check separation equipment. Check for excessive dust load in gas stream and for purge-line malfunctioning. Add alkali for pH modification. Install abrasion-resistant liners in high-wear zones if liquor modifications are not practical.
c. Excessive gas velocity.	c. Check throat pressure-drop and reduce to design point.
Erratic automatic-throat operation	
a. Prime-mover malfunction.	a. Remove from service, repair or replace. Most throats can be held in a fixed position close to design pressure-drop by mechanical means during this procedure.
b. Sensor signal incorrect.	b. Check sensor taps on vessel for solids buildup. Check transmission tubing for liquid buildup or air leaks. Clean or repair sensor.
c. Transmitted signal incorrect.	c. Clean or repair sensor. Check instrument air-supply pressure and filters. Check tubing for leaks. Check positioner filter and connections. (Clean instrument air is critical here.) Thoroughly clean positioner internals and check freeness of operation.
d. Damaged damper disk-mechanism.	d. First make external inspection of drive train. If damaged area is not observed, shut unit down and make internal inspection using a throat-actuator manual override. Check for packing damage and excessively tight packing gland.

Erosion protection

In the high-velocity venturi throats, at tangential liquor and gas zones, and near spray areas, consideration must be given the possibility of erosion, as well as the combined effect of corrosion-erosion. This becomes of primary importance when gas-stream solids are abrasive and the liquor being recycled is acid. For these operating conditions, design options such as use of silicon carbide and alumina wear-surfaces must be considered. Fig. 2 shows Hastelloy alloy C high-velocity spray nozzles that were used for handling dilute sulfuric acid and abrasive solids. Service life of these nozzles was approximately two weeks. The abrasive solids in the liquor continuously eroded the corrosion-resistant skin of the alloy so that corrosion and erosion worked together to cause a catastrophic failure. Nozzles of silicon carbide have replaced alloy nozzles for this kind of service and, in most instances, have provided years of troublefree service.

Access to equipment

If critical areas are inaccessible, developing problems may go undetected during normal inspections. Access doors, sight ports, or both, should be provided to allow inspection of wear and corrosion zones, as well as areas where there is solids buildup potential. When downtime must be kept at a minimum, fast-opening doors with reliable latches should be considered.

For many applications, it is reasonable to make spray headers removable during operation, thus allowing for a quick check of these critical components. Shutoff valves and elbow connections directly upstream of each header will simplify removal.

Solids accumulation

Another consideration in scrubber design is the prevention of plugging and blockage of small-clearance openings such as spray heads, liquor pipes and tray perforations. Interference with normal gas and liquor flow patterns can result in high-temperature and corrosion damage to local areas of the scrubber and, in some cases, to downstream equipment. Spray nozzle openings on slurry service should be above 1/2-in. dia. In some cases where the solids concentrations are high or the solids are coarse and sticky, minimum openings must be increased to 1-in. dia. or larger. Low-pressure, open-pipe feeds are often preferred over sprays for scrubber-liquor injection where high solids concentrations or high temperature is anticipated.

Tray plugging occurs when solids in the incoming gas and recycle liquor build up around and eventually bridge the tray perforations. The basic causes of plugging are improper design of tray-flushing sprays, plugged flushing nozzles, low tray-liquor flow, sticky solids in the gas stream, and chemical scaling (usually in the form of calcium salts).

Selection of a mist eliminator is usually a compromise between maximum performance and minimum maintenance. Generally, the smaller the open area that the eliminator presents to the gas stream, the higher the mist-removal efficiency and the greater the possibility of plugging. Mesh-type mist-eliminators must be avoided on applications where there is anything more than a trace of solids in the gas or liquor. Parallel-blade chevron designs

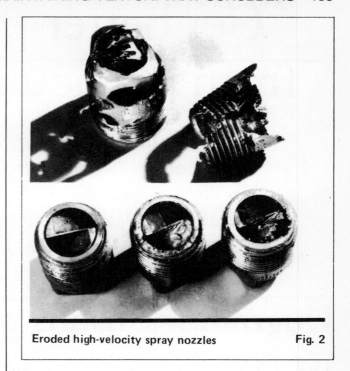

Eroded high-velocity spray nozzles Fig. 2

are often indicated where sticky materials or high dust-loads are involved. Chevron units require flush sprays above or below the blades (or both) on many applications, and flushing liquor should be clean.

Open-vane separators are a good choice for many scrubber units up to a maximum diameter of 14 ft. Open cyclonic separators will not plug up, since they have no internals, but their efficiencies are below those of impingement-type entrainment separators.

Preventive maintenance

When a scrubber is properly designed for its specific application, and unexpected conditions do not turn up in actual operation, it should provide many years of trouble-free service. Table I gives a typical preventive-maintenance schedule for a combination venturi-tray scrubber. The appropriate intervals between inspections differ substantially from job to job, and the intervals indicated are approximate minimums and maximums. It is suggested that preventive-maintenance checks start with the minimum intervals and be gradually increased toward the maximum ones. Following a guide of this type will further assure a troublefree and extended service life.

The author

William J. Kelly is project manager for Swemco Inc., 470 Park Ave. South, New York, NY 10016, telephone (212) 725-5450. He previously worked as product manager of cooling towers for Blazer Corp., and as a scrubber sales engineer for Peabody Engineering. After receiving a B.S.M.E. degree from New York State University, he spent several years as U.S. Merchant Marine officer, and as a Navy nuclear test engineer.

Get better performance from particulate scrubbers

The following two articles tell how to obtain improved product recovery and pollution control from particulate scrubbers. The first suggests ways to raise collection efficiency, cut entrainment, and lessen opacity. The second gives tips on pinpointing the causes of poor scrubber performance.

Upgrading existing particulate scrubbers

The author looks at several options for enhancing performance: adjustments to process conditions, greater power input, changes in internal designs, and alternative mechanisms of collection.

*Seymour Calvert,** Air Pollution Technology (APT), Inc.*

□ What can you do if your particulate scrubber is not performing well enough to meet air-pollution-control requirements and you do not want to invest in a new system?

This question may arise for a number of reasons. Often, new regulations demand better performance than an existing scrubber can provide. In other instances, a new scrubber system does not have the efficiency it was designed for, and attempts by the vendor and the owner to adjust it have failed. There are also cases in which performance has deteriorated over the years for reasons that are unclear to the operator; a similar situation crops up when corrosion or erosion have severely damaged the system, and the operator has to decide whether to replace it as is or with some alternative form.

Furthermore, scrubber efficiency can be satisfactory but other characteristics might not be. High power consumption is one trait that is becoming increasingly undesirable. Often the system may be performing acceptably on an existing process, but a capacity expansion is planned and the ability of the system to handle a higher gas flowrate comes into question.

Whatever the situation, it can be profitable to look into upgrading the existing system rather than replacing it with an entirely new one. Modification will generally involve less disruption of plant facilities, little or

*To meet the author, see p. 153.

Originally published October 24, 1977

154

no new space requirements, fewer delays because of long delivery times, and easier training periods for operating personnel since they are already familiar with much of the equipment.

Define the problem

The starting point in evaluating options for scrubber upgrading is to set forth the general facts. We may summarize the most important classes of information as follows:

1. What criteria must be met and in what way are they not being satisfied? Particulate-emission restrictions can be expressed in terms of particle concentration in the gas, mass rate of particle emission, opacity of the plume, and effect on pollutant concentration in ground-level air. Other factors that come into play are power needs, gas-flow capacity, turndown capability, and operation and maintenance limitations.

2. What is the nature of the emission and its source? Has the source operation changed during the period in question?

3. What are the configurations and dimensions of the scrubber system ancillaries? These include ducts, dampers, fans, the liquid-handling system and the stack.

Why is the limit being exceeded?

The next step is to close in on the specific nature of

the problem, which generally falls into one of several categories:

1. The particulate emission-rate or concentration is too high because collection efficiency is too low.

2. The particulate mass-emission-rate or concentration is too high because of mist entrainment leaving the stack.

3. The opacity limit has been exceeded even though the emission rate or concentration limitations may have been met.

4. Particulate emission levels are being met, but other system characteristics are unsatisfactory.

Measurements involved in obtaining this kind of information are specialized and will demand the assistance of competent, experienced people.

It is essential that particle-size distribution and concentration in the inlet and outlet gas streams be known. The most important instruments used in finding these are cascade impactors, for measuring particulate-size distribution, and filters, for pinpointing particle concentration. Cascade impactors fractionate particles into categories according to their aerodynamic particle size (see *Chem. Eng.*, Aug. 29, p. 56). The mass of each of these size fractions is determined by weighing quantities that are on the order of a few milligrams and less. Many factors can seriously bias the sample taken into the cascade impactor, and other factors can influence the accuracy of the measurements. Sampling and size analysis can require several days' work by an experienced crew.

Although particle concentration can be fixed solely from cascade-impactor measurements, it is generally wise to obtain confirmation by using filters. The filters can not do size fractionation, but will generally give a more reliable measurement of the total particle mass than will the cascade impactor. A substantial literature on particle sampling and sizing exists [7].

Liquid-entrainment rate and droplet-size distribution in the gas leaving the scrubber can be measured by means of small cyclone separators, or specially designed cascade impactors. Because the droplet size is generally large compared to that of particles entering and leaving the scrubber, entrainment sampling is more subject to bias than is particle sampling. Evaporation or condensation can also give rise to errors in the size distribution and in the amount of liquid entrainment. Considerably less work has been done on entrainment measurement than on particle size determination, but there are a few helpful documents in the literature [4,5].

Opacity must be determined by trained observers. As will be discussed later, opacity is related to measurable quantities such as particle-size distribution and concentration [8], but air pollution rules generally do not permit the use of theoretical computations. Certification of plume watchers is usually done by air-pollution-control agencies at "smoke schools," and requires from one to three days of training.

Measurement of scrubber-system parameters such as gas and liquid flowrates, temperatures, pressures, compositions, and power input can normally be made by plant personnel. Where gas flowrates are not known, it may be necessary to determine them by means of pitot-tube traverses. It is important to record the hu-

midity, temperature, pressure and density of the gas during these velocity measurements.

Particle collection efficiency

In most cases, one will look at the possibilities for raising particle collection efficiency or cutting power needs, no matter what the specific difficulty may be. These changes are worth trying because they can be done in steps and may be carried only as far as appears worthwhile. The first point to consider is whether the scrubber in question is achieving its maximum potential.

If the scrubber is not performing up to par and if "par" would be satisfactory, the next step is to look into operational changes. If the scrubber is performing up to par and if par is not satisfactory, one should look at alternative mechanisms. Some things that can be done short of replacing the scrubber with a totally new system are:

■ Change the particle size distribution or concentration through source operation changes.

■ Put more power into the existing scrubber to obtain higher efficiency.

■ Install a different type of scrubber within the existing shell.

■ Augment scrubber efficiency with flux-force/condensation (F/C) features.

Scrubber potential

Scrubber performance can be predicted in most cases. The APT cut/power plot [2], provides a useful tool for relating scrubber performance to pressure drop. The cut diameter, d_{PC}, is the aerodynamic diameter of the particle that would be collected at 50% efficiency. From the cut diameter, one can predict the efficiency at other particle sizes. Knowing the inlet particle-size distribution, one can then compute the overall collection efficiency that would be obtained at a given scrubber pressure drop or power input.

Performance can also be predicted by applying basic principles of particle technology. One general method for developing mathematical models for predicting efficiency is the unit mechanism approach [3].

Below-par performance

Should the scrubber's performance fall below par, it becomes necessary to track down the cause on an individual case basis. There are some common problems to look for. Prominent among these is excessive entrainment. This will show up as a high particulate loading in the scrubber outlet.* In order to determine whether this high loading results from entrainment, one should look for evidence of moisture in the duct or on the sampling apparatus. If this is observed, special sampling and analytical techniques can distinguish between the particle carried over by entrainment and that passing through because of low primary collection efficiency.

Air leaks into the scrubber or ductwork on induced-draft systems can lead to an increase in fan horsepower and/or a decrease in the amount of gas drawn through the scrubber. Old systems and those in which severe corrosion has occurred will often develop significant

*Measured by filter sample, using EPA Method 5.

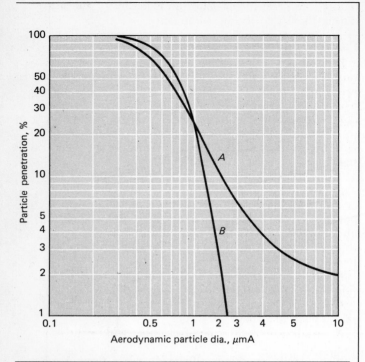

In venturi unit, higher liquid-to-gas ratio improves collection of large particles   Fig. 1

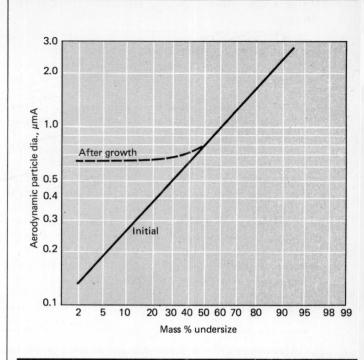

Particle growth resulting from water-vapor condensation reduces power demand   Fig. 2

Nomenclature

c	Mass concentration, g/cm^3 or gr/ft^3
C'	Cunningham correction factor, dimensionless
d_p	Particle physical diameter, μm
d_{pa}	Aerodynamic particle diameter, $\mu mA \equiv$ $d_p(C'\rho_p)^{1/2}$
d_{pg}	Geometric-mean particle diameter, μmA
d_{PC}	Performance cut diameter, μmA
Pt	Penetration, fraction (c_o/c_i) or percent
μm	Micrometer (micron)
μmA	Aerodynamic diameter $\equiv d_p(C'\rho_p)^{1/2}$, $\mu m(g/cm^3)^{1/2}$
ρ_p	Particle density, g/cm^3

Subscripts

i	Inlet
o	Outlet

leakage. Corrosion and erosion can also deteriorate scrubber internals so that the power is not being used effectively.

Sometimes, too low a liquid rate is used, under the assumption that efficiency depends only on pressure drop. This assumption is approximately true if there is sufficient liquid. The definition of sufficient liquid rate depends on the type of scrubber and on the particle size distribution of the particulate to be collected. To illustrate this point, Fig. 1 shows a plot of particle penetration as a function of particle size for a venturi scrubber operating at a given pressure drop and two different water-to-gas ratios. Curve A represents a ratio of 5 gal of water/cfm of gas; curve B, 15 gal/cfm. In both cases,

the pressure drop is 10 in. H_2O, and in both cases the cut diameter is roughly the same. However, the higher liquid-to-gas ratio provides a much better collection efficiency (lower penetration) for larger particle sizes than does the lower ratio. This is because the latter does not provide a sufficient droplet concentration to completely sweep the gas stream.

Alternatives

When a scrubber is already operating at its maximum potential, or when this level of performance is not satisfactory, the search for alternatives must begin. It is always advisable to take a careful look at the option of changing the source operation or the method of transporting the gaseous emissions from the source. For example, if one can increase particle size by decreasing the maximum temperature reached in a furnace or kiln, the change can substantially reduce power requirements. Often it is possible to decrease the quantity of gas needing treatment by means of careful ducting, and perhaps by recycling the gas stream.

If an increase in power consumption can be tolerated in order to lower capital investment, it may be possible to enhance the capability of an existing scrubber. The manner will depend on the scrubber. For a gas-atomized spray scrubber, such as a venturi, a higher gas velocity or liquid-to-gas ratio will result in a higher collection efficiency. Tray-type scrubber performance can be improved by the use of more trays or of smaller perforations on the trays. Spray scrubber performance can be improved by using a higher liquid-to-gas ratio or higher liquid pressure.

Another alternative is to choose a different type of

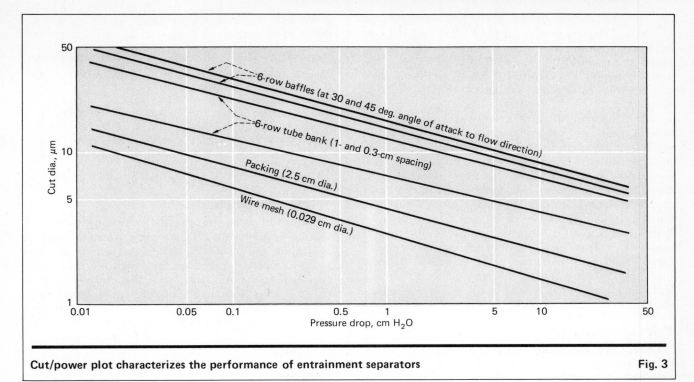

Cut/power plot characterizes the performance of entrainment separators   **Fig. 3**

scrubber that can reach a lower cut-diameter, or that displays better efficiency for a given pressure drop. It may be possible to replace the existing scrubber internals with those of another type and thus save the cost of a new shell, foundations, etc.

Water vapor condensation can enhance performance by increasing the mass of the particles, thereby bringing greater deposition forces to bear on them. Fig. 2 shows the effect of condensation on aerodynamic size. The straight line on the log-probability plot of particle diameter vs. mass percent of particles under that size applies to the original size distribution, whereas the dashed line represents a particle-size distribution accompanying condensation of water vapor. If a venturi scrubber were used to obtain 85% collection efficiency, a pressure drop of about 75 in. H_2O would be required for the particle size distribution, whereas only about 30 in. H_2O would be needed if condensation and particle growth took place.

Electrostatic charge may also be used to enhance the scrubber's performance. The charge may be applied to the particles, the drops, or both. Efficiency characteristics are essentially the same as for wet electrostatic precipitators.

Entrainment separation

Mist eliminators or entrainment separators are generally necessary to prevent undesirable emissions of liquid droplets from the scrubber. An unfortunate consequence of thorough and vigorous liquid-gas contacting in the scrubber is that some liquid is atomized and carried out of the scrubber by the gas that has been cleaned. The liquid entrainment, or mist as it is commonly referred to, will generally contain both suspended and dissolved solids.

In many cases, excessive entrainment imposes a limitation upon scrubber capacity. That is, while the scrubber itself might be capable of handling a larger gas flowrate, the rate at which entrainment becomes excessive will dictate a limit on capacity.

Separation principles

The flowrate and droplet-size distribution of entrainment depend greatly on the nature of the gas-liquid contacting arrangement within the scrubber. Droplets entrained from the scrubber contacting zone may be separated from the gas by a number of mechanisms; the initial collection of these droplets is sometimes referred to as primary collection.

After initial collection, the droplets coalesce into a sheet of liquid, unless the velocity of impact is so high that some of the drops shatter. In a properly designed and operated entrainment separator, liquid should drain freely. Capacity of the separator is limited by the reentrainment and/or flooding that occurs when the gas velocity ranges too high for a given liquid flowrate.

Equipment types

Apparatus used for entrainment separation can be grouped into categories according to the mechanism of operation:

Gravitation sedimentation. Within the scrubber and its outlet ducting, sedimentation is always active and important. However, discrete entrainment separators using sedimentation rarely follow a scrubber.

Centrifugal deposition. Cyclone separators of various designs are commonplace. Radial baffles and other types of guide vanes can be installed to induce a rotary motion of the gas stream within the scrubber shell. Zigzag-baffles, chevrons, corrugated sheets and similar devices force one or more abrupt changes in the gas flow direction. Droplets deposit on the baffles, and the collected liquid film either runs down the baffles (if the major axis is vertical) or drips off as large drops (if the axis is horizontal). A directional change in the gas duct can likewise cause considerable deposition of large droplets.

Inertial impaction. Beds of massive packing—such as saddles, rings and other elements—are used in either vertical or horizontal gas-flow configurations. Fibrous packings can also be used in beds comparable to massive-packed systems. Knitted-wire mesh, screens, metal wool, felted fibers, open-celled polymeric sponge, and glass fiber have been used for this purpose. Other impaction devices employ banks of round tubes, streamlined struts, and other shapes in vertical and horizontal orientations. Trays (perforated, valve, impingement and others) have been used for special purposes. Because these are scrubbing devices, they generate entrainment; their use after another type of scrubber may be redundant and should be carefully assessed.

Primary efficiency

Overall collection efficiency for droplets is dependent upon primary collection and reentrainment rates. In most cases, overall efficiency is low at low gas velocities because primary efficiency is low. At higher velocities, efficiency increases. However, at still higher velocities, reentrainment begins and causes overall efficiency to fall, even though primary efficiency remains high.

Methods for predicting primary collection efficiency for various types of separators were reported by the author and coworkers [4,5,6]. A concise representation of the primary efficiencies of several types of separators can be shown using the same cut/power relationship used for characterizing particle collection efficiencies in the scrubber itself. This relationship is given by a plot of the droplet diameter collected at 50% efficiency (i.e., the cut diameter) against the gas-phase pressure drop, or power input for the separator (Fig. 3).

Plots shown in Fig. 3 are based on design equations and experimental correlations. Curves are given for baffles at two angles of attack to the flow direction, tube banks with two different spacings between tubes within a row, packing of one particular size, and knitted mesh with a certain wire diameter.

Reentrainment

The liquid and gas flow-capacities of an entrainment separator are limited by reentrainment. Experimental data on collection efficiency for several types of separators have been taken by several investigators (see Ref. 5 for a bibliography).

The rate of reentrainment depends on separator geometry, gas velocity, liquid flowrate, and separator orientation. Increasing the gas or liquid flowrate will increase the rate of reentrainment. Liquid drainage is best when gas flow is horizontal and collection surfaces are near-vertical; also, with this configuration, reentrainment occurs at higher flowrates than for horizontal elements.

Some approximate values of gas velocity (based on an empty cross-section) at the onset of entrainment are given here to illustrate the range possible for well-designed equipment at moderate liquid loadings:

Separator	Gas velocity, ft/s
Zigzag with upward gas flow and horizontal baffles	12–15
Zigzag with horizontal gas flow and vertical baffles	15–20
Cyclone (inlet gas velocity)	100–130
Knitted mesh with vertical gas flow	10–15
Knitted mesh with horizontal gas flow	15–23
Tube bank with vertical gas flow	12–16
Tube bank with horizontal gas flow	18–23

Solids deposition

Industrial experience with entrainment-separator fouling and plugging, and some quantitative experimentation on suspended solids deposition [5], have yielded empirical guidelines for design. Vertical collection surfaces stay cleaner than horizontal ones, due to better liquid drainage. Intermittent washing with sprays is beneficial, but the details of the washing system and procedure depend on the specific case. Precipitation scaling must be controlled through the system's chemistry.

Opacity

Opacity refers to the appearance of a smoke plume or cloud when viewed by a human observer. It can be graded quantitatively in accordance with a scale that is memorized by the observer during a training procedure. Regulations based on opacity are favored by air-pollution-control agencies because they are easy to enforce and because they relate to the principal source of complaints from citizens.

Black-smoke opacity is rated according to a Ringelmann number scale of 0 to 5, corresponding to visibility reduction of 0 to 100% in 20% steps. Because black smokes absorb light, their appearance results from attenuation of light that would otherwise be transmitted through the plume. White smokes do not absorb as much light as black, and their appearance is strongly influenced by light scattering. Depending upon the background against which white plumes are observed, their appearance varies greatly.

Opacity contributors

Several factors govern the visual opacity of a plume. Foremost is the effective particle surface area available for light scattering. The effective area depends on the amount of particle surface area, and on a light-scattering coefficient that varies with particle size and shape, index of refraction, and wave length of the light. The amount of particle-surface facing the light path through the plume depends on particle size, concentration and density, and on the length of the light path.

For white plumes, the light-scattering angle is extremely important, whereas for black plumes it is not. The scattering angle is the angle between the direction of the light path and a line from the observer to the plume. The smaller the scattering angle, the more nearly the observer is looking into the light source (generally, the sun). The light-scattering angle depends on the angle of the sun above the horizon (and therefore

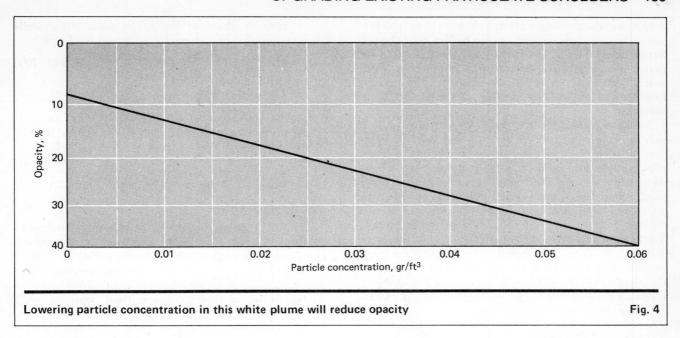

Lowering particle concentration in this white plume will reduce opacity   Fig. 4

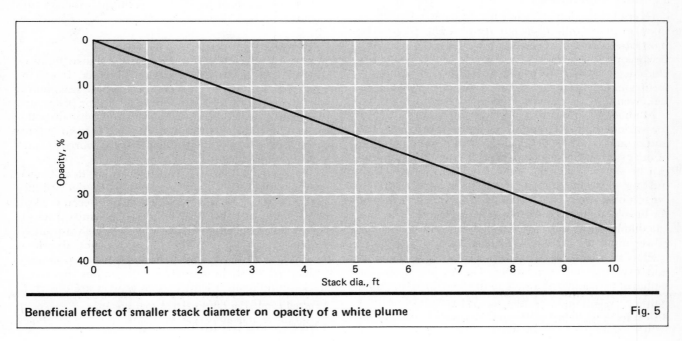

Beneficial effect of smaller stack diameter on opacity of a white plume   Fig. 5

on the time of day and season), on the horizontal offset angle between the observer's position and the position he would have if the sun were directly at his back, and on the ratio of the distance between the observer and the smokestack to the elevation of the plume above the observer's position.

Reducing opacity

Because of the several factors involved, there is more than one way to reduce opacity. The most obvious is to lower the particle concentration. The nature of the light-scattering phenomenon is such that $\ln(100 - \%$ opacity) is a linear function of the product of stack diameter and particle concentration. This relationship holds at all opacities for black plumes, and up to about 40% opacity for white plumes. Fig. 4 shows a plot of opacity vs. particle concentration based on this relationship for a hypothetical case in which the stack diameter is 5 ft, particle mass-median dia. is 1 μm, geometric standard deviation of particle-size distribution is 3, particle density is 2.5 g/cm^3, sun angle is 50 deg above horizon, observer offset angle is 30 deg, and the observer is standing 2 stack heights from the plume. The curve shows that for a particle concentration of 0.035 gr/acf (and an index of refraction of 1.5), the opacity of a white plume viewed against a blue sky would be 20%. To get 10% opacity, the concentration would have to be reduced to 0.016 gr/acf.

Another approach for reducing opacity would be to decrease stack diameter. Fig. 5 presents a plot of opacity

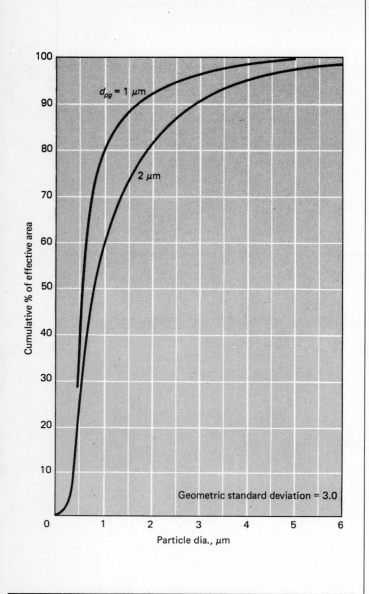

Cutting concentration of particles having diameters in the midrange lessens opacity   **Fig. 6**

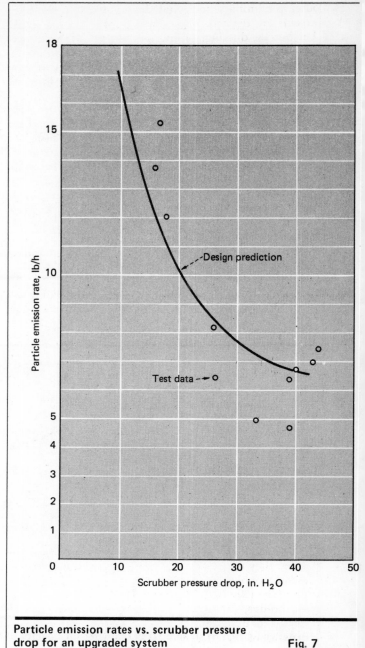

Particle emission rates vs. scrubber pressure drop for an upgraded system   **Fig. 7**

vs. stack diameter for the conditions given above, with the same particle concentration of 0.035 gr/acf. It can be seen that to reduce opacity to 10%, stack diameter would have to be about 2.3 ft.

In some cases, it may be possible to change particle size distribution by means of a process change. Fig. 6 plots cumulative percentage of total, effective, light-scattering cross-section vs. particle dia. for particle size distributions having mass median diameters of 1 μm and 2 μm and geometric standard deviation of 3. For the larger size distribution, only 10% of the light-scattering area arises from particles smaller than 0.4 μm, and another 10% from those larger than 3 μm. Cutting concentration in the mid-range of particle diameters would have a great effect on plume opacity.

Application of upgrading techniques

The potential for scrubber system upgrading is illustrated by a project carried out by Air Pollution Technology, Inc., for an inorganic chemical producer.

Five scrubbers handling from 40,000 to 50,000 cfm each were made obsolete within a few years of their installation by the advent of more-stringent air pollution rules. Their emission rates had been running 30 to 60 lb/h at a scrubber pressure drop of 14 in. H$_2$O. The new regulations required approximately 18 lb/h or less. The company was considering the installation of new venturi scrubbers with 40-in. H$_2$O pressure drop, after attempts to adjust the existing scrubbers failed.

A scrubber upgrading program was initiated, and

after preliminary tests and design study, a modified system was devised. This system utilized the old scrubber bodies, ducting, piping, pumps, etc. The changes required addition of a new pretreatment section, scrubber internals, and a high-efficiency entrainment separator. It was also necessary to replace the existing fan with one capable of higher pressure drop.

Results obtained with the modified system are illustrated in Fig. 7, a plot of emission rate against scrubber pressure drop. The company elected to operate the scrubbers at around 20 in. H_2O pressure drop in order to provide reserve capacity for future needs. The modified scrubber system was installed for considerably less money than the new venturi scrubber system would have cost, and the horsepower requirement was roughly half that which would have been required for the new venturi scrubber system.

Scrubber system upgrading can pay off in many instances because small-particle technology is complex and its principles are not well-understood by many engineers. Careful study can indicate whether the scrubber was improperly designed in the first place, or whether conditions have changed so that something better can be devised.

References

1. Calvert, S., *J. Air Pollution Control Assn.,* Vol. 24, 1974, p. 929.
2. Calvert, S., *Chem. Eng.,* Vol. 84, Aug. 29, 1977, p. 54.
3. Calvert, S., Goldshmid, J., Leith, D., and Mehta, D., "Scrubber Handbook," Natl. Tech. Info. Service, Springfield, Va., PB 213-016, 1972.
4. Calvert, S., Jashnani, I., Yung, S., and Stalberg, S., "Entrainment Separators for Scrubbers—Initial Report," NTIS, PB 241-189, 1974.
5. Calvert, S., Yung, S., and Leung, J., "Entrainment Separators for Scrubbers—Final Report," NTIS, PB 248-050, 1975.
6. Calvert, S., Yung, S., and Barbarika, H., "Entrainment Separators for Scrubbers," paper presented at EPA/APT Second Fine Particle Symposium, New Orleans, May 1977.
7. Harris, D. B., "Procedures for Cascade Impactor Calibration and Operation in Process Streams," EPA 600/2-77-004, 1977.
8. Weir, A., Jones, D. G., Papay, L. P., Calvert, S., and Yung, S., "Factors Influencing Plume Opacity," *Environmental Science and Technology,* Vol. 10, 1976, p. 539.

Troubleshooting wet scrubbers

Improper gas-liquid traffic, fouling, and mist carryover can hamper the ability of packed beds and low- and high-energy devices to control emissions.

William Gilbert, Croll-Reynolds Co.

□ Even though wet scrubbers are relatively simple devices, they require proper care to ensure long service life and troublefree operation. In this discussion, some commonly used scrubbers will be divided into three basic groups, according to their maintenance needs: packed-bed units, and low- and high-energy devices. Nearly all of these scrubbers involve two distinct zones of operation—a contacting zone where the vapor or particle is captured, and a disengaging zone where the liquid is eliminated from the gas.

Packed scrubbers use various internal arrangements. The most common is a countercurrent tower arrangement. This type of unit is particularly well-suited for high-efficiency gas absorption. It is also chosen for dust collection when the dust is soluble. Other units operate in a cross-flow manner for extremely soluble gases or because of restricted space availability. Although theoretically possible, very few packed beds are operated in a cocurrent fashion. This would require an additional vessel for separation.

Low-energy scrubbers as defined here include a wide variety of designs, typically using from 0.75 to 3 hp per 1,000 cfm. Many preformed-spray scrubbers, centrifugal and baffle-type units, self-induced spray devices, and some mechanically aided scrubbers fall into this category.* They are most often selected to recover dust or combinations of dust and gases. Low-energy scrubbers normally handle particles down to approximately 3 microns as an average size. Since many of these devices employ a cocurrent flow pattern, an additional separator vessel is necessary to disengage entrained liquid from the gas.

High-energy scrubbers include some plate-scrubbers and many gas-atomized spray units, such as venturi, orifice and rod-bank scrubbers. Often these devices feature a venturi-throat design, in which the liquid enters by a crossflow arrangement. The high gas velocity relative to that of the liquid atomizes the liquid; the high throat-velocity creates considerable turbulence, which promotes contacting. This type of unit is most often used for dust collection only, although some gas absorption also occurs. Its widest application is for capturing very fine particulates (less than 3 microns). A cyclone separator or similar arrangement usually accompanies this type of scrubber. Mesh type mist eliminators or chevrons may also be specified.

Problems with wet scrubbers normally manifest themselves as either high utility needs or low removal efficiencies. Other areas of consideration involve abra-

*See p. 154 for further description of the devices listed here.

Originally published October 24, 1977

sion or corrosion (see pp. 159-162 for further discussion of these latter trouble-areas.)

Packed towers

One of the most common difficulties noted with packed towers is high pressure drop on the gas side. As the liquid rate increases, the pressure drop will increase until, at the floodpoint, the liquid will actually begin to form a layer above the packing. Typical packed towers for air pollution control are designed to operate at values far below the floodpoint. Many are equipped with spray nozzles that make it extremely difficult to overload the tower with liquid.

The first step in avoiding flooding rates is to measure either the inlet or discharge liquid flows. If a spray nozzle is used, its liquid pressure should be checked, as well as its condition. If a trough or weir-type distributor is used, there may be little restriction on liquid flow through the unit.

A quick check for high liquid rate can be obtained by turning off the liquid temporarily. If the pressure drop immediately falls to the dry value, then the liquid rate is in excess of design. If the pressure drop remains high, other problems, such as fouling or plugging, are present.

Assuming that the liquid rate is correct, a check should be made of the gas rate through the unit. As this rate increases at any fixed liquid rate, the pressure drop through the column will increase. It should also be noted, of course, that the efficiency of the tower will be decreased by excessive gas rates.

An S-pitot tube or similar device can be used to obtain stack velocity readings, needed for calculating the total volume leaving the column. Exit-gas volume can differ significantly from that of the inlet. Whenever possible, inlet measurements should be taken to ensure that the gas volume is equal to or less than the design volume. High inlet temperatures, or changes in water content, can alter the volume rate considerably.

Changes in gas composition from the design values can also alter the pressure drop through the column. As gas density increases, the pressure drop will tend to increase. If the tower was originally designed for a substantial percentage of soluble or condensible components, the equivalent volume of noncondensibles will cause a greater pressure drop.

Next to improper gas or liquid rates, the presence of insoluble solids becomes the greatest concern. Solids or gases that precipitate within the tower can cause plugging in the mist eliminator, packed bed, or support plates. These items can be visually inspected. Normally, packing plugs at the top or bottom, and not elsewhere along the bed. Plugging results from inlet material accumulating on the bottom of the packing, or from liquor recirculated to the top of the bed.

Since many packed towers use injection-molded plastics, warnings should be posted against steam cleaning. Many problems in tower operation can be traced to a maintenance program involving extremely hot solutions used to clean the internals. Although the packing may not actually melt if such solutions are sprayed on the bed, they can cause deformation because of the weight of the packing resting on the bottom layer.

In scrubbers having fiberglass-reinforced plastic (FRP) walls, any incipient corrosion normally attacks the inner resin layer. Exposed fiberglass mat then begins to shred. If such material is picked up in the recycle loop, it can cause plugging at the top of the packed bed. Besides raising the pressure drop, this is a sign that more-serious structural damage will occur to the tower. The cause should be pinpointed immediately. If a particular area is affected, care should be taken during repairs to eliminate the contributing conditions. If these are due to chemicals not previously suspected, additional washdown sprays may prove to be a temporary answer. Serious damage may call for alternative materials of construction.

A previous operational history helps but is not always available. Repeated upsets can shift the arrangement of the internals. When packing is installed randomly, it is normally loose. Gas velocities through such an arrangement are so low that they do not cause problems under most conditions. During startup or upset, the gas flowrate can suddenly rise and fluidize part of the packing, resulting in a non-level bed. With a spray-type or weir-type distributor, the liquid will then be unevenly distributed. The result: channeling of the gas and liquid with low collection or gas-absorption efficiencies. A quick visual check of the tower packing will determine whether leveling of the packing is necessary.

Mist carryover usually follows high gas or liquid rates, or a defective mist eliminator. Normally, a high pressure drop will also ensue. In some cases, a temporary upset may force the mist eliminator out of its normal position within the column. Visual inspection to ensure that no gaps appear between the various sections of the separator is advised.

Sometimes, a low gas velocity can cause carryover. Mist eliminator systems are normally designed for minimum velocities of 3 ft/s. If the spray system in the column is not properly designed, it may create a certain amount of fine mist and induce carryover through the mist eliminator at gas velocities below the initial design point. For this reason, a low nozzle pressure is strongly recommended. If an existing tower has a high nozzle pressure (over 25 psig), changing the nozzle may eliminate carryover problems.

If high gas velocity is the culprit, it may be possible to adjust the inlet gas rate. If not, substitute an alternative tower packing to reduce the overall pressure drop. For example, a more-open-design 2-in. packing may replace existing 2-in. saddles. Although the surface areas may be similar, and the overall efficiency the same, pressure drop can be reduced to original design conditions. Care should be taken when considering such substitutions that efficiency is not sacrificed.

Packed towers can operate over a wide range of gas rates. The maximum rate is normally the design condition. As the rate decreases, the effective liquid-to-gas ratio will increase. This means that the efficiency will remain constant or actually increase slightly. At some point, efficiency will begin to decrease again. This occurs when pressure drop through the column becomes so low that the gas is not forced to distribute itself across the entire packed bed. Such a condition often results at values of 30% of the design gas rate.

If wide variations in the gas rate are anticipated, the

manufacturer of the column should be consulted to assure that it can handle these variations. A spray-type distributor is strongly recommended for such service. In trough- or weir-type distributors, the gas velocity and accompanying turbulence normally ensures good distribution of liquid. But as gas velocity decreases, the distribution of the liquid can be disturbed. In existing columns, a spray-type distributor and a mist eliminator section can be added where variations in the gas rate are causing problems with efficiencies.

Low-energy scrubbers

Ejector venturis are representative of a variety of designs for low-energy scrubbers. Such units are used for both absorption and dust collection. Energy consumed by any of these devices remains about the same for a given dust-collection efficiency. In most cases, energy needs vary more with the nature of the dust.

Problems with plugging do not usually occur in open designs—often used whenever particle agglomeration can take place within the throat of a unit. Other difficult applications crop up with subliming materials and with gases, such as sulfur dichloride, that decompose. Where particulates can form within the unit, washdown systems are often employed.

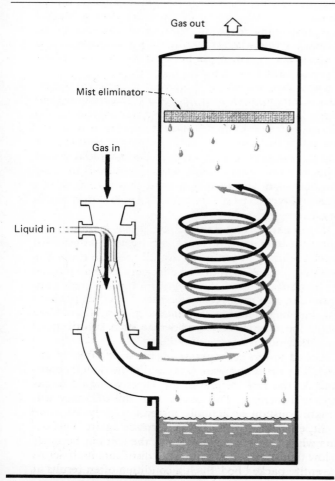

Venturi scrubber equipped with cyclone separator and mesh mist-eliminator   Fig. 1

(labels in figure: Gas out, Mist eliminator, Gas in, Liquid in)

Operating conditions should be checked against the design conditions. In a recycle loop, if the makeup rate is too low, concentration of contaminants will increase above design conditions, resulting in poor absorption. Low makeup rates can also concentrate solids in the recycle loop. This can lead to serious abrasion problems caused by salting out of solids from the solution.

Energy input

If the liquid flow or a mechanical rotor is used to introduce energy, the gas flowrate can be allowed to vary over a wider range than if energy is introduced via the gas. In the latter case, some mechanical method must be devised to ensure that the energy becomes available for removing the dust.

For example, an ejector venturi uses the liquid to draw in the gas. At zero draft, the unit might typically use 60-psig liquid with $L/G = 40$ gpm/1,000 cfm. If the gas flowrate drops to one half the design value, the liquid pressure remains the same while the L/G rises to 80 gpm/1,000 cfm. The energy of the gas has effectively increased, implying that the dust collection efficiency should equal or exceed the design level.

Alternatively, a low-pressure-drop standard venturi scrubber, as shown in Fig. 1, relies on the gas to create a high velocity through the throat. This results in a substantial pressure drop in high-energy scrubbers, but it can be as low as 6 to 10 in. At that range, this unit is effectively a low-energy scrubber. Since the unit relies on the pressure drop across the throat, cutting the gas volume will reduce the gas velocity through the throat, decreasing performance. Adding a restriction to the throat restores the original velocity and maintains good scrubbing effectiveness.

Whenever a pressure drop becomes too high, the same procedures should be applied to low-energy scrubbers as to packed towers. Check the gas rate, liquid rate, and gas composition before proceeding further. At this point, the methods for reducing a high pressure drop differ somewhat. For example, in an ejector venturi, a low liquid pressure or flowrate could be responsible for a low draft (negative pressure drop) or higher pressure drop than design. A high gas rate could still cause a high pressure drop. In units that rely on the gas to create the necessary scrubbing action, either a high gas rate or a high liquid rate will raise pressure drop.

In dust collection, concern must be focused on the particle size distribution and the loading of dust. This is analogous to gas composition and gas concentration in gas absorption. Collection efficiency is affected by high dust loading or by smaller-than-anticipated particle size. As particle size falls, efficiency will tend to drop for any amount of energy. Fig. 2 illustrates typical energy requirements for various particle sizes.

Samples of the particle-size distribution are best left to professional testing agencies or an in-house expert. Total mass measurements, which yield the dust loading, may be made at the site. Most scrubbers are insensitive to dust loading, so that minor changes will not appreciably diminish performance. This does not mean that a unit designed for 1 gr/ft³ can handle 5 or 6 gr/ft³, but it does permit 1.2 gr/ft³ without appreciable change in performance. If inlet loadings much higher than design levels are encountered, a precleaner or spray of some type may help to relieve the problem. The original manufacturer should be consulted.

Changes in operating conditions can often result in changes in the dust loading. If the unit is being used to control dust from plant exhaust fans, it may be possible to relieve the problem through better housekeeping. Manways that are not in use should be closed when vessels are charged, and ventilation fans should be set to provide adequate gas flow without pulling excessive dust away from charging operations.

It may also be possible to adjust throughput, reaction rates or other process conditions so as to minimize emission problems. In some cases where emissions are expected to stay at the higher levels, it may be necessary to up energy input. In an ejector venturi an increase in liquid pressure or flowrate can compensate for changes in dust loading or composition. In a standard venturi or impingement-type design, a higher liquid-to-gas ratio or gas-side pressure drop may compensate for the increased duty.

Separator arrangements

Separator designs vary greatly. For most low-energy applications, the separator vessel stands apart from the scrubber body. This allows for some variation in arrangement of the units, and often the separator vessel is also used to store liquid for recycling. Separators for low-energy scrubbers generally follow either gravitational or impingement designs.

For gravitational separation, the simplest device would be a very large settling chamber. However, unless the liquid spray exiting the scrubber is very coarse, such devices are normally impractical. Modified designs that impact the liquid on a surface to create larger droplets are often used.

If pressure drop goes too high in the separator, plugging can be suspected. In some impingement units, simple plates used for mist elimination offer a surface for solids accumulation. Such devices should always be installed so that they can be easily removed for cleaning. If solids do collect, access should be provided so that settled material can be flushed out. In large vessels, a manway suffices for this purpose; a large-diameter cleanout connection on the bottom proves helpful. On smaller vessels, the washdown can often be accomplished through the gas outlet. A spool piece in the exit ductwork allows access for a hose to wash down the internals.

Other devices that use impingement techniques include mesh mist-eliminators and chevron-type separators. These devices create a tortuous path for the exiting gas and are often matched with scrubbers that produce a moderately fine spray. The liquid droplets impinge upon either small wires or flat plates. Larger droplets then fall off the pad due to their greater size. While such units provide good efficiency, they do present a risk of plugging. This risk can be reduced with a chevron type, which permits high efficiency at pressure drops of 0.5 to 1 in. H_2O. Chevrons have an open design that can handle larger solids and are easier to clean.

If a high pressure drop persists, a check should be made of the gas velocity, and the pad should be visually inspected for plugging, which most often occurs on the bottom face. In many cases, the pad can be steam-cleaned if made of stainless steel or similar metal construction. Care should be taken with nonmetallic pads to avoid damage in cleaning.

If carryover becomes a problem, inspect the unit for possible gaps. In most applications, one must insert mesh mist eliminators through a manway or similar opening. Although the pads are manufactured for an

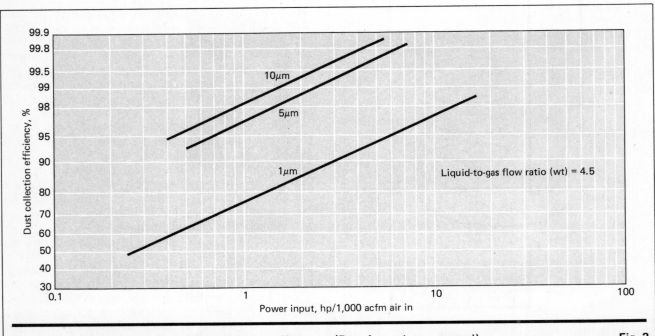

Particles smaller than expected will compromise efficiency. (Data from ejector venturi)   Fig. 2

exact fit to the tank, there have been cases where they were cut in the field for easier installation. This can create gaps during service, since the pads must support additional weight from the liquid draining off. Gas blowing through can force the mesh apart at the seams. If this happens, light line or wire can be used to tie the pads in place.

Chevrons are subject to the same concern over fit. Again, it is suggested that these sections be wired in place to reduce the possibility of a section lifting out of place and allowing carryover. Because these units work by impingement, there is a minimum as well as maximum velocity for good operation. On a typical mesh pad, the design velocity may be 5 to 9 ft/s. A minimum of 3 ft/s is often suggested. Low velocities do not create difficulties, since they generate coarser sprays that can be separated via gravitational effects. If carryover is anticipated at low gas flowrates, a smaller pad may be installed, or part of the existing pad can be blocked off to boost velocity.

High-energy scrubbers

High-energy scrubbers follow designs similar to those for more-coarse dust collection. The energy range generally varies from 3 to 6 hp/1,000 cfm. The designs most commonly employed use high gas velocities to atomize the liquid (Fig. 1). Gas-side pressure drop ranges from 15 to 45 in. H_2O for many applications.

Because the unit has relatively open design, it normally does not plug. High pressure drops indicate either that gas volume is greater than anticipated or that liquid rate is higher than design. A flowmeter check or pressure-gage check (if spray nozzles are used) will usually point up this problem. Gas composition or gas temperatures can also affect volume throughput.

These units are used more for dust collection than for gas absorption. The effects of high dust-loading and particles smaller than anticipated are very similar to those already discussed for low-energy scrubbers. Because of the fine size of particles being collected, characterization of the dust becomes difficult. Overall inlet- and exit-loadings can be measured using standard EPA methods, but more-detailed information regarding particle-size distribution is beyond normal maintenance procedures.

With most venturi scrubbers, a low pressure drop is of more concern than a high one. Low pressure drops will tend to diminish collection efficiency, and can result from low gas rate or low liquid rate.

In some cases, the liquid distributor may be plugged.

This can happen with units using weirs, spray nozzles, or even open designs. If collected material is insoluble and recirculating, it can block such devices. Constant flow prevents such troubles.

If the unit is shut down frequently, it is possible to dry out dissolved solids within the liquid distributor. Over a period of time, this can lead to problems. If the unit must be operated in this manner, it should be flushed every time it is shut down.

If variations in gas flowrate through a venturi are expected to be large, a variable-throat design is often used; it is then necessary to adjust the throat area to maintain a high velocity for scrubbing. If the gas rate fluctuates by less than ±25%, a fixed-throat design can often be substituted.

With a variable-throat design, care should be taken to ensure that the actuator is operating properly. If the actuator seizes, it will cause difficulties with pressure drop through the unit. Seizure generally occurs when solids build up on the shaft, or when corrosion causes the operator to malfunction. The shaft should be disturbed occasionally even if the gas flow appears to be relatively constant. This can be done by shutting off the fan temporarily. By forcing the shaft to move rapidly, a workman can remove buildup without dismantling the system. Manually adjustable throats should also be worked occasionally, so as to avoid freezing.

The separator most often coupled with high-energy venturis is a cyclone device. Carryover can accompany gas rates lower than the maximum design value, necessitating the installation of a chevron or mesh-type mist eliminator at the top of the vessel. The cyclone design will be maintained, but the final polishing removal of mist will be achieved by the impact-type separator. This setup works well whenever gas flowrates are expected to fluctuate.

The author

William Gilbert is Division Manager, Air Pollution Control Equipment, for Croll-Reynolds Co., 751 Central Ave., Westfield, NJ 07091. Previously with Betz Environmental Engineers and Schutte & Koerting Co., he has authored a number of papers on particulate scrubbers. He holds an M.B.A. from Seton Hall University and a B.S. in chemical engineering from the University of Pennsylvania.

How to choose a

Large cyclone collects dust from coal grinder; smaller one separates product in a pneumatic conveying system

Christian Doerschlag, Doerschlag Co., and Gerhard Miczek, G. Miczek Co.

☐ Since the first cyclone design was patented in 1886, manufacturers and researchers have accumulated so much data on these separation devices that the engineer who must choose one is confronted with a multitude of designs and claims for performance. The various accessories available for pressure recovery, reduction of pressure drop, and efficiency improvement can both confuse and frustrate the engineer. Indeed, comparing cyclone efficiencies as they are given by various makers can be downright misleading unless the basis for comparison is clearly defined.

Cyclone design

As used in industry, the term "cyclone" includes mechanical collectors, without moving parts, that use centrifugal force to separate particles from a gas stream. A basic cyclone equipped with tangential entry is shown in Fig. 1. Its principal components are the inlet section, a cylindrical barrel, a conical transition from barrel to dust outlet, a gas outlet-pipe, and a dustpot below the dust outlet.

Operation of the cyclone is very simple. The dust-laden gas stream channels tangentially into the cyclone barrel, which imparts a spinning, vortexed flow-pattern to the gas-dust mixture. Centrifugal force separates the dust from the gas stream, and the dust travels first to the walls of the barrel, then down along the conical section to the dust outlet. The spinning gas also travels down along the wall toward the apex of the cone, but reverses direction in the center of the cone, and leaves the cyclone through the gas outlet-tube at the top.

As will be shown later, the relationships between the various cyclone components play an important role in the overall collection efficiency of a particular design, and changing only one or two dimensions will affect both capital and operating requirements.

Cyclone units range from the "run of the mill" model, which can be specified from a textbook calculation, to the high- and ultrahigh-efficiency models, which to meet performance standards demand detailed knowledge and experience in cyclone design.

In the quest to improve collection efficiencies, researchers have tried many variations of the basic cyclone form. Cyclone barrels (separation chambers) have been built with both cylindrical and tapered sides; tangential inlets have been designed with both square

Originally published February 14, 1977

cyclone dust collector

Greater emphasis on dust collection—for both pollution control and product recovery—has enhanced the attractiveness of this low-maintenance, low-cost gas cleaning device.

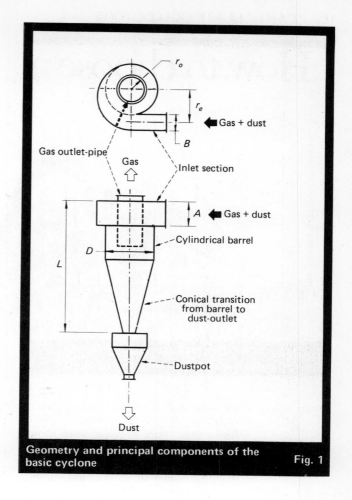

Geometry and principal components of the basic cyclone **Fig. 1**

and round cross-sections; and axial inlets equipped with spinner vanes have also been employed. Various forms of guide vanes have been installed in gas outlets, and flat plates—as well as guide vanes—have been mounted in dust outlets.

Many of these variations are, in fact, useful in improving the performance of multicyclones (a number of cyclones operating in parallel). However, even in recent times, there have been few modifications that have improved upon the collection efficiency of the basic cyclone.

Efficiency and capacity

In many instances, a cyclone's performance is thought satisfactory only because its true efficiency is not critical, yet still exceeds the minimum required for the separation. In such cases, deviations in cyclone efficiency are smaller than deviations arising from the process itself, and do not appreciably affect product quality. Rough-cut separation of coarse dusts and pre-collection are examples of applications where low efficiencies can be tolerated.

In this analysis of cyclone performance, therefore, we will eliminate the "run of the mill" cyclone and concentrate on the models denoted as high-efficiency. Irrespective of manufacturers' claims for high, very-high, or ultrahigh-efficiency performance, all data should be reduced to a common base to allow a fair comparison.

Some of the primary considerations in any evaluation of cyclones are: gas volume; cyclone body diameter; pressure drop; and particle size-distribution.

Cyclones are generally designed to meet limits on efficiency and power consumption; remaining design variables can be related to these factors.

Various theories hold that in a mixture of dusts there is a limit-particle size above which all particles are collected, and below which they are lost. Theoretically, the limit particle is that which moves in a circular path of approximately the same diameter as the gas outlet-tube and whose movement is balanced by the centrifugal forces of rotation and the drag forces that act on the particle. Particles smaller than the limit size are drawn into the inner swirl and are swept out of the cyclone; larger, heavier particles are thrown outward against the cyclone wall and are collected.

To calculate the limit-particle size, we can use, for example, the formula developed by Feifel [2]. Drawing also on Stokes' law to describe the settling velocity of a dust particle, we can relate the limit-particle diameter to various dust and gas properties, as well as the cyclone dimensions:

$$d_{theor.} = C_o h^{-1/4} D^{1/2} Q_s^{1/4} \mu^{1/2} \gamma_d^{-1/2} \qquad (1)$$

$$h = 5.21 \Delta P / \gamma_G \qquad (2)$$

and
$$Q_s = 4Q(1/h)^{1/2}/\pi D^2 \qquad (3)$$

167

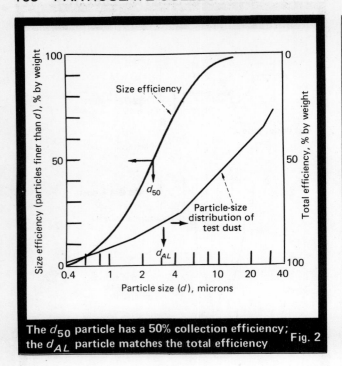

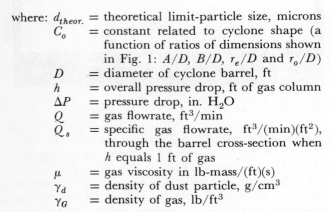

The d_{50} particle has a 50% collection efficiency; the d_{AL} particle matches the total efficiency   Fig. 2

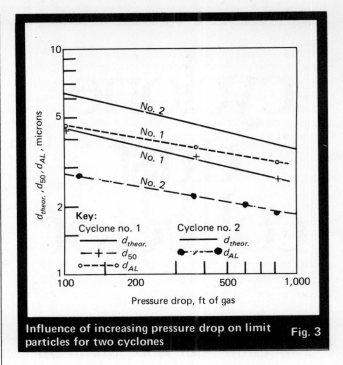

Influence of increasing pressure drop on limit particles for two cyclones   Fig. 3

where: $d_{theor.}$ = theoretical limit-particle size, microns
C_o = constant related to cyclone shape (a function of ratios of dimensions shown in Fig. 1: A/D, B/D, r_e/D and r_o/D)
D = diameter of cyclone barrel, ft
h = overall pressure drop, ft of gas column
ΔP = pressure drop, in. H_2O
Q = gas flowrate, ft^3/min
Q_s = specific gas flowrate, $ft^3/(min)(ft^2)$, through the barrel cross-section when h equals 1 ft of gas
μ = gas viscosity in lb-mass/(ft)(s)
γ_d = density of dust particle, g/cm^3
γ_G = density of gas, lb/ft^3

Certain restrictions limit the usefulness of this formula. First, the factor C_o must be defined. Theoretically, C_o can have any value between zero and infinity. However, we have found that practical values range between 2.7×10^3 and 4.2×10^3. This means that the maximum ratio of C_os for two cyclones of different geometries is normally 1.0 to 1.55. Furthermore, in experiments, we could not detect any measurable difference in the limit-particle size that resulted from variations of the factor C_o over the range mentioned. These results indicate that the shape of the cyclone (without considering its length, L) has little influence on the collection efficiency as long as other parameters in Eq. (1) *do not change*.

(Shape of cyclone is here defined as the position and configuration of the tangential inlet and the diameter of the gas outlet-tube. The effects of C_o on efficiency will be neglected in this discussion.)

In evaluating measured results, we can define the limit particle as that dust particle (described by its equivalent diameter) that has a 50% collection efficiency (cut point d_{50}), or that corresponds in size to the intersection of the measured total efficiency with the particle-size distribution curve (average limit-particle, d_{AL}) as shown in Fig. 2. Both definitions are inaccurate since the first neglects the usual unsymmetrical form of the size vs. efficiency curve, resulting in a more-favorable limit particle; whereas the second is based on a value that varies with the shape of the particle-size distribution curve, resulting in a less-favorable limit particle [3]. A typical example is shown in Fig. 2.

The influence of three important variables—overall pressure drop, barrel diameter, and specific flowrate—was assessed from measurements taken from accurate laboratory tests. In most of the tests, the effects of the three variables were measured independently of each other [3,4,5].

Pressure drop and specific flowrate

Fig. 3 shows the influence of the pressure drop, h (feet of gas), on the limit particle for two different cyclones. Limit-particle size falls with rising pressure drop, but the performance of cyclone no. 2 betters the theoretical prediction.

Fig. 4 shows the change in size of the limit particle with respect to cyclone diameter. Again, the measured values follow theoretical expectations.

Fig. 5 shows measured limit-particle sizes for six cyclones of different specific gas flowrates. Measured values have been corrected to a common reference basis, using a cyclone diameter of 1 ft, a gas viscosity of 1.2×10^{-5} lb-mass/(ft)(s), and a dust density of 2.5 g/cm^3. Shape factors (C_o) ranged between 2,700 and 4,700, and no corrections were made for C_o.

A theoretical relationship between d_t and Q_s was then calculated for $C_o = 3.5 \times 10^3$ and $L/D = 4$. As shown in Fig. 5, the experimental results for this case deviate considerably from theory, indicating that the influence of specific gas flowrate is much greater than assumed by Feifel.

What conclusions can now be drawn by comparing theory and experimental results? First, we find that the limit-particle size diminishes (raising collection efficiency) with $h^{1/4}$ and increases (dropping collection efficiency) with $D^{1/2}$, if Q_s is constant. With the same cyclone shape, therefore, we improve efficiency by choosing a large pressure drop (h). At the same time, for a given gas flowrate, cyclone diameter (D) declines, also enhancing efficiency. In practice, pressure drop is limited by abrasion and by energy needs.

A further conclusion is that for two cyclones having the same geometry, the smaller one will be more efficient.

Fig. 5 also shows that *a cyclone with a small specific gas flowrate is much more efficient than a cyclone with a larger one as long as cyclone diameter, pressure drop, gas viscosity and dust density are not changed.*

Up to now, we have kept the barrel diameter, D, constant. If we now take the gas flowrate, Q, to be constant—i.e., we want to build a cyclone for a certain gas flowrate—then the data in Fig. 5 will change slightly. Since the specific flowrate, Q_s, is related to D (Eq. 3), we alter Q_s if we alter D. Hence, experimental values of $d_{theor.}$ at different Q_s must be corrected for a cyclone diameter that corresponds to the assumed value of Q.

According to theory, $d_{theor.}$ at constant Q is independent of Q_s (and also D). This we expect from combining Eq. (1) and (3):

$$d_{theor.} = C_o \left[\frac{Q}{\pi/4} \right]^{1/4} h^{-3/8} \mu^{1/2} \gamma_d^{-1/2} \qquad (4)$$

In Fig. 6, theoretical and experimental values of $d_{theor.}$ have been compared for a constant Q. The curve showing experimental values is somewhat flatter than Fig. 5, but still relatively steep. This means that for a constant gas flowrate, Q, and pressure drop, h, a smaller-diameter cyclone with a high Q_s will hardly be as efficient as a larger-diameter one with a low Q_s.

Feifel's formula does not recognize cyclone length.

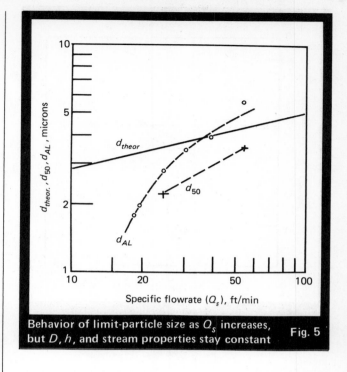

Behavior of limit-particle size as Q_s increases, but D, h, and stream properties stay constant **Fig. 5**

From experience, however, we know that efficiency improves with length when inlet and outlet dimensions are constant. Fig. 7 shows an example of the influence of relative length ratio (L/D) on overall efficiency. Ratios above 5 cannot usually be justified economically.

Parallel cyclones

Collection efficiencies discussed so far apply to individual cyclones with totally sealed dust chambers. With higher gas volumes, however, the dimensions of such a cyclone become so great as to cause a progressive reduction of efficiency. To obtain an acceptable efficiency, we would have to compensate by selecting a smaller specific gas flowrate or a higher pressure drop. A smaller

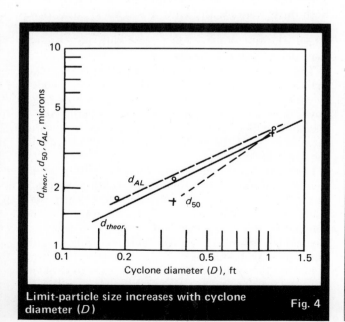

Limit-particle size increases with cyclone diameter (D) **Fig. 4**

Nomenclature

A, B = inlet dimensions, ft
$d_{theor.}$ = theoretical limit-particle size, microns
C_o = constant related to cyclone shape (a function of ratios of dimensions shown in Fig. 1: A/D, B/D, r_e/D and r_o/D)
D = diameter of cyclone barrel, ft
h = overall pressure drop, ft of gas column
ΔP = pressure drop, in. H_2O
r_e = mean radius of inlet section at point of stream entry, ft
r_o = radius of gas outlet-pipe, ft
Q = gas flowrate, ft^3/min
Q_s = specific gas flowrate, ft^3/(min)(ft^2), through the barrel cross-section when $h = 1$ ft of gas
μ = gas viscosity in lb-mass/(ft)(s)
γ_d = density of dust particle, g/cm^3
γ_G = density of gas, lb/ft^3

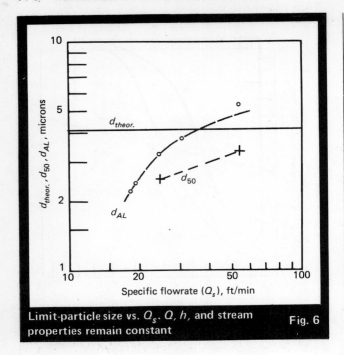

Limit-particle size vs. Q_s, Q, h, and stream properties remain constant Fig. 6

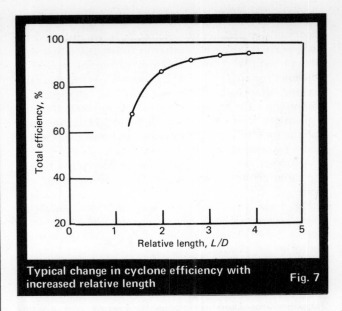

Typical change in cyclone efficiency with increased relative length Fig. 7

specific gas flowrate will only extend the dimensions of the collector, whereas a higher pressure drop will result in greater abrasion and energy needs.

For these reasons, the designer, noting that a smaller-diameter cyclone achieves a higher collection efficiency, may elect to arrange a number of small-diameter cyclones in parallel instead of building one large-diameter cyclone. Since it is not practical to provide each of these small cyclones with its own sealed dust-chamber, the dust outlets of all cyclones are connected to a common dust hopper.

If the gas pressure at the cyclone dust outlet is not identical for each cyclone, secondary circulation will develop through the dust hopper from one cyclone to another. Those cyclones that take in gas through their dust outlet perform at lower collection efficiency and may cease to collect any dust at all.

To obtain the same pressure conditions in each of the parallel cyclones, we would have to provide exactly the same dimensions in each, and the same pressure at all inlets and behind all outlets. In addition, the dust concentration and size distribution for each cyclone would have to be the same. However, since none of these stipulations can be met in practice, it is not surprising that we can find multicyclones operating with efficiencies lower than if only one cyclone had been used.

By modifying the dust discharge of the cyclone, the above-mentioned disadvantages can be reduced. Fig. 8 [5] shows the influence of backflow (air admitted through the dust outlet), on collection efficiency for two cyclones. As shown, one cyclone is only slightly affected by ΔQ, whereas total efficiency rapidly drops for the other cyclone as ΔQ increases.

Size efficiency and total efficiency

The total efficiency of a cyclone (i.e., its ability to recover from a dust a certain percentage by weight) cannot be predicted from the limit-particle calculation. For

this purpose, the engineer needs to know the pertinent size-efficiency curve (or fractional efficiencies), which can only be found by actual measurements. These efficiencies correspond to particle size—or to a narrow band of particle sizes—usually measured in the lab. For particles below 40 microns, a test dust containing a continuous range of sizes is used. Size efficiencies are evaluated by comparing the size distribution in the feed dust with that of the recovered dust, and adjusting for the measured total efficiency.

Such an experimental result should be unequivocal for a given cyclone feed gas, pressure drop, temperature, and dust gravity. Unfortunately, this is not so. Size efficiencies depend still further upon surface characteristics of the dust particles. A sticky dust, even when well dispersed into the air stream, shows better efficiencies than a dust that is completely nonsticky.

Fig. 9 shows two size-efficiency curves measured under the same conditions with two different dusts [5]. Both dusts were fly ash. Dust A was sticky, dust B was nonsticky. The air going through the cyclone had a relative humidity below 40%, and both dust samples were dried at 300°F before use.

The ratio of cut points (limit-particle sizes) is greater than 2. It is easy to see from Fig. 5 that this difference in cyclone quality corresponds to a change in specific gas flowrate of 20 to 50 ft/min. Strictly speaking, size or fractional efficiencies can be compared only when measured under the same conditions for dusts having the same makeup.

In practice, the cyclone usually collects dusts at efficiencies corresponding to nonsticky conditions (Fig. 9, dust "B"). Now if the size efficiencies have been established using a sticky test-dust, then the total efficiency calculated for an installation, especially when the cleaned gas is hot, is overrated, and performance guarantees cannot be fulfilled.

For total-efficiency calculations, we need to know not only size efficiencies, but also the particle-size distribution of the dust being collected. Herein, however, lies another pitfall. Even if the specifications are based on

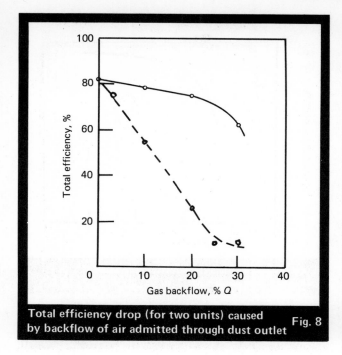

Total efficiency drop (for two units) caused by backflow of air admitted through dust outlet   Fig. 8

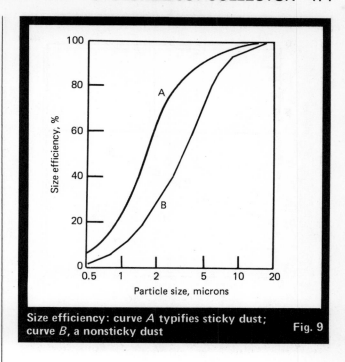

Size efficiency: curve *A* typifies sticky dust; curve *B*, a nonsticky dust   Fig. 9

good measurements and the size distribution is known for the appropriate range, it is still important to scrutinize the measurement method used. Data obtained from different methods can be very different indeed. Fig. 10 and 11 compare size distribution curves [6] established by sedimentation in liquid, and by the Bahco method (Fig. 12).* Depending on test procedure, the latter method may give variable results.

It is often assumed that the total efficiency calculation is correct if the particle-size distributions used in calculating size and total efficiencies have all been measured by the same technique. This assumption is not always correct.

Size efficiency curves shown in Fig. 13 were all measured for a test cyclone under the same working conditions. The first two curves were established by using the Bahco method. A coarse test dust was used in case no. 3, a fine test dust for no. 4. Curve no. 5, established by sedimentation, was identical for both dusts.

These curves show that size efficiencies measured by Bahco are not independent of dust particle-size distribution. Moreover, total collection efficiencies calculated by the Bahco method, and by liquid sedimentation (e.g., for the dust in Fig. 10) can be quite different, as shown in Fig. 13.

Cyclone applications

Because of numerous misapplications, lack of understanding, and stricter environmental demands on particulate collection, the cyclone has lost much of its attraction and is often looked upon as something from the past. Often, a cyclone may have been installed at a time when the only objective was to collect a portion of escaping dust that could be profitably returned to the process, while leaving the finer dust to escape. With time and the advent of environmental restrictions, even some of the finer dust must now be collected, but since

*See ASME Power Test Code No. 28.

the cyclone as originally designed is not capable of meeting the new demands, it is replaced with a more rigorous (and more costly) collector such as a cloth filter or wet scrubber.

Many such applications could be satisfied by either replacing the original cyclone with a properly designed, more-efficient cyclone, or placing a better cyclone in series with the existing unit. In many instances, the new performance requirements can be met while retaining a relatively simple piece of equipment requiring low maintenance.

Cyclones *can be successfully used* in applications where:
☐ Dust must be collected in dry form.
☐ Temperatures are high.
☐ Dust concentrations are high.
☐ Gas is under high pressure.
☐ Dust or gas becomes corrosive when wet.

Cyclones *should not be specified* for conditions where:
■ Dust will adhere to cyclone and dust hopper walls because of its surface properties, or because temperatures drop below the gas dewpoint.
■ Dust is very fine, below 1 to 5 microns, depending on the dust density, and the gas flowrate.

Cyclones are favored for applications where the collected dry dust is valuable. Instances occur in the food industry where the desired product is a powder, and contamination with the minute fibers of a cloth filter cannot be tolerated, or where retention of the collected powder on walls or in the cloth filter-material would generate a health hazard.

Cyclones are constantly used as the lowest-cost collector in all those applications where only a small portion of the dust to be collected is below 5 microns. In groups, or as multicyclones, these devices are often used as a first-stage collector in large modern plants. In case the dust concentration proves too large for cloth filters, electrostatic precipitators, or even wet collectors, cyclones are provided for precollection. It may also be

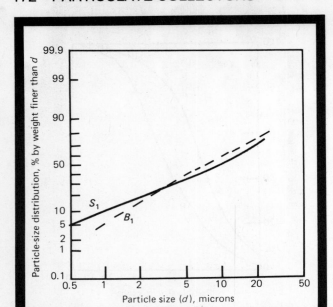

Size distribution for fine dust. S_1 from sedimentation in liquid; B_1 from Bahco test · Fig. 10

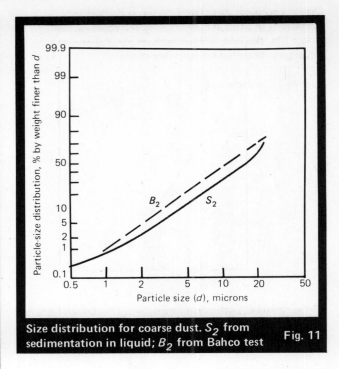

Size distribution for coarse dust. S_2 from sedimentation in liquid; B_2 from Bahco test · Fig. 11

necessary to protect expensive wet collectors from excessive abrasion, and in such instances, cyclones offer low-cost protection.

A special area of application for cyclones is the cleaning of very hot gases that have high dust loads. The units are built from heat-resistant materials and are enclosed in a refractory-lined vessel. For cyclones of larger diameter, it is possible to line internal surfaces with layers of insulating, abrasion-resistant refractory.

One such application is the use of cyclones in separating valuable catalyst from gas streams during fluid catalytic cracking. In order to reach the rather high collection efficiencies commonly required for such plants, where initial dust concentrations can reach 7,000 grains per ft^3 of gas or more, cyclones are strung together in two or three stages.

Other similar applications include recovery of usable products in operations such as drying and calcining; collection of iron oxide dust in ore beneficiation processes; and various gas-cleaning operations connected with fluid-bed incinerators.

Fig. 10–13 courtesy Fisher Klosterman, Inc.

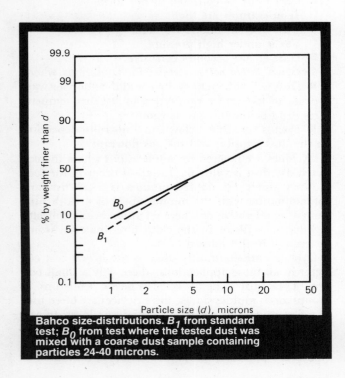

Bahco size-distributions. B_1 from standard test; B_0 from test where the tested dust was mixed with a coarse dust sample containing particles 24-40 microns.

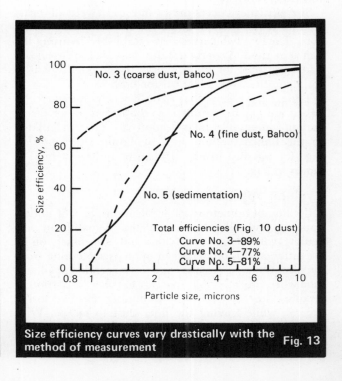

Size efficiency curves vary drastically with the method of measurement · Fig. 13

Cyclone recovers dust from drying and classifying operations in plant manufacturing refractories. For longer life, the unit is lined with an abrasion-resistant refractory mounted on steel hex-mesh

For high-pressure gas applications, conventionally designed cyclones can be enclosed by a pressure vessel, with the cyclone being exposed only to the operating differential pressure drop plus some safety margin. Cyclones have been designed for removing sand and grit from very high-pressure natural-gas wellheads, and operate quite effectively enclosed in a pressure vessel. Collected particulate is removed through letdown valves operating on a continuous basis. Other high-pressure applications where cyclones seem to be the only suitable collectors crop up in various coal-gasification systems.

Construction materials

As with other pieces of equipment, materials for cyclone construction depend to a high degree on what the application is, what choice of materials is possible, and how much the installation can cost. Cyclones have been built from practically every construction material available.

For normal applications, mild carbon steel is the first choice for low cost and reasonable life. Depending on the type of dust and operating conditions, such cyclones can last for decades, or can wear out in a matter of days. Cyclones fabricated from mild steel, and sized conservatively for low inlet velocities to reduce abrasion, are still operating on coal-fired boiler flue gases after 27 years of service. To overcome the action of erosive dusts, harder metals, such as abrasion-resistant steels, are used, or the cyclone can be lined with a thin layer (approx. $\frac{3}{4}$ in. thick) of abrasion-resistant refractory. If the collected dust is abrasive, and refractory lining must be ruled out because product degradation cannot be tolerated, units can be designed larger than normal to minimize the erosive grinding action of the dust. Such cyclones are naturally more costly but have their proper place in special applications.

When a large number of identical cyclones of moderate size are needed (6 to 24-in. barrel diameter), it is good practice to cast them from white iron, which has excellent abrasion resistance and can improve life expectancy over mild steel several times.

Corrosive atmospheres demand higher-class materials, such as stainless steel for food processing, or clad steels to reduce the cost of larger units requiring corrosion control. Small cyclones are often machined from a solid block of metal, or cast in various types of plastic materials.

Economics

The selection of a cyclone collector is based on the total efficiency required, which in turn depends on the dust particle-size distribution and the fractional efficiency of the cyclone.

The economic choice yields the lowest expenditure per unit time, within the limits of cyclone collection ability. Capital costs per unit time are based on the total investment cost and life expectancy of the installation, adjusted for interest costs. For operating costs, we usually do not include more than the cost of power, since for smoothly operating installations, maintenance is negligible.

As previously shown, the collection efficiency (or the limit particle) is influenced by several variables. We can now select these to yield the lowest overall cost per unit time for any required collection efficiency. To do this, we can adjust the following:

1. Specific gas flowrate (Q_s).
2. Number of units—those in parallel will be smaller.
3. Pressure drop.
4. Length of cyclone.

Since it is possible to develop a number of cyclone designs where the length of the cyclone is at a cost

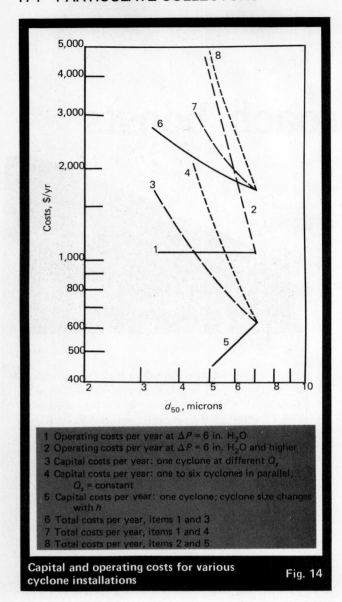

1 Operating costs per year at $\Delta P = 6$ in. H_2O
2 Operating costs per year at $\Delta P = 6$ in. H_2O and higher
3 Capital costs per year: one cyclone at different Q_s
4 Capital costs per year: one to six cyclones in parallel; Q_s = constant
5 Capital costs per year: one cyclone; cyclone size changes with h
6 Total costs per year, items 1 and 3
7 Total costs per year, items 1 and 4
8 Total costs per year, items 2 and 5

Capital and operating costs for various cyclone installations   Fig. 14

flowrate is variable, improved collection efficiency can be obtained with the lowest cost by selecting a single cyclone for the total gas flow. Less favorable is the selection of several smaller cyclones to handle the same gas flow.

If a higher pressure drop is selected to boost collection efficiency, operation of the cyclone becomes less economical.

To meet a required collection efficiency, the project engineer should start his calculations first for a single cyclone. If this is not adequate, even with the lowest specific gas flowrate, then several parallel cyclones can be substituted, using the lowest specific gas flowrate. If this measure is still inadequate, then the pressure drop can be raised.

The above procedure is an ideal simplification of the selection process, since there are other factors specific to the installation that will influence the design of the lowest-cost solution, such as:
- Available space.
- Allowable abrasion.
- Availability of intermediate sizes.
- Fabrication and transport difficulties.

Data shown in Fig. 14 account only for the actual cyclones and do not include costs for the overall facility, such as manifolds and hoppers, piping, and installation.

References

1. Foerderreuther, C., Wirkungsweise von Staubsichtern und Staub-abscheidern, *Arch. Waermewirtschaft*, Vol. 9, pp. 323–328, 1928.
2. Feifel, E., Zyklon Staubabscheider: Der Zyklon als Wirbelsenke, *Forschung*, Vol. 9, pp. 68–81, 1938.
3. Miczek, G., Comparative Studies of Cyclone Collectors, *Staub*, Vol. 20, pp. 9–13, 1960.
4. Miczek, G., Collection Limitations of Cyclone Dust Collectors, *Almanac 2*, The Czechoslovak Scientific-Technical Soc., 1957.
5. Miczek, G., Unpublished Measurements, 1951–1952.
6. Miczek, G., Unpublished Measurements, 1970–1972.

optimum with respect to specific gas flowrate, we need concern ourselves only with the first three factors.

Cost optimization—an example

Annual expenditures have been worked out for the following cyclone installation:

Air flowrate	10,000 SCFM
Cyclone pressure drop	6 in. H_2O, or higher
Economic life of cyclones	5 yr
Material of construction	mild steel, $\frac{3}{16}$ in.
Interest on capital	8%
Operating costs (power only)	0.015 $/kWh
Fan efficiency	60%
Operating hours per year	6,000

Fig. 14 summarizes cyclone investment costs, operating costs and total costs per year for a range of limit-particle sizes, or cut points. We have calculated three cases, always keeping two variables constant but changing the third. For the case where the specific gas

The authors

Christian Doerschlag is owner of the Doerschlag Co., 519 S. Otterbein Ave., Westerville, OH 43081, an engineering company for process systems and equipment. Before establishing his own firm he was Vice-President, Engineering, of American Van Tongeren Corp., Columbus, OH. Mr. Doerschlag holds a bachelor's degree in mechanical engineering from Ohio State University and is registered as a P.E. in Ohio. He is a member of the American Soc. of Mechanical Engineers and the Air Pollution Control Assn.

Gerhard Miczek is owner of G. Miczek Co., Dust Collectors and Industrial Gas Cleaning, 1501 Cowling Ave., Louisville, KY 40205. He has worked as a scientific researcher in institutes in Czechoslovakia, and as a research and project engineer in Austria. In the U.S., he has been employed by Fisher Klosterman, Inc. as a research engineer, and by American Van Tongeren Corp. Mr. Miczek holds a degree in mechanical engineering from CVUT-Prague, Czechoslovakia.

New design approach boosts cyclone efficiency

Cyclones, if more precisely designed, could do a far better job of collecting particles than has been assumed. This procedure converges to optimum designs with a minimum number of trial-and-error calculations.

Wolfgang H. Koch, *Amoco Chemicals Corp.* and
William Licht, *University of Cincinnati*

Originally published November 7, 1977

How to design more-efficient cyclone particle collectors has attracted renewed interest. Tightened emission regulations have impelled process engineers to exploit the capabilities of cyclones to the fullest. If better designed, cyclones could control particle pollution much more efficiently. Furthermore, improved cyclone performance could greatly reduce the need for more-complicated secondary cleaners.

The cyclone's simplicity of construction, low energy requirements, and ability to operate at high temperatures and pressures make it attractive for cleaning up gases for processes: for example, coal gases for chemical feedstock, and gases from fluidized-bed combustion for running gas turbines. Because cyclones can be easily cleaned and kept sanitary, they are especially valuable in the food processing industry, such as in the recovery of spray-dried products.

New approach valuable for design estimates

Previous attempts to predict cyclone collection efficiencies have been only moderately successful. A number of proposed models [4,5,13] have relied on the approach of Rosin, Rammler and Intelmann, first published in 1932 [9].

A new theoretical approach developed by Leith and Licht, which predicts grade efficiency from physical-properties data and cyclone design ratios, also predicts collection-efficiency curves that agree well with published data [6]. However, this model does not apply at extremely high inlet velocities or volumetric flowrates. It predicts a limiting efficiency of 100% as the inlet velocity approaches infinity. Because particles reentrain at high velocities, collection efficiency actually reaches a limiting value of less than 100%.

Investigating the conveyance of particles in horizontal pipes, Zenz established an experimental model for particle saltation in terms of Reynolds number and drag coefficient [16]. Extending the correlation to cyclone design applications, the Kalen and Zenz model predicts an optimum value for the inlet velocity, and the velocity at which particle reentrainment becomes significant [3].

The approach presented combines the theoretical collection-efficiency model of Leith and Licht with the saltation velocity correlation proposed by Kalen and Zenz. The combination of these models results in a cyclone design approach that will converge to an optimum design within a minimum number of trial-and-error calculations. A method is presented for graphically determining the optimum inlet velocity and the cyclone diameter for a desired separation.

This new approach should be most valuable for an initial design estimate, which can be made by an individual having little experience in the design of air-pollution-control equipment. Although most engineers who routinely perform such design work have access to elaborate design simulation programs, an occasional need to design cyclones would not justify the work necessary to develop an elaborate program. The approach presented is also intended to eliminate the need for such a program.

Calculate collection efficiency

The theoretical approach to calculating cyclone collection efficiencies advanced by Leith and Licht is based on the concept of continual radial backmixing of the uncollected particles, coupled with the calculation of an average residence time for the gas in a cyclone having a tangential inlet [6]. Table I shows a typical cyclone with all the necessary dimensions and gives values of the relative dimensions for several typical designs. The derivation of the pertinent design equations is given by Leith and Licht [6].

Cyclone "grade," or "fractional," efficiency may be calculated by Eq. (1):

$$\eta_i = 1 - \exp\left\{-2\left[\frac{G\tau_i Q}{D_c^3}(n+1)\right]^{0.5/(n+1)}\right\} \quad (1)$$

In Eq. (1):

$$\tau_i = \rho_p(d_{p_i})^2/(18\mu) \quad (2)$$

$$G = 8K_c/K_a^2 K_b^2 \quad (3)$$

The vortex exponent, n, may be calculated as a func-

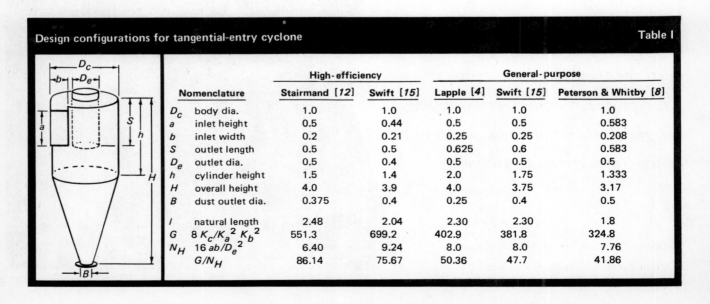

	Nomenclature	High-efficiency		General-purpose		
		Stairmand [12]	Swift [15]	Lapple [4]	Swift [15]	Peterson & Whitby [8]
D_c	body dia.	1.0	1.0	1.0	1.0	1.0
a	inlet height	0.5	0.44	0.5	0.5	0.583
b	inlet width	0.2	0.21	0.25	0.25	0.208
S	outlet length	0.5	0.5	0.625	0.6	0.583
D_e	outlet dia.	0.5	0.4	0.5	0.5	0.5
h	cylinder height	1.5	1.4	2.0	1.75	1.333
H	overall height	4.0	3.9	4.0	3.75	3.17
B	dust outlet dia.	0.375	0.4	0.25	0.4	0.5
l	natural length	2.48	2.04	2.30	2.30	1.8
G	$8K_c/K_a^2 K_b^2$	551.3	699.2	402.9	381.8	324.8
N_H	$16\,ab/D_e^2$	6.40	9.24	8.0	8.0	7.76
	G/N_H	86.14	75.67	50.36	47.7	41.86

Design configurations for tangential-entry cyclone   Table I

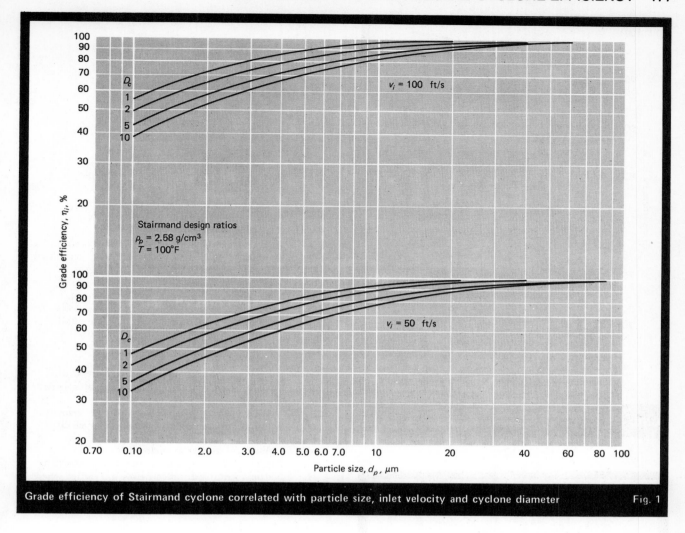

Grade efficiency of Stairmand cyclone correlated with particle size, inlet velocity and cyclone diameter   Fig. 1

tion of cyclone diameter, with D_c in feet, and temperature, T, in °F [1]:

$$n = 1 - \left[1 - \frac{(12D_c)^{0.14}}{2.5} \right] \left[\frac{T + 460}{530} \right]^{0.3} \quad (4)$$

The cyclone configuration factor, G, is a function only of the configuration and is specified by the seven geometric ratios that describe its shape. The procedure for calculating G is outlined later in the first sample calculation. Although it is well known that collection efficiency increases with dust grain-loading, this model does not account for the additional variable. It estimates performance conservatively, being based upon light grain-loading.

With the foregoing relationships, grade efficiencies may be calculated as a function of cyclone design parameters and operating conditions. An overall efficiency may be found for a given particle distribution via Eq. (5):

$$\eta_T = \Sigma m_i \eta_i \quad (5)$$

This model is consistent with experimentally observed trends. It predicts gains in efficiency with increasing values of particle density, inlet velocity and

Standard aerosol particle-size distributions [11]   Table II

Particle size,	Percent by weight less than			
µm	Superfine	Fine	Coarse	Example
150	—	100	—	—
104	—	97	—	—
75	100	90	46	94
60	95	80	40	92
40	97	65	32	86
30	96	55	27	79
20	95	45	21	67
15	94	38	16	58
10	90	30	12	44
7.5	85	26	9	34
5.0	75	20	6	22
2.5	56	12	3	8

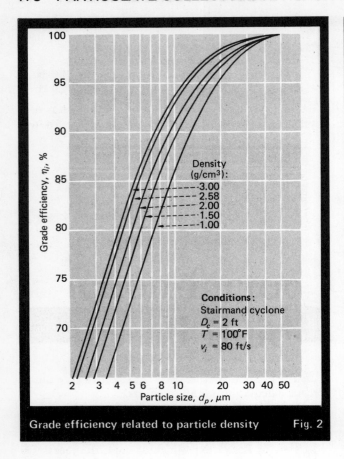

Grade efficiency related to particle density   Fig. 2

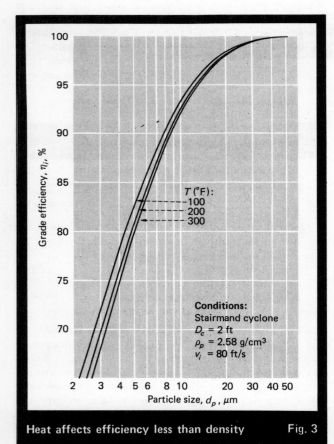

Heat affects efficiency less than density   Fig. 3

Nomenclature

B	Cyclone dust-outlet dia., ft
D_c	Cyclone dia., ft
D_e	Cyclone gas-outlet dia., ft
G	Cyclone configuration factor
H	Cyclone height, ft
K_a	a/D_c
K_b	b/D_c
K_c	Cyclone volume constant, Eq. (18)
N_H	No. of inlet velocity heads, Eq. (14)
N_o	No. of cyclones in parallel
ΔP	Pressure drop, in. H_2O
Q	Total gas flowrate, actual ft^3/s
Q_i	Flowrate to each cyclone, actual ft^3/s
S	Gas outlet length, ft
S_Δ	Characterizes relative ease of conveyance, a measure of effective particle size and size distribution
T	Temperature, °F
V_H	Volume below exit duct (excluding core), ft^3
V_{nl}	Volume at natural length (excluding core), ft^3
V_s	Annular volume above exit duct to middle of entrance duct, ft^3
a	Inlet height, ft
b	Inlet width, ft
d	Dia. of central core at point where vortex turns, ft
d_p	Particle size, equivalent aerodynamic spherical dia., ft
f_ρ	Density correction factor for saltation velocity
f_T	Temperature correction for saltation velocity
g	32.2 ft/s^2
h	Cylindrical height of cyclone, ft
i	Subscript denotes interval in particle size range
l	Natural length (distance below gas outlet where vortex turns), ft
m_i	Mass fraction of particles in size range designated by i
n	Vortex exponent
r	Effective radius, ft, Eq. (10)
v_i	Inlet velocity, ft/s
v_s	Saltation velocity, ft/s
Δ	$[3\mu^2/4g(\rho_p - \rho_f)\rho_f]^{1/3}$, ft
η_i	Grade efficiency for particle size at mid-point of interval i, %
η_T	Overall efficiency, %
μ	Fluid viscosity, lbm/ft-s
ρ_f	Fluid density, lbm/ft^3
ρ_p	Particle density, lbm/ft^3
τ	Relaxation time, s, Eq. (2)
ω	Eq. (8), ft/s

cyclone body height, and declines in efficiency with increasing fluid viscosity, cyclone diameter, outlet diameter and inlet width. Fig. 1, 2 and 3 show typical calculated grade-efficiency curves for the Stairmand design (Table I) as a function of particle size, inlet velocity and cyclone diameter, as well as the dependence of the grade efficiency on particle density and fluid temperature.

Although increases in efficiency with higher particle density can be determined directly from Eq. (1) and (2), the effects of temperature are somewhat more complicated. A temperature rise increases the fluid viscosity,

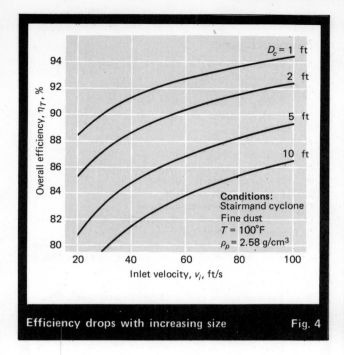

Efficiency drops with increasing size   Fig. 4

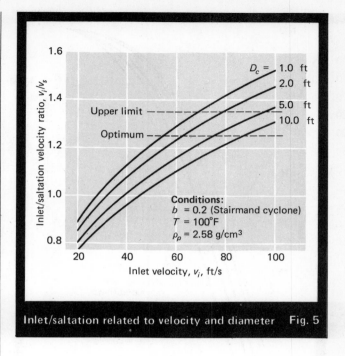

Inlet/saltation related to velocity and diameter   Fig. 5

which lowers the magnitude of the relaxation time, τ. The vortex exponent also decreases. The net result is lower grade efficiency.

Fig. 2 and 3 show that the effect of temperature on efficiency is far smaller than that of particle density. The magnitude of both effects decreases with increasing particle size. Fig. 4 reveals the calculated overall collection efficiency of a Stairmand cyclone treating Stairmand's "fine dust," which is described in Table II [11].

Saltation velocity

The concept of saltation arises in predicting the conveyance of solids in horizontal pipes or conduits. Saltation velocity may be defined in several ways, two of which apply to cyclone design:

1. The minimum fluid velocity necessary to prevent the settling out of solid particles carried in the stream.

2. The velocity necessary to pick up deposited particles and transport them without settling.

For this study, the first definition will be used, although the second one is needed in order to predict an upper limit for the cyclone inlet velocity. Zenz has shown experimentally that the velocity given by the second definition differs from the first by a factor of 2 to 2.5. He developed a correlation relating the saltation velocity in pipelines to particle and fluid properties [16]:

$$v_s/\omega = (\text{constant})(d_p/\Delta)^{S_\Delta} \qquad (6)$$

In Eq. (6), the saltation constant, S_Δ, depends on particle size distribution and particle shape, and ω and Δ are functions of fluid and particle properties. Experimental data indicate a strong dependence on particle geometry.

To apply the saltation concept and correlation to cyclone design, Kalen and Zenz made the following assumptions [3]:

1. There is no slippage between fluid and particles.

2. The cyclone inlet width is the effective pipe diameter for calculating saltation effects.

3. Grain loading is less than 10 grains/ft³.

4. The diameter effect on the saltation velocity is proportional to the 0.4 power of the inlet width.

In view of Zenz's experimental data, the fourth assumption represents an approximation that becomes more accurate with increasing particle sphericity. Expressing particle size as equivalent aerodynamic spherical diameter should increase the model's accuracy.

To approximate the fluid path inside a cyclone via a coiled pipe model, Kalen and Zenz developed an empirical correlation for the saltation velocity [3]:

$$\frac{v_s}{\omega(\text{No. of } gs)^{1/3}} = 5.19b^{0.4} \qquad (7)$$

In Eq. (7):

$$\omega = [4g\mu(\rho_p - \rho_f)/3\rho_f^2]^{1/3} \qquad (8)$$

$$\text{No. of } gs = v_i^2/32.2r \qquad (9)$$

$$r = 1/2(D_c - b) \qquad (10)$$

Defining the inlet width in terms of a dimensionless size ratio b/D_c, and substituting Eq. (8), (9) and (10) into Eq. (7):

$$v_s = 2.055\omega\left[\frac{(b/D_c)^{0.4}}{(1 - b/D_c)^{1/3}}\right]D_c^{0.067}v_i^{2/3} \qquad (11)$$

Thus, saltation velocity is a function of particle and fluid properties as well as cyclone dimensions.

Kalen and Zenz have shown that maximum cyclone collection efficiency occurs at $v_i/v_s = 1.25$ [3], and Zenz has determined experimentally that reentrainment occurs at $v_i/v_s = 1.36$ [16]. Fig. 5 shows v_i/v_s ratios calculated from Eq. (11) as a function of inlet velocity and cyclone diameter for an inlet width ratio of $b/D_c = 0.2$. The commonly accepted inlet-velocity working range of 50 to 90 ft/s, depending upon cyclone diameter, is

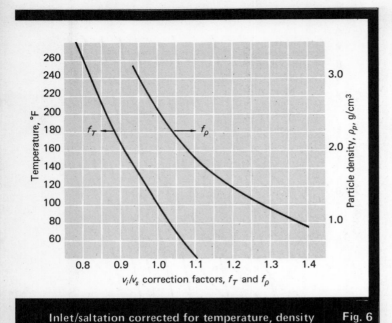

Inlet/saltation corrected for temperature, density   **Fig. 6**

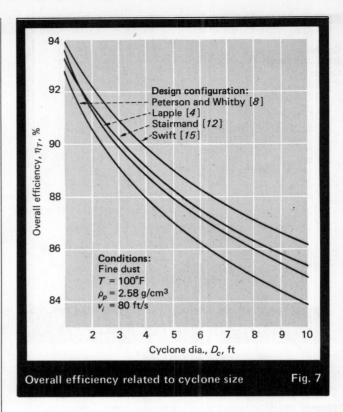

Overall efficiency related to cyclone size   **Fig. 7**

consistent with these results. Fig. 6 gives correction factors for the v_i/v_s ratio for temperature and particle density, using $T = 100°F$ and $\rho_p = 2.58$ g/cm³ as basis. Correct v_i/v_s ratios may be calculated with Eq. (12):

$$v_i/v_s = f_T f_\rho (v_i/v_s)_{\text{base}} \qquad (12)$$

The correction factors f_T and f_ρ are for temperature and density, respectively.

Estimate pressure drop

A correct estimate of the pressure drop across a cyclone is necessary, in addition to collection efficiency, so that cost effectiveness may be calculated. A magnitude of 10 in. H₂O, or less, is a generally acceptable operating range. Correlations for pressure drop predictions have been largely empirical.

A review of pressure drop theories by Strauss [14] and a comparison by Leith and Mehta [7] revealed that the approach of Shepherd and Lapple [10] is the simplest and is of comparable accuracy to the more-complex ones. This correlation estimates pressure drop in cyclones as follows:

$$\Delta P = (\rho_f v_i^2/2g)(N_H/\rho_{\text{H}_2\text{O}}) \qquad (13)$$
$$N_H = K(ab/D_e^2) \qquad (14)$$

The number of inlet velocity heads, N_H, depends upon only three of the seven geometric design ratios. If no inlet vane is present, $K = 16$; with a neutral inlet

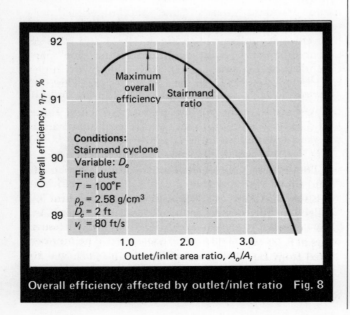

Overall efficiency affected by outlet/inlet ratio   **Fig. 8**

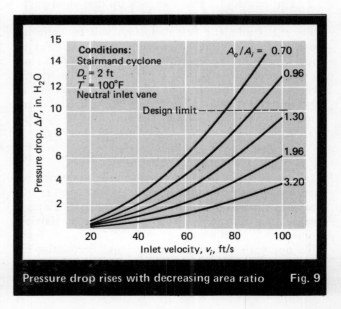

Pressure drop rises with decreasing area ratio   **Fig. 9**

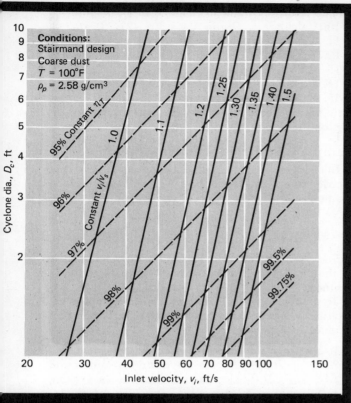

Stairmand cyclone design optimized for coarse dust   Fig. 10

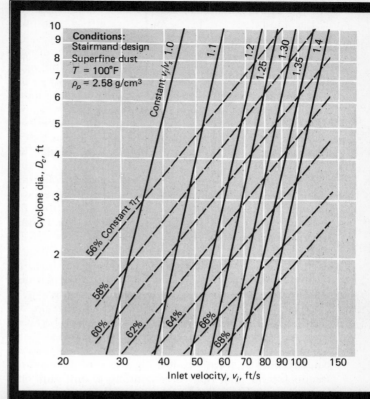

Stairmand cyclone design optimized for superfine dust   Fig. 12

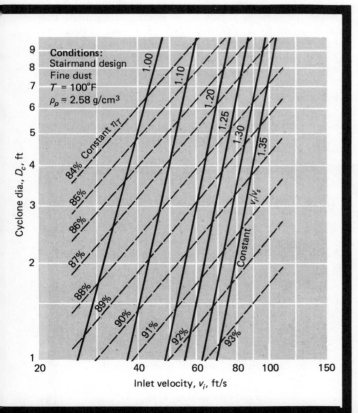

Stairmand cyclone design optimized for fine dust   Fig. 11

vane, $K = 7.5$. Eq. (14) may be simplified by the appropriate conversion factors:

$$\Delta P = 0.003\rho_f v_i^2 N_H \qquad (14a)$$

Because pressure drop strongly depends on inlet velocity, it becomes obvious that high velocities cause not only reentrainment but also excessive pressure drop.

Comparing cyclone designs

So as to compare accepted cyclone designs, Stairmand's standard "coarse," "fine" and "superfine" particle distributions [11] were tested with the Stairmand high-efficiency cyclone [12]. Fig. 4, 10, 11 and 12 were prepared from the results. Additionally, the "fine" particle distribution was tested with the Swift high-efficiency design [15] and the Lapple [4] and Peterson and Whitby [8] general-purpose designs. The design configurations for the cyclones tested are listed in Table I, and particle size distributions for the standard aerosols are presented in Table II and Fig. 13.

Fig. 7 yields overall efficiency as a function of cyclone size for the four cyclone designs collecting "fine" dust. As would be expected from Eq. (1), collection efficiencies depend upon the magnitude of the constant G. Therefore, the obvious optimization strategy would be to maximize the magnitude of G. However, pressure drops in a cyclone must also be considered. The number of inlet velocity heads, N_H, is directly proportional to pressure drop and is a function of the cyclone design parameters. Maximizing the ratio G/N_H would appear to be a far better optimization strategy in selecting a design configuration. Because Table I values show the Stairmand design to be the optimum, it has been used in further calculations.

Calculating cyclone efficiencies as a function of outlet/inlet area ratio, Strauss [14] and Kalen and Zenz [3] determined that collection efficiency increases with a decreasing area ratio. Using the Leith and Licht model, with Stairmand's design configuration (Table I), we found that efficiency reaches a maximum and declines on either side of the optimum with decreasing as well as increasing outlet/inlet area ratios. The results are shown in Fig. 8.

Note that the magnitude of the decrease in collection efficiency at lower than optimum outlet/inlet ratios is insignificant. Analysis of the model indicates that the volume at natural length, V_{nl}, exhibits maximum/minimum behavior, indicating that this effect will be observed only in cyclones of natural length, $l < (H - S)$. Additional information on the cyclone volume effect may be found in the sample calculations.

Because the number of inlet velocity heads, N_H, is inversely proportional to the outlet/inlet area ratio, the ratio has a significant effect on pressure drop. This effect has not been previously discussed. Fig. 9 shows that the magnitude of ΔP increases appreciably with decreasing outlet/inlet area ratios and indicates why design for a ratio greater than that corresponding to optimum efficiency may be desirable.

Cyclone design procedure

To facilitate the preliminary estimate of design parameters, overall collection efficiency for the three standard dusts were calculated for the Stairmand cyclone design. The resulting data are shown in Fig. 10, 11 and 12 as inlet velocity vs. cyclone diameter at constant overall efficiency and constant inlet/saltation velocity ratio.

These steps are recommended for the design of a cyclone (the data needed are the physical properties of the particles and gas, temperature, particle size distribution and volumetric flowrate):

1. Calculate the required overall efficiency.
2. Compare particle size distribution to Stairmand's three standard aerosols, and pick the standard that most closely resembles the actual aerosol (Fig. 13 and Table II).
3. Correct overall efficiency to a basis of 100°F, $\rho_p = 2.58$ g/cm^3, using Fig. 2 and 3, and the mass median particle diameter. Correct the optimal ratio, $v_i/v_s = 1.25$, for actual operating conditions, using Fig. 6 and Eq. (12).
4. From Fig. 10, 11 or 12, depending on the appropriate standard aerosol, and with the η_T and v_i/v_s calculated in Step 2, find D_c and v_i.
5. Pick cyclone design ratios in accordance with constraints (see following sample calculation).
6. Calculate $Q_i = abv_i$ (volumetric flowrate per cyclone).
7. Calculate number of cyclones required in parallel (adjust to nearest integer); $N_o = Q/Q_i$.
8. Calculate v_s with Eq. (11) or read from Fig. 5.
9. If $v_i/v_s > 1.35$ (adjusted), recalculate design ratios and v_i. (Note: For cost optimization, it may be neces-

sary to reduce N_o, and increase D_c or v_i, or both.)
10. Calculate n, G and τ via Eq. (2), (3) and (4).
11. Calculate grade efficiency via Eq. (1).
12. Repeat Steps 10 and 11 for other particle sizes.
13. Calculate overall efficiency via Eq. (5).
14. Compare results with required efficiency.
15. Calculate pressure drop via Eq. (14a) or read from Fig. 9. If necessary, make adjustments in A_o/A_i, v_i, D_c, or the design ratios, or in all. Return to Step 4.

Sample calculation: configuration factor

In order to adequately describe a cyclone, seven geometric ratios must be specified in terms of the cyclone's diameter:

$$\frac{a}{D_c}, \frac{b}{D_c}, \frac{D_c}{D_c}, \frac{S}{D_c}, \frac{h}{D_c}, \frac{H}{D_c}, \frac{B}{D_c}$$

For simplicity, only the main dimension will be used in the following equations. A cyclone diameter of 1 is assumed.

To obtain a workable design, the dimensionless geometric ratios cannot be chosen arbitrarily. Certain constraints must be observed:

1. $a < S$ (to prevent short-circuiting).
2. $b < 1/2(D_c - D_e)$ (to avoid sudden contraction).
3. $S + l \leq H$ (to keep the vortex inside the cyclone).
4. $S < h$.
5. $h < H$.
6. $\Delta P < 10$ in. H$_2$O.
7. $v_i/v_s \leq 1.35$ (to prevent reentrainment).
8. $v_i/v_s \simeq 1.25$ (for optimum efficiency).

After the design ratios have been determined, the configuration factor, G, must be calculated in order to obtain grade efficiencies from Eq. (1) [6]:

1. Calculate natural length, l:

$$l = 2.3D_e(D_c^2/ab)^{1/3} \qquad (15)$$

For $l < (H - S)$, calculate the cyclone volume at the natural length, V_{nl}:

$$V_{nl} = \frac{\pi D_c^2}{4}(h - S) + \left(\frac{\pi D_c^2}{4}\right)\left(\frac{l + S - h}{3}\right)$$
$$\times \left(1 + \frac{d}{D_c} + \frac{d^2}{D_c^2}\right) - \frac{\pi D_e^2 l}{4} \qquad (16)$$

In Eq. (16),

$$d = D_c - (D_c - B)[(S + l - h)/(H - h)] \qquad (16a)$$

For $l > (H - S)$, calculate cyclone volume below the exit duct, V_H:

$$V_H = \frac{\pi D_c^2}{4}(h - S) + \left(\frac{\pi D_c^2}{4}\right)\left(\frac{H - h}{3}\right)$$
$$\times \left(1 + \frac{B}{D_c} + \frac{B^2}{D_c^2}\right) - \frac{\pi D_e^2}{4}(H - S) \qquad (17)$$

2. Calculate the cyclone volume constant, K_c, using V_{nl} or V_H:

$$K_c = (2V_s + V_{nl,H})/(2D_c^3) \qquad (18)$$

In Eq. (18),

$$V_s = [\pi(S - a/2)(D_c^2 - D_e^2)]/4 \qquad (19)$$

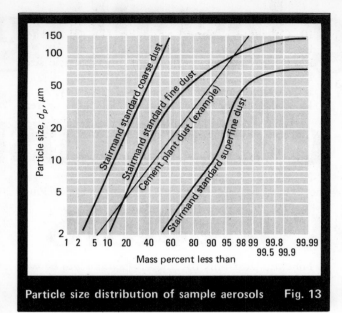

Particle size distribution of sample aerosols   Fig. 13

3. Letting $K_a = a/D_c$ and $K_b = b/D_c$, the configuration factor can be calculated with Eq. (3).

Note that $l < (H - S)$ for all the design configuration factors tested and given in Table I. In rare cases, the natural length of a cyclone was found to be much larger than $(H - S)$. Depending on the relative magnitude of the difference $l - (H - S)$, it may be possible to obtain a negative value for V_H. Aslami and Licht have recently proposed a change in Eq. (16) and (17) that modifies the volume of the central cyclone core by considering it to be conical rather than cylindrical [2].

Sample problem: cement dust cyclone

A particle size distribution of cement dust emitted from a Portland cement kiln is given in Table II and Fig. 13. Design a cyclone as a precleaner for the following conditions:

Particle density	1.5 g/cm^3
Air rate	159,600 actual ft^3/ton feed
Dust rate	230 lbm/ton feed
Feed rate	5 ton/h
Air temperature	250°F
Air pressure	Atmospheric
New plant emission limit	0.30 lbm/ton feed

The overall collection efficiency, η_T, of 99.87% that is required exceeds the capability of cyclone collection.

Therefore, the cyclone should be designed for maximum η_T.

Fig. 13 indicates that the aerosol may be approximated by fine dust.

Mass median dia., $dp_{50} = 12\mu m$.

Temperature correction for $\eta_T = 0.8\%$ (Fig. 3).

Density correction for $\eta_T = 3.0\%$ (Fig. 2).

Correction for $v_i/v_s - f_T = 0.81 f_\rho = 1.195$ (Fig. 6).

Corrected $v_i/v_s = 1.21 < 1.25$ (too low); adjust uncorrected value to 1.29.

Using Fig. 11, it becomes obvious that high collection efficiencies can only be achieved with small cyclones. Because the total volumetric flowrate is large (222 actual ft^3/s), arbitrarily set a limit of four cyclones in parallel. For $v_i/v_s = 1.29$, take $\eta_T = 90.5\%$ (corrected 86.7%), then find from Fig. 11: $D_c = 2.67$ ft, and $v_i = 73$ ft/s.

Use Stairmand design ratios (Table I).

As $Q_i = 53.3$ actual ft^3/s, $N_o = 4.16$; adjust to 4. Adjust $D_c = 2.75$ ft, $Q = 55.3$ actual ft^3/s ($v_i = 73$ ft/s).

Variables are within established limits: $n = 0.623$, $G = 551.3$ (Table I).

Using seven particle sizes, add their corresponding mass fractions: $\eta_T = 84.6\%$; Eq. (2), (3) and (4).

The dust rate for second-stage cleaning: 35.5 lbm/ton feed.

Final-stage cleaning efficiency required: 99.1%.

$\Delta P = 3.1$ in. H$_2$O.

References

1. Alexander, R. M., *Proc. Austral. Inst. Min. Met.* (N.S.), Vol. 152, 1949, p. 202.
2. Aslami, M. A., and Licht, W., Paper 33a, AIChE 81st Natl. Meeting, Kansas City, Apr. 1976.
3. Kalen, B. and Zenz, F. A., AIChE Symposium Ser., Vol. 70, No. 137, 1974, p. 388.
4. Lapple, C. E., Air Pollution Engr. Manual (J. A. Danielson, ed.), U.S. Dept. of Health, Education, & Welfare, Public Health Service Pub., No. 999-AP-40, 1967, p. 95.
5. Lapple, C. E., and Shepherd, C. B., *Ind. & Engr. Chem.*, Vol. 32, 1940, p. 605.
6. Leith, D. and Licht, W., AIChE Symposium Ser., Vol. 68, No. 126, 1972, p. 126.
7. Leith, D. and Mehta, D., *Atm. Environ.*, Vol. 7, 1973, p. 527.
8. Peterson, C. M. and Whitby, K. T., *ASHRAE J.*, Vol. 7, No. 5, 1965, p. 42.
9. Rosin, P., Rammler, E. and Intelmann, W., *Z. Ver. Deut. Ing.*, Vol. 76, No. 18, 1932, p. 433.
10. Shepherd, C. B., and Lapple, C. E., *Ind. & Engr. Chem.*, Vol. 31, 1939, p. 972.
11. Stairmand, C. J., *Filtration & Separation*, Vol. 7, No. 1, 1970, p. 202.
12. Stairmand, C. J., *Trans. Inst. Chem. Eng.*, Vol. 29, 1951, p. 356.
13. Stern, A. C., Kaplan, K. J., Bush, P. D., Cyclone Dust Collectors, *API Report*, 1955.
14. Strauss, W., "Industrial Gas Cleaning," Pergamon Press, N.Y., 1966, Chap. 6.
15. Swift, P., *Steam Heating Engr.*, Vol. 38, 1969, p. 453.
16. Zenz, F. A., *Ind. & Engr. Chem. Fund.*, Vol. 3, 1964, p. 65.

The authors

Wolfgang H. Koch is a research engineer with Amoco Chemicals Corp., Naperville, Il 60540, where he works in process design and economics. He was previously employed in the electroplating industry in the U.S. and abroad. Holder of a B.S. in chemical engineering from Rutgers University and M.S. and Ph.D. degrees in chemical engineering from the University of Cincinnati, he is a member of AIChE and the American Electroplaters Soc.

William Licht is professor of chemical engineering at the University of Cincinnati, Cincinnati, OH 45221. Formerly head of the Dept. of Chemical and Metallurgical Engineering, he has also been a visiting professor at the University of Minnesota. Active as a consultant, he has served as chairman of the Air Pollution Board of the City of Cincinnati. His research and consulting interests include modeling and design of duct collectors, fluidized-bed technology and gas purification. He holds Ch.E., M.S. and Ph.D. degrees from the University of Cincinnati. A Fellow of AIChE, he is also a member of the Amer. Pollution Control Assn., Amer. Soc. for Engr. Educ. and Sigma Xi.

Selecting, installing cyclone dust collectors

Here is a rundown on the nuts-and-bolts aspects of cyclone dust collectors: design variations, materials of construction, maintenance problems, energy requirements and ancillaries.

Theodore I. Horzella, *Enviro Energy Corp.*

and maintaining

☐ Cyclone dust collectors have a relatively high collection efficiency and can handle gases containing high dust concentrations. So, they have become the workhorses of the chemical process industries.

While the cyclone dust collector will rarely meet air-pollution-control codes when used alone, it should always be evaluated as a precleaner to be followed by other collectors.

Using a cyclone precleaner will not necessarily reduce the size of the secondary dust collector, but it will very often both simplify the design and increase the reliability of the secondary collector and accessory equipment.

An excellent example of the value of a cyclone dust collector is found in connection with fluidized-bed dryers and calciners. In such operations, most of the material processed is suspended in the gas stream. In a typical case, a fluidized-bed dryer used for drying coal in a coal-preparation plant caused about 50% of the coal processed to be suspended in the dryer exhaustgases. This represented about 100 tons/h of coal. A cluster of six cyclone collectors operating in parallel with a pressure drop of 3 in. W. G. (water gage) was capable of removing 99.5% of the suspended coal. The remaining 0.5% or ½ ton/h was then removed by a wet gas scrubber.

In a sintering operation that uses a baghouse as the final air-pollution-control equipment, cyclone collectors remove in excess of 80% of the suspended dust, and effectively protect the bags from undue wear caused by the heavy concentration of solids.

Cyclone collector designs

We prefer to group the various cyclone-dust-collector designs into two broad categories—*axial* and *tangential* gas entry.

The *axial-gas-entry cyclone* (Fig. 1) is best constructed by using castings for both the inlet vanes and the tubes. Hence, the axial-entry design is commercially available in only a few diameters, because casting patterns are costly to design, make and maintain. In addition, several hundred vanes and tubes are cast per run, and inventories must be limited. This results also in a limitation on materials of construction.

For all practical purposes, most manufacturers limit their stock to two or three diameters and only one or two alloys.

The *tangential-entry cyclone* (Fig. 1) is best produced by using fabricated plate. Fabricated-plate cyclones can be manufactured in every size and shape and lend themselves well to application of internal refractory linings.

Thus, the materials of construction required for the application will have a decisive influence on the design selected.

Another factor that should be kept in mind is that axial-entry cyclones are limited to dusts that can be

Originally published January 30, 1978

handled without undue wear or danger of plugging the required inlet vanes.

Axial-entry cyclones have a gas capacity about twice that of wraparound-entry cyclones of the same diameter.

The shape of the inlet vane will influence both the pressure drop and the dust collection characteristics of the cyclone.

Manufacturers of axial-entry cyclones have developed various higher-efficiency vanes in an attempt to bring their equipment closer to air-pollution-control code requirements. The gas velocities passing the entry vanes are relatively high, on the order of 5,000 to 7,000 ft/min, as compared with 2,400 to 4,200 ft/min experienced with wraparound or tangential-entry cyclones. This high-velocity gas carrying entrained dust particles requires hard alloys for the vanes, to give them an acceptable life. As the vanes wear, their efficiency lessens, and they must be periodically replaced. However, properly selected equipment will last more than five years.

Typical materials:

	Approximate Brinell hardness	Relative life
Grey iron	180	1
Chromehard	325	1.5
White iron	500	2

Depending on dust loading and dust abrasiveness, the equipment manufacturer will select a suitable vane and tube material. He will also limit the gas-entry velocity, thus reducing the gas-handling capacity of a given diameter. The dust-collection efficiency is, of course, affected by the gas-entry velocity.

For example, a typical axial-entry cyclone with a nominal 9-in. dia., operating at a pressure drop of 3-in. W.G., and handling a dust having a specific gravity of 2.3 and consisting of about 50% particles under 10 μm will perform as shown in Table I.

The same gas-flow equations as for wraparound and tangential cyclones apply. Capacity varies depending on the design of the vanes and the cylinder.

Dust-collection efficiencies expected of axial-entry cyclones will vary not only depending on the vane and the cylinder shapes but also on the roughness of their surfaces. However, for preliminary estimates and for dusts having a specific gravity in the range of 2 to 3, Fig. 2 is applicable.

The graph has a wide range. For example, with 50% of the dust under 10 μm, collection efficiency may range between 73 and 85%.

One manufacturer consulted gave a range of efficiencies between 70 and 86% for cylinder diameters between a nominal 9 and 24 in. However, the same manufacturer would guarantee the efficiencies of each diameter

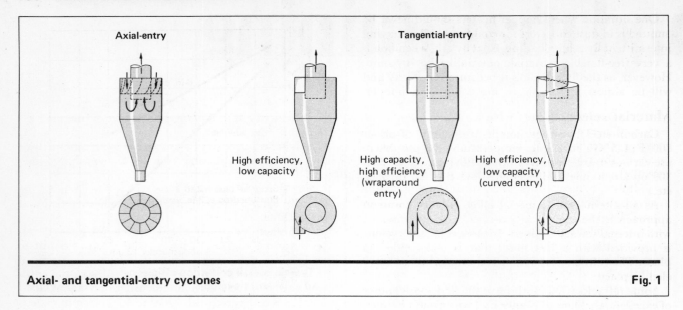

Axial- and tangential-entry cyclones   **Fig. 1**

and vane design provided the gas velocities were maintained within specified limits.

Materials of construction

Generally, cyclone dust collectors operate dry. There are some exceptions, especially where the dust collected is tacky and tends to build up inside the cyclone-walls—for example, in collecting the urea dust produced when drying urea prills. Here, water or a urea solution is sprayed into the cyclone cylinder, and the urea dust is collected as a solution.

However, because dry operation is most common, corrosion is a minor consideration in the selection of materials of construction for cyclones. Normally, the main factors are:

- Temperature of the gas.
- Abrasiveness of the dust particles.
- Protection of the dust from contamination.
- Corrosiveness of the gas.

In reviewing the various processes, we will find that the materials of construction must be selected not only for normal operation but with special attention paid to startup, shutdown, or when there is a process upset or other emergency situation.

The most critical factor in startup and shutdown of the process is the possibility of moisture condensation on the internal surfaces of the cyclone. Condensation

will invariably start corrosion and initiate dust agglomeration and buildup.

Tangential-entry cyclones are commercially available in diameters ranging from about 4 to 120 in. Capacities of such cyclones range from about 30 to 130,000 ft^3/min. For larger gas flows, two or more cyclones are grouped in clusters.

The gas-handling capacity of these cyclones is expressed by the equation

$$\text{Gas flow} = KD^2$$

where gas flow is expressed in ft^3/min., and D in ft. The constant, K, depends on the shape of the tangential gas entry and the proportions of the cyclone body.

With inlet gas velocities of 3,000 ft/min, K is 300 for the high-efficiency tangential-entry cyclone, and 900 for the high-volume (wraparound) tangential-entry type.

Fig. 3 shows two typical designs of tangential-entry cyclones and their fractional collection efficiencies.

In most processes involving the drying of solids, the exhaust-gas temperature is maintained as low as the drying characteristics of the solids will permit, so as to obtain the best thermal efficiency.

In such cases, the engineer should take precautions to avoid reaching the dewpoint and thus causing condensation of moisture:

1. Preheat the system before the feed of material is started.

2. Continue the supply of hot gases after shutdown until the system has been completely drained of dust and water vapor.

3. Insulate the ductwork, cyclone and hopper.

4. Provide artificial heating of the hopper, preferably by installing electric heaters or steam tracing before applying insulation.

Often the above first three steps will prevent the condensation of moisture and, thus, the danger of corrosion. However, experience shows that Step 4—the addition of hopper heaters—is a most wise investment. It will both eliminate corrosion and minimize the chance of material buildup in the hoppers.

Performance of axial-entry cyclones			Table I
	Vane A	Vane B	Vane C
Gas flow, actual ft^3/min	580	680	760
Inlet gas velocity, ft/min	5,740	5,440	4,940
Collection efficiency, %	87	81	77
Relative life expectancy	1	1.06	1.17

One situation where hopper heaters should never be omitted is in dust collectors for coal-fired boilers having intermittent fly ash unloading. Coal fly ash, when hot, is a very free-flowing material, not unlike hot gypsum. However, as the fly ash cools it becomes very tacky and will be almost impossible to move out of a hopper.

Material selection

Carbon steel has a gas-temperature limit of about 800°F (425°C). For higher temperatures, it is possible to use various stainless steels or other alloys, e.g., Type 304, 309 or 310 stainless, Hastelloy Alloys B or C, Inconel, etc.

As an alternative to special alloys, a very common approach is the use of carbon-steel cyclones provided with internal refractory liners. In most cases, the cyclone is provided with a hex mesh that is tack-welded in place. The refractory material is then gunited or troweled in place.

Some refractory materials have the added advantage of excellent abrasion resistance and will remain in operating condition for extended periods of time.

Abrasiveness of the dust is a major problem in mineral processing and metallurgical operations. It is also found in handling of catalyst vehicles, such as alumina used in petroleum refining operations. For such applications, the cyclone can be fabricated of heavier gage or abrasion-resistant metal, or it can be provided with wear liners.

Many times, the use of highly-abrasion-resistant refractory materials, which also stand up to higher operating temperatures, is the most practical approach to abrasion. Most cyclone manufacturers prefer to have the refractory liner installed in the field, and this requires bringing in specialized equipment and experienced labor.

Protection of the dust from contamination is particularly critical in food processing and in the production of chemicals in which traces of iron oxide could cause discoloration. Materials often used for prevention of contamination are the various grades of stainless steels, aluminum, alloy steels, Corten, etc.

Often, special materials of construction are required because the equipment must be periodically cleaned with steam and detergents, as in some food and pharmaceutical process operations.

Corrosiveness of the gas itself, as stated earlier, is not a common problem with cyclones, except when proper operating procedures are not followed.

The use of stainless steels will usually increase the initial cost of the cyclone by a factor of 3 as compared with carbon-steel construction. For this reason, it is often best to use carbon steel and provide interlocked automatic controls to minimize or eliminate faulty startup and shutdown procedures.

Equipment costs

A major cost in the fabrication of a cyclone is that of labor. The axial-entry cyclone (usually consisting of cast tubes and vanes inside a fabricated housing) requires less labor, especially for larger gas flows (in excess of 100,000 actual ft³/min).

However, the cast tubes and vanes limit their appli-

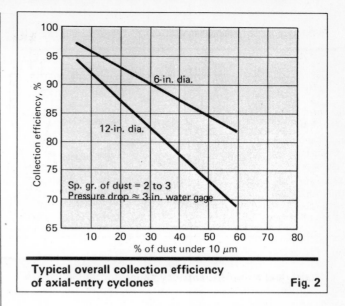

Typical overall collection efficiency of axial-entry cyclones **Fig. 2**

cation to those cases where their resistance to temperature, dust abrasiveness, and corrosion are compatible with the process.

Another factor that limits the application of axial-entry cyclones in the chemical process industries is the danger of plugging of the gas-entry vanes. These vanes are more sensitive to high dust loadings and tacky dusts than are the tangential- or wraparound-entry cyclones.

(We point to the above factors because first cost may

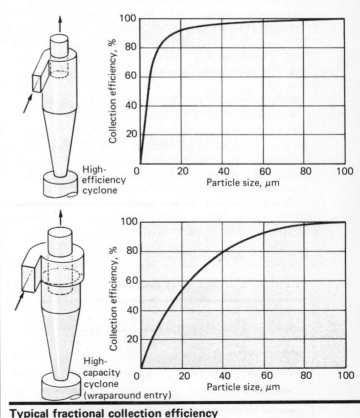

Typical fractional collection efficiency of tangential-entry cyclones **Fig. 3**

Relative costs of various cyclones			Table II
	Gas throughput, actual ft^3/min		
	10,000	50,000	100,000
Axial-entry cyclone	5,000	10,000	20,000
No. of tubes/dia.	18/9	28/12	56/12
Tangential-entry cyclone			
Diameter	52	80	80
Number	1	2	4
3/16 in. thick carbon steel, cost, $	5,000	20,000	40,000
3/16 in. thick 304 stainless, cost, $	12,000	50,000	100,000
3/16 in. thick carbon steel, 4-in. thick refractory, cost, $	7,000	30,000	60,000

be the last thing to consider in equipment selection.)

Table II lists the first cost of cyclones selected to handle three gas volumes, without consideration of collection effectiveness, dust loadings, or abrasion resistance.

The costs are intended only as a relative guide. The thickness of the material, the finish of the welds, and the type of refractory will all have a substantial effect on such costs.

Maintenance costs of cyclones are negligible, and a 20-yr life is frequently attainable. Wear will occur, especially at the entry section of tangential-entry cyclones and at the vanes of axial-entry cyclones. With refractory-lined cyclones, there may be spalling and occasional fallout of part of the refractory, especially on the horizontal top section and the gas outlet tube.

In general, the main reason for wear and tear will be abrasion, depending on the dust handled and lack of control in operating temperature. Hence, an "average" cost for maintenance is difficult to establish.

Energy requirements

Cyclone collectors are normally selected for pressure drops on the order of 3 to 6 in. W.G. At these drops, power requirements are about 1 to 1.5 hp per 1,000 std. ft^3/min of dry gas. Lower pressure-drops are experienced only with electrostatic precipitators, which normally operate at drops under 1 in. W.G.

A clear-cut power-consumption evaluation is not easy to make, since the cyclone dust collector will normally require secondary collectors to meet emission-control regulations.

We recommend evaluating the complete system, which may consist of:

1. Cyclone + electrostatic precipitator.
2. Cyclone + wet scrubber.
3. Cyclone + baghouse.
4. Cyclone + cyclone in series.

1. Cyclone + electrostatic precipitator combinations are used to reduce very heavy dust loadings. This type of combination will be dictated by the precipitator manufacturer and is rather rare in the chemical process industries.

2. Cyclone + wet-scrubber combinations are of special interest to the chemical engineer because the cy-

clone substantially reduces the amount of slurry that has to be processed in the system. In most industrial processes, a wet scrubber will operate with slurries containing from 5 to 15% solids in suspension.

A wet scrubber handling 50,000 actual ft^3/min of a gas containing 10 grains/actual ft^3 of dust will require a slurry-handling capacity of about 50 gpm if the slurry concentration is maintained at 15%.

The same scrubber provided with a cyclone precleaner will reduce the dust loading by about 80%, to 2 grains/actual ft^3 entering the scrubber. With this dust loading, the slurry-handling capacity will be reduced to about 10 gpm.

The additional power required for the cyclone, about 50 hp, will be insignificant compared with the cost of handling the additional 40 gpm of slurry, which may require large settling tanks, filters, centrifuges, chemical additives, etc.

Another case where the cyclone + wet-scrubber combination contributes to substantial energy savings is found in the operation of coal- or hog-fuel-fired boilers. Some of the fuel may not be completely burned and can be collected in the cyclone for return to the boiler. The finer dust (mostly completely burned fuel), which is not collected in the cyclone, goes to the scrubber.

3. Cyclone + baghouse combinations are more difficult to evaluate. As a rule, the manufacturer of baghouses will select the gas velocity across the filter cloth (usually stated as air-to-cloth ratio) with due consideration to the dust loading. The selection is not as clearly defined, and personal judgment or commercial considerations may lead the baghouse manufacturer to select higher or lower air-to-cloth ratios.

Most baghouses operate with pressure drops on the order of 5 to 9 in. W.G. Adding a cyclone as preliminary collector increases the system pressure drop by about 50%. Unless the reduction in dust loading accomplished by the cyclone results in a higher air-to-cloth ratio, there will be little advantage to installing such a device. Some advantages—not easily established—include the spark-arrestor protection that a cyclone can provide in applications involving furnaces. Some protection against damage to the bags by large particles of dust (which would be removed from the gas stream before it enters the baghouse) is also afforded by the precleaning cyclone.

Higher dust loadings require bag cleaning at more-frequent intervals. This has a bearing on bag longevity. Those baghouses using compressed air for cleaning of the bags will experience a reduction in the amount of compressed air used and, therefore, will benefit from the precleaning cyclone. For greatest reduction in compressed-air consumption, the baghouse must be provided with an automatic control to both measure the pressure drop and activate the cleaning mechanism as required.

An accurate energy-consumption and total-usage-cost evaluation for the cyclone + baghouse combination is indeed difficult. The engineer will have to rely heavily on both his judgment and advice from the baghouse manufacturer.

4. The combination of two cyclones in series is used at times. Most of a very heavy dust loading is removed in

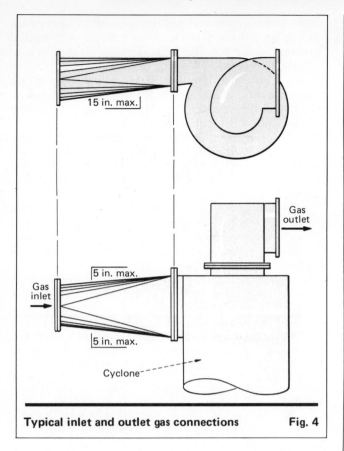

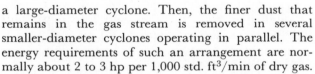

Typical inlet and outlet gas connections   **Fig. 4**

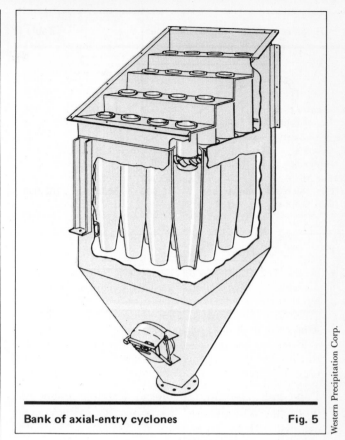

Bank of axial-entry cyclones   **Fig. 5**

Western Precipitation Corp.

a large-diameter cyclone. Then, the finer dust that remains in the gas stream is removed in several smaller-diameter cyclones operating in parallel. The energy requirements of such an arrangement are normally about 2 to 3 hp per 1,000 std. ft³/min of dry gas.

Because of the uncertainty of the overall collection efficiency of such systems, two-stage cyclone dust collectors are rarely used.

Auxiliary equipment

The auxiliary equipment required for the operation of a cyclone will consist essentially of:

> Ductwork to and from the cyclone.
> Gas-flow control dampers.
> Dust-removal valves.
> Induced-draft fans.
> Insulation.
> Hopper heating devices.
> Hopper rapping or vibration devices.
> Hopper-dust-level measuring devices.
> Hopper inspection doors.
> Pressure-drop measuring devices.
> Hopper ventilation and baffles.

Ductwork design

Proper ductwork design is an important element in the final dust-collection efficiency of a cyclone.

If the cyclone consists of a single tube, it is sufficient to provide an inlet conversion section from the usually round duct to the cyclone's rectangular entry, as shown in Fig. 4.

The outlet duct can be either a continuation of the cyclone outlet tube or a volute, as shown in Fig. 4. Many times, instead of the volute shown, a rectangular box is suitable.

If the cyclone consists of more than one tube, such as shown in Fig. 5, the inlet and outlet ducts will have a definite influence on performance. It is most important to keep in mind that an even gas distribution to all of the tubes will ensure highest collection efficiency, by maintaining proper gas inlet-velocities.

The layout of the equipment sometimes makes it close to impossible to furnish a straight inlet duct long enough to ensure proper gas distribution. In such case, gas deflectors in the duct or manually adjustable dampers in each tube's gas entrance will permit some adjustments during the original startup period. They can be adjusted by measuring and equalizing the pressure drop across each tube. This is not a simple task, but it is a practical approach.

If the gas flow is not constant, the collection efficiency of the cyclone will be affected, and it may be necessary to stop the gas flow to one or more of the tubes. This is normally accomplished with remotely operated manual dampers. These dampers are shut off one at a time to maintain the proper gas flow to the other tubes. An alternative to this is to bleed ambient air (or to recirculate air from the cyclone exit) into the cyclone inlet duct. This approach permits constant gas flow and proper collection efficiency. One drawback is that the gas temperature will vary depending on the amount of ambient air bled into the system. This could cause

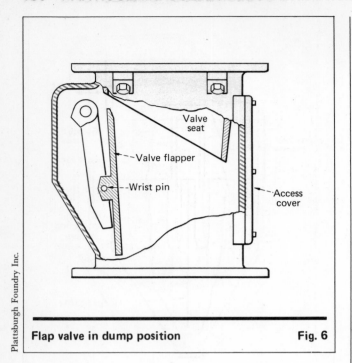

Plattsburgh Foundry Inc.

Valve
seat

Valve flapper

Wrist pin

Access
cover

Flap valve in dump position   **Fig. 6**

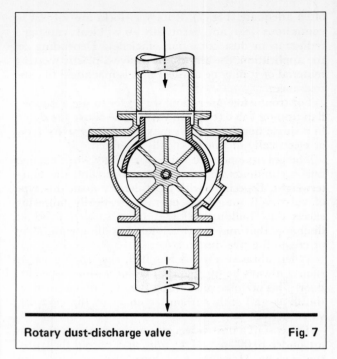

Rotary dust-discharge valve   **Fig. 7**

problems due to moisture condensation or overloading of the induced-draft fan motor. Careful design will eliminate such dangers.

In the case of axial-vane-inlet cyclones, which are normally 6 to 12 in. dia., a number of tubes are installed in a common housing. (See Fig. 5.)

The gas volume entering those tubes near the gas entry will be substantially larger than the volume entering the tubes in the back. A collector 6 tubes deep may be rated at a gas-handling capacity per tube about 20% higher than if the collector had 15 tubes in depth.

If the gas flow is not constant, the multiple-tube collector of the axial-gas-entry type must be designed in several compartments, each with a damper to shut off the gas flow, to maintain proper gas flow to the tubes in service.

Another extremely important factor in the design of the inlet ductwork to cyclones is the maintenance of proper gas velocities to prevent the settling of dust in the ducts themselves.

Dust settling can be extremely serious. Combustible dust, such as incompletely burned coal fly ash, may smolder in the duct and overheat the metal. Once dust has settled, it will often agglomerate and, after full gas flow is reestablished, the gas velocity may be inadequate to reentrain the dust. Thus, the pressure drop through the system will increase, and the induced-draft fan may be incapable of handling full gas flow.

At times, agglomerated dust is reentrained, but large agglomerates may plug the inlet vanes of axial-entry cyclones or the dust discharge valves.

Dampers

The gas-flow-control dampers described with the ductwork design above are usually of the single-blade design and operate either fully open or closed. Remote manually controlled operation, either with a hand

chain or with electric or pneumatic operators, is recommended.

In the selection of damper location, it is most important to avoid the creation of dead spots where dust will accumulate. Even if the location of the damper is carefully selected, provision should be made for access doors both upstream and downstream. These doors are indispensable both for inspection and to remove dust deposits.

Dust-removal valves

Dust-removal valves or airlocks are extremely important in the proper operation of cyclone collectors. Dust may be removed from the hoppers, either continuously or intermittently. However, entrance of ambient air through the discharge valve will cause a substantial reduction of dust collection efficiency.

The dust valve is continuously subject to the abrasive action of the dust and often it also operates at high temperatures. For intermittent discharge of dust, a well-designed slide gate—or better yet, a seat and flap valve as shown in Fig. 6—will provide reliable service. If the dust is allowed to accumulate to an excessive height, there is always the danger of agglomeration; consequently, intermittent dust-discharge should be used only with dusts that will freely flow from the hopper.

With intermittent discharge, there is also always the danger that the operator may forget to unload the hopper. If the dust level reaches the bottom of the cyclone, dust will be reentrained in the gas stream, and complications may develop. If there is a secondary dust collector capable of handling the added load, the problem may be minor. However, if the induced-draft fan is located downstream of the cyclone, permanent damage to the fan wheel and housing may result from the heavy dust load in the gas stream.

For continuous dust discharge, a rotary airlock is

often adequate (Fig. 7). Rotary airlocks are subject to continuous wear, and eventually air will leak, causing a reduction in dust collection efficiency. Depending on the application, the airlock may have a plastic wearing material or it may be all steel that is machined to close tolerances.

For troublefree operation, we prefer to use a double-flap tipping valve (two flap valves, one above the other, in a single housing) which may be of the gravity type, or electrically or pneumatically operated.

The gravity-operated valve opens only when enough dust accumulates above the flap to overcome the counterweight. Especially with light-density dust, this type of valve will sometimes operate erratically, allowing excess dust buildup followed by a sudden floodlike discharge that may overload dust-handling equipment or create fugitive dust problems.

When abrasive dust is handled, the tipping valves should always be of the type having a side inspection door. This permits quick inspection and replacement of the flaps and seals without removal of the complete valve body.

The cost of a dust valve, especially when gas volumes are under 10,000 actual ft^3/min, may exceed the cost of the cyclone. However, a cyclone with an improperly operating dust valve will be inefficient. There is absolutely no justification in cutting costs when furnishing a dust valve.

Induced-Draft Fans

While the induced-draft fan is not an accessory to the cyclone itself, one must remember that the cyclone in most processing applications will allow the passage of substantial amounts of uncollected dust. Unless the fan is located downstream of the secondary dust collector, it must be selected so that it will be able to operate with dust-containing gas. Abrasion or dust deposits on the fan may cause an imbalance of the wheel and a loss of fan efficiency.

In most cases, the selection should be of a radial-blade wheel, proper shaft seals, and quick-opening inspection door. The housing should be split so that the fan wheel can be replaced. Proper roof clearance should be provided for removal and installation of the wheel. The fan should be provided with a damper to prevent overloading of the motor during cold startup.

Insulation

If the cyclone must handle gases above 150°F (65°C), it is necessary to provide insulation for personnel protection, to prevent heat losses and, as discussed previously, to prevent condensation of moisture.

Insulation can be installed either in the shop or in the field and the decisive factors will be size of the equipment, accessibility, and any need for field welding.

Costs of field insulation will vary, depending primarily on accessibility of the equipment.

Hopper heaters

Hopper heating devices should be used whenever there is a danger of moisture condensation inside the hopper walls. The installation and cost of electric heating strips is not high and will eliminate many operating

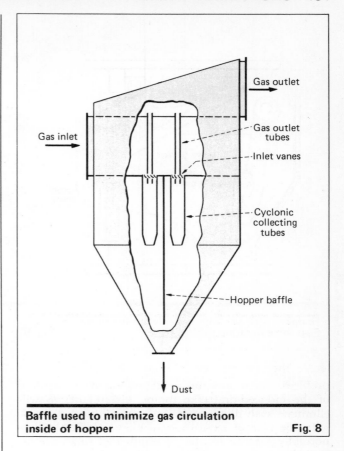

Baffle used to minimize gas circulation inside of hopper   Fig. 8

problems, especially when the equipment is outdoors. Strip heaters can automatically switch on and off, and use only an insignificant amount of power.

Steam heating is also practical, but only if steam is readily available, which is not usually the case near the cyclonic dust collectors.

Hopper rappers

Hopper rappers or vibrators are a necessary evil in a few applications. A proper hopper slope, without internal obstructions for the dust flow, plus heating and insulation, will normally eliminate the need for rappers.

In doubtful cases, where experience is meager with a particular dust, we recommend providing anvil plates welded to the hopper walls. These plates should be covered with insulation, but properly marked. If rappers become necessary, they can then be added with minimal field expenditures.

Measuring hopper dust-levels

Hopper dust-level measuring devices are always recommended when manual dust unloading will be applied. They are also recommended when the degree of dust loading is yet unknown or when dust loading will fluctuate. There are several commercial level-measuring instruments, and careful selection is required. Corrosiveness and temperature are critical.

Inspection doors

Hopper inspection doors are too often ignored by the engineer. Usually, the only consideration given is to

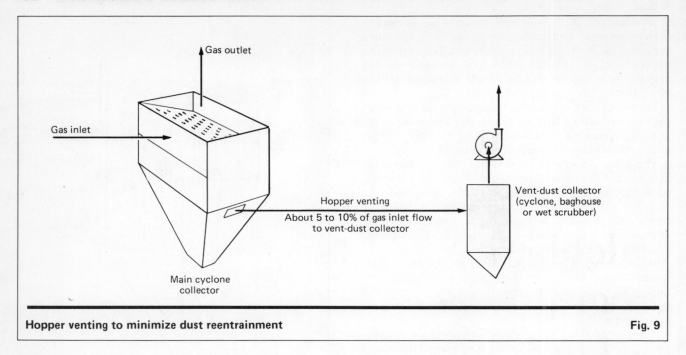

Gas outlet

Gas inlet

Hopper venting

About 5 to 10% of gas inlet flow
to vent-dust collector

Vent-dust collector
(cyclone, baghouse
or wet scrubber)

Main cyclone
collector

Hopper venting to minimize dust reentrainment   Fig. 9

locate them where they are accessible. However, an inspection door must also be gastight, because a leaky one will reduce collection efficiency and cause corrosion. The inspection door must be provided with a safety latch that will prevent the accumulated dust—often at very high temperature—from showering the operator as the door is opened. Many serious accidents have occurred when unsafe doors were unlocked and an uncontrolled flow of dust left the hopper.

Properly designed hopper access-doors often cost a substantial amount and there is a tendency to skimp on them.

Measuring pressure drop

Pressure-drop measurements are always a good indication of the amount of gas flow, or the possibility of a plugged cyclone collector. A simple U tube with metal tubing connecting it with the cyclone inlet and outlet costs little but will be a very effective maintenance tool.

A slow change in pressure drop under steady-state process conditions may anticipate buildup of dust, and corrective measures can be taken well before a serious problem develops. (A log of pressure drops may be maintained right next to the U tube.)

Hopper ventilation and baffles

When a number of cyclones discharge dust into a common hopper (especially when using cyclones of the axial-entry type), there is a difference in absolute pressure between the cyclones near the gas entry duct and

the cyclones to the rear of the common hopper. This differential pressure causes circulation of gases from the tubes to the hopper and from the hopper to those tubes that are under a greater negative pressure. As gases from the hopper move into the tubes, they reentrain dust, which goes into the exhaust gases.

The problem can be partially corrected by installing baffles inside the hopper, as shown in Fig. 8. These baffles somewhat reduce the internal hopper circulation. If about 5% to 10% of the total gas entering the dust collector is removed from the hopper, as shown in Fig. 9, and handled through a secondary cyclone, scrubber or baghouse, the dust collection efficiency can be increased by about 5%. In some cases, this additional efficiency has permitted cyclone systems to meet air pollution codes.

The author

Theodore I. Horzella is vice-president of Enviro Energy Corp., Suite 220, 16161 Ventura Blvd., Encino, CA 91436. He received his B.S. in chemistry from Santa Maria University, Chile, and his M.S.Ch.E. from Iowa State University, Ames, Iowa. He has also done graduate work in business administration at the University of California. He is a member of AIChE, Sigma Xi, and the American Management Assns.

Calculator program solves cyclone efficiency equations

A procedure for predicting the fractional efficiencies of cyclones appeared on pages 175-183. This article presents a calculator program for solving the equations needed to calculate the efficiencies.

Yatendra M. Shah and *Richard T. Price*, *Pedco Environmental, Inc.*

☐ The procedure described in the article beginning on p. 175 for determining cyclone fractional (grade) efficiency is based mainly on cyclone dimensions and the flow characteristics of particle-laden gases. It involves progressively solving several equations and using the results to calculate fractional efficiency. The equations are complex, containing exponential and logarithmic functions.

This calculator program for finding cyclone efficiencies for different input variables greatly reduces calculations and improves accuracy. Although designed for Texas Instrument calculator TI-59 and printer

PC-100A, the program can be adapted for other programmable calculators with minor modifications.

Table I gives the input variables for the efficiency equations. These are either used directly in the final grade efficiency equation, or their derivatives are required for solving it.

The equations for calculating cyclone fractional efficiencies are:

For l—Natural length (the distance below the gas outlet where the vortex turns), ft:

$$l = 2.3D_e(D_c{}^2/ab)^{1/3} \qquad (1)$$

Originally published August 28, 1978

Input variables for efficiency equations — Table I

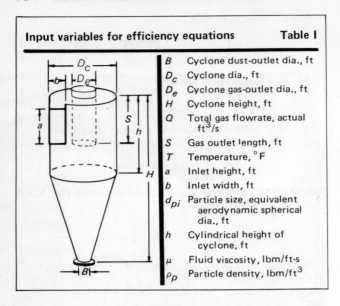

B	Cyclone dust-outlet dia., ft
D_c	Cyclone dia., ft
D_e	Cyclone gas-outlet dia., ft
H	Cyclone height, ft
Q	Total gas flowrate, actual ft³/s
S	Gas outlet length, ft
T	Temperature, °F
a	Inlet height, ft
b	Inlet width, ft
d_{pi}	Particle size, equivalent aerodynamic spherical dia., ft
h	Cylindrical height of cyclone, ft
μ	Fluid viscosity, lbm/ft-s
ρ_p	Particle density, lbm/ft³

Overall structure of cyclone-efficiency program — Table II

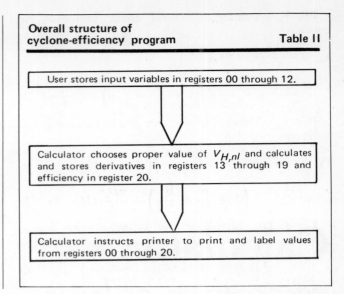

User stores input variables in registers 00 through 12.

Calculator chooses proper value of $V_{H,nl}$ and calculates and stores derivatives in registers 13 through 19 and efficiency in register 20.

Calculator instructs printer to print and label values from registers 00 through 20.

Detailed listing of cyclone-efficiency program

Location	Code	Key	Location	Code	Key	Location	Code	Key	Location	Code	Key	Location	Code	Key	Location	Code	Key	Location	Code	Key	Location	Code	Key	Location	Code	Key	Location	Code	Key
000	02	2	035	11	11	070	02	2	105	07	07	140	53	(	175	23	lnx	210	05	5	245	04	04	280	04	04	315	69	Op
001	93	.	036	95	=	071	93	.	106	33	x²	141	43	RCL	176	54	)	211	55	÷	246	43	RCL	281	95	=	316	06	06
002	03	3	037	42	STO	072	05	5	107	54	)	142	11	11	177	45	yˣ	212	53	(	247	01	01	282	69	Op	317	02	2
003	65	×	038	15	15	073	54	)	108	55	÷	143	75	−	178	53	(	213	43	RCL	248	95	=	283	06	06	318	03	3
004	43	RCL	039	43	RCL	074	54	)	109	53	(	144	53	(	179	02	2	214	17	17	249	69	Op	284	01	1	319	02	2
005	05	05	040	14	14	075	65	×	110	01	1	145	43	RCL	180	94	+/−	215	85	+	250	06	06	285	06	6	320	03	3
006	65	×	041	32	x⇄t	076	53	(	111	08	8	146	01	01	181	65	×	216	01	1	251	01	1	286	01	1	321	69	Op
007	53	(	042	43	RCL	077	55	÷	112	65	×	147	55	÷	182	53	(	217	54	)	252	04	4	287	07	7	322	04	04
008	43	RCL	043	15	15	078	53	(	113	43	RCL	148	02	2	183	53	(	218	54	)	253	01	1	288	69	Op	323	43	RCL
009	04	04	044	22	INV	079	43	RCL	114	00	00	149	54	)	184	43	RCL	219	54	)	254	04	4	289	04	04	324	08	08
010	33	x²	045	77	CE	080	12	12	115	54	)	150	54	)	185	19	19	220	54	)	255	69	Op	290	43	RCL	325	95	=
011	55	÷	046	11	A	081	85	+	116	54	)	151	65	×	186	65	×	221	95	=	256	04	04	291	05	05	326	69	Op
012	53	(	047	43	RCL	082	04	4	117	95	=	152	53	(	187	43	RCL	222	42	STO	257	43	RCL	292	95	=	327	06	06
013	43	RCL	048	15	15	083	06	6	118	42	STO	153	43	RCL	188	18	18	223	20	20	258	02	02	293	69	Op	328	02	2
014	01	01	049	77	CE	084	00	0	119	18	13	154	04	04	189	65	×	224	22	INV	259	95	=	294	06	06	329	00	0
015	65	×	050	12	B	085	54	)	120	53	(	155	33	x²	190	43	RCL	225	52	EE	260	69	Op	295	03	3	330	02	2
016	43	RCL	051	01	1	086	55	÷	121	02	2	156	75	−	191	10	10	226	04	4	261	06	06	296	05	5	331	03	3
017	03	03	052	75	−	087	05	5	122	65	×	157	43	RCL	192	65	×	227	01	1	262	02	2	297	03	3	332	69	Op
018	54	)	053	53	(	088	03	3	123	43	RCL	158	05	05	193	53	(	228	04	4	263	00	0	298	03	3	333	04	04
019	54	)	054	53	(	089	00	0	124	04	04	159	33	x²	194	43	RCL	229	01	1	264	01	1	299	69	Op	334	43	RCL
020	45	yˣ	055	01	1	090	54	)	125	55	÷	160	54	)	195	17	17	230	69	Op	265	04	4	300	04	04	335	09	09
021	53	(	056	75	−	091	45	yˣ	126	53	(	161	85	+	196	85	+	231	04	04	266	69	Op	301	43	RCL	336	95	=
022	01	1	057	53	(	092	93	.	127	43	RCL	162	02	2	197	01	1	232	43	RCL	267	04	04	302	06	06	337	69	Op
023	55	÷	058	53	(	093	03	3	128	01	01	163	65	×	198	54	)	233	00	00	268	43	RCL	303	95	=	338	06	06
024	03	3	059	01	1	094	54	)	129	33	x²	164	43	RCL	199	55	÷	234	95	=	269	03	03	304	69	Op	339	03	3
025	54	)	060	02	2	095	54	)	130	65	×	165	16	16	200	53	(	235	58	Fix	270	95	=	305	06	06	340	04	4
026	95	=	061	65	×	096	95	=	131	43	RCL	166	54	)	201	43	RCL	236	04	04	271	69	Op	306	01	1	341	03	3
027	42	STO	062	43	RCL	097	42	STO	132	03	03	167	95	=	202	04	04	237	95	=	272	06	06	307	06	6	342	04	4
028	14	14	063	04	04	098	17	17	133	33	x²	168	42	STO	203	45	yˣ	238	69	Op	273	01	1	308	03	3	343	69	Op
029	61	GTO	064	54	)	099	43	RCL	134	54	)	169	19	19	204	03	3	239	06	06	274	06	6	309	03	3	344	04	04
030	13	C	065	45	yˣ	100	06	06	135	54	)	170	01	1	205	54	)	240	02	2	275	01	1	310	69	Op	345	43	RCL
031	43	RCL	066	93	.	101	65	×	136	65	×	171	75	−	206	54	)	241	00	0	276	05	5	311	04	04	346	10	10
032	08	08	067	01	1	102	53	(	137	53	(	172	53	(	207	45	yˣ	242	01	1	277	69	Op	312	43	RCL	347	95	=
033	75	−	068	04	4	103	53	(	138	89	π	173	01	1	208	53	(	243	03	3	278	04	04	313	07	07	348	69	Op
034	43	RCL	069	55	÷	104	43	RCL	139	65	×	174	22	INV	209	93	.	244	69	Op	279	43	RCL	314	95	=	349	06	06

For V_{nl}—Volume at natural length (excluding the core), ft^3:

$$V_{nl} = \frac{\pi D_c^2}{4}(h - S) +$$

$$\frac{(\pi D_c^2)}{4}\frac{(l + S - h)}{3}\left(1 + \frac{d}{D_c} + \frac{d^2}{D_c^2}\right) - \frac{\pi D_e^2 l}{4} \quad (2)$$

For V_H—Volume below exit duct (excluding the core), ft^3:

$$V_H = \frac{\pi D_c^2}{4}(h - S) + \left(\frac{\pi D_c^2}{4}\right)\left(\frac{H - h}{3}\right) \times$$

$$\left(1 + \frac{B}{D_c} + \frac{B^2}{D_c^2}\right) - \frac{\pi D_e^2}{4}(H - S) \quad (3)$$

For d—Dia. of central core at point where vortex turns, ft:

$$d = D_c - (D_c - B)\left(\frac{S + l - h}{H - h}\right) \quad (4)$$

For n—Vortex exponent:

$$n = 1 - \left[1 - \frac{(12D_c)^{0.14}}{2.5}\right]\left[\frac{T + 460}{530}\right]^{0.3} \quad (5)$$

For τ_i—Relaxation time, s:

$$\tau_i = \rho_p(d_{pi})^2/(18\mu) \quad (6)$$

For G—Cyclone configuration factor (specified by the geometric ratios that describe the cyclone's shape):

$$G = 8K_c/K_a^2 K_b^2$$

Substituting values of K_a, K_b, and K_c from the Koch and Licht article gives:

$$G = \{2[\pi(S - a/2)(D_c^2 - D_e^2)] + 4V_{H,nl}\}\frac{D_c}{a^2 b^2} \quad (7)$$

In equation (7), use V_H when $S + l > H$, and V_{nl} when $S + l \le H$.

Table III

Location	Code	Key	Location	Code	Key	Location	Code	Key	Location	Code	Key	Location	Code	Key	Location	Code	Key	Location	Code	Key	Location	Code	Key	Location	Code	Key
350	03	3	385	98	Adv	420	06	06	455	06	06	490	09	09	525	54	)	560	53	(	595	33	x^2	630	75	−
351	06	6	386	02	2	421	98	Adv	456	98	Adv	491	75	−	526	54	)	561	43	RCL	596	65	×	631	43	RCL
352	03	3	387	00	0	422	02	2	457	06	6	492	43	RCL	527	54	)	562	14	14	597	43	RCL	632	09	09
353	06	6	388	02	2	423	00	0	458	01	1	493	11	11	528	75	−	563	85	+	598	14	14	633	54	)
354	69	Op	389	07	7	424	03	3	459	01	1	494	54	)	529	43	RCL	564	43	RCL	599	65	×	634	55	÷
355	04	04	390	69	Op	425	01	1	460	07	7	495	85	+	530	05	05	565	11	11	600	89	π	635	53	(
356	43	RCL	391	04	04	426	69	Op	461	69	Op	496	33	x^2	531	33	x^2	566	75	−	601	55	÷	636	43	RCL
357	11	11	392	43	RCL	427	04	04	462	04	04	497	53	(	532	65	×	567	43	RCL	602	04	4	637	08	08
358	95	=	393	14	14	428	43	RCL	463	43	RCL	498	53	(	533	43	RCL	568	09	09	603	95	=	638	75	−
359	69	Op	394	95	=	429	17	17	464	20	20	499	43	RCL	534	15	.15	569	54	)	604	42	STO	639	43	RCL
360	06	06	395	69	Op	430	95	=	465	65	×	500	08	08	535	95	=	570	55	÷	605	16	16	640	09	09
361	03	3	396	06	06	431	69	Op	466	01	1	501	75	−	536	42	STO	571	03	3	606	61	GTO	641	54	)
362	07	7	397	98	Adv	432	06	06	467	00	0	502	43	RCL	537	16	16	572	54	)	607	00	00	642	54	)
363	03	3	398	02	2	433	98	Adv	468	00	0	503	09	09	538	61	GTO	573	65	×	608	51	51	643	54	)
364	07	7	399	03	3	434	03	3	469	95	=	504	54	)	539	00	00	574	53	(	609	76	Lbl	644	95	=
365	69	Op	400	03	3	435	07	7	470	69	Op	505	55	÷	540	51	51	575	01	1	610	13	C	645	42	STO
366	04	04	401	06	6	436	02	2	471	06	06	506	03	3	541	76	Lbl	576	85	+	611	43	RCL	646	13	13
367	43	RCL	402	69	Op	437	04	4	472	98	Adv	507	54	)	542	12	B	577	43	RCL	612	04	04	647	61	GTO
368	12	12	403	04	04	438	69	Op	473	98	Adv	508	65	×	543	53	(	578	13	13	613	75	−	648	00	00
369	95	=	404	43	RCL	439	04	04	474	98	Adv	509	53	(	544	43	RCL	579	55	÷	614	53	(	649	31	31
370	69	Op	405	15	15	440	43	RCL	475	91	R/S	510	01	1	545	04	04	580	43	RCL	615	53	(			
371	06	06	406	95	=	441	18	18	476	76	Lbl	511	85	+	546	33	x^2	581	04	04	616	43	RCL			
372	98	Adv	407	69	Op	442	95	=	477	11	A	512	43	RCL	547	65	×	582	85	+	617	04	·04			
373	98	Adv	408	06	06	443	69	Op	478	89	π	513	02	02	548	89	π	583	43	RCL	618	75	−			
374	02	2	409	98	Adv	444	06	06	479	55	÷	514	55	÷	549	55	÷	584	13	13	619	43	RCL			
375	00	0	. 410	04	4	445	98	Adv	480	04	4	515	43	RCL	550	04	4	585	33	x^2	620	02	02			
376	01	1	411	02	2	446	02	2	481	65	×	516	04	04	551	65	×	586	55	÷	621	54	)			
377	06	6	412	04	4	447	02	2	482	53	(	517	85	+	552	53	(	587	43	RCL	622	65	×			
378	69	Op	413	02	2	448	02	2	483	43	RCL	518	43	RCL	553	43	RCL	588	04	04	623	53	(			
379	04	04	414	69	Op	449	02	2	484	04	04	519	02	02	554	09	09	589	33	x^2	624	53	(			
380	43	RCL	415	04	04	450	69	Op	485	33	x^2	520	33	x^2	555	75	−	590	54	)	625	43	RCL			
381	13	13	416	43	RCL	451	04	04	486	65	×	521	55	÷	556	43	RCL	591	54	)	626	11	11			
382	95	=	417	16	16	452	43	RCL	487	53	(	522	43	RCL	557	11	11	592	75	−	627	85	+			
383	69	Op	418	95	=	453	19	19	488	53	(	523	04	04	558	85	+	593	43	RCL	628	43	RCL			
384	06	06	419	69	Op	454	69	Op	489	43	RCL	524	33	x^2	559	53	(	594	05	05	629	14	14			

Operating instructions for cyclone-efficiency program — Table IV

Step	Procedure		Enter		Press		Display
1.	Partition registers		3	2nd	Op		
			17				719.29
					CLR		0
2.	Store program:	Run card side 1					1
					CLR		0
		Run card side 2					2
					CLR		0
		Run card side 3					3
					CLR		0
		Run card side 4					4
					CLR		0
3.	Store input variables		μ	STO	0	0	μ
			a	STO	0	1	a
			B	STO	0	2	B
			b	STO	0	3	b
			D_c	STO	0	4	D_c
			D_e	STO	0	5	D_e
			ρ_p	STO	0	6	ρ_p
			d_{pi}	STO	0	7	d_{pi}
			H	STO	0	8	H
			h	STO	0	9	h
			Q	STO	1	0	Q
			S	STO	1	1	S
			T	STO	1	2	T
					CLR		0
4.	Reset calculator				RST		0
5.	Run program				R/S		

* For subsequent runs, start at 3; if any of the variables do not change, these need not be reentered.

Sample output run of cyclone-efficiency program* — Table V

Sample printout	Actual symbol	Register number
0.0000	UU	μ	00
4.5000	—A	a	01
2.5330	BB	B	02
1.8960	—B	b	03
6.3330	DC	D_c	04
3.7920	DE	D_e	05
62.4300	RP	ρ_{pi}	06
0.0000	DP	d_{pi}	07
26.3330	HH	H	08
8.5520	—H	h	09
516.7000	QQ	Q	10
3.4480	SS	S	11
110.0000	TT	T	12
4.3014	—D	d	13
14.6101	—L	l	14
22.8850	HS	H-S	15
209.4317	VV	$V_{nl,H}$	16
0.7276	—N	n	17
0.0003	TI	τ_i	18
89.7282	GG	G	19
63.3051	%E	η_i	20

*The input variables and output values are fixed to four decimal places; values of μ and d_{pi} in this sample run are 1.28×10^{-5} and 3.281×10^{-5}, respectively.

that the program has been recorded on the magnetic cards beforehand, and the calculator has been just switched on.

A sample output run is shown in Table V. The single-spaced items are input variables. The calculated derivatives and efficiency are double-spaced and follow the input.

For η_i—Fractional efficiency:

$$\eta_i = 1 - exp\left\{ -2\left[\frac{G\tau_i Q}{D_c{}^3}(n+1) \right]^{0.5/(n+1)} \right\} \quad (8)$$

The program is structured for storing the input variables in data registers 00 through 12. The derivatives of these input variables and the value of the efficiency are stored in registers 13 through 20. The input and output values can be recalled whenever necessary. The effect of an individual input variable on the efficiency can be examined by storing the changed value in the assigned register; other variables need not be reentered.

The overall structure of the program is summarized in Table II. The detailed listing of the program is presented in Table III. The program requires register partitioning of 720 program registers and 30 data registers. The calculator should be placed in 719.29 mode by pressing ⬚3⬚ ⬚2nd⬚ ⬚Op⬚ ⬚1⬚ ⬚7⬚.

Operating instructions for the program are given in Table IV. The presentation of the instructions assumes

The authors

Yatendra M. Shah (Pedco, 11499 Chester Rd., Cincinnati, OH 45246) has compiled and analyzed cost data for pollution control systems, and performed system analyses of computer programs on costs of coal-cleaning plants, flue-gas desulfurization systems and precipitators. He was project manager for the preparation of a manual for enforcement of New Source Performance Standards for coal-preparation plants and of compliance schedules and CPM networks for emission systems. Holder of B.S. and M.S. degrees in mechanical engineering from Bombay University, he is a registered engineer and a member of ASME.

Richard T. Price is involved in analyzing the control of air pollution at copper smelters and in the design of equipment to reduce dust emissions. As project manager, he was responsible for analyses of converting installations from oil and gas to coal firing. His other projects have included comparing air-pollution control options for boilers, and determining the compliance of generating stations to sulfur dioxide and particle-emission standards. Holder of a B.E. in chemical engineering from Youngstown State University, he is studying for an M.S. in environmental engineering at the University of Cincinnati.

How To Specify Pulse-Jet Filters

Pulse-jet fabric filters are becoming increasingly popular for collecting dusts from industrial processes. Here is a guide to the design considerations that will lead to top performance.

TED M. RYMARZ, Flex-Kleen Corp.

When specifying fabric-filter installations for air-pollution control, engineers today increasingly select the type that uses pulsed jets of air for bag cleaning. In fact, demand for these jet filters grows even though they are initially more expensive than units that depend on mechanical shaking for bag cleaning.

The popularity of pulse-jet filters is based on several inherent advantages that can mean lower overall costs. However, to make the most of potential savings, engineers should be sure that the filters are properly designed and installed. In the following discussion, we will cover the key factors in successful use of pulse-jet filters.

Higher Ratios, Fewer Bags

Jet cleaning allows use of felted fabrics at a high air-to-cloth ratio. This means that fewer bags can handle the same volume of gas at a given pressure drop. Typically, pulse-jet units operate at air-to-cloth ratios that are two to four times higher than those of conventional fabric filters cleaned by mechanical agitation.

Further, it is not necessary to divert flow from bags while they are being cleaned. This eliminates the need for extra bags where full flow must be continuous.

Originally published March 31, 1975

The ability to meet performance requirements with fewer bags is, in itself, often enough to offset the higher cost of pulse-jet units. Additionally, the flange-to-flange cost of any kind of fabric filter is only about 15% of the total expense for purchasing, installation, operation and maintenance. We find that there are very significant cost considerations apart from the initial outlay for equipment between the gas inlet and outlet flanges.

The pulse-jet design saves on installed cost because it does not require dampers and multiple inlets that are needed where flow must be shut off for cleaning. Maintaining the cleaning system is simple since there are no moving parts except solenoid and diaphragm valves, which can be serviced from the outside while the unit continues operation. Finally, longer bag life—two to three times that of mechanical-shake units—gives a double savings because the labor involved in removing a filter bag and installing a new one often equals the price of the bag.

While all pulse-jet fabric filters are alike in principle, they can vary considerably in pollution-control capability, reliability and costs, depending on how well they are engineered and built to fit the application. Among the important variables are gas volume, pressure and temperature; characteristics of the dust to be collected; and physical design and construction.

Size Generously

Because the cost of pollution control varies with the volume of gas treated, some engineers tend to skimp on gas volume by basing it on ideal conditions. Instead, volume for filter sizing should be estimated generously. There is little one can do to increase filter capacity once the installation is in place, short of expanding it to hold more bags. The volume estimate, therefore, should consider not only the nominal gas flow but also the maximum that might occur during upsets.

A factor should be included to cover the probability that wear and tear will gradually increase air leakage into the flue system, putting a greater load on the filter. And, of course, the likelihood of eventual source changes should be recognized at the time the system is specified. If possible, space for expansion should be provided for at the time of the initial installation.

Gas temperature may be the single most important factor in day-to-day operations because bag life can be seriously reduced when temperatures exceed the fabric's practical limit. It is not enough in the design stage to simply indicate normal gas temperature. One should estimate the frequency, extent and duration of unusual surges.

Natural fibers are seldom used for pulse-jet filter bags. The bags are made of a needled-felt fabric, which consists of a felted-fiber batt that is punched into an open woven backing. The most common fibers, and their nominal continuous operating limits in °F., are:

Polypropylene	180
Acrylic	250
Polyester	275
Nomex (a Du Pont nylon)	400

Glass fibers, which can operate up to 550°F., have not been used in needled-felt form. However, development work indicates that there may be some pulse-jet applications for them.

BAGFREE WALKWAY requires a larger housing but greatly simplifies the task of changing filter bags

Provide Dust Details

The filter manufacturer should be given complete information on the nature of the dust—quantity, smallest particle size, and size distribution. Of course, he also needs to know the type of operation, and the makeup of the particulates. Two especially important considerations are the dust's tendency to absorb moisture (which might cause plugging) and the degree of bridging in the hopper. A 60-deg hopper angle is typical, but the maker can provide a steeper slope if he knows in advance that the dust will not flow readily at 60 deg.

Bulk density is another important factor, particularly if the dust is lightweight. Comparing bulk density with specific gravity gives a measure of the void fraction in the bulk dust.

If the buyer sends a dust sample to the filter manufac-

How the Pulse Jet Works

The housing of a pulse-jet fabric filter is divided by a plate into a bag compartment and a clean-gas plenum. The divider plate contains openings, usually fitted with venturis, at each bag location.

Bags, held open by an interior framework called the cage, are clamped to the underside of the divider plate with their open ends up. The clamp seals the top of the bag so that only gas that passes through the fabric can flow upward into the plenum. Dust remains behind on the outside of the bag.

Bags are cleaned by the action of compressed air rather than mechanical shaking. Solenoid-operated air valves, controlled by an adjustable sequence timer, admit pulses of compressed air through a distribution sys-

tem in the clean-air plenum to orifices above the bags. As a pulse of air "jets" down into the bag, it sets up a shock wave that flexes the cloth and dislodges the accumulated dust cake, which falls into a hopper. There are no moving parts inside the collector housing.

A key advantage is that cleaning takes place while dusty air continues to flow into the filter compartment. This means it is not necessary to divert flow from bags being cleaned—and not necessary to provide extra bags to maintain full capacity during cleaning.

Pulse-jet filters can be operated at high air-to-cloth ratios, and have relatively low pressure drop through the fabric. These advantages add up to reduced space requirements and lower operating costs.

turer, the sample should be supplied with details on where and how it was collected. A scoop of coarse material taken from a conveyor belt probably will not represent the dust floating off at a transfer point.

Fine particles require relatively low air-to-cloth ratios. For example, submicron solids from a metallurgical furnace might call for a ratio as low as 2.5:1, while for coarse particles from a sanding operation the ratio could be as high as 30:1. Naturally, most applications fall between these extremes. The point to remember is that the proper air-to-cloth ratio depends on particle size and distribution. Trying to cut corners by specifying too high a ratio will produce an excessive pressure drop across the filter. This is a very important factor in pulse-jet applications because the effectiveness of the cleaning method is influenced by pressure drop.

Other Considerations

Physical design of the filter will be affected by the pressure difference between the system and ambient air, and whether the higher pressure is internal or external. As one would expect, several design decisions depend on whether the unit will be indoors or exposed to wind, precipitation and temperature extremes.

A pulse-jet filter must be supplied with air that is compressed to within 10 to 15% of the design pressure, which is usually 90 to 100 psig. At lower pressures, the cleaning action will not be fully effective, and gas pressure drop across the filter will be excessive.

The required operating pressure must be available at the filter itself, allowing for pressure drop between the compressor and filter. However, if the design pressure cannot be assured, a somewhat lower figure can be tolerated if the filter manufacturer is aware of the problem in the design stage. Of course, compressed air should be clean and dry. A moisture trap or air dryer in the supply system is strongly recommended.

When specifying insulation, both internal and external temperatures have to be considered. To avoid serious plugging of the fabric, it is very important that internal gas temperature be kept above its dew point. At the same time, exterior temperature should not be excessive. The Occupational Safety and Health Administration now specifies limits on equipment surface temperatures where workers may come in contact with the equipment.

Beyond the basic design considerations, there are options that should be discussed in advance with the filter supplier. In general, one must reach a balance between the ready availability and lower cost of a standard model, versus a longer delivery time and higher first cost of special options that may be essential for maximum efficiency and performance.

Safety, Convenience Options

If a dust or gas presents an explosion hazard, certain design provisions are mandatory. Other safety features may not be so obviously necessary, but should be investigated. For example, fixed exterior platforms may be preferred instead of erecting ladders or temporary scaffolding when access to the filter is necessary. Similarly,

STANDARD PANELS that are factory assembled cut field-construction time and reduce labor costs.

interior grids are not normally included in the base cost, although they eliminate the need for laying planks or some other kind of temporary supports.

Other frequently used safety options include level indicators that detect excessive accumulation of dust in hoppers, and zero-speed switches that signal stoppage of the hopper-emptying conveyor.

Custom design features may be selected to accommodate an unusual layout. Others may be desirable to reduce servicing costs. For example, the most economical designs fit as many filter bags as possible into the housing. Without room to enter the housing, workers must remove outer bags to gain access to those inside. Depending on the anticipated frequency of bag changing, it might be advantageous to have bagfree walkways. This may increase the size of the filter box by 30%, although the effect on overall costs is not great.

Another matter for discussion is the question of field assembly versus full or partial shop assembly. Even large

Keys to Filter Performance

- Size generously to allow for maximum flow that could occur during upsets.

- Be sure to consider the frequency, magnitude and duration of unusual temperature surges, which can seriously affect the life of filter bags.

- Provide the filter supplier with full details on composition of dust to be collected.

- Check the structural design for proper stability and mechanical fit between components.

- Follow a regular schedule for bag replacement to avoid an outage at an inconvenient time.

EXPOSURE TO WIND and weather in rooftop installation calls for careful engineering of filter supports.

fabric filters usually can be shipped in subassemblies that reduce onsite labor and installation time.

Construction Tips

Many construction details have a substantial effect on performance and reliability. A solid-state timer, for instance, probably will be more reliable than electromechanical types. It is important to have high-quality tubing connecting the jet-pulse diaphragm and the solenoid pilot-valve so that the tubing can withstand rough treatment and outdoor environments.

The air-jet system plays a critical role because cleaning action suffers if the jet does not discharge down the full length of the bags. Pipes that distribute compressed air to the bags should be heavy enough (schedule 40) so they will not distort, and should have orifices located precisely over the bags.

Bent cages that allow filter bags to touch each other or the collector structure can cause rapid bag wear. Welds should not leave rough spots on the cage exterior because these also provide points for accelerated wear.

Since rust can roughen the cage, stainless steel should be considered to prevent corrosion, a particular problem when the equipment is shut down frequently or over weekends. Where the products of corrosion could contaminate the dust (food processes, for example), we recommend stainless steel for all dust-contact parts below the tubesheet.

The means of suspending the filter bag and cage is another key component. It should be strong, corrosion resistant, and designed so that correct installation of the cage and bag can be easily handled. The suspension often incorporates a venturi to enhance the air-jet action. If a bag ruptures, which occurs even in the best units, dust can be picked up by the jets and blown at high speeds through the venturis. A cast-aluminum venturi has sufficient thickness to withstand considerable abrasion; light-gauge spun venturis wear out much sooner.

Watch Structural Design

Too often, the structural aspects—fit and stability—do not receive enough attention. As an example of fit, the air-lock valve and conveyor system must be matched to each other and to the filter hoppers. Not only should the physical dimensions agree, but also the capacity to remove dust as fast as the filter collects it.

Although it seems obvious that structural support is not a place for cost-cutting, fabric filters have been known to sag, settle, and even be blown over by the wind.

With panelized construction, virtually any size of filter can be assembled simply and economically by bolting together flanged sections. This works especially well in a confined area. Possible problems are panels without adequate rigidity that may eventually crack, and flanges that will not readily mate for gastight assembly because of distortion during shipping or handling.

Replacing Bags

Bag life in a properly engineered filter should be at least one year. Often it is several years. But sooner or later, bags must be replaced, and it is vital that the new ones have the exact diameter of the originals in order to obtain a tight seal. They should also have enough slack to flex properly during the cleaning pulse.

Bag seams should be sewn with a double lock-stitch so they cannot unravel, and it is good practice to have a reinforcing stitch angling across the top of the vertical seam. The long seam must be straight to make the bag perfectly cylindrical—otherwise the air pulse may not flex the bag enough to clean it thoroughly. Fabric type should never be changed without first consulting with the filter manufacturer.

A regular schedule of bag replacement is better than waiting for failure. Ruptured bags can force a lengthy outage at an inconvenient time. A design that provides top openings makes replacement easier because sometimes a faulty bag leaves a telltale streak of dust in the clean-air plenum. Usually, locating a ruptured bag among many in the dusty interior of a filter is quite time-consuming.

Meet the Author

Ted M. Rymarz is manager of applications research for Flex-Kleen Corp. (subsidiary of Research-Cottrell Inc.), 222 S. Riverside Plaza, Chicago, IL 60606. He is responsible for application of pulse-jet filters and related research and development activities. He joined the company in 1970 as assistant chief engineer after working in fine-particle research at the IIT Research Institute. Mr. Rymarz received a B.S. in chemical engineering from the University of Illinois in 1952.

Cooling Hot Gases Before Baghouse Filtration

Gas filtration through fabric bags is often needed for pollution control or product recovery. But high temperature can destroy the bags. Here is how to choose the cooling method that best fits your situation.

PAUL VANDENHOECK, Combustion Equipment Associates, Inc.

Successful operation of arrays of cloth dust-arresters (baghouses) in filtering high-temperature gases requires that the gases be properly cooled prior to filtration. Otherwise, all bags are liable to be destroyed, representing the loss of one-third the cost of the dust arrester and probably entailing an operational shutdown until the bags are replaced and the cause of their loss remedied. (Permissible maximum temperatures range from 180 F. for cotton bags to 550 F. for glass-fiber bags.)

The three basic methods used for the cooling of high-temperature gases are: dilution, radiation, and evaporative cooling.

Dilution

The simplest method of cooling exhaust gases is to dilute them with ambient air. Ambient air is generally added near the baghouse, provided good mixing of air and gases can be assured. This can be accomplished by positioning the fan, which propels gas through the system, between the ambient-air inlet and the baghouse. Or, if a cyclone collector is employed to eliminate sparks from the gas stream before it reaches the baghouse, positioning the cyclone between the ambient-air inlet and baghouse will assure good mixing.

Ambient air can be admitted through a simple Y-shaped fitting installed in the gas duct. A damper in the leg of the Y that opens to atmosphere can be automatically controlled to admit the proper amount of air required to bring the gas temperature to the desired predetermined value. (A temperature sensor installed at the collector inlet controls a motor-operated damper that ad-

Originally published May 1, 1972

mits more or less air as required to maintain the gas stream at the predetermined temperature.)

Dilution is almost always uneconomical because the volume of ambient air required to achieve the desired cooling is so great that a substantially larger baghouse is required to handle it all (Fig. 1).

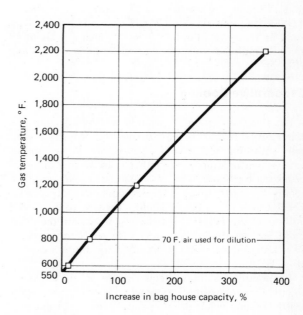

ADDED CAPACITY needed in baghouse when hot gases are cooled by dilution with ambient air—Fig. 1

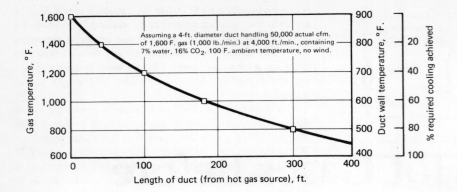

Assuming a 4-ft. diameter duct handling 50,000 actual cfm. of 1,600 F. gas (1,000 lb./min.) at 4,000 ft./min., containing 7% water, 16% CO_2. 100 F. ambient temperature, no wind.

RADIATION effectiveness in cooling hot gases— Fig. 2

Radiation

Radiation cooling transfers heat from the gas stream to the atmosphere via the duct walls; as the gas travels toward the dust arrester. It is to take advantage of radiation cooling that the ambient-air intake is located near the baghouse when dilution cooling is used. As a rule of thumb, radiation cooling provided by 100 ft. of duct conveying 1,600 F. gases to the baghouse can take care of one-third to one-half of the total cooling requirements. Since 100 ft. is an average distance between the hot-gas source and the collector, substantial cooling is economically accomplished. When further radiation cooling is desired, additional lengths of duct are calculated on the basis of heat to be removed from the gas under regional weather conditions.

Radiation-cooling ducts may be in the form of vertical U's that may be as much as 40 ft. or more high. Although radiation cooling does not require an enlarged baghouse, it does have economic limitations. Radiation cooling of gases that are above 1,600 F. entails exotic construction materials and is uneconomical. Cooling of gases below 1,000 F. requires substantial surface areas, lengthy duct runs, and increased fan horsepower, which quickly offset the economic advantages of a minimum-size baghouse (Fig. 2).

Evaporative Cooling

Evaporative cooling is accomplished by injecting fine water droplets into the gas stream—the droplets evaporate, absorbing heat from the gas. Spray nozzles are located either in a quench chamber or in ducting that precedes the baghouse. The gas is cooled quickly and in a relatively small space. Greater baghouse capacity is needed than for radiation cooling because of the added volume of the evaporated (gaseous) water—on the other hand, 90% less baghouse capacity is needed for evaporative cooling than would be required for dilution cooling (Fig. 3).

Evaporative-cooling systems require careful engineering for proper operation. Systems for cooling gases below 800 F. require water droplets smaller than 125 microns. There are three classes of spray nozzles that can produce droplets within this category. They are: high-pressure-

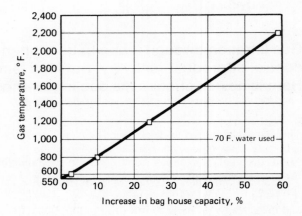

70 F. water used

BAGHOUSE capacity increase needed when hot gases are cooled by means of evaporative cooling—Fig. 3

water nozzles, high-pressure air- or steam-atomization nozzles, and centrifugal spinning nozzles. High-pressure-water nozzles are common at present, but high pressure air- or steam-atomization nozzles are newly being adapted for gas-cooling applications. Centrifugal spinning nozzles are not appropriate, due to required enclosure configuration and to maintenance necessitated by particulate matter in the gas stream.

High-Pressure-Water Nozzles—These inject water at 150 to 300 psi. through a small orifice. Each nozzle sprays but a small quantity of water, and some 20 or more nozzles are employed to achieve desired cooling. Water can be properly atomized only at the nozzle's rated flow capacity; the nozzles cannot be modulated. Therefore, nozzles are operated in banks, and varying spray requirements are met by turning appropriate banks of nozzles on and off. Hence, the resulting system cannot provide closely graduated cooling temperatures because cooling temperatures change in steps dependent on the number of nozzle banks in operation.

From a maintenance standpoint, the water supply must be free of particulate matter. Drinking water can contain sufficient solids to erode a nozzle orifice within one week. Even when high-purity water is available, and the nozzles are protected by adequate strainers, weekly

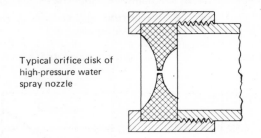

High-pressure air or steam atomization nozzle

Typical orifice disk of high-pressure water spray nozzle

COMPARISON of designs of nozzles used for water-spray, and for high-pressure air (or steam) atomization—Fig. 4

inspection is required to check for erosion and for nozzle plugging from pipe-scale deposits, corrosion, and other solids accumulation. When nozzles are off, they should be continuously purged with compressed air to prevent plugging and corrosion by deposits from the gas stream. Valving can be arranged so that when water to the nozzles is turned off, compressed air is automatically turned on, reaching the nozzles through the same piping as the water.

The banks of nozzles are usually installed in the well of a vertical spray chamber with a suitable access door and internal or external scaffolding for nozzle inspection and maintenance. The chamber is of large diameter to provide sufficient volume to retain the water droplets for a period long enough to complete their evaporation. A typical quench chamber exhausting 25,000 actual cfm. at 500 F. might be 9 ft. dia. by 30 ft. high.

Existence of the chamber leads to efforts to "kill two birds with one stone," and the chamber is often required to act as a crude spark arrester or mist eliminator, rather than being followed by a mechanical dust collector, which is completely effective in removing sparks and moisture from the gas stream. Absence of a mechanical collector following a spray chamber can result in bag damage due to sparks and water-droplet carryover. It can also lead to serious fan abrasion by coarse dust particles when the fan is located between the spray chamber and the baghouse.

High-Pressure Air- or Steam-Atomization Nozzles— These (Fig. 4) provide superior spray droplets of less than 50 microns, and require virtually no maintenance. This type of nozzle is predominantly used in spraying heavy fuel oils with entrained solid contaminants into furnaces. It sprays water even better and is now being adapted to evaporative cooling of gases. It has a striking advantage in that only one nozzle is required. As a result, it can be installed right inside the gas duct, along its center axis. No quench chamber is needed, provided the duct is long enough to furnish the required retention time. The single nozzle will spray any flow of water to meet gas cooling demands. This includes properly spraying the maximum amount of water that might be required, down to 10% of this amount.

The drawback of this nozzle is its compressed air or steam consumption, which can be as high as 1 lb. of 50-100 psi. air or steam for every 5 lb. of water atomized. When the quantity of water to be atomized is under 10 gpm., plant compressed air or a separate compressor is often the most economical propellant unless steam is readily available.

Evaporative Cooling Expertise

Safe, dependable, low-maintenance evaporative cooling requires more knowhow than simply spraying water inside a duct or chamber. All the water injected into the gas stream must be evaporated before the gases reach the baghouse. Otherwise, metal parts may be corroded and gas passages in the bag fabric become plugged as dry dust being collected is turned to sludge. Failure to evaporate injected water can result in buildup of nonflowing sludge in the duct or baghouse hopper, requiring continued maintenance in shoveling it out. Spray water must not contact the walls of the duct or the spray chamber. Otherwise, metal distortion would occur or, when lined chambers were used, refractory would spall. Lining may be required to protect the metal from high-temperature gases, corrosive gases, or both.

Evaporative cooling of a hot gas stream down to a set temperature requires the injection of a precise amount of water consistent with the gas mass treated. Liquid-travel distance required for complete evaporation after injection is dependent on water droplet size and gas temperature, and can be calculated. However, in designing system components, particularly quench chambers, to be sure that all moisture will be evaporated before the gas stream reaches the baghouse, experience is required. Such factors as low gas-velocity and incomplete mixing of gas and water do not permit quench-chamber evaporation within theoretically calculated intervals. Accordingly, larger chambers are required than theory alone would indicate.

Radiation-Cooling Expertise

Since radiation cooling involves a lengthy duct between the source of the hot gas and the baghouse, particular care must be taken to avoid duct abrasion and material buildup within the duct.

Abrasion is caused by impingement of particles on the duct surface. Removal of coarse dust particles from the gas stream is one step in minimizing abrasion. This can be accomplished by passing the gas stream through a low-velocity chamber, impingement device or gravitational inertial separator located near the gas source.

Avoiding turns in the duct is the second step. Little abrasion is experienced in straight runs, provided particles remain suspended and there is no gas turbulence. Bends in the duct will be subjected to great wear unless adequately protected. This is best accomplished where feasible by letting the particulate impinge on a buildup of its own material. Squared-off outer corners of bends will accomplish this (material will build up in the corners thus formed).

To avoid material buildup along the bottoms of straight runs, at least a minimum gas velocity should always be maintained, keeping in mind that gas volumes, and therefore velocities, change with temperature. Minimum velocities required will range from 3,500 ft./min. to 6,000 ft./min., depending on the particulate and duct configuration. Difficulties can be expected if gases from more than one source are manifolded to a common duct, unless balancing problems can be eliminated.

Overcooling

Gases must not be cooled below their acid or water-vapor dewpoint. Otherwise, moisture will condense on dust-arrester components, corroding metal parts and plugging fabric. The end-result will be equipment down-time and high maintenance costs.

Dust-Arrester Protection

A few seconds of runaway temperature can cause thousands of dollars worth of damage to dust-collection equipment. Provisions must be made to instantly isolate the baghouse from the gas stream in the event of equipment, utility or process failure. In the case of equipment or utility failure, interlocks can shut off the flow of gases or cause them to bypass the baghouse.

Instantaneous protection against high temperatures due to process failures is more difficult to assure. Thermocouple sensors have a built-in time response lag to temperature change. This can be accentuated by dust buildup on the sensing element. For practical purposes, however, thermocouples are the only method of initiating protective actions based on temperature changes. Therefore they are used in shutting off the gas flow or to divert gases away from the baghouse.

However, heat sinks may be provided between the hot gas source and the dust arrester to reduce temperatures long enough for the thermocouples to activate shutoff or bypass valves. The heat sink also gives up heat to the gas stream when sudden gas-temperature drops are encountered, preventing excessive cooling of the gas stream.

Pitfalls to Avoid

The technology for cooling high-temperature exhaust gases by U-tube radiation cooling or by air dilution is relatively simple and well within the capabilities of most baghouse suppliers. However, these approaches were developed during times when labor and material costs played less-critical roles than they do today in the all-important profit making picture.

The theoretical economies to be realized through evaporative cooling are appreciated. Unfortunately, the technology and expertise are limited. Completely suc-

cessful installations are the exception rather than the rule.

Close coupling of the water-injection area to the baghouse is the most expensive mistake frequently made. It requires that all components work perfectly, and leaves no room for individual component failures, process upsets or less-than-perfect maintenance. The result is periodic loss of all bags.

Inadequate bag protection from moisture carry-over due to faulty water-injection equipment, and from sparks from the process, is the next source of costly mistakes, paid for in high maintenance and bag replacements.

Other common complaints of improperly designed evaporative cooling systems are: excessive fan erosion due to coarse particle carry-over in positive-pressure systems; need for frequent nozzle inspection; material buildups; metal distortion due to heat; and refractory failures.

Cutting Costs

Since air-pollution-control equipment does not contribute to profits, the tendency is to reduce its capital cost to a minimum. Maintenance, however, is expensive, and the approach, therefore, should be to cut overall costs without sacrificing quality and reliability.

Steam or air atomizing has built-in cost savings. Atomization requires less retention time and the use of only one nozzle, instead of six to eight banks of three or more high-pressure nozzles each. This enables a straight section of stainless-steel duct followed by a 100% effective mechanical cyclone-collector to replace the expensive, low-velocity, bulky quench tower required with high-pressure water-spray cooling. Also eliminated is a brick-lined duct from the gas-emission source to the quench tower.

Maximum use of unlined stainless-steel ductwork permits substantial use of radiation cooling, resulting in savings of 20 to 30% on baghouse fan power and capital costs. It also provides a substantial heat sink between the cupola and baghouse to ensure a fail-safe system against surges of high-temperature gases. The mechanical collector further guarantees complete protection against water-droplet, spark and abrasive-particle carryover. This claim cannot be met by the best of quench towers.

A well-designed atomizing system taking full advantage of all available cooling and gas handling techniques can be provided at competitive first-cost prices, and offers substantial operating and maintenance savings.

Meet the Author

Paul Vandenhoeck is manager of the Process Systems Dept., Combustion Equipment Associates, Inc., 555 Madison Ave., New York, NY 10022. He is a specialist in wet-scrubber, mechanical, electrostatic-precipitator and bag-filter equipment for industrial air-pollution control. He received his B.S. from Yale University, and did graduate work in mechanical engineering at Svenska Kullager Fabriken Aktiebolaget, Gothenburg, Sweden. He is a member of the American Inst. of Mining and Metallurgical Engineers, and the Foundrymens Soc.

Baghouses: separating and collecting industrial dusts

The selection and operation of baghouse collectors is mainly an art. Here is practical information on filter fabrics and cleaning; baghouse design, operation and servicing; and methods for protecting the system from fire and explosion.

Milton N. Kraus

☐ The separation of dusts from industrial-gas streams may often be done by using filters made of natural or synthetic fibers. These filter elements are bag-shaped, and placed in structural enclosures called baghouses.

In addition to supporting the filter elements, the baghouses contain baffles for directing airflow into or out of these elements, equipment for cleaning the fabric, and a hopper for collecting and discharging the dust.

Baghouses are used to control dust nuisance whenever dust-laden air must be discharged to the atmosphere, or to recover valuable dust from a process venting system. Sizes may range from small bin-venting filters to large multicompartment filters that receive dust from an extensive system of exhaust ducts.

Dust filtration and collection is an art. The sizing and selection of dust filters are based on (a) past experience and (b) actual tests that use specific fabrics at specific dust-to-air loadings. Because of the many types of collectors and filter fabrics, combined with numerous material properties that affect filtration, the selection of a filter should be made by actual test at the design load, using the least-costly fabric for the requirements.

Separation principle for filters

Dust filtration in baghouses is accomplished by passing the dust-laden air through a filter fabric that is formed into cylindrical tubes or oblong bags. As the air passes through the fabric, the dust is retained on the surface and in the interstices of the yarn strands, building up a filter cake that also acts as a filter medium. In order to reduce the resistance to airflow (so as not to affect the flowrate), the filter cake must be periodically dislodged. The method of cleaning the fabric is the basic difference between various filters.

The greater the area of filter cloth provided for a given dust loading on a specific baghouse, the longer it takes for a filter cake to build up that will affect the airflow. In comparing the area of filter fabric offered by different vendors, we should consider the different methods of cleaning the fabric. The criterion should be a constant pressure drop across the fabric for a specified airflow rate and a specified dust loading, rather than the term "air-to-cloth ratio."

If checked dimensionally, this term is the face velocity of the air (ft/min) through the effective area of the fabric for a given type of filter. The air-to-cloth ratio only has significance when comparing the performance of a particular manufacturer's line of baghouses when handling different materials. The cleaning method establishes whether a baghouse is suitable for continuous or intermittent service.

Properties of filter fabrics

In order to function properly, a filter fabric must have the following properties:

Permeability—The fabric must be sufficiently porous to

Originally published April 9, 1979

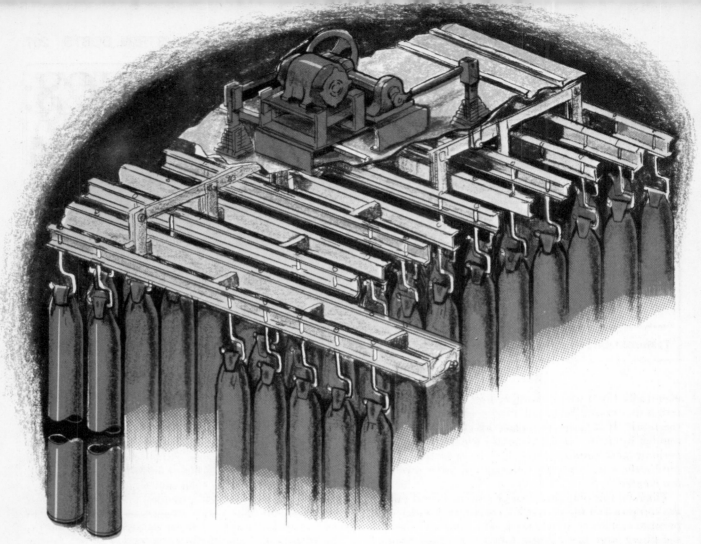

permit a satisfactory flow of air. The permeability of a fabric is generally stated as the clean airflow in (ft³/min)/ft² of fabric, at a pressure differential of 0.50 in. water column, as determined by the Frazier test.* However, the Frazier permeability alone is no guide to fabric selection because materials having the same permeability may not have the other characteristics required to successfully filter a specific material. Also, the filter cake formed on the fabric surface will reduce airflow.

Mechanical strength—The fabric must resist the tensile forces caused by the operating pressure differentials, by mechanical shaking during cleaning, and by pulsing during reverse airflow. It must withstand abrasion where it is clamped to tubesheet ferrules, where it is supported on metal grids or retainers, and where it is subjected to the impact of the filtered materials.

Solids retention—Fabric construction must be open enough to prevent the accumulation of fines in the interstices of the yarn, yet tight enough to prevent fines from blowing through the fabric. This may seem anomalous, but the filter cake on the surface of the fabric also acts as a filter medium and narrows the interstices.

Corrosion resistance—The fabric must resist attack and weakening due to chemical action between the fibers

*ASTM Standard D-737, Standard Method of Test for Air Permeability of Textile Fabrics, American Soc. for Testing and Materials, Philadelphia, Pa.

and the filtered materials, especially if moisture is present due to condensation.

Heat resistance—From some processes, the fabric must resist high-temperature exhaust gases. Each fabric has a definite temperature limit beyond which it will tend to disintegrate.

Cleanability—The fabric must have a surface texture that is conducive to rapid release of the filter cake during cleaning. This, too, may seem anomalous, since the surface must also retain the cake. However, the forces involved during cleaning are usually the airflow in the opposite direction to normal airflow, and mechanical flexing of the fabric. The fabric should have a high rate of electrostatic-charge dissipation so as to shed charged dust particles.

Dimensional stability—The fabric must resist stretching or shrinking that would affect its permeability. Many untreated fabrics tend to increase in permeability after laundering.

The properties of commonly used filter fabrics are shown in Table I.

Fabric construction

A knowledge of the construction of filter fabrics will indicate how well the preceding property requirements can be met. However, final selection should be made by test and by an economic study that weighs first cost against probable maintenance expense. All fabrics will

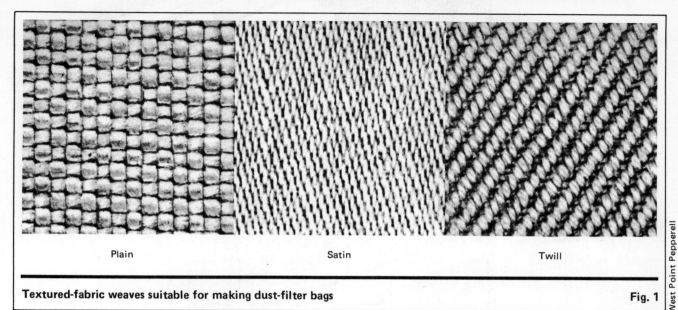

Plain   Satin   Twill

Textured-fabric weaves suitable for making dust-filter bags   Fig. 1

eventually blind when handling fine materials, or materials that crystallize or polymerize when embedded in the fabric. It is better to replace all cloth in a filter at regular intervals (as determined by experience or by periodic permeability tests) than to operate the unit until airflow is drastically reduced and other problems are created.

Filter fabrics may be felted or woven. Felted fabrics are composed of fibers that are compressed under high pressure and are relatively thick. The pressure enmeshes the fibers, and in so doing forms a strongly bonded sheet of material having complex labyrinthine interstices through which air can pass in three directions. Synthetic monofilament fibers cannot form a strong bond between themselves and are usually pressed over a woven sheet of scrim fabric for strength. Wool fibers are naturally barbed and have a stronger bond than synthetics. Most felts can absorb moisture due to capillary

action between the fibers—even though the fibers themselves are moisture resistant.

A simple test for relative absorption of moisture is to stand equal lengths of fabric on edge in a jar containing a shallow layer of water. After several minutes, the samples should be withdrawn and the height to which the water has risen in each sample noted.

The thickness of felt provides for maximum dust impingement. The many changes in direction due to lateral airflow traps the finest particles. Inspection (a) by ultraviolet light of the cross-section of felt fabrics that have handled dust containing fluorescent materials, and (b) by microscopic examination of the cross-section of felts handling other types of dust, shows that the particles rarely penetrate through more than one third of the felt thickness. While there is a reduction in air permeability, this is not as much as would occur in a woven fabric having the same quantity of trapped dust per square inch of filter area.

Felted fabrics are used for maximum product recovery where filtered air is recirculated or discharged into a work space. Felts are primarily used in collectors having a high face velocity for airflow (i.e., high air-to-cloth ratio).

Woven fabrics are composed of twisted yarns that are woven into geometric patterns having various spacings between the yarns, and with a specific surface finish that is designed to retain or shed a filter cake, depending upon the application. The permeability of these fabrics depends on the type of fiber, the tightness of twist and the size of the yarn, the type of weave (geometric pattern), the tightness of weave (thread count), and the type of surface finish. Correct coordination of all these factors for a successful filter application cannot be done by abstract analysis. The fabric must be tested in a dust collector considered suitable for a particular application on the basis of operational and economic factors. Woven fabrics have a low ratio of weave openings to yarn area and require a surface filter cake to retain the dust. This limits the face velocity for airflow

Properties of common filter fabrics			Table I
Fabric	Temperature limitation, °F	Chemical resistance	
		Acids	Alkalis
Cotton	180	Poor	Good
Wool	200	Good	Poor
Nylon	200	Poor	Good
Polypropylene	200	Excellent	Excellent
Orlon	260	Good	Fair
Dacron	275	Good	Fair
Dynel	160	Good	Good
Glass	550	Excellent	Poor
Nomex	400	Fair	Good
Polyethylene	165	Good	Good
Teflon	400	Excellent	Excellent

Relative ratings: Excellent, Good, Fair, Poor.
Source: Publications of the Industrial Gas Cleaning Institute, Inc., Alexandria, Va.

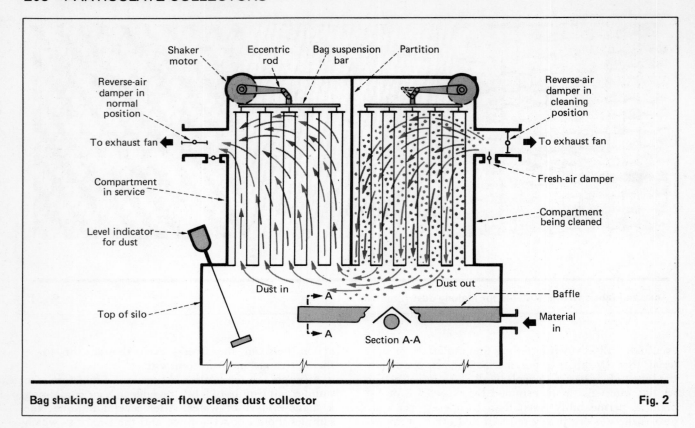

Bag shaking and reverse-air flow cleans dust collector   **Fig. 2**

to low values [1.5 to 3 (ft³/min)/ft²] in order to prevent dust from blowing through the openings and destroying the filter cake.

The permeability of woven fabric is very dependent on its construction. A review of textile terminology will prove useful in understanding the effects of yarn design and fabric weaving on permeability. The yarn may be made from:

1. Staple—the term for short fibers of cotton or synthetic material—drawn into parallel strands and twisted into yarn by spinning.

2. Continuous filaments of synthetic material, made by extruding a solution through a finely perforated nozzle and then spinning the filaments into yarn.

The degree of twist of the yarn affects permeability. A tight twist resists penetration by dust particles, and a loose twist permits retention of dust, or "blinding" of the yarn. Filament yarns are stronger than staple ones. Spun yarns have more flexibility than filament yarns and more resistance to repeated flexing.

Weaving is the interlacing, at right angles, of a series of yarns to form a textured fabric. The basic weaves are plain, twill and satin (Fig. 1). In the plain weave, each crosswise or filling yarn passes alternately over and under one warp or lengthwise yarn. Each warp yarn passes alternately over and under each filling yarn. This type of construction makes the tightest weave; by varying the thread count, the weave may be made as open and porous as required.

In the twill weave, the filling yarn passes under one warp yarn, then over two or more warp yarns. In each succeeding row, the filling thread moves one warp either left or right, forming the diagonal lines that iden-

tify a twill weave. Twills have fewer interlacings than plain weave and, depending on the thread count, greater permeability.

In the satin weave, the filling yarn passes over one and under four or more warp yarns in the next row. This weave has few interlacings, spaced widely but regularly. It has a smooth surface and greater permeability than the other weaves. Cotton fabric in this weave is known as cotton sateen. Regardless of weave, the size of yarn and the number of warp and filling yarns per inch (thread count) greatly affect permeability. A low thread count will permit higher airflow rates at low pressure drop, whereas a high count will trap more fine particles and produce a higher pressure drop.

The woven fabric may have various finishes. Cotton fabrics may be preshrunk to maintain dimensional stability. They may also be napped in order to hold a filter cake. (Napping is the process of scratching the cloth surface to raise the fibers and form a soft pile.) Synthetic fabrics may be heat-set to obtain dimensional stability and a smoother surface with uniform permeability. All fabrics may be silicone-sprayed to obtain abrasion resistance, improved cake release, or low moisture absorbency.

Many other fabrics are available for special applications. Selection of these should be made only after determining that the common fabrics are not suitable for the proposed service, and after discussing the application with specialists in filter-media firms.

Example of performance

The following example illustrates the unpredictability of fabric performance. Specifications for the original

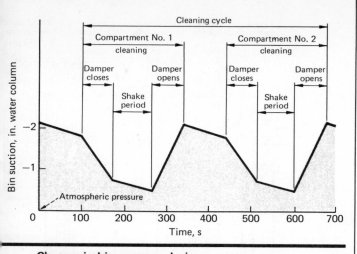

**Changes in bin pressure during
cleaning of two-compartment filter**   **Fig. 3**

by clamps or snap rings, should have a reinforced cuff.
All hems and seams should have two rows of stitches,
with edges folded or treated to prevent fraying.
Spreader rings should be enclosed in a pocket that is
doubly reinforced inside and outside the bag.

The suspension end of hung bags should be doubly
reinforced and folded over to form a loop for hanging
on a wide suspension hook, or fitted with a metal grom-
met if hung from an eye bolt or hook bolt. Cuffs of
oblong bags that are hooked over support pins should
be fitted with metal grommets at the hook locations.
Where bags must be grounded, a copper wire should be
sewn the full length of the bag, with sufficient lead to
clamp to a tube ferrule.

Filter cleaning

The shape of the filter element, its method of support
and the normal direction of airflow through the filter
will determine the method of cleaning. Normal airflow
will hold the dust cake against the fabric. Stopping the
flow of air will tend to release the dust cake. Simultaneously flexing the element, or reversing the airflow, will
ensure release of the cake so that it can drop into a
hopper.

Flexing can be induced (1) mechanically, by motor-
driven crank mechanisms that oscillate closed-end tube
elements, or (2) pneumatically, by reverse airflow for
flow-through or closed-end tube elements.

On flow-through elements that collect dust on the
inside surface, spreader rings may be used to form nodes
for more-effective reverse-air flexing and to prevent
collapse of the elements. (A collapsed element would
prevent the dust from falling into the hopper.) On
closed-end tubular elements that collect dust on the
inside surface, a top-suspension tensioning device is
depended upon to prevent collapse of the elements
during cleaning and to hold their surfaces taut to form
the dust cake during normal airflow.

Envelope, oblong or tubular elements that collect
dust on their external surfaces are generally placed over
metal retaining cages or grids that keep the fabric taut
so as to retain the dust cake and maintain the internal
open area required for clean-air flow. During cleaning,
reverse airflow flexes the fabric away from the retainer
grids; restoring normal airflow whips the fabric
against it.

On collectors where normal airflow through the fabric is not stopped, an extra-high reverse airflow is
needed to dislodge the dust cake. This may be done by
using high-pressure air jets or separate air-reversal
blowers.

Cleaning methods for baghouses

Commercial baghouses use one or more of the cleaning methods previously discussed. These methods may
be further described as follows:

Shaking only is used on unit filters having woven bags
where service can be interrupted long enough for cleaning—usually 3 to 4 min. The bags may be (1) envelope
type, mounted over a metal spacer, with the closed end
connected to a motor-driven shaker grid and tensioning
bar, and the open end fastened to a clean-air plenum
wall; or (2) tubular type, with the closed end suspended

bag material as furnished by a filter manufacturer are
compared to a replacement material supplied six years
later by the same manufacturer for the same filter and
the same service.

	Original	Replacement
Fabric manufacturer	ABC Co.	XYZ Co.
Material	Orlon	Orlon
	filament	staple
Thread dia., in.	0.0085	0.024
Thread count, yarns/in.	70 x 68	90 x 60
Weave	Twill	Satin
Permeability, (ft^3/min)/ft^2	20 to 25	43
(Frazier)		

Despite the apparent dissimilarity of these fabrics,
they gave the same performance. In this case, filtration
and pressure drop were evidently a function of the type
of filter cake built up on the fabric during service rather
than a function of the fabric.

The fabrics are formed into filter bags having configurations to suit the collector. The bags may be cylindrical tubes open at both ends for attachment to top and
bottom tubesheets using band clamps or snap rings, and
fitted with anticollapse spreader rings to permit the
released dust to fall out. Or, the bags may be cylindrical
tubes closed at the top to permit suspension from a
system of shaker bars, and open at the bottom for
fastening to a tubesheet. Or, they may be cylindrical
tubes closed at the bottom and open at the top for
installation over a wire retaining-cage suspended from a
tubesheet.

The bags may be oblong—in one version, closed at
one end, open at the other for slipping over a retaining
cage (like a pillowcase); in another version, closed at
one end for top suspension, with several cylindrical
openings at the lower end for clamping to a tubesheet.

In any bag design, the following details should be
considered:

Thread used to sew the bag should be of the same
material as the bag, or compatible with it. The open
ends of the bag, which are attached to tubesheet ferrules

from motor-driven shaker bars, and the open bottom end fastened to a tubesheet for the entry of dust-laden air. The cleaning cycle may be initiated by (a) a timer set for a specific cleaning period, (b) a pressure-differential sensor, or (c) a manual stop-button connected to a time-delay relay. In this method, the exhaust fan stops, allowing the internal pressure to reach atmospheric, and the shaker motor operates for a preset interval. After cleaning, operation of the dust collector may be restored automatically or manually, depending upon the service.

Shaking with reverse airflow is used with either type of bag described for unit filters. However, for continuous service, the filter elements must be distributed among two or more compartments, and the exhaust fan must remain in operation during the cleaning cycle. Each compartment is fitted with its own shaker mechanism and a set of reverse-air dampers. Each set comprises an outlet damper, ducted to the exhaust fan, and a smaller damper opening directly to atmosphere. A program timer permits cleaning the bags in each compartment successively, with the time interval between cleanings adjusted to suit the service.

During cleaning, the timer activates the reverse-air damper drive to close the clean-air outlet damper to the fan, and simultaneously opens the atmospheric-air inlet damper. Suction from the active compartments draws air through the bags in the reverse direction. Limit switches make sure the dampers are correctly positioned and start the shaker mechanism. After the shake period, the reverse-air damper drive opens the clean-air outlet damper and closes the atmospheric-air inlet damper, reactivating the compartment. During the cleaning cycle, a certain amount of dust is picked up by the reverse airflow and deposited on the bags in the active compartments.

Fig. 2 shows a two-compartment filter used for venting a silo receiving material from a pneumatic conveying system. Fig. 3 shows the variation in bin pressure during the cleaning cycle for this collector.

Reverse-airflow cleaning is used for all types of filter elements (tubular or envelope, woven or felt materials, dust inside or outside of the bag, elements distributed between several compartments or all in one housing). There are many arrangements for creating reverse airflow—all intended to enable continuous service. Several common arrangements are:

1. A compressed-air nozzle at the clean-air discharge of each bag injects high-pressure air into the tube in the reverse direction from normal airflow in pulses of short duration (1/10 to 1/25 s) for systems having closed-end tubular bags supported on metal retainer cages suspended from the clean-air plenum tubesheet. The pulses are controlled by external timers that operate special air-supply solenoid valves, each of which serves a group of bags. The time between pulses is adjustable, but is factory-set for the intended service. See Fig. 4.

2. A variation of this method is to divide groups of bags (including the clean-air plenum) into zones by means of internal partitions. The clean-air plenums are manifolded into an outlet duct via poppet valves. Each zone is equipped with a single pulse valve that supplies compressed air to the group of bags. During the clean-

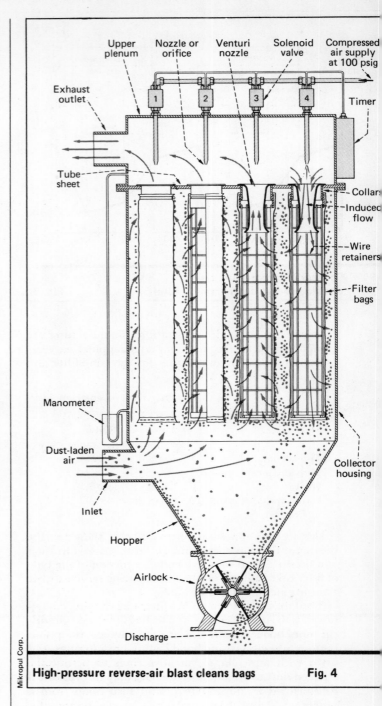

High-pressure reverse-air blast cleans bags   **Fig. 4**

Mikropul Corp.

ing cycle, the poppet valve in the zone to be cleaned closes, stopping airflow through the zone. The pulse valve opens for 0.1 s, admitting a burst of air into the plenum to clean the bags. The poppet valve then automatically reopens, putting the cleaned zone back on-stream. This operation is repeated in each successive zone until all bags in the collector have been cleaned. See Fig. 5.

3. A motor-driven plenum chamber receiving air from a separate blower delivers reverse air to closed-end oblong bags that are supported as in the previous arrangement but not zoned. The plenum rotates and blows reverse air into the clean-air discharge end of each bag as its nozzles travel over the bags. Timing is

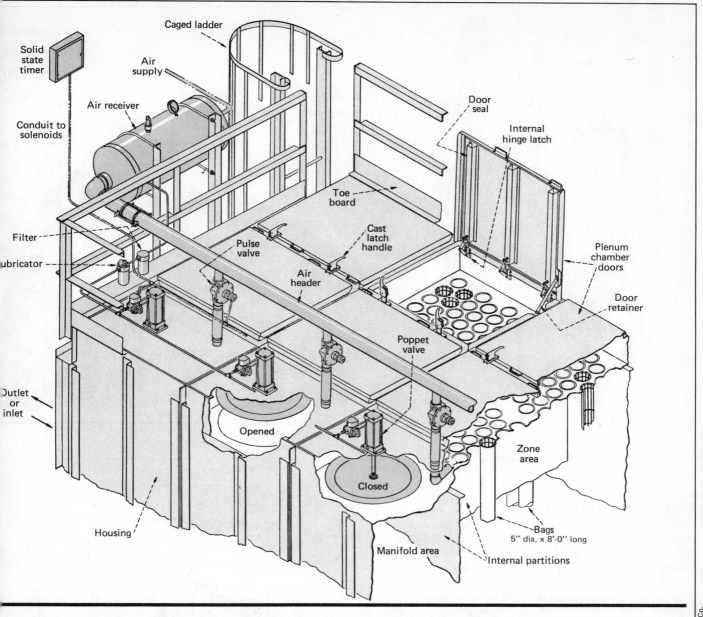

Partitioned collector uses a pulse valve to admit reverse air for cleaning all bags in zone   Fig. 5

Fuller Co.

determined by the speed of rotation of the plenum chamber. See Fig. 6.

4. A counterweighted frame carries a series of blow-rings that encircle each filter tube. The frame is driven slowly up and down by chains supported from overhead shafts and sprockets. Air is continuously supplied to the blow-rings by an external blower or fan connected by a hose, or a specially designed air-supply connection to a manifold mounted on the blow-ring carriage. See Fig. 7.

5. A traveling air-plenum, moving horizontally, is positioned over the clean-air discharge of each vertical line of envelope-type filter bags. Diaphragms on the traveling plenum seal off adjacent rows of bags. Reverse air may be supplied in two ways: (a) atmospheric air may be drawn through the filter bag in the reverse direction from normal airflow by the operating suction

of the exhaust fan, or (b) clean air may be blown back into the bags from a separate blower mounted on the traveling plenum. See Fig. 8.

Design requirements for filters

Manufacturers offer standard designs for baghouses. However, such designs may have to be radically modified to suit the user's requirements. These may be fixed by the characteristics of the material entering the baghouse, and by the operational needs of the duct system to which the baghouse is connected. Before preparing a specification for a baghouse, consideration must be given to design details such as: housing construction, physical location, servicing provisions, method of operation, and protection from fire and explosion.

Blank covers are provision for future filter elements

**Rotating plenum provides reverse-air flow
to clean the bags in cylindrical collector**

Fig. 6

Housing construction

Baghouses may operate under positive or negative pressures. Baghouses that receive air from a fluidized-bed process or from a positive-pressure pneumatic conveying system generally operate under positive pressure. The pressure is limited by the pressure drop across the bags because discharge is directly to atmosphere.

Most baghouses operate under negative pressure, which imposes stringent design requirements for the bag enclosure walls.

Although the pressure drop across a baghouse may be 3 to 6 in. water column, the negative pressure imposed by the exhaust fan is determined by the duct-system requirements that may be five to ten times those of the pressure drop. If the air inlets to the baghouse are blocked by powder or closed inlet dampers, the suction may increase to values well above the normal operating level. The user must determine the maximum suction that may be imposed on the baghouse, and seek a unit that is rated for this condition.

If the dust-air mixture can become explosive, the baghouse walls must also resist a positive pressure that is determined from dust-explosion tests. These fix the vent area in the walls, so as to limit pressure buildup. The user must specify the allowable positive pressure for baghouse design.

Cylindrical baghouses are simpler to design for high suction or high pressure than are flat-sided enclosures.

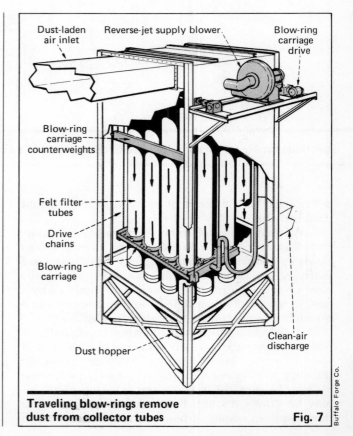

**Traveling blow-rings remove
dust from collector tubes**

Fig. 7

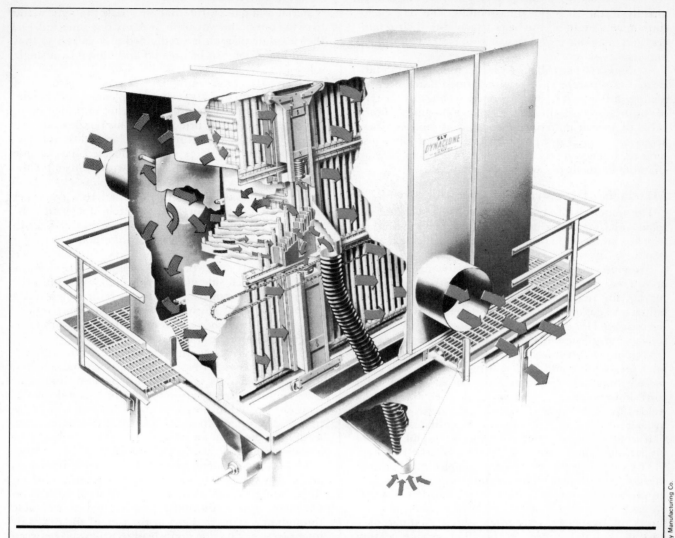

Traveling plenum moves horizontally to remove dust from envelope-type bags   **Fig. 8**

W.W. Sly Manufacturing Co.

Standard suction ratings for cylindrical units range from 15 to 160 in. water column, depending on diameter. Rectangular or square baghouses have ratings that range from 15 to 30 in. water column. These units require external battens and cross-bracing to stiffen them. Multicompartmented units have interior walls that require special bracing to resist differential pressures and stress reversal during cleaning operations. Tubesheets are usually made of heavier plate than are the walls, and are fitted or formed with stiffeners because the tubesheets not only serve to support the bags and retainers but also act as walkways during tube replacement. Distortion of tubesheets can cause top-suspended bag retainers to touch each other and retain dust. Some baghouses are only available as cylindrical units; others may be square, rectangular or cylindrical, depending on size and application.

Fabrication of baghouses

The enclosures for housing and supporting the bags may be made in several ways:

1. Modular design—A complete, factory-assembled unit consisting of bolted or welded components may include the dust hopper, supporting legs, cleaning accessories, timer, internal baffles, bags, etc. Installation only requires rigging into location atop a base to hold down the legs before connecting to the inlet and exhaust systems, and utilities. A group of parallel-connected units may be used to get additional capacity. The parallel connections may be internal, using common clean-air or dusty-air plenums; or external, using ductwork manifolds attached to inlet and outlet connections on each module. Module sizes are limited to those that can be transported over the road. The largest (12 to 14 ft square in cross-section) would require a low-bed trailer.

2. Factory-welded subassemblies—The bag housing, inlet and outlet plenum chambers, dust hoppers, etc., are separately fabricated to form the largest subassemblies that can be shipped over the road. Connections to adjacent components may be prepared for bolting or welding at the jobsite. The method of assembly depends on the materials of construction and location of the baghouse. Field welding is precluded where bags are already installed in the subassembly. The bag enclosures may contain all the bags attached to their retainers, and may have cleaning accessories factory-piped and factory-wired.

3. Knocked down—All structural components are fabricated as separate panels, with formed flanges for assembly in the field by bolting or by welding. This type of construction is limited to very large baghouses fabricated of special materials or located at jobsites where access by crane is not possible. The flanged joints also serve as stiffeners on the flat surfaces of hoppers, plenum chambers, tubesheets and bag enclosures.

Materials of construction and unit size

The materials of construction and the sheer size of some units may preclude complete factory assembly, which is preferred in order to eliminate leakage and fix responsibility.

Baghouses requiring hot-dipped galvanized surfaces or internal linings of rubber, polyurethane or other plastics for resisting corrosion or abrasion will need bolted and gasketed joints because welding will destroy the coating. Hot-dipped galvanized sections are limited to sizes large enough to fit into available dip tanks.

Welding of galvanized surfaces is possible, but special joint preparation and field coating the weld with compatible materials are required in order to maintain corrosion resistance. Larger galvanized sections may be welded by using lower-cost electrogalvanized sheet steel instead of hot-dipped, in order to avoid the size limitation imposed by dip tanks. However, special welding preparation and coating of the weld will be required.

The weight of the zinc coating on electrogalvanized sheets should equal that normally specified for hot-dipped galvanizing (a coating of not less than 1 oz/ft^2 on each surface). Complicated surfaces such as tubesheets on baghouses requiring interior coatings may have to be stainless steel or Hastelloy because coating these surfaces may not be possible. All bolted joints will require gasketing or the application of a sealing compound to the mating surfaces. These materials should be compatible with the interior coating and the dust entering the collector.

The large size of a unit may require fabrication and shipment knocked-down in order to erect the baghouse in areas inaccessible to a crane. Such areas could be (a) building interiors, (b) yards surrounded by high buildings, or (c) roofs loaded with other equipment. In many cases, a crane could lift the components to a peripheral location, from where they could be trolleyed on a cable to the final position.

Access for servicing

The location of access doors is principally determined by the method of grouping the bags and attaching them to their supports. The closed-end tubular bag and its metal retainer suspended from the clean-air plenum tubesheet is the most common design because it permits the closest grouping of bags in a given space. Access is through hinged doors located entirely across the clean-air plenum, atop the collector. The bag and its retainer are simply dropped through the tubesheet and fastened to it by clamps, snap rings or twist locks, using the bag

material as a gasket. Internal air-nozzle piping is made to swing out of the way or to be easily disconnected. Or, the clean-air plenum is subdivided and valved so that each subdivision can be isolated and pulsed by a single solenoid valve.

Where the bags must be serviced regardless of weather conditions, the clean-air plenum can be designed as a walk-in housing requiring only one door, but only on collectors fitted with an air nozzle over each bag. Where bags and retainers are clamped from below the tubesheet, the doors must be located on the side walls of the unit so that bags are within 30 in. of the opening to allow proper takeup of clamps.

If adjacent walls or equipment interfere with proper location of a side door, the available doors must be used, necessitating the removal of a group of bags in order to reach those furthest from the door.

Bags suspended from shaking bars, with their open ends fastened to a bottom tubesheet, are usually grouped on each side of a narrow walkway within the clean-air compartment that is reached through a single door in the wall. During rebagging, the most remote bags must be attached and tensioned. Then, work must progress toward the walkway.

Envelope-type bags on metal spacer frames are installed in horizontal rows, with their open ends clipped to the vertical slots of a clean-air plenum, and their closed ends hooked to a support frame. Several rows of bags are located in tiers, one above the other. All bags in a compartment are accessible from two doors—one leading to the clean-air plenum, the other leading onto a walkway in the dusty-air chamber.

Long, large-diameter bags that are open at both ends are usually fastened to a bottom tubesheet that is accessible through a side door at the bottom of each compartment, and to the top tubesheet that is accessible through one or more doors in the roof of the dusty-air inlet plenum. The side doors lead to walkways between the bags. Some designs limit the number of rows of bags on either side of the walkway in order to improve access for clamping and to avoid damage to the bags.

All doors should be loosely hinged, and clamped or latched in such a manner that all surfaces of the gasket are sealed airtight. Gaskets should be closed-pore sponge rubber (40 Durometer hardness) that is cemented into a confining recess around the door or around its frame. The rubber and its cement should be compatible with each other and with the dust to which they are exposed. Latches or clamps should be so designed that they can be opened manually by operators without using a wrench or special tool. The frames of all doors should extend beyond any external insulation so that they can be flashed properly. The inside surface of vertical or sloping doors on the dusty side of the collector should be flush with adjacent walls in order to prevent dust buildup on ledges.

Platforms, railings and ladders should be furnished for reaching all elevated equipment and all access doors. These should be built in accordance with applicable safety codes. Rooftop equipment should be accessible for servicing from within the peripheral handrailing. Ladders should have safety cages starting 7 ft above floors or platforms, and should be equipped with

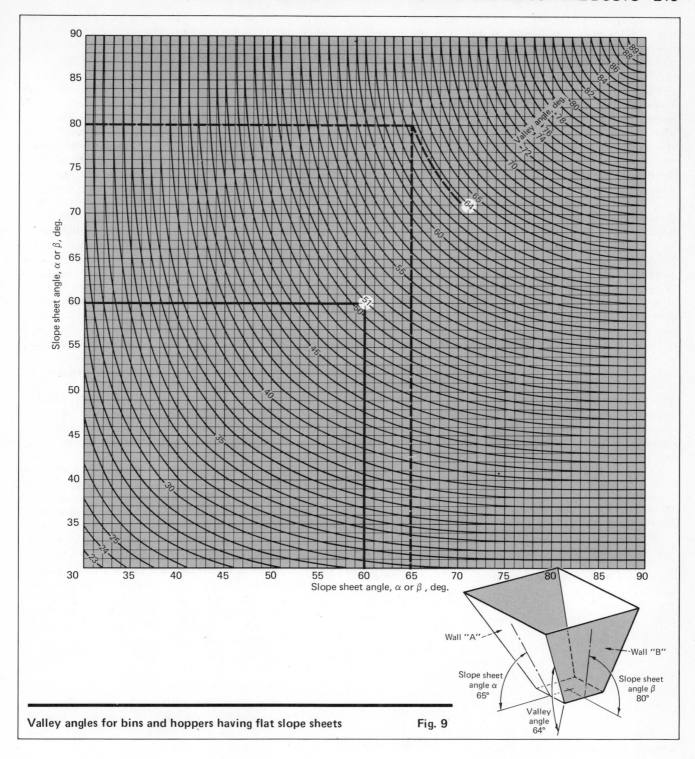

Valley angles for bins and hoppers having flat slope sheets   **Fig. 9**

grab rails and, for greater safety, with narrow, nonskid treads in place of rungs. Platforms should be made from serrated grating so that dust cannot accumulate and slipping is prevented.

Dust hoppers and dust discharge

The material removed from the bags may be dropped directly into a bin or silo if the baghouse is used for venting that bin or silo. Otherwise, the material is dropped into a collecting hopper that is shaped to direct the flow to the dust outlet usually fitted with an airlock-type discharge valve.

On multicompartment units, each compartment should be fitted with its own dust hopper, to permit its isolation during cleaning or maintenance. To eliminate a multiplicity of airlock discharge valves, the bag compartments may discharge into a common V-shaped hopper equipped with a screw conveyor that delivers the dust to a single discharge valve. Such a screw conveyor must operate at the baghouse pressure. Therefore, the conveyor must be equipped with effective seals at the ends of the trough and at the bearings, which should be outboard of the hopper. The conveyor must also be able to handle the collected dust at the proper rate.

The flow properties of the collected dust should be known, since they may be quite different than the properties of the material from which they originated. The slope angles of the dust hopper may have to be steeper than those of the storage silos of the original material because (1) fine dusts tend to pack more readily than coarse materials, (2) the dust may contain moisture picked up in the conveying air or formed by condensation within the collector, (3) collected material should be removed as rapidly as it is deposited on the hopper walls. Steep slope angles increase the overall height of the baghouse and may limit the space available for installing a discharge valve. However, failure to provide the proper slope will create problems after installation.

A conical hopper attached to a cylindrical baghouse presents no problem other than the increase in height required to obtain a steep slope. Attachment of a square, rectangular or conical hopper to a square or rectangular baghouse requires study—especially of the valley angles formed by the slope sheets of inverted pyramidal hoppers, or by transitions to conical hoppers. These valley angles should be at the minimum value required to obtain material flow. If the wall-slope angles to the horizontal are known, the valley angles may be obtained from Fig. 9, or from the formula from which the chart is derived:

$$\cot^2 \theta = \cot^2 \alpha + \cot^2 \beta$$

where θ is the valley angle, and α and β are the slope angles between the horizontal and Wall A and Wall B of the hopper, respectively, as shown in the sketch in Fig. 9.

Provision of the proper valley angle at an intersection does not ensure material flow, because moist materials can bridge across the intersection. To minimize bridging, the intersection should be formed to as large a radius as possible when joining the flat slope-sheets internally.

On baghouses cleaned by reverse-air blast (using compressed air), the slope-sheet angles can be reduced by a few degrees because the air blast helps to propel the collected powder down the slope.

Materials may still hang up even though the proper slope angles are provided on dust hoppers, because of process upsets, changes in moisture content, etc. To ensure flow, mounting pads should be installed on the slope-sheets for attaching mechanical or pneumatic vibrators for inducing flow. The vibrators should be placed at the centroid of each surface. The flow inducers should be operated only when needed. Cylindrical hoppers require only one pad, located at the centroid of the projected surface.

The dust hopper and its supports should be designed on the basis that it could become full of material and may also have to support a surcharge. This situation may arise if the baghouse becomes full of material because of a broken bag, a malfunction of the dust-

discharge valve, or stoppage of a downstream dust conveyor.

Outdoor installations

Since many exhaust systems draw air from within warm buildings, or handle hot materials from processes or handle materials containing moisture, condensation is a possibility within the baghouse. This can occur on baghouses located indoors or outdoors, particularly in winter. Some manufacturers' guarantees are automatically voided if condensation occurs inside the unit.

In such situations (especially outdoors), the baghouse should be insulated. Insulation should be of sufficient thickness to prevent condensation at the lowest outdoor temperature recorded in a given location. All doors, dust hoppers and projecting stiffeners should also be insulated. Outdoor insulation should be protected with aluminum or stainless-steel jacketing, applied so as to shed water. Flashings of the same material should be formed and fitted around all access openings. In some cases, removable covers containing the insulation may be fitted over doors to permit easy access.

If a baghouse does not require a separate enclosure, a watertight housing should be provided around dust hoppers, dust-discharge valves, and other dust-recovery equipment located below the baghouse to permit access for servicing during bad weather. Hinged or easily removable weathertight hoods should be furnished to protect all external operating mechanisms.

All electrical conduit to timers and actuators mounted on the baghouse should be watertight. Pneumatic actuators, solenoid valves, shaking mechanisms, and air piping to them, must also be insulated or traced, to prevent condensation of moisture and its subsequent freezing, which will prevent operation of the accessories.

If baghouses handling hot gases are shut down for the weekend, they become especially susceptible to condensation problems that prevent startup. Such units should be equipped with an independent heater and a pressurizing fan designed to maintain the normal operating temperature within the unit during shutdown. A gravity-closed damper on the discharge of the main exhaust fan will permit retention of the heated air. Baghouses operating under pressure should have a weathertight cover or an automatic backdraft damper at the air outlet to prevent intake of cold air during shutdowns.

Protection against fire

When hazardous materials are handled, protection must be provided to eliminate or minimize hazards. Some materials may only require fire protection; others, fire and explosion protection.

The best procedure for determining the hazards in handling a particular dust with unknown characteristics is to submit test samples to the fire-insurance underwriter's laboratory or to a commercial testing laboratory that has facilities for conducting:

1. Combustion test—Ignition by match flame or Bunsen burner to determine whether the material is self-extinguishing, or whether the flame will spread.

2. Spontaneous-combustion test—Exposing material to high temperature for a prolonged period (several hours). The material is contained in a crucible immersed in a hot salt bath. Moist materials should be tested at their usual moisture content, and again after drying.

3. Autoignition-temperature test—Same as the preceding test, but raising the temperature continuously to determine the temperature at which ignition occurs. Evaporation of water from a moist sample may exclude air, so that the autoignition point could be on the high side.

4. Dust-explosion test—Material is placed in a bomb having a variable vent opening, mixed with compressed air, and then ignited by an electric spark. The pressure developed is measured and reported for two conditions—(a) unvented, and (b) for a specific ratio of vent area to bomb volume. Materials known to be explosive should also be tested for rate of pressure rise in addition to the pressure developed from an explosion.

Where possible, any baghouse subject to fire or explosion should be located outdoors, and away from important equipment.

Fire protection may be provided by a built-in sprinkler system. The sprinkler heads would be rated about 25° (F) below the temperature limit of the bags if no other protection were available; and 100° (F) above this limit, if early-warning protection were installed. The sprinklers would be connected to a building sprinkler system. The heads are usually located where they can cover most of the bag area. Additional heads may be needed to cover the dust hoppers. Because of difficulty in placing heads in some baghouses, assistance of the fire underwriter's engineering department should be obtained.

Opening of the sprinklers would be indicated by the water-flow detector and alarm on the building sprinkler system. However, a local alarm to warn system operators may also be required. The local alarm would be a fire detector within the baghouse—the detector being set at the same temperature as the sprinkler head. Provision must be made to dispose of the water released by the sprinklers. When the baghouse is outdoors, a nonfreeze sprinkler system will be needed. A dry system, using air in the sprinkler lines, is expensive and requires considerable maintenance. A system filled with a nonfreeze solution may foul the bags when the heads are released. Hence, the bags may require replacement even if they are not burned.

Several alternatives are available for sprinklers:

1. An inert-gas (Halon) injection system may be installed. When actuated, it will exclude or reduce the oxygen content in the baghouse so that combustion cannot be sustained. The sensor is a rate-of-rise temperature detector, set at the temperature limit of the bags, that operates a control unit to electro-explosively rupture a diaphragm to release the inert gas into the baghouse. The gas storage vessels are located outside the collector. If the system is actuated, there is no mess to clean up except char from the fire. The control unit can also initiate other functions such as closing dampers, stopping fans and sounding alarms.

2. A steam-smothering or a water-deluge system that is actuated manually or by a rate-of-rise temperature detector uses open nozzles, located in the same manner as sprinkler heads.

3. A manually operated deluge system can be de-

vised for collectors having reverse-jet air manifolds by connecting a water header to the air manifold pipes at the ends opposite the air header. When water is turned on, the air pulses speed up wetting of the bags. The timer can be temporarily switched to pulse all nozzles simultaneously, and pulse time could be increased to obtain faster water distribution.

Protection against explosion

Explosion protection may be provided by installing an inert-gas suppression system or explosion doors.

The suppression system is similar to that for fire protection, except that a highly sensitive pressure detector initiates action of the control unit. The pressure setting is determined from data for the rate of pressure rise obtained during the bomb test. The baghouse may be isolated so as to contain the inert gas without loss by the use of fast-acting inlet and outlet dampers, or by shutting down the exhaust fan and rapidly closing an outlet damper.

The inert-gas suppression system detects the pressure rise caused by an explosion, and acts in milliseconds to stop pressures from reaching values that would rupture the bag housing. Because of the short duration of an explosion, and the dependence on proper functioning of all control equipment, a backup system of explosion doors may be desirable.

Explosion doors are generally installed in baghouse walls on the dusty-air side. They may be hinged doors with special, spring-loaded, quick-releasing latches. They may be rupture panels made of frangible materials such as aluminum foil, lead sheets, polyethylene or Teflon. They may be laminates of metal and plastic designed to rupture along prescored lines to obtain full opening of the panel. They may be loose doors held in place by their own weight or by suction.

Latched or loose explosion doors on outdoor baghouses must have rain hoods built around them. Weighted doors, unattached by any other means, should have chains connected to them, which are fastened to an internal structural member, so that the doors do not become missiles when blown open.

Every explosion is rapidly followed by an implosion that can be as damaging to the structure as the explosion. Explosion doors that are pivoted or hinged should be designed to prevent automatic reclosing. Such doors should be reclosed manually.

The concluding installment of this two-part series on baghouse collectors, to appear in the issue of Apr. 23, will deal with the selection and testing of filters, measuring performance, and acceptance testing of baghouses. A reprint notice for both parts will be carried at the end of the article in that issue.

The author

Milton N. Kraus, 12 Chris Ave., Hillsdale, NJ 07642, recently retired from Colgate-Palmolive Co., Jersey City, N.J., where he was supervisor of project engineering. Previously, he worked as a supervisory marine engineer in design at the New York Naval Shipyard and became a licensed marine engineer with the U.S. Maritime Service. He also served as division engineer for dairy processing plants and as mechanical superintendent for a power-plant construction firm. He has a B.M.E. from Pratt Institute. He is a licensed professional engineer in New York and New Jersey, and a member of ASME and Tau Beta Pi.

Baghouses:
selecting, specifying and testing industrial dust collectors

The operating parameters for dust removal with filter bags determine the proper removal system. Here are practical guidelines for evaluation.

Milton N. Kraus*

☐ To obtain a baghouse collector that will efficiently remove industrial dusts from gas streams requires: an appropriate filter system; either a pilot-plant test on the unit to be considered, using a sample of the dust to be separated, or a pilot-plant test in the field, using the actual dust-laden stream; a method for measuring the system's performance variables; and, finally, acceptance testing of the installed unit.

Filter selection

Selection starts with submitting a detailed specification to dust-collector vendors, prepared on the basis of information obtained from process engineers and that presented in the preceding article. This information may be summarized in tabular form (see Fig. 1 on p. 222), with details amplified in a separate procurement specification. Tabulated data should be grouped under several headings:

1. Process specifications will cover all information regarding the dust to be handled and the dust-collection system that will deliver material to the baghouse. The dust data should include all physical and chemical properties and characteristics that may affect the filter fabric and its cleaning, or the recovery of dust from the baghouse. These data include density, particle-size distribution, temperature, moisture content, pH or corrosiveness, tackiness, toxicity, abrasiveness and electrostatic charge.

Data for the dust-collection system should include: the quantity and nature of the conveying medium, its temperature and moisture content, source of the dust,

anticipated dust load, allowable pressure drop across the baghouse, and the baghouse's maximum design pressure.

2. Mechanical specifications will list the known or desired details of the baghouse, such as materials of construction, size and location of access openings, method of fabrication, fire and explosion protection, method of dust removal, need for insulation, access platforms, structural supports, painting requirements, exhauster details and performance characteristics. Spaces should be left for insertion of unknown data by the vendor.

3. Electrical specifications will include tabulation of available electrical services for power and control, types of enclosures required for timers, starters, relays, etc., and motor-drive data.

4. Remarks will contain information not tabulated under other headings, such as special construction details, and information that would help the vendor in preparing the bid.

The procurement specification should accompany the tabulated data sheet (such as Fig. 1). A detailed description of the desired construction for each baghouse component should be given insofar as possible. These components should include bag enclosure, bags, dust hoppers, bag-cleaning mechanisms and lubrication requirements. Overall performance requirements for the baghouse would be stated, allowing the vendor to propose any deviations from the detailed specifications that he believes would improve the design.

Analyzing the bids

Each vendor submits a bid that may conform to the procurement specification but will be sized according to

*To meet the author, see p. 231

Originally published April 23, 1979

the vendor's experience. Bids should first be analyzed on the basic criteria set by the purchaser as being desirable for handling the entering dust. Size may be investigated later, or may be proven by pilot-plant test.

A typical list of desirable requirements for a hot, tacky dust:

- Absence of moving mechanical parts to be cleaned.
- Cleaning of filter media by reverse airflow.
- Loosened dust flows in same direction as normal airflow. (This prevents redeposit of dust on cleaned filter fabric. On some designs, this directs the dust into the dust hopper.)
- Minimum amount of ancillary equipment.
- Minimum requirements for utilities.
- Minimum floor space.
- Modular, all-welded design to resist the specified minimum pressure (inches of water column).

These criteria should be revised to suit the characteristics of other types of dust. Obviously, not all baghouses will meet the requirements.

In addition to a tabulation of comparative data for bid analysis (Table I), a separate list (Table II) should be made to show how the bids conform to the selection requirements.

From these tabulations, selection may be narrowed to one or more vendors whose equipment meets most of the criteria. If more than one vendor qualifies, the submissions must be investigated more closely by considering price, delivery, guarantees, availability of a test unit, etc.

For hot, tacky dust, all baghouses depending on mechanical components that could be fouled if a bag were to break were eliminated. Even if a bag did not break, fine dust could build up and smear the mechanisms and affect their operation. Such baghouses can only be considered where savings in first cost and maintenance costs justify the risks.

Equipment guarantees

A materials-and-workmanship guarantee is standard with most vendors, and lasts for a period of one year from date of shipment. Because construction of the baghouse and the system it serves may extend over a long period of time, the purchaser should request an extension of time for this.

The vendor should guarantee that baghouse emissions will comply with the pollution-control laws applicable to the baghouse location. The purchaser's acceptance of the baghouse could be based on acceptance of the installation and issuance of an operating permit by the environmental protection agency having jurisdiction, or by test. The recommended test procedure is that established by the Industrial Gas Cleaning Institute [1].

Filter testing: vendor's pilot plant

Testing of a baghouse is time-consuming, and there is a tendency to bypass this phase of equipment selection and rely on a vendor's application experience in sizing a particular filter. However, an experienced vendor will request that a quantity of the dust (about several cubic feet) be sent for test in a pilot plant. The unit would be equipped with brand-new filter tubes of the vendor's choice, based on an analysis of the dust characteristics

by the vendor's laboratory. A typical test procedure would be:

1. A volumetric feeder is calibrated by setting it for a high discharge rate, and weighing the test material discharged over a period of time.

2. A blower exhausts air from the baghouse, and a calibrated venturi tube in the suction line meters the airflow.

3. The feeder is set to discharge dust into the filter's inlet connection at the desired rate. The discharged powder is dispersed by an air jet in order to simulate plant conditions.

4. The cleaning-cycle programmer is set at the vendor's recommended rate, and operating utilities for the baghouse turned on.

5. The filter is operated for at least eight hours in order to build up a filter cake on the bags. However, if the differential pressure across the bags were to build up to the vendor's standard operating-pressure differential prior to this time, the dust feeder would be recalibrated to a lower rate and the airflow reduced accordingly.

6. The standard pressure differential is then approached and maintained (by trial and error) for various rates of dust loading and airflow. Each combination of dust loading and airflow is maintained for several hours.

During this test, the data are plotted as dust loading in (grains/ft³ air)/ft² cloth area. Typical curves for various materials are shown in Fig. 2. These are called loading curves, and are only useful for sizing the unit on test when handling the tested dust.

The dust-loading test is usually run at ambient-air temperatures. When the baghouse is operated at a higher air temperature, the filter rate must be corrected. The correction factor is obtained from Fig. 3, which is a curve for constant-volume flow. In effect, this factor increases cloth area. It compensates for the increased air viscosity at high temperatures. (The viscosity of air increases almost in direct proportion to the increase in temperature.) In general, the relationship of the correction factor for the filter rate is:

$$\frac{F_1}{F_2} = \frac{\phi_1}{\phi_2} \qquad (1)$$

where F_1 is the filter rate at a given temperature and a pressure drop of 3.5 in. water column,* F_2 is the corrected filter rate, ϕ_1 is the filter-rate correction factor for ambient temperature during test, and ϕ_2 is the filter-rate correction factor for the proposed operating temperature.

After the filter rate is corrected for temperature, a separate calculation must be made to allow for expansion of air at the higher temperature, so as to obtain the actual filter rate at the face of the fabric during regular operation.

The cloth area for a full-size baghouse, based on test data, is calculated from:

$$A_C = V_T/F_2 F_a \qquad (2)$$

where A_C is total area of cloth, ft²; V_T is total volume of gas stream at operating temperature, ft³/min; F_2 is

*Mikropul Corp. standard, Ref. 3.

Comparative data for the analyses of bids, and selection of a dust collector

| | Housing | | | | Filter-bag data | | | | | | |
Make of collector	Maximum design vacuum, in. water column	Modular design	Standard assembly	Standard material	Standard size	Medium	Dust cake on bag	Fastening Top tube-sheet	Bottom tubesheet	Bags cleaned by	Cleaning controlled by
A	30	Square	Welded	Stainless steel	4½-in. dia.	Felted	Outside	Clamped to cage	—	High-pressure air jet in each bag	Timer—operating solenoid valves
B	17	Rect.	Welded	Carbon steel	5¾-in. dia.	Felted	Outside	Clamped to cage	—	High-pressure air jet in each bag	Timer—operating solenoid valves
C	17	Square	Bolted	Carbon steel	5½-in. dia.	Felted	Outside	Clamped to cage	—	High-pressure air jet in each bag	Timer—operating solenoid valves
D	20	Rect.	Welded	Galvanized	8-in. dia.	Woven	Inside	Pressed-in collar	Clamped	Air suction into bag housing	Timer—activating air-reversal dampers
E	15	Square	Welded or bolted	Carbon steel	6-in. dia. or 11-in. dia.	Woven	Inside	Snapring on chain-hung cap	Clamped	High-pressure air jet in each bag plus air suction into housing	Timer—activating air-reversal damper and solenoid valves
F	40	Cyl.	Bolted	Carbon steel	Oval	Felted	Outside	Clamped by cage	—	Low-pressure air blast into each bag	Continuously rotating plenum, and cam track
G	12	Rect.	Bolted	Carbon steel	Envelope	Woven	Outside	Hooked over pins on vertical mullions		Air suction into bag housing	Continuously traveling suction plenum
H	16	Square	Bolted	Carbon steel	12-in. dia.	Felted	Inside	Clamped	Clamped	Traveling air blow-ring on each bag	Low-pressure air blow-case traverses bag length. Differential-pressure switch starts case
I	20	Cyl.	Bolted	Carbon steel	9-in dia.	Woven	Inside	Snap ring	Snap ring	Traveling air blow-ring on each bag	Low-pressure air blow-case traverses bag length continuously

corrected filter rate; and F_a is an application factor from Table IV.

Estimating the filter rate

A preliminary estimate of the test-filter rate can be made by using a filter rate guide developed by one manufacturer for application to pulse-jet or blow-ring baghouses that have continuously cleaned felt-filter media [2]. Such baghouses maintain a pressure differential of 3.5 in. water column at design flow and loading conditions. The guide consists of five factors, multiplied together to arrive at a filter rate. These are:

Material factor, F_m, from Table III allows for the properties and characteristics for a material that would affect its tendency to form a filter cake during normal airflow and release it during the cleaning cycle for the bags by reverse airflow.

Application factor, F_a, from Table IV allows for the type of process delivering dust to the baghouse and for upsets that cause variations in the dust load.

Temperature effect, ϕ, from Fig. 3 compensates for in-

creased air viscosity at high temperatures, as previously explained.

Particle-size factor, F_f, from Table V allows for the fact that fumes and fine dusts tend to blind fabrics more rapidly than do coarse particles.

Dust-load effect factor, F_d, from Fig. 4, which is a typical performance curve, is applicable only to this type of baghouse operating at constant differential pressure.

An actual pilot-plant test eliminates the factors for material, particle-size and dust-load effects, but the temperature and application factors must be obtained, as described above, in order to determine the required cloth area.

The preliminary determination of filter rate can be used to roughly size a baghouse for cost-estimating purposes. For baghouses using woven fabrics that require cleaning at less-frequent intervals, the filter rate established by the above guidelines can be divided by five. The resulting rate, however, is only applicable for dust loads within the capability of such baghouses, and for cost-estimating purposes only.

Table I

General		
Airflow, path sequence	Motors required to operate	Remarks
Dust hopper, tubes, top plenum	None	
Dust hopper, tubes, top plenum	None	
Dust hopper, tubes, top plenum	None	Requires removal of air manifolds to replace bags
Top plenum, tubes, top of bag compartment	1-Rapper 2-Timers	
Inlet plenum, dust hopper, tubes, bottom of bag compartment	None	Cloth area of 6,500 ft² and less has welded construction and 6-in.-dia. bags
Side of bag compartment, tubes, top plenum	1-Plenum drive 2-Blowers	Cylindrical units must be paralleled for large capacities
Side of bag compartment, baffle, envelopes, clean-air chamber	1-Plenum drive	
Top plenum, tubes, bottom of bag compartment	1-Blow case 2-Blowers	Blower air is recycled
Top plenum tubes, center of bag compartment	1-Blow case 2-Blowers	Cylindrical units must be paralleled for large capacities

Filter testing: field pilot-plant

Some vendors can supply a test unit for installation at a prospective user's plant. This method of evaluation should only be undertaken under one or more of the following circumstances: (1) when there is no precedent for filtering a specific material; (2) when a filter design is deemed novel and the user has had no previous experience with it; (3) when the filter unit must be evaluated under a variety of plant-operating conditions or when handling a variety of materials; or (4) when determination of the effect of weather conditions on filtration is desired for a proposed outdoor installation.

A test unit is generally supplied on a rental basis, with an option to purchase. The accrued rental charges could be applied to the purchase price of the test unit or of a larger unit. The agreement with the vendor should provide for: (1) freight, handling and installation expense, usually borne by the user; (2) services of an installation supervisor, if one is desired; (3) allocation of reconditioning expenses for damage caused beyond

normal wear during a test; (4) responsibility for loss of the unit from any cause; (5) access to the test area by the vendor's representatives and furnishing test data to them; (6) schedule of charges and method of billing; (7) statement of credits allowed upon return of the test unit; and (8) a list of auxiliary equipment furnished with the test unit.

Installation of a field test unit at an operating plant usually requires that a small percentage of dust-laden air be drawn from an existing exhaust system to match the test unit's capacity. This requires insertion of a special nozzle into a straight section of the existing duct—not just a "tee" or lateral connection. The open end of the nozzle should face upstream and be centered so as to receive a representative sample of the dust load.

Air drawn through the test filter should enter the sampling nozzle at the velocity that normally exists in the cross-section from which the sample is drawn. This procedure is called isokinetic sampling. The exhaust fan for drawing this sidestream through the test unit must be sized for the airflow required to obtain the isokinetic velocity at a suction pressure low enough to overcome (a) the suction resistance in the main duct plus (b) the estimated resistance of the test filter and ductwork.

Some form of shutoff gate should be placed in the suction line, close to the test unit, so that the unit may be opened for inspection and modification without affecting flow in the main duct, and that the unit may be shut down without causing reverse airflow through its exhaust fan. If tests above the isokinetic velocities are planned, the exhaust-fan's drive should be able to increase the fan's speed in order to obtain higher airflows.

Collected dust from the unit should discharge through an airlock into a drum for weighing. If no airlock is used, the dust may flow into a drum through an airtight drum cover, so that the drum is maintained at the same negative pressure as the test unit. However, every time the drum has to be emptied, weighed or inspected, the unit must be shut down. Provision should be made for periodic disposal of collected dust. Otherwise, the dust may back up and fill the baghouse, voiding the test data.

A typical outdoor test-setup for a unit handling hot detergent dust is shown in Fig. 5. If hot or moist dust is to be processed, provision should be made to completely insulate the unit. A heated, plywood enclosure should protect the dust hopper and collecting drum, but insulation and electrical tracing of the hopper may still be required.

On units supplied with compressed air for cleaning, the air lines and solenoid valves should be electrically traced. All electrical heat tracing should be thermostatically controlled to prevent overheating and burning of the collected dust. A large temperature differential between the air-inlet and air-outlet connections may indicate an air leak into the collector, or the need for additional insulation.

Measuring performance

Means for measuring performance should be installed and data taken for airflow rate, dust load, temperature, pressure, and timing of cleaning cycles.

Airflow rate—Install a calibrated venturi meter or

EQUIPMENT DATA SHEET

DUST COLLECTORS

PURCHASE SPECIFICATION NO.

DATA SHEET NO.

PROJECT TITLE		
PROJECT LOCATION		DATE PREPARED
PREPARED BY	REVISIONS	DATE

QUANTITY		ITEM NO. AND TITLE
SERVICE CONDITIONS ☐ INDOOR ☐ OUTDOOR ☐ CONTINUOUS ☐ INTERMITTENT		QUOTATIONS SHALL BE IN ACCORDANCE WITH PURCHASE SPECIFICATION NO.

PROCESS SPECIFICATIONS

DUST DATA	MATERIAL	LB. PER HOUR	TEMPERATURE
	APPARENT DENSITY	TRUE DENSITY	ABRASIVENESS

SIEVE ANALYSIS

CONVEYING AIR DATA	QUANTITY	TEMPERATURE	MOISTURE %	MAXIMUM DESIGN PRESSURE, SUCTION INCHES WATER COLUMN
	ALLOWABLE PRESSURE DROP INCHES WATER		ACTUAL PRESSURE DROP INCHES WATER	

MECHANICAL SPECIFICATIONS

MANUFACTURER	MODEL OR TYPE	SIZE

FILTER HOUSING	MATERIAL	GAGE	NO. OF COMPARTMENTS
	ACCESS DOOR SIZE	ASSEMBLY ☐ ALL WELDED ☐ BOLTED PANELS	☐ UNITIZED COMPARTMENTS

FILTER CLOTH	MATERIAL	WEAVE	
	TUBULAR DIAMETER	OTHER	AIR PERMEABILITY (FRAZIER

CLEANED BY ☐ REVERSE AIR FLOW _____ CFM   ☐ MECHANICAL SHAKERS   ☐ OTHER

TOTAL FILTER AREA		NET AIR TO CLOTH RATIO
GROSS _____ SQ. FT. NET _____ SQ. FT.		_____ CFM PER SQ. FT.

PLENUM CHAMBER	MATERIAL	INLET SIZE	EXPLOSION DOORS SQ. FT. PER COMPART.
DUST HOPPER(S)	MATERIAL		GAGE
FILTER COMPARTMENTS PER HOPPER	VOLUME	SHAPE ☐ INVERTED CONE ☐ INVERTED PYRAMID	ANGLE OF SLOPE SHEET
MINIMUM VALLEY ANGLE	ACCESS DOOR SIZE	SIZE OF DISCHARGE CONNECTION ☐ FLANGED ☐ OTHER	

DUST DISCHARGE VALVE		RAPPERS ☐ REQUIRED ☐ NOT REQUIRED
TYPE	FURNISHED BY	

INSULATION	☐ REQUIRED ☐ NOT REQUIRED	FURNISHED BY	FOR SPECIAL REQUIREMENTS SEE REMARKS
FAN	FURNISHED BY _____ CO. ☐ SEE FAN DATA SHEET ATTACHED	WITH COLLECTOR	TYPE OF HOUSING ☐ DUST TIGHT ☐ WEATHER PROOF ☐ NONE
ACCESS PLATFORMS	☐ REQUIRED ☐ NOT REQUIRED	FURNISHED BY	FOR SPECIAL REQUIREMENTS SEE REMARKS
STRUCTURAL SUPPORTS	☐ REQUIRED ☐ NOT REQUIRED	FURNISHED BY	CO. DWG. NO.
PAINTING	☐ NOT REQUIRED	☐ SHOP COAT	☐ OTHER (SEE REMARKS)

ELECTRICAL SPECIFICATIONS

MOTOR DRIVES	TYPE	SERVICE PHASE _____ CYCLES _____ VOLTS
	STARTERS FURNISHED BY	BAG SHAKER(S) _____ RAPPER(S) _____ FAN

CONTROLS	SERVICE PHASE _____ CYCLES _____ VOLTS	INTERLOCKS
	TIMER CABINET—NEMA TYPE	TIME CYCLE

CONTROL TRANSFORMERS FURNISHED BY

REMARKS

Purchase specification for dust collectors

Fig. 1

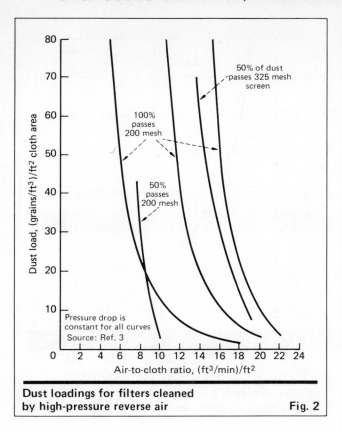

Dust loadings for filters cleaned by high-pressure reverse air   **Fig. 2**

(Chart labels: Dust load, (grains/ft³)/ft² cloth area vs. Air-to-cloth ratio, (ft³/min)/ft². Curve labels: "50% of dust passes 325 mesh screen"; "100% passes 200 mesh"; "50% passes 200 mesh". Note: "Pressure drop is constant for all curves Source: Ref. 3")

orifice in the clean-air discharge duct, with the required lengths of straight duct coming before and after the device. Or, install two connections in a straight section of duct for making a pitot-tube traverse (Fig. 5). These connections should be at right angles to each other and at the same cross-section of the duct.

When testing, obtain the airflow rate before, during and after cleaning the bags, if possible. By measuring the pressure differential across the exhaust fan, an approximate check of airflow may be made by referring to the fan characteristic curve of static pressure vs. air volume for the operating speed of the unit.

When the fan's characteristic curve rises and then drops, so that it is uncertain which of two identical static pressures has been read, the motor horsepower should be measured (using clamp-on electrical measuring instruments for quick readings). The horsepower vs. air volume curve should then be used to determine on which side of the characteristic curve the static-pressure reading belongs. (The horsepower-air volume curve is usually a straight line, increasing proportionally with capacity.)

During a test, a flow measurement higher than that required to obtain isokinetic velocity indicates the presence of an air leak in the test equipment, usually at the dust-discharge connection.

In a filter using reverse flow of atmospheric air for cleaning, the reverse-airflow rate should be obtained by installing a calibrated orifice at the reverse-air inlet and measuring the pressure drop across it.

Dust load—Install a platform scale close to the dust-collecting drum (Fig. 5). Weigh the load in the drum at regular intervals and use the weights of filled, sample-sized containers of known volume to determine average powder density.

Temperature—Install thermometer connections at the baghouse inlet and outlet. The thermometers can be of the indicating or recording type. A glass-stem indicating thermometer inserted in a cork-stoppered hole in the duct may suffice. The sensing bulb should be cleaned after each use. Wet-bulb readings on the dust duct must be taken by rapidly inserting a glass-stem

				Auxiliary			
Make of collector	Moving parts	Reverse air-flow only	Dust follows normal airflow	equipment required	Utilities required	Floor space	Housing per specification
A	None	Yes	No	Timer	High-pressure air, control a.c.	Minimum	Yes Stainless-steel material
B	None	Yes	No	Timer	High-pressure air, control a.c.	Minimum	Special
C	None	Yes	No	Timer	High-pressure air, control a.c.	Minimum	Special
D	Damper	Yes	Yes	Hopper, rappers, timer	High-pressure air, control a.c., rapper motors	Multiple compartment	Special
E	Damper	Yes	No	Timer	High-pressure air, control a.c., rapper motors	Multiple compartment	Special
F	Rotating plenum	Yes	No	Reverse-air blower	Blower and plenum motors.	Minimum	Special
G	Rolling plenum	Yes	No	None	Plenum motors	Minimum	Special
H	Traveling blow case	Yes	Yes	Reverse-air blower, limit controls, reversing starters	Control a.c., blower and case drive-motors	Minimum	Special
I	Traveling blow case	Yes	Yes	Motors		Minimum	Special

Conformity of bids to selection requirements for dust collector   **Table II**

thermometer with a wetted wick over the bulb, and holding it in place until the temperature reaches its lowest level. The test opening should be sealed during this reading. The wick should be cleaned and rewetted before each reading.

Wet- and dry-bulb readings of the ambient air surrounding the test unit should be recorded at the time other data are taken, by using a sling psychrometer.

Pressure—Install manometer connections in the inlet and outlet ducts, close to the filter, to measure the pressure differential across the unit. Also, install connections on the clean and dirty sides of the baghouse at the tubesheet to get the pressure differential across the bags.

If factory-installed connections are provided, make sure that they are not blocked by paint or galvanizing metal, and that a cotton filter is installed in the dirty-air side. When testing, take readings of the pressure differential across the bags and across the baghouse just before a cleaning period starts, and just after it ends.

On baghouses using compressed air for cleaning, install a pressure gage on the air-supply manifold. Observe this gage regularly to ensure that proper cleaning pressure is maintained.

Timing of cleaning cycles—Before starting a test, operate the program timer, and check the time between closing and opening of the air-reversal dampers and the starting and stopping of shakers in each compartment of multicompartment units equipped with bag-shaking and reverse-airflow dampers. Check the time between the termination of the period in one compartment and the start of the cleaning cycle in the next compartment, as set up in the timer.

On units having reverse-airflow pulsing valves, time the period between successive blows or pulses, and if possible, the duration of the airflow.

The cleaning-cycle time is the total elapsed time required to clean all the bags in a filter once.

Regular inspections of baghouses

In addition to obtaining data at the baghouse, a complete record of process operations that exhaust into it should be obtained. Operating conditions, rates, downtime, upsets, etc., of the process should be correlated with events occurring at the baghouse. The baghouse should be opened for inspection at regular intervals as determined by process operations—at least once per week. The bags should be inspected for:

1. Permanent coating of dust that cannot be removed by cleaning.

2. Plastering of moist material on one side of each filter element.

3. Hardness of the coating for evidence of condensation.

4. Thickness of the coating along the length of the bag.

5. Color of the coating compared to the color of the collected dust.

6. Condition of the fabric, opening of the weave or felt, tears, wear due to rubbing or flexing, etc.

Baghouse interiors should be inspected for buildup of powder on the walls of the bag housing and dust hopper—especially at corners and ledges, at baffles opposite

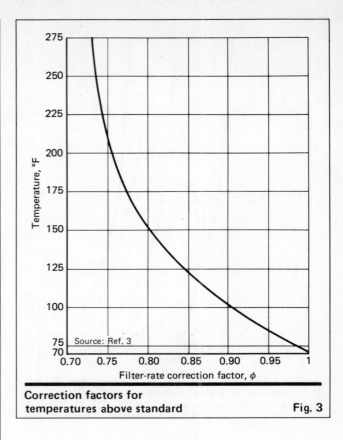

Correction factors for temperatures above standard   **Fig. 3**

the inlet, at valley angles, and at dust outlets. Weekend or nightly condensation can cause buildup at dust outlets. A sample of any builtup material should be collected and analyzed for particle size and composition. This may reveal an unusual concentration of an ingredient that has been scalped off the material in the process, and should not be in the baghouse.

After several hundred hours of operation, a number of bags should be removed from strategic locations within the baghouse and tested for air permeability. The test should be made on sections cut from the top, center and bottom of each bag. Permeability of the used bag should be compared with that of a new one.

These data would be used to estimate the life of the bags or to determine the frequency of washing or dry-cleaning needed. The method of cleaning must be determined by trial and error to suit the filter fabric and the materials lodged in it. A cleaned bag should also be tested for air permeability in the same manner as a dirty bag.

Acceptance testing of installed filters

The information presented here applies mainly to field-assembled units that arrive at the jobsite knocked-down or as subassemblies. This information may be modified as necessary for factory-assembled units. Testing for acceptance can only start after completion of all mechanical and electrical work. A typical procedure for such testing is to:

1. Inspect the baghouse and all system ductwork for structural integrity.

2. Check the direction of rotation of all motor-driven equipment—such as fans, conveyors, cleaning mecha-

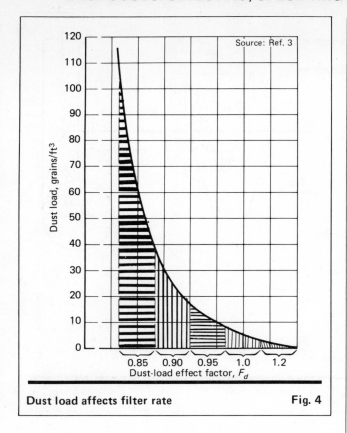

Dust load affects filter rate   **Fig. 4**

done before connecting the piping to air-cylinder operators, solenoid valves, and measuring instruments.

6. Connect all compressed-air piping, blanking off any possible escape openings. Check the piping for air leakage by applying full pressure, closing the air-supply valve, and observing a test gage for loss of pressure.

Rapid loss of pressure indicates a large leak that should be sought for by following the length of the piping and listening for escaping air. Slow loss of pressure indicates a small leak that should be sought by applying soapy water to each joint, after isolating the piping from the main air-supply valve. (A leaking supply valve can falsely indicate a well-functioning system because it replenishes the leaking air.)

7. Check the operation and sequencing of each program timer. Follow up by checking the functioning of the cleaning mechanisms controlled by the timers. For example, reverse-air dampers should operate freely and seat tightly. Check timing between opening, closing and cycling of the dampers between compartments; check operation of the linkage to shaking mechanisms and the duration of the shaking period; and check that solenoid valves are pulsing in proper sequence and are set for the recommended off-time. Observe the action of poppet valves on units equipped with air blast into the plenum.

8. Check carriage level, tracking of air-supply hose, and the limit-switch operation for carriage reversal, on filter units having blow-ring carriage drives.

9. Check operation and starting sequence of dust-discharge valves and conveyors.

10. Check settings of pressure and temperature switches, alarms, etc. Check the scale ranges of indicating instruments, including manometers so as to prevent blowout of manometer liquid.

Inspection for structural integrity

Testing for housing tightness should be made before applying the insulation and jacketing. This is most important on outdoor installations, where rain can enter the housing. A preliminary inspection of both the

nisms, and airlock dischargers—by actually observing the motor shafts.

3. Lubricate all motor-driven equipment, such as gear reducers, crank arms, and pivots.

4. Compare motor-starter overload-relay ratings with motor-nameplate ratings to ensure proper matching, so as to suit starter location.

5. Blow out dirt, water and scale from all compressed-air pipelines, and make certain that air-line lubricators are filled with the proper oil. This should be

Material characteristics affect filter rate through fabric-bag collectors				Table III
		Factor, F_m		
15*	**12**	**10**	**9**	**6†**
Cake mix	Asbestos	Alumina	Ammonium	Activated carbon
Cardboard dust	Buffing dust	Carbon black, fine	phosphate	Carbon black,
Cocoa	Fibrous and	Cement	fertilizer	(molecular)
Feeds	cellulosic	Ceramic pigments	Coke	Detergents
Flour	materials	Clay and brick dusts	Diatomaceous	Fumes and other
Grain	Gypsum	Coal	earth	dispersed
Leather dust	Lime, hydrated	Fluorspar	Dyes	products direct
Sawdust	Perlite	Kaolin	Fly ash	from reactions
Tobacco	Salt	Limestone	Metal powder	Powdered milk
	Sand	Rock dust, ores	Metal oxides	Soaps
	Sandblast dust	and minerals	Pigments	
	Soda ash	Silica	Plastics	
	Talc	Sugar	Resins	
			Silicates	
			Starch	
			Stearates	

*In general, any physically and chemically stable materials.
†Includes solids that are unstable due to hygroscopic nature, sublimation or polymerization.
Source: Ref. 3.

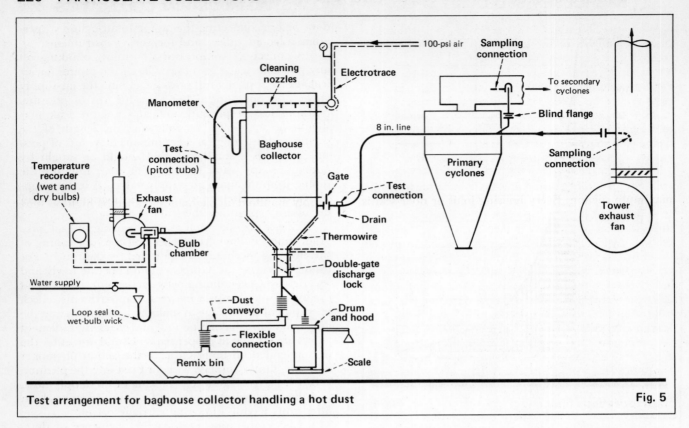

Test arrangement for baghouse collector handling a hot dust   Fig. 5

clean and dusty sides of the baghouse interior should be made to ensure removal of all construction materials and debris.

The bags should be checked for (1) proper assembly to tubesheets; (2) proper tension at both high and low positions of the shaker mechanisms, when furnished; (3) proper suspension, so that the bags do not touch each other or the baghouse walls. Interior walkways and baffles should be in place.

Access doors should be checked to ensure (1) that gaskets are in place and mate with the seating surfaces when the doors are closed and (2) that fastening or latching devices are in place, lubricated, and operating freely. Inspect walls for evidence of daylight from un-

used bolt-holes, unconnected piping, open joints or pinholes.

Fire detectors, sprinklers and deluge nozzles should be in place. Fusible sprinkler-head systems should be checked for tightness before closing and sealing the housing. (This should be done before bag installation, so that a leak will not damage the bags.)

An exterior inspection should be made after the baghouse access and inspection doors are closed and latched. All bolted joints on panels, dust hoppers and connecting ductwork should be inspected for unused bolt-holes (have proper bolt installed), alignment (mismatched holes should be redrilled, and unused holes or spaces filled in), and presence of sealing compound or a gasket between mating parts.

On outdoor suction systems, the housing can be checked for tightness by starting the fan and pulling maximum suction on the housing, using clean air. Inlet dampers can be partially closed to obtain this condition and reduce airflow so as not to overload the motor.

Application factor	Table IV
Process	**Factor, F_a**
Nuisance venting	1.0
Relief of transfer points, conveyors, packing stations, etc.	
Product collection	0.9
Air-conveying venting mills, flash dryers, classifiers, etc.	
Process-gas filtration	0.8
Spray dryers, kilns, reactors, etc.	
Source: Ref. 3	

Particle-size factor	Table V
Size	**Factor, F_f**
Over 100 micron	1.2
50 to 100 micron	1.1
10 to 50 micron	1.0
3 to 10 micron	0.9
Under 3 micron	0.8
Source: Ref. 3	

While the housing is under suction, a hose should be played on all access doors, and on all bolted and field-welded joints exposed to the weather, taking care to avoid direct impingement into the openings on reverse-air dampers. Obviously, all outdoor electrical systems should be sealed before this test is done.

After hosing down the equipment, each compartment and plenum chamber is entered and inspected for evidence of water, which is then traced to its source. The causes of leaks, if any, are then eliminated.

On indoor suction systems where hosing is impractical, the fan can be started and the system set to obtain maximum suction. Every housing joint or door seal is then inspected by listening or feeling for airflow into the unit. Each suspected leak is then marked for later correction. Although minor leakage is tolerable, a process upset or fan failure may cause positive pressure inside the baghouse, and spew dust into the room.

On positive-pressure systems, the fan is started and clean air directed into the baghouse. Housing joints and door seals may then be inspected for leaks. Testing for leaks on the clean-air side of positive-pressure baghouses is unnecessary, and on many large baghouses impractical. The best way to check for leaks is by brushing all the joints and sealing surfaces with soapy water, especially where the escaping dust is hazardous. Bag connections to the tubesheets must be tight because leakage goes directly to atmosphere.

After these checks have been made, the exhaust fan may be started. Then, the clean airflow through each equipment connection or exhaust hood is adjusted to obtain the required intake velocity for dust pickup. The exhaust-fan motor should be frequently checked during this period for overheating or other evidence of overloading.

Overloading may be caused by (a) excessive airflow due to the low resistance of new bags; (b) the blower, on a positive-pressure system, acting as a booster; or (c) an oversized fan. In the first case, the exhaust-fan discharge should be temporarily throttled until a filter cake builds up and the pressure drop in the baghouse becomes constant. A larger motor and speedup of the fan may be necessary where the positive pressure in a baghouse, bin or receiver must be made negative. A permanent damper may have to be installed on the discharge of an oversized fan.

Starting up a collector

The operating sequence for a typical dust-collection system is:

1. Start dust-removal equipment.

2. Start bag-cleaning equipment. When starting up with brand-new filter bags, delay starting the timers for the cleaning system until the anticipated pressure differential is attained. If this differential cannot be reached because of operating conditions, the timer may be started, and the system operated at a lower pressure differential.

3. Start the exhaust fans. Reduce initial airflow to new bags until filter cake is formed, then increase airflow to full load. On a unit that handles hot dust, preheat the baghouse before allowing dust to enter, in order to prevent condensation.

4. Observe the pressure differential across the bags. If this increases above a predetermined limit, make the following investigations:

a. Check for overloaded filter bags by manually advancing the cleaning cycle, and observe how this affects pressure drop. If overloading is apparent, shorten the cleaning cycle, or lengthen the cleaning time, or do both. On a unit operating intermittently, the exhaust system may have to be shut down at regular intervals to permit cleaning, pending determination of a means to increase filter capacity or reduce dust load.

b. Check the compressed-air pressure at the air-supply manifold on baghouses cleaned by high-pressure reverse-air jets. If too low, check for clogged air filters or for insufficient air capacity to accommodate the added load.

c. Recheck rotation of the exhaust fan. If rotation is reversed, a fan sometimes appears to be operating satisfactorily simply because it still discharges air and creates suction—especially with flat-bladed fans.

d. Check the cleaning sequence between compartments to see if they are being cleaned in too rapid succession. The hopper may be overloaded with dust if two or more bag compartments discharge into it.

5. Observe the discharge duct for evidence of dust. A light plume is normal on new bags until the filter cake forms. The plume should cease after several hours of operation. If it persists or becomes heavier, check the clean-air side or clean-air plenum for powder buildup. If buildup is present, recheck the tightness of all bag connections to the tubesheet.

Tests for emissions

Most environmental protection codes provide a temporary operating period to allow for evaluation and testing of the baghouse before issuing an operating certificate. During this period, the purchaser or vendor should sample the emissions by using an established procedure acceptable to the local environmental protection agency (see Ref. 1). Emissions from the unit should be less than the amount originally reported in the application to construct and install the dust-control system. The inspector for the code-enforcement agencies should be invited to witness the tests in order to expedite approval if the emission rates are acceptable.

References

1. IGCI Test Procedure for Determining Performance of Particulate Emission Control Equipment, Industrial Gas Cleaning Institute, Inc., 700 N. Fairfax St., Alexandria, VA 22314.
2. Frey, R. E. and Reinauer, T. V., New Filter Rate Guide, *Air Engineering*, Apr. 1964.
3. Bulletin RD-1, Reference Data and Tables, Mikropul Corp., Summit, NJ 07901.
4. Kraus, M. N., "Pneumatic Conveying of Bulk Materials," Ronald Press, New York, 1968.

Selecting and Specifying Electrostatic Precipitators

Industrial electrostatic precipitators are complex devices. There are many added-cost features that will pay off in better operation and lower maintenance, but are likely to be omitted in the low-bid specification. Also, proper erection and inspection procedures are vital if you expect to receive trouble-free service and high efficiencies.

GILBERT G. SCHNEIDER, THEODORE I. HORZELLA, JACK COOPER and PHILIP J. STRIEGL, Enviro Energy Corp.

There are many details that you must be aware of if you want to select and specify precipitators intelligently.

For example, inclusion of many specialized design-features will enable the precipitator to be erected easily and to be operated and maintained with the fewest problems. But since most precipitators are bought on a bid basis, these features are likely to be omitted (to provide for the lowest cost), unless you have specified that they must be included.

Careful attention to detail during the erection of the precipitator will pay dividends during startup, and in later operation. Here, again, you must know which problems must be avoided.

One of the things that complicates the purchase of electrostatic precipitators is that there is much "art" involved in the selection of the equipment by the vendor. This selection relies more on experience with previously sold precipitators than on solid engineering data and cal-

Originally published May 26, 1975

culations. Depending on the supplier's experience bank, it is perfectly reasonable for an engineer to receive vendors' bids that, for the same gas flow, vary in size by factors of two or more.

In the design of this type of equipment, it should be noted that size increases directly with gas volume for a constant efficiency but increases exponentially as efficiency requirements rise. That is, costs increase exponentially with increase of efficiency.

Before we go into details of the precipitators, let us discuss the particulate-containing gases that will be going through them.

Particle Size of the Pollutant

Particulate air-pollution-control problems involve particles under 100 microns (μm) in size. Although the particles are not spherical, the particle size is expressed as

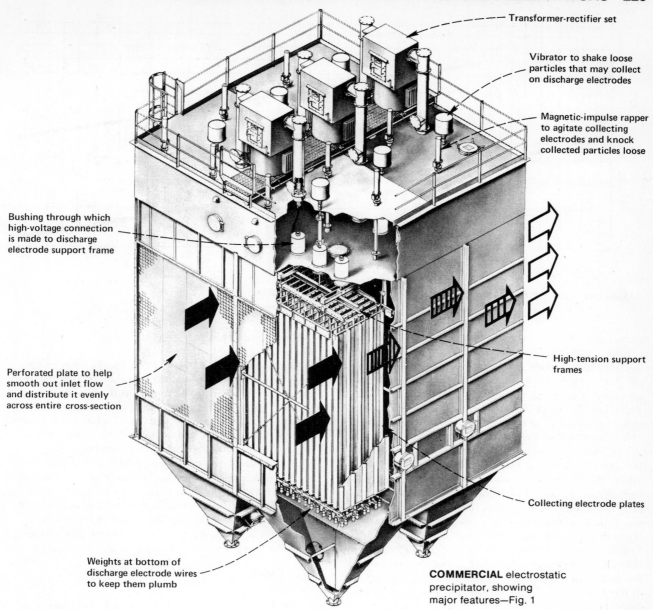

Transformer-rectifier set

Vibrator to shake loose particles that may collect on discharge electrodes

Magnetic-impulse rapper to agitate collecting electrodes and knock collected particles loose

Bushing through which high-voltage connection is made to discharge electrode support frame

High-tension support frames

Perforated plate to help smooth out inlet flow and distribute it evenly across entire cross-section

Collecting electrode plates

Weights at bottom of discharge electrode wires to keep them plumb

COMMERCIAL electrostatic precipitator, showing major features—Fig. 1

the diameter of an equivalent sphere that would follow the settling rate of Stokes' law.

Particles are often described, depending on their size or nature, as:

- Dust—Particles from 0.1 to 100 microns (μm).
- Mist—Liquid droplets suspended in a gas.
- Fume—Solid particles or liquid droplets that are formed by condensation from a vapor.

In most industrial applications, particle size interests us only insofar as it affects the capability of the air-pollution-control equipment. With the electrostatic precipitator, as with other industrial air-pollution-control devices, the larger particles are easier to collect (except when the particle is large but extremely fluffy).

There is a basic difference between the electrostatic precipitation principle and the mechanical methods (used in centrifugal separation, wet scrubbing and gas filtration). In the precipitator, the electrical forces are ap-

plied only to the suspended particles. In the mechanical methods, the complete gas stream is subject to externally applied forces, resulting in a much higher consumption of energy for the collection process. The size of the precipitator will be affected by the particle size, but the energy consumption for the collection process will remain almost constant.

Properties of the Particles

Certain physical and chemical properties of the particles are important because they affect the properties of the agglomerates that result when the particles reach a collecting surface.

For example: Mists will form liquid droplets that flow by gravity into the hopper. Some metallurgical fumes, such as zinc and lead oxides, form low-bulk-density layers that will break (during rapping of the collection

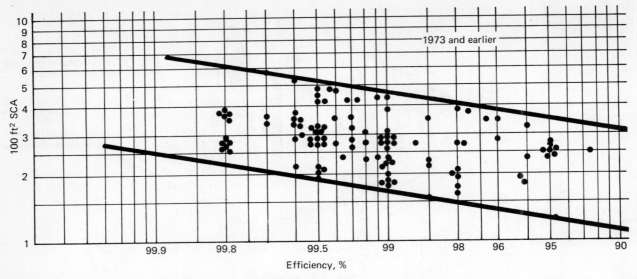

VARIATION in size (SCA) supplied for any specified efficiency (coal-fired boilers)—Fig. 2

plates) into fluffy agglomerates, which float in the gas stream. Other fine dusts, such as cement, form a relatively dense agglomerate that quickly falls into the hopper during rapping. Finally there are dusts, such as those produced in no-contact (low odor) kraft recovery boilers, which are very tacky and difficult to remove by rapping; special attention must be given to a proper rapper system to minimize operational problems.

The electrical conductivity of the particle is most important. (However, in the field of electrostatic precipitation, the reciprocal property, "resistivity," is used for the sake of numerical convenience.) Nonconductive or high-resistive particles, as they deposit on the surface of the collecting electrode, form an electrical insulating layer that prevents the movement of ions to ground. Thus the flow of current from the discharge electrode to the collecting electrode is reduced, and the voltage differential is increased until sparking occurs. A resistivity of the particles above 2×10^{11} ohm-cm is generally considered the limit for proper electrostatic precipitation.*

Excessive conductivity (or low resistivity) of particles, such as carbon black and other carbonaceous materials, causes the particles to immediately lose their charge as they contact the collecting electrode. Thus the particles are not retained on the collecting electrode surface, but rather reenter the gas stream.

Conditioning To Modify Particle Resistivity

As White points out in "Industrial Electrostatic Precipitation," "adaptation of conditioning methods to practical situations requires a broad knowledge of the basic principles and contingent factors to obtain useful results."

Conditioning to modify particle resistivity may consist of:

■ Addition of chemicals, or water vapor, to the gas stream.

■ Modification of the material producing the dust.

■ Change of the gas temperature.

More often than not, a combination of these methods will be chosen.

Addition of chemicals (or water vapor), even in very small amounts, has shown remarkable effect on the resistivity of the particles. Ammonia, sulfuric acid, sodium chloride, sulfur trioxide, and other substances, added in trace amounts, have improved the operation of some precipitators from an unacceptable performance to one that meets air-pollution-control requirements.

In most cases, the conditioning chemical is adsorbed on the surface of the particles. However, several cases of conditioning of weak basic particles by using both strong acids and water vapor suggests that the acid is adsorbed on the particle surface, and that water vapor is then absorbed. A similar situation occurs when conditioning weak-acid particles with a strong base and water vapor.

Gas Flow

Gas flow is critical in the design and operation of an electrostatic precipitator. The basic principle on which the precipitator works—the migration of minute particles to the collecting electrode—involves a finite length of time. If the gas velocity in any of the passages around the collecting electrodes exceeds the design gas velocity, some particles will not have adequate time to reach the collecting electrode.

After the particles have been deposited on the collecting electrodes, they are made to fall into the hoppers by rapping the electrodes. During this operation, good gas-flow patterns are critical to avoid reentrainment of particles in the gas stream. This reentrainment is possibly the most devastating effect caused by poor gas-flow.

At times, excessive gas velocity—due to unbalanced gas flow—may cause reentrainment from the hoppers. Uneven gas flow may result from part of the gas moving

*The resistivity of various dust particles, under operating variables such as gas temperature, gas moisture, and chemical composition of the gas, is presented in tabular and graphical form in "Industrial Electrostatic Precipitation," by H. J. White, Addison-Wesley, Reading, Mass., 1963. Because the resistivity is affected by so many variables, we do not attempt to give such information in this article.

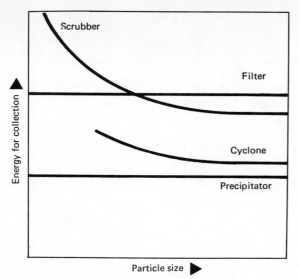

ENERGY consumption of dust collectors—Fig. 3

through the hopper space rather than between the collecting plates. This gas "sneakby" through the hoppers will result in very poor performance. With today's requirements of efficiencies above 99%, a 1% sneakby would make it impossible to meet the required efficiency. Baffles in the hoppers, when properly designed and installed, will usually correct such gas-flow imbalance.

Modification of the materials that produce the dust has been successfully applied to coal-fired boilers. Coal that contained natural conditioning agents (such as sulfates) was mixed with low-sulfur coal of high resistivity. The resulting mixture produced a dust that could be effectively collected in an electrostatic precipitator.

Changing the temperature of the particulate-containing gas is a widely accepted practice. In general, temperature in the 200 to 400° F of range will produce high-resistivity dust when natural conditioning agents are absent. (This is typical of low-sulfur-coal-fired boilers.)

One answer is to locate the precipitator ahead of the heat exchanger that heats incoming air (and, consequently cools the hot gas). The gas-flow upstream of the air heater (at about 700° F) represents about 1.5 times the volume of gas flow downstream of the air heater (at about 300°F). It would appear that a larger precipitator would be required to handle the gases at higher temperature (higher volume). However, in actual practice (with low-sulfur coals), because of the high-resistivity problems in the 200 to 400°F range, it is the smaller gas volume that requires the larger precipitator.

It should be kept in mind, though, that the "hot" precipitator (upstream of the air heater) may not necessarily be the universal answer for low-sulfur coal. Some low-sulfur coals contain other impurities that may have a conditioning effect, thus making the "cold" precipitator (downstream of the air heater) the more economical selection.

This should warn prospective users of precipitators that dusts from untested sources should be tried out in a pilot precipitator or, if feasible, in a full-scale existing installation. Such test work will support a final decision on a "hot" or "cold" precipitator.

The best precipitator will function no better than the gas-flow distribution allows. The flue design, both upstream and downstream of the precipitator, has a direct effect on gas distribution. Transitions and turning vanes must be designed in accord with principles developed in wind-tunnel tests. Diffusion plates or screens at the precipitator inlet will convert a flow of great turbulence into a multitude of small turbulences immediately downstream of each opening of the plate. These small turbulences will promptly fade out, causing a nonturbulent flow through the precipitator. Diffusion devices are usually perforated plates that are provided with rappers to dislodge any dust buildup.

Selection of the Precipitator

Precipitators are selected using the Deutsch equation or some of its modifications, and applying numerous experience factors.

$$\eta = 100 \cdot [1 - e^{-wA/V}]$$

Where:
η = collection efficiency, %; w = migration velocity, cm/s; A = area of the collecting electrodes, ft²; V = gas flow, thousand ft³/min; e = base of natural logarithms.

The units in the above equation are used by U.S. manufacturers of precipitators to express the migration velocity in whole numbers. The above equation is used to calculate the collecting area (A) when the gas flow (V) and the collection efficiency required (η) are known. The equipment designer will select a migration velocity (w) from his experience file. This migration velocity will vary, depending on operating conditions, source of the dust, and temperature and chemical composition of the gases. The ranges of such variations are too large for us to present meaningful data for the chemical engineer.

The migration velocity (w) is the average rate at which particles are charged, and conveyed to the collecting electrode (where they lose their electrical charge and are removed into the hoppers). Those particles that are not collected, or which are reentrained into the gas stream, are factored-in in this migration velocity. At the present state of the art, these factors are determined by experience.

"Experience" factors, such as particle size, particle agglomeration, gas velocity, gas temperature, gas composition, and chemical and physical properties of the particles, are all grouped into this so-called migration velocity.

Also included as experience factors are design peculiarities typical to the specific application, such as configuration of the ductwork, use of distribution plates, the type of process producing the particulates, and many other subjective-judgment elements the designer wishes to incorporate.

It is easy to understand that today's precipitator design is strongly tinted by the experience, and subjective decisions, of the individual selecting the equipment.

Attempts to quantify the experience factors and to incorporate them in computer programs are underway by numerous engineers and scientists. The precipitator in-

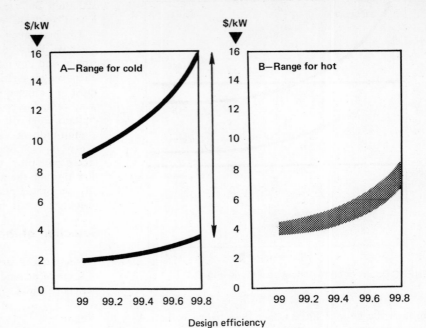

USER'S cost of erected precipitators of various design efficiencies. Variation in particle resistivity makes for a much wider cost range for cold precipitators—Fig. 4

dustry, however, has been very slow in employing the computer in this important task; hence, the selection of equipment is made today in basically the same manner as in the early 1940s. Most of the computer studies have come from outside the electrostatic-precipitator industry. The greatest difficulty in applying the resources of electronic data processing to the selection of precipitators is the lack of complete field-test results covering all of the variables that affect the basic concept of migration velocity. The designers of electrostatic precipitators use the following parameters as guidelines in equipment selection:

Dry Precipitators—Face velocity (velocity of the gas across the precipitator) is kept in the range from 1 to 15 ft/s, depending on the dust source.

Aspect ratio (collecting-plate total depth divided by collecting-plate height) is kept above about 1.0.

The electrical system is arranged in bus sections, each bus section representing any portion of the precipitator that can be independently energized.

The number of fields (number of bus sections arranged in the direction of gas flow) are calculated thusly: As a rule of thumb, manufacturers will use 1 field for up to 90% collection efficiency, 2 fields for up to 97%, 3 fields for up to 99% and 4 or more fields for efficiencies above 99%.

The number of cells (number of bus sections arranged in parallel) are established so that if any field is shorted out, the overall precipitator efficiency will not fall below the specifications.

Often, more than one cell will be energized from a common high-voltage electrical set (or source). However, more than one field in any one cell should not be energized from the same high-voltage electrical set, since a short would affect more than one field in the same cell,

causing a substantial reduction in collection efficiency. In general, one high-voltage electrical set is used for up to 25,000 ft² of collecting surface. About 55mA are supplied per 1,000 ft² of collecting surface.

Wet Precipitator—The wet precipitator is ideally suited for removing acid mists or other materials that can be collected as a liquid solution or suspension. The wet precipitator is also useful for either high- or very-low-resistivity dusts, provided that the process is not affected by such wet collection of the dust. There are cases where the dust can be removed from the collecting surface only by wet washing of the surface, and here the wet precipitator is the only answer—unless alternate methods of collection (such as fabric filters) are used.

The wet precipitator may be designed similarly to a dry one, but furnished with water sprays for continuously, or periodically, washing the collecting surfaces and the discharge electrodes. More often, though, the wet precipitator is designed as a series of pipes in parallel, each containing a single discharge-electrode in its center. Gases flow vertically through each pipe, and the collected mist flows down the pipe surface. Intermittent washing or flushing is usually required. As in the dry precipitator, the basic design parameters vary with each application.

BASIC STRUCTURAL CONSIDERATIONS

Structural engineering and design considerations are frequently overlooked by the engineer who specifies and buys electrostatic precipitators. Such details are perhaps as important as the design of performance parameters if the machine is to operate reliably for its expected life. It is often presumed that the manufacturer's experience and engineering capability should be accepted at face value. Consequently, evaluations of proposals are based essentially upon migration velocities, face velocities, quantity of collecting surface, electrical supply, sectionalization, number of rappers, and the like.

Of course these criteria are extremely important in comparing competitive designs. And, in most cases, the manufacturer's structural engineering capability is sufficient. Still, it is prudent to evaluate the structural design—at least, beyond the details of casing thickness, design pressure and temperature, and construction materials (which are the items usually included in proposals).

Hence, to provide a more complete picture before final evaluation, you should thoroughly examine (a) the supplier's standard fabrication and erection dimensional-tolerance system; (b) his procedures for qualifying his subcontractors; (c) his quality-control and inspection procedures in the shops and in the field; (d) the caliber of the individuals designated as construction supervision or advisers and service engineers, and (e) even the workload in his organization to assess the level of competence that will be applied to your precipitator.

This section of our article is not intended as a construction or design manual for electrostatic precipitators. Each manufacturer has his standards, design philosophy and "track record." Certainly, we do not wish to evaluate here the relative merits of the designs offered. Rather, what follows is an attempt to provide the user's buyers, engineers and operators with a guide that will enable them to examine and discuss the equipment with the manufacturers more knowledgeably. Assurance that the electrostatic precipitator will be designed and erected so that it will operate both reliably, and as intended, should be based upon considerably more than merely pages of legalistic terms, conditions and warranties. (It is now generally accepted that low price and "paper" guarantees do not generate the best investment in pollution-control equipment.)

In the competitive atmosphere of a "bid business," a manufacturer will generally propose only the equipment and features that he believes necessary and sufficient to meet the contract requirements. Although he will strive to point out his "features" at sales presentations, he will rarely offer designs that, in his judgment, will make his price noncompetitive. He will try to avoid increases in proposal and engineering costs that will be caused by deviations from his standards, which, after all, were developed largely as a result of commercial aspects and his knowledge of operating experience.

Therefore, a plant engineer must either carefully express his needs in the specification, recognizing the impact on his cost or, alternatively, evaluate proposals with extreme care.

Catastrophic structural failures have been rare (although they have occurred). What one must be wary of, usually, is the subtle structural problem. One such problem might be lack of provision for expansion, possibly stemming from a temperature assumption that allows no margin (causing excessive deflection of the substructure or the interior precipitator beams and columns). There are many other such problems, even insufficient attention to fabrication and erection tolerances (which will result

in misalignment and other operating difficulties after the precipitator is onstream).

Under the pressure of plant operation, with the natural reluctance to shut down units because of high cost, the true causes of malfunction are extremely difficult to ascertain, and repairs are expensive in both time and money. Even more disturbing is that the symptoms are often the result of several deficiencies that overlap and tend to mask each other. Fortunes have been spent by manufacturers and plants in making corrections and modifications in one shutdown after another, frequently without any benefit.

Problem prevention usually requires no more than a professional attention to detail, proceeding from a knowledgeable understanding of functional requirements, in logically following the design, fabrication and erection phases of the precipitator. The cost of this review will be far less than that of even a few days of forced shutdown.

The Support Structure

Use of standard AISC (American Institute of Steel Construction) allowable stresses and deflections for

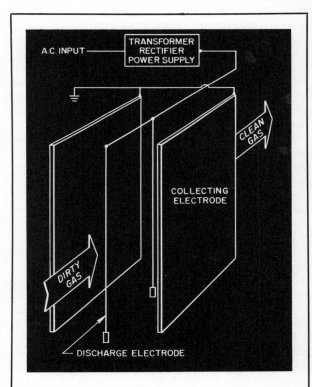

Basics of Electrostatic Precipitators

Particles in the dirty gas entering a precipitator are charged by the discharge electrode. The particles then migrate to the collecting electrodes (collecting plates), where they adhere and lose their charge. The particles are generally dislodged from the collecting plates by vibrators or rappers that are attached to the plates, and fall into hoppers below the plates. In some cases, however, it is necessary to wash the particles off; liquid particles may drip off by themselves.

structural-steel design may not provide a rigid enough "platform" to support the multiple columns of a precipitator. Some deflection is inevitable, but it is the relative deflections between support points that must be examined for the effects on internal alignment.

In an attempt to minimize these relative deflections, manufacturers generally place high restrictions on the allowable deflections, frequently leading to the need for massive structural girders. Additional structural columns to grade should be studied as an alternative (to reduce the support of the precipitator by members in flexure).

Because the casing expands in the plan view with increasing temperature, it is necessary to provide for sliding at the support points, with resulting frictional forces. This friction causes torsion on the structural member (since the sliding takes place across its top flange), and consequent bending in the precipitator columns. Also, the vertical loads will be eccentric to the steel member in either the hot or cold position (or both), depending upon the detail dimensioning. Most designers will offset the precipitator columns relative to the steel for half the calculated expansion.

The hoppers (with insulation and lagging), moving with the casing, must clear the flanges of the support girders in both the cold and hot positions. To provide the space required, in view of the large girders (usually needing rather wide flanges), the precipitator columns are extended—in effect raising the casing in relation to the steel. The bending moments caused by friction forces at the base are thus increased and the columns must be checked carefully. Errors occur here frequently because of poor coordination between designers of the precipitator and those who design the structural steel.

To reduce the frictional resistance, Lubrite (or other friction-reducing plates) are interposed between the structural steel and the precipitator-column base. When these are used in the design, one should carefully check allowable bearing-pressures and dimensional details in the hot and cold positions. Installation must be supervised to assure that the sliding surfaces are horizontal.

Fabrication and erection tolerances usually dictate that "shim packs" be installed between the precipitator base and the structure, thereby allowing for vertical adjustment of the casing. Logically, the base of the precipitator should be detailed "high" in the event that the structural tolerance is on the "plus" side at erection—otherwise, shims would be of no value.

Perhaps some adjustable "jackbolt" provision could be made so that shims could be fitted after the loads have been applied rather than as they are commonly used today—that is, as a tool to overcome tolerances "to a level" after erection of the steel.

To control expansion movements, most precipitators are anchored to the structure at one column base, with all others allowed to grow radially from the one point. If all the column bases are bolted to the steel, it is necessary to make provision for movement by using slotted holes. Considering fabrication and erection tolerances and the possibility of overtightened bolts, many designs incorporate shear-bar guides and stops at a limited number of columns along the two lines that radiate from the anchor column parallel to the structural steel.

These bars both control the expansion and transmit lateral shear at the base (from wind or seismic loads, which may be in excess of frictional loads). These shear bars must be located accurately and must be deep enough to allow for the maximum shim pack. Therefore, they should be placed after the casing is installed. Admittedly, access to the shear bars at this time is somewhat difficult and, occasionally, an erector may place the shear bars prematurely. This is another reason for careful field quality-inspection.

Casings

Functionally, the precipitator casing both houses and supports the collecting surfaces and discharge systems, forming a gastight enclosure between the inlet and outlet plenums or flue connections. Therefore, the casing is subjected to, and must be designed for:

- Static or dead loads of all components, including any equipment located on the roof, superstructure weights, hoppers, and dust loads.
- Roof live-loads and snow-loads.
- Loads and movements imposed by connecting flues.
- Wind and seismic loads.
- Internal gas pressure (or vacuum).
- Dynamic loading imposed by vibrators and rappers.

Additionally, the casing will be subjected to the elevated temperature of the flue gas; hence, the design must provide for the casing's overall expansion. Superimposed upon this expansion relative to the surrounding structure, there are also the differential expansions between components, and the consideration of thermal gradients in specific members as the stresses and deformations are influenced by end-restraints (connections).

The structural components of the casing are:

Upper Beams—These beams support the collecting surfaces and discharge systems, transformer and other electrical equipment on the roof, as well as the roof casing itself, and part of the superstructure and other loads. Because these beams are somewhat shielded from the gas stream, they are subjected to temperature differentials between bottom and top, particularly during the transient states of startup and shutdown.

Columns That Support the Upper Beams—These columns are subjected to lateral loads: i.e., pressure from wind that is transmitted through the casings, seismic forces, and flue loads. Since the columns form part of the heated enclosure, the bases must be permitted to "slide" relative to the cold steel substructure. Therefore, any friction loading results in additional column stresses.

In multicell or multichamber precipitators, interior columns are required to reasonably limit the spans of the beams. In the direction of gas flow, an interior partition ties the columns together, acting as a diaphragm. Frequently, this partition is a double wall having insulation between two panels to provide a heat break in the event one chamber is shut down. Certain designs employ bracing in lieu of panels. Placed at ninety degrees to the gas flow, the columns are either designed to be self-supporting for the full height, or use K-bracing of one form or another.

Some casings are developed as horizontal panels, with

the "columns" formed by massive vertical "stiffeners."

Lower Beams—A series of lower beams (and baffles) tie the columns together, supporting hoppers in most cases.

Casing—The casing consists of stiffened panels designed to transmit pressure and wind loads to the columns. Also, external loads exerted by plenums and flues (or stub stacks) may be transmitted through the stiffeners of the casing sheets.

Roof—The casing roof consists of secondary members and a cover plate that supports the discharge system (through electrical insulators), transformer-rectifier sets, control panels, etc.

Stresses in the Structure

We must readily appreciate the possibilities of thermal stresses in a structure of the foregoing type, resulting from differential heating and cooling of the elements.

Aside from the potential for structural failure, particularly at connections, excessive deformation of the casing structure may cause misalignment of the discharge and collecting systems. If such deformation is purely elastic, internal inspection of the precipitator in the cold condition will reveal nothing. Therefore, the best protection is a rigorous and thoroughly understood design analysis, paying particular attention to deflections and thermal stresses.

It would be well to become aware of the manufacturer's design criteria prior to award of the contract, and to include a review of all stress analysis in parallel with the approval of drawings during the engineering phase of the contract.

The increased allowable stresses for A-36 steel over A-7 has tended to "lighten" structures. However, the modulus of elasticity (governing deflections) has not increased and, therefore, deflections are greater in many modern designs. As the temperature increases, the modulus decreases, so that deflections in the hot condition should be carefully examined.

Fortunately, most precipitators operate at gas temperatures below 800°F, so the phenomenon of creep rarely needs be considered. However, because of (1) marked reductions in allowable stresses, (2) the possibility of graphitization, and (3) decreased oxidation corrosion resistance in carbon steel (A-36) at temperatures in excess of about 750°F, manufacturers look to other steels for construction.

Because of availability and cost, there is a tendency in the direction of certain proprietary materials such as Corten or Mayari R as opposed to "pedigreed" alloys; i.e., those recognized by the ASME codes as acceptable for elevated temperatures.

We would recommend to the plant engineer that he carefully discuss these materials with the metallurgists employed by the suppliers. In our opinion, there is little or no structural improvement made through the use of these materials (in place of A-36) up to gas temperatures of 800°F. Actually, designs to higher stress levels, a reason for selecting Corten A, again may result in greater deflections of members.

The primary benefit of using these proprietary steels may lie in some increased resistance to certain corrosive media at low temperatures and during shutdown. Note that although the gas temperature may wisely be used as a design temperature, the metal temperatures of main structural members should be about 50 deg lower.

Inspection and maintenance of both collecting and (most particularly) discharge sytems, is (in many designs) via key-interlocked doors through the roof casing (usually one door per dust-plate section). The operator has to crawl under casing stiffeners and over and around suspension hardware. The floor of the crawl space is usually the discharge-electrode support system, with the implications of scraped shins and sore knees. Specifying minimum clearances will eliminate the competitive tendency among manufacturers to reduce casing, rapper shaft, and suspension hardware costs by providing uncomfortably low headroom. Walkways and access doors between fields are a worthwhile investment for inspection, cleaning and general maintenance of the precipitator internals.

Miscellaneous

Pains must be taken during erection of casing, collecting and discharge systems to assure alignment. However, because of the effects on deflections of the structure, final alignment should be checked after all the loads have been imposed on the precipitator. These loads must include attachment of both the inlet and outlet flues, which often impose significant forces and moments on the casing. Frequently, the precipitator erection is "completed" by one contractor, with another installing the flues at a later date. Normal fabrication tolerances may require considerable "jacking" of the materials in order to make welded connections. This process may seriously impair alignment of internals.

It is preferable that alignment be rechecked after a short period of initial operation, allowing the system a sort of "shakedown run" through a few temperature cycles, because of the possibility that some weld connections may yield plastically.

Designs that separately support inlet and outlet plenums to relieve the casing of the loads must include supports that compensate for expansion of the precipitator relative to those supports. Pressure loads on the plenum faces may transmit torsional loads onto precipitator columns, depending upon the connection design.

Roofs, Penthouses and Superstructures

Insulator-Compartment Designs—Some designs provide separate insulator compartments, with the roof casing covered by insulation, and with a walking-surface deck plate. This deck plate will be cold relative to the casing. It must be adequately supported, either through rigid insulation or metal framing, and sloped for drainage (so as to prevent water and ice accumulation).

Clearances must be provided for the movements of insulator compartments, rapper-shaft sleeves and any other equipment or equipment supports that will move as the casing and precipitator structure expands.

Designs for any of the deck plate's metallic supports

that are attached to the casing must be examined for expansion provision. The criticality increases with higher gas temperatures and larger precipitator size. The exposed deck plate must be maintained watertight to prevent damage to insulation.

Penthouse—Penthouse designs extend the casing over the precipitator to form single large insulator compartments, eliminating the need for bus ducts.

The operating temperature of the penthouse is controlled by radiation and convection from the precipitator roof, conduction through the casing, and the flow of pressurizing air required to purge insulators. Heat losses depend upon the method of insulation. Frequently, the insulation is attached to the underside of the penthouse roof and to the inside of the walls so that the roof is cold steel, presenting a walking surface.

The structure is then required to absorb the expansion differential between the precipitator and the penthouse, with reactions transmitted into the casing, with a potential for causing distortion and misalignment.

Rapper shafts extending through the height of the penthouse, with rappers mounted above the roof, must be protected against any forces that may cause misalignment or binding.

If pressure testing of the precipitator shell along with the flue system is desired, it is important that the penthouse design-pressure be specified. Frequently, penthouses are designed of thinner materials than the precipitator casing. At best, pressure tests of precipitators are difficult and costly because of the many penetrations that must be temporarily sealed.

Weather Enclosures—Superstructures, or weather enclosures, may be provided to protect personnel and equipment; this will be money well spent, since it will assure maintenance inspections on a routine basis. These structures are cold, so connections to the precipitator must transmit vertical and lateral loads while also allowing for differential expansion.

Under weather enclosures, the precipitator roof is an insulated walking surface but the surface need not be weathertight (i.e., a sealed deck plate is not necessary). However, in some monolithic castable designs that fail to provide for differential movements in the casing, the penetrations and the castable do not last very long. Fissures, and the ultimate development of rubble, allow radiation and connection from the casing, which will make the area extremely uncomfortable for maintenance personnel.

Weather enclosures must be properly ventilated for both the comfort of personnel and the protection of electrical equipment and controls that are located on the precipitator roof. Undersized roof vents or gravity ridges may be less costly than liberally sized louvered blowers located along the walls; but in the long run, the operating difference will be appreciated.

Hoppers

Whether suspended from the casing or supported directly on the substructure that is interposed between the casing and the steel, hoppers are required for both collection and temporary storage of the dust. The simplest and most common hopper is pyramidal, converging to a round or square discharge. Frequently, the hoppers are baffled at the division between two dust-plate sections, to prevent gas bypassing the treater.

In certain applications, where the dust is to be removed by screw conveyors, the pyramid converges to an elongated opening along the length of the conveyor. In others, where the dust agglomerates in a sticky fashion (e.g., salt cake) and has a tendency to build up on any sloping surface, the hoppers are eliminated and the casing is extended down to form a flat-bottomed box under the precipitator. The dust is removed by drag conveyors that cover the entire bottom plate.

For many applications, however, hoppers slopes will shed the dust (provided the angles are steep enough). The valley angle (the angle between the corner of the hopper and a horizontal plane) should be checked as the governing minimum slope. Although valley angles as low as 52 deg have been used successfully, for some processes, 60 deg is an often-specified minimum.

Hoppers must be kept clean and dry. Therefore, although many designs do not require vibrators (they are both costly and require maintenance), it may be prudent to examine past experience in operating plants of the same or similar process and install mounting provision for vibrators anyway (to avoid later costly removal of insulation and lagging if operation shows the need for vibrators).

Moisture-laden dust that hits cold steel hoppers, has a tendency to stick. Therefore, insulation of hoppers is vital. However, this is sometimes not sufficient, and additional heating of the hoppers may be advisable, albeit expensive.

Installing hopper heating after operation is a costly construction affair involving building of scaffolding, and removal and replacement of the insulation and lagging. Therefore, in marginal cases, provision might be made in the design of the hopper and insulation to allow for future installation of strip heaters. Other common methods of hopper heating are attachment of mineral-insulated heating cables or by installing steam tracing. These devices are best applied at initial erection.

The discharge of the hopper should be as large as practical and the inner surfaces free of all projections and rough welds. Internal ladder rungs, attached by welding to hopper walls, in addition to presenting projections for dust buildup, provide a hazard to personnel in the event of weld failure caused by corrosion or external vibrators.

When a baffle extends too far down into a hopper, there is a danger of its acting as a "choke," causing bridging between the baffle and one or both sides of the hopper. Stopping the baffle a liberal distance (say 2 ft) clear of the sloping hopper wall should not, usually, allow gas bypassing. A gas sweep under a baffle of this type, considering the pressure drop of the turn, is possibly (and even probably) a symptom of poor gas distribution to the precipitator (that is, a downward jet at the entrance).

Access to hoppers should be via external, key-interlocked doors. Bolt-on doors through baffles should be avoided because of the dangerous possibility of dust accumulation on the far side of the door. Liberal "poke-hole" ports should be provided to allow for clearing a blockage at the discharge.

Overfilling of hoppers is a cause of precipitator problems. Dust buildup is capable of lifting discharge systems, shorting out sections, and frequently causing electrode breakage.

Level alarms are extremely valuable, provided they are kept in working order. Too often, because they are located near the top of the hoppers (even so high as to place them above the bottom of the structural steel supporting the precipitator), they are inaccessible for periodic inspection and maintenance. Also, the temperature of the atmosphere in this confined area may be sufficient to cause the alarm mechanism to fail; this is a point for

critical review of details prior to installation of the instrument.

Hopper capacity should be carefully checked to provide a reasonable time for minor maintenance of the dust-removal system.

With certain types of dust, either those that are pyrophoric or contain high levels of carbon, the danger of fire increases if hoppers are allowed to overfill and smolder, up to the levels where a spark from the precipitator may ignite the mixture. The smoldering material, itself forming clinkers, may structurally damage the hopper without actually breaking out in flame. Poorly placed (or faulty) level alarms cannot be the primary cause of the damage, but the preventive alarm function is one that is certainly

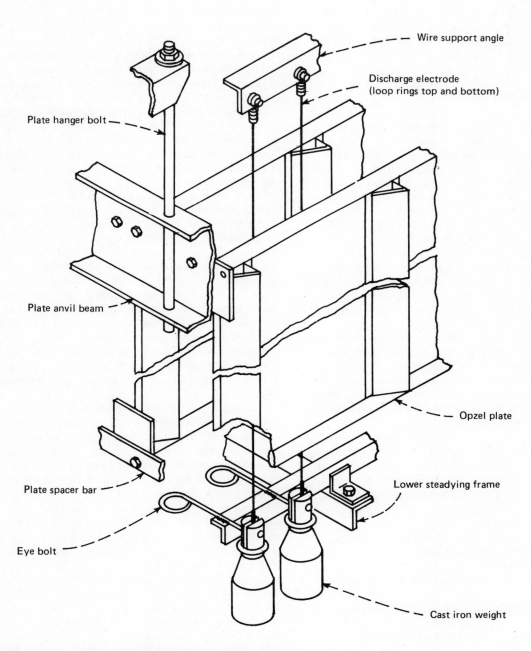

DETAIL of wire-weight system precipitator electrode construction—Fig. 5

well worthwhile designing and maintaining properly.

Anywhere from 60% to 70% of the dust will be removed by way of the inlet hoppers. However, in the event of inlet-field failure, the dust load will be transferred to the next hopper downstream, and so forth. This point is particularly important in sizing both discharge systems and conveyors. As for conveyor selection, the Conveyor Equipment Mfrs. Assn. places a manual at the disposal of the plant engineer. Just a few points are worth mentioning here. There is no better assurance of performance than a liberally sized, low-speed conveyor, that normally operates at 15-45% of the trough loading, depending upon the dust. The loading should be based upon the lowest anticipated dust density.

Alignment of the conveyors is important, and to a great extent it will depend upon the alignment of hopper connections. Because of the difficulty in erecting multiple hoppers to close alignment tolerances, field-adjustable flange connections are recommended. Also, provision for expansion between hopper connections and conveyor troughs must not be overlooked.

Discharge Systems

There are two basic designs of discharge-electrode systems offered today by manufacturers.

Wire-Weight System—The wire-weight system consists of individual electrode wires suspended from an upper support-frame. The wires are best shrouded in some fashion to prevent arcing to exposed, sharp, ground edges, or where the electrical clearance is reduced by passing the tops and bottoms of the collecting dust plates. The wires are held taut by suitable weights suspended from their bottoms. The weights in turn are spaced by a guide frame. The frame must be stabilized against swinging, an action that may be generated mechanically by the gas stream, through "electrical wind," by an improperly functioning automatic voltage-control, or some combination of these.

Commonly, the stabilization is accomplished by trusses extending from the upper support-frame to the guide frame. Rapper energy, transmitted through the trusses, aids in keeping the lower guide-frames clean.

Guide frames of a design that permits dust buildup high enough to raise the weights may cause slackening of the wires, arcing, and eventual wire failure.

Guide frames that are stabilized by ceramic, or other, insulators from the casings or hoppers can be a maintenance problem. Dust buildup on the insulators during operation, although resistive in some cases, presents a source of leakage to ground. Moisture gathered during shutdown (or low-load operation) might lead to complete failure of an insulator.

Electrode wire failure will be virtually nil, assuming:

■ Reasonable care, during erection, in alignment of the casings and surfaces.

■ A well-designed support, guide and stabilizer system.

■ Reliable, properly adjusted automatic voltage-controls.

■ Good operating maintenance of the dust-handling system.

Unfortunately, there are many precipitators in which the foregoing preconditions have not been met; repeated wire failure (really the symptom, not the problem) has been the culprit in the eyes of the operators.

Rigid Wire Frame—The rigid wire frame was furnished by U.S. suppliers prior to 1950 and then virtually abandoned (in favor of the wire-weight design) because of its many reliability and operating problems. Recently, U.S. licensees of foreign manufacturers have reintroduced frame electrodes to this country, and there are quite a few modern installations both in operation and on order (as there are wire-weight designs in European plants).

The rigid frame requires a high degree of quality control, both in fabrication and erection, and is intrinsically more costly. Replacement or repair is an expensive, time-consuming undertaking. It is about the same as attempting to replace a dust-collecting plate.

At lower temperatures, up to 400°F, warpage of the frames is uncommon, but for operating temperatures above 400°F, or with cyclical operation, potential deformation of the frames becomes seriously undesirable.

The rigid frame employs wider gas lanes, or ducts, to provide electrical clearance between the frame and the dust-collecting plate. This leads to larger casings to house the required surface areas.

It is important that the engineer be fully aware of the differences and the requirements of each design philosophy in detail, so that he avoids incorrect evaluations of one versus another.

The erection sequence usually consists of casings and hoppers first, followed by collecting surfaces and, then, discharge systems. If the casings are not erected to true dimensions, plumbed vertically, and square cornered in the plan view, attempts will often be made to compensate during the installation of collecting surfaces, i.e., using guides that should be free of frictional loads as "jacks," and so forth. Then, the discharge system, which should hang freely, is stabilized in an offset position to maintain, as best as possible, the wire-to-plate centers. This kind of construction will most probably prove an operating headache from the first day on; again, this is a vital reason for in-depth, step-by-step quality control and inspection, *regardless* of the pressure of construction schedules.

Discharge systems are supported from the casing through standoff electrical insulators. These must be kept in a clean, dry condition during operation, to prevent dust- or moisture-coatings from accumulating, because such coatings present a leakage path to ground. Wet accumulations can be very common during shutdown as the moisture in the gas condenses.

Warmed, filtered pressurizing-air supplies must be adequately designed to prevent such a problem. The system must provide distribution to a multiplicity of insulators, none of which may be allowed to "starve" because of disproportionate flow. This design problem is similar to that of balancing an air-conditioning system. Some method for checking distribution should be provided to the operators. Maintenance routines of changing filters and checking heater elements also should be established as soon as the system is operational.

Most commonly used electrical insulators lose dielectric strength as temperature increases. Although the maximum temperature varies with the insulating material, 400°F is a probable limit. Therefore, it is necessary that electrical insulators be isolated thermally from hot gases. The purge-air system normally suffices, but insulators mounted on hot casing-steel may be affected by conduction, at least for several inches along the length. Fortunately, most electrical insulators retain structural strength under higher temperatures and also act somewhat as thermal insulators, so that, if the electrical path is long enough, the effect of the conducted heat is limited to a short distance up the insulator.

Collecting Surfaces

The keys to any successful collecting surface configuration are:

Dust-Plate Trueness—It is important to ensure trueness of the individual dust plate, i.e., it must be free of kinks or excessive "oil canning." This trueness will depend upon care in fabrication, the packaging for shipment, and how the surfaces are stored and handled in the field. Most manufacturers will supply complete procedural details if requested to do so.

Dust-plate bundles should be stored on edge, on closely-spaced, level dunnage that has been positioned to take the loads directly from the shipping frames on the bundles. If stored for long periods, bundles should be protected from weather.

Most damage occurs during unpackaging and raising. Supervisory experience is the best guide. Although many foremen are thoroughly competent, having been involved in the erection of other precipitators, it is wise not to leave this operation solely to the devices of the workmen. Removing a damaged dust plate after the unit is "buttoned up" is difficult and costly. Close inspection as dust plates are installed will pay dividends.

The effect of wind on deformation of surfaces during raising is minimized by dust-plate packaging designs that permit installation of entire bundles into the casing shell prior to their being opened.

Depending upon orientation of the gas inlet, high winds passing through during construction—in the direction of gas flow—may damage plates, and this may go unnoticed if inspection is not constant.

Ruggedness of the Support System (and Its Dimensional Tolerances)—This system supports the dust plates and, in many designs, must transmit rapper energy to them. Be aware that a manufacturer's "standard" design may not be sufficiently rugged for all types of rappers. The design must allow for alignment adjustment at erection and should also allow for readjustment, if necessary, after shakedown operation. Particular attention should be paid to the effects of vibration and impact loading (notch sensitivity) at all welded points.

Alignment—Sufficient, adequate spacers must be provided to maintain alignment, while also allowing for possible temperature variation between dust plates (and, of course, temperature differentials between the casings and dust plates).

Often, the surfaces are guided at the bottom by using interior plates whose primary function is to provide a gas

TYPICAL transformer-rectifier (T-R) set for powering precipitator—Fig. 6

baffle. Such baffle plates may be somewhat thin, and may expand more than exterior casings; hence they frequently distort (or buckle). If this happens, the collecting surfaces may be pulled out of alignment through the guides.

Rapper Anvils—Rapper anvils that are attached to either dust-plate supports or rapper header beams require particular attention because of the duty to which they are subjected. Since alignment is extremely critical, designs that permit bending of flanges, or other local deformations, must be questioned.

Alignment Tolerances—Beware of "tolerances" given on drawings to govern alignment unless there are full explanations of:

■ How materials may be installed to those tolerances.

T-R sets

Rapper

Pipe and guard run

T-R SET, rappers and other items of precipitator equipment mounted on the roof—Fig. 7

■ How inspection and checking is intended. Inspection by "eyeballing" is always somewhat subjective.

■ How adjustment is provided to meet the requirements.

Baffles—The baffles between the extreme outer dust plates and the casings are necessary to prevent bypassing of untreated gases, because there are no discharge electrodes in this space.

Many designs call for such baffles to be installed after the surfaces are in and aligned. Unless there are walkways between fields (and the baffles are thus made accessible), such installation is almost impossible because of workspace limitations. Erectors often add these baffles to the casing plates while they are still on the ground, or before the dust plates are installed. The result is frequently a less-than-desirable closing of the space. Even inspection is difficult. Therefore this feature is best discussed with the manufacturer during the design stage.

Rappers

Rappers are used to remove dust from the collecting and discharge surfaces, and their effectiveness and reliability are vital. The types generally furnished are:

■ Electromagnetic impulse, either single or multiple.
■ Electric vibrators.
■ Pneumatic impulse.
■ Various mechanical hammers. These are usually associated with foreign designs but are sometimes furnished by others for special applications.

Each manufacturer has developed rapper applications for compatibility with his suspension system and rapper schedule (number of surfaces per rapper), based upon his experience and tests. Generally, pneumatic rappers will impart more energy than either electromagnetic rappers or electric vibrators, and will remove tenacious dusts more readily. However, it is important to be certain that all the hardware in the system is designed to withstand such high-energy forces. Changing from electrical vibrators to pneumatic rappers (in an attempt to improve operation) without also "beefing up" the hardware has led to structural failure.

Mechanical hammers are frequently very effective. But moving parts in the dirty gas stream become a maintenance problem. Any repairs will require shutdown of a chamber or system.

In the final installation, it is important that there is no binding of rapper shafts against casings so that the energy is directed where it belongs—to the surfaces to be cleaned. The design of rapper shafts through penthouses should be examined thoroughly with regard to the expansion differential between the penthouse and the casing.

Rapper controls should be readily adjustable for intensity, sequence and cycle time. Optimizing these for best rapper performance with the aid of a dust density meter in the outlet or stack) is a good practice.

It is prudent to check—at least once a week, but preferably once a day—that all rappers and the controls are in working order. A manual control that allows operation of one rapper at a time, coupled with an indicator panel in the main control room that monitors the automatic control, is an excellent feature. (Indicators that simply monitor power to a rapper may not tell the full story of the operation.)

Equipment on the Roof

The first-cost economics of locating automatic voltage-controls and rapper controls on the precipitator roof, or "operating floor," are attractive. However, the reliability of control elements that are subjected to the hot and generally dusty atmosphere makes it well worth evaluating the additional wiring and conduit cost that results from locating the controls in the main air-conditioned control room.

In engineering a precipitator, manufacturers may not provide a detailed layout of all of the components on the roof, thereby leaving conduit or piping runs to a contractor's judgment.

Frequently, the problems of close spacing of equipment, and of access to doors and cabinets, do not become evident until after erection. Further, the contractor tends to install the lowest-cost system possible in keeping with his specifications. This may result in an overall layout that is not conducive to good maintenance. The best protection is a concise specification and a detailed drawing study.

Although infrequent, transformer-rectifier (T-R) sets do sometimes fail. The methods of removal and replacement should be thoroughly understood and agreed upon. Again, although the first cost may be higher, transformer rectifiers located on a separate platform—with good access for removal—could be a sound investment. A spare T-R set, in storage on the platform, would make for minimum downtime after a failure.

A kilovoltmeter in the control panel of the T-R set is more indicative of what is happening in the precipitator than is a voltmeter on the transformer's primary side. When a kV meter is not provided, T-R sets may be equipped originally with provision for attaching such a meter for checking and test purposes.

Gas Distribution, Gas Proportioning, and Flues

Uniform gas distribution, with the gases entering perpendicularly to the face of the precipitator, is important to proper operation. High velocity jets may either cause erosion of dust from collecting surfaces or permit volumes of gas to move through the machine relatively untreated.

Flow-model studies (and the distribution devices installed—most often one or more perforated plates) are not always effective in developing the desired distribution. However, flow-model studies will generally not be conducted unless specified by the customer although they are at least qualitative indicators of what is going on. Unfortunately, most flow devices installed are fixed by design, so changes or adjustments require costly shutdowns.

The velocity of the gas entering the precipitator is so low, and the areas are so broad, that a pitot-tube traverse to check velocity is impractical. Instead, the most commonly used tool is the hot-wire anemometer. It would be worthwhile to allow a period in the construction schedule of two weeks to one month to allow for conducting tests and making adjustments before operation. Most programs have a time schedule that precludes this step as a luxury. If these adjustments are neglected, an equivalent period may have to be spent in making the changes after startup.

There are devices such as the "Konitest" that may be used effectively, but they require that specially designed hardware be installed initially in order to make them practical. Attention must be paid to flue designs to avoid both the close coupling that makes distribution more difficult, and surfaces that may allow for gross dust accumulation. Of what value is a flow-model study if the dust builds up for several feet on the bottom of a flue?

Any distribution device must be kept clean through adequate rapping.

Multiple-chamber precipitators require some means for gas proportioning. These are most commonly louver dampers at the outlet. Guillotine shutoff dampers at the inlet should not be used for proportioning since they tend to destroy proper gas distribution to a chamber.

During the initial layout phase of a project, attention must be given to the adequate location of test ports for both inlet and outlet sampling. After locating the ports in a section of the flue where a reasonably uniform velocity profile may be expected, it is advisable to provide proper platforms and weather protection for the test crews.

ECONOMICS

Installed Cost

The installed costs of electrostatic precipitators vary considerably, depending upon construction location, whether they are new or retrofitted, and on the season of the year. The fob. cost, however, is more-or-less predictable—it is basically a function of the area of collecting surface provided.

After the collecting-surface area has been determined (as described on p. 97), the packaging configuration is established. The package is then subjected to various constraints, such as the length-to-height ratio (L/H), the contact time, the number of fields, the gas velocity, etc. If the relationships between the quantity of gas and the package meet the criteria of past experience, that particular geometry will be priced and offered to the customer. The specific geometry is also affected by a supplier's standard modules for length, height and width.

Variations in Installed Cost

The limitations of the selection criteria as influenced by suppliers' experience, plus variations in packaging geometry, result in installed costs that vary by several hundred percent for similar applications.

Electrostatic precipitation is a mature technology wherein the fundamental principles have not changed. The changes presently occurring are in the refinement of application, such as attempts to maximize corona power, improve gas distribution, optimize rapping techniques and prevent reentrainment.

The fundamental change that has taken place in the recent past is that the technology is being required to operate at its boundary conditions, which results in exponential changes in costs. For example, older performance curves that related efficiency with cost were relatively flat and predictable in the efficiency range for which most precipitators were bought (90 to 98%). But these curves became asymptotic as the efficiency passed 99%. In these days of EPA and state regulations, one seldom sees precipitator specifications calling for less than 99% efficiency, and this is the area where prior knowledge is practically nonexistent both for performance and for cost data. This requirement imposes a need to completely understand the effects of independent variables on electrostatic precipitator efficiency.

Cost Effectiveness of Efficiency Models

The current state of the art in predicting efficiencies from the independent variables may show a 1% difference between the predicted and observed values. This is a good statistical fit, but that 1% difference at 96-97% efficiency level translates into an almost 10% difference in the precipitator's fob. cost. The same 1% difference in fit about the 99% efficiency point can result in a 25% differential in cost.

Thus, to be cost effective, there must be significant improvement in the currently used efficiency models. The current sizings generally contain a plethora of contingency and safety factors to ensure compliance with the customer's requirements. This is because the penalties for being wrong are so high that they can bankrupt a small company and seriously damage the profitability of even the very large companies. These factors have resulted in a marked tendency toward overkill in precipitator design, and most users have had to fund for significant added costs for pollution control over what they had expected.

Cost Effectiveness of Precipitator Geometry

The specific geometry of the enclosure for the collecting surface that has been specified by the efficiency model, affects the installed cost.

One precipitator supplier can produce a relatively small precipitator in 35 different "standard" combinations, to meet various application criteria. The selection of the least-cost configuration in conjunction with an im-

proved efficiency model is necessary to provide a truly cost-effective precipitator to the users.

Operating Costs

The variables that have an effect on installed cost also can affect operating costs. The major electrical requirements are a function of the design power-density (watts per square foot of collecting surface). Power requirements range from 0.00019 kW/actual cfm to 0.00040 kW/acfm, according to a recent study of TVA installations. Additional operating costs can be incurred, depending upon the specific installation requirements. Typical requirements are as follows:

- Rapper system—1 kVA/rapper panel.
- Control and signal power—0.25 kVA/T-R control panel.
- Insulator-compartment vent system—4 kVA/compartment.

These are generally ignored in cost comparisons, as they are not significant when compared to the corona power requirements. Another general guideline is that the annual operating costs are 10% of the installed cost.

Maintenance Costs

Costs of maintaining a precipitator are influenced by relative size, efficiency requirements and design parameters. A review of TVA's costs shows a range of $0.01 to $0.03/actual cfm of gas treated. Typical items of maintenance are rappers, rapper anvils, electrode wires, ash-handling-system parts, curtains and electrical controls.

Meet the Authors

◄ **Gilbert G. Schneider** is Executive Vice-President, Enviro Energy Corp., Suite 220, 16161 Ventura Blvd., Encino, CA 91436. An expert in air-pollution control, he has been sales manager for the Western Precipitation Div. of Joy Manufacturing Co. He has a B.A. in mathematics from Wofford College, S.C., and a B.Ch.E. from Rensselaer Polytechnic Institute, Troy, N.Y. He is a guest lecturer at University of Southern California and is a member of AIChE.

Theodore I. Horzella is Vice-President of Enviro Energy Corp. He received his ► B.S. in chemistry from Santa Maria University, Chile, and his M.S.Ch.E. from Iowa State University, Ames, Iowa. He has also done graduate work in business administration at the University of California. He is a member of AIChE, Sigma Xi, and the American Management Assns.

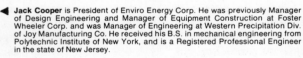
◄ **Jack Cooper** is President of Enviro Energy Corp. He was previously Manager of Design Engineering and Manager of Equipment Construction at Foster Wheeler Corp. and was Manager of Engineering at Western Precipitation Div. of Joy Manufacturing Co. He received his B.S. in mechanical engineering from Polytechnic Institute of New York, and is a Registered Professional Engineer in the state of New Jersey.

Philip J. Striegl is Vice-President of Enviro Energy Corp. He was formerly associated with Allis-Chalmers Corp. and Joy Manufacturing Co. plants. He has a B.S. in metallurgical engineering from Illinois' Institute of Technology, and has done graduate work in business administration at Marquette University and the University of Wisconsin.

Electrostatic precipitators in industry

Here is an overview of the subject, touching on theory, design, sizing, controls, component reliability, efficiency, the upgrading of old equipment and the retrofitting of new, and the conditioning of gases.

Robert L. Bump, Research-Cottrell, Inc.

☐ For over half a century, electrostatic precipitation has been the method of choice to control particulate emissions at industrial installations ranging from cement plants and pulp and paper mills to oil refineries and coke ovens. In most cases, the particulates to be collected are by-products of combustion. In others, they are dust, fibers or other small solids from a production process.

In the past decade, precipitator design—spurred by increasingly stringent emission regulations—has advanced at an especially rapid rate. Efficiencies and availability records not considered possible a few years ago are now routinely achieved. Although the geometrically accelerating cost curves associated with higher and higher purity standards are well known, the cost of precipitators has not risen exponentially as might have been expected. Why not? This article explores some of the reasons.

General precipitator design

A modern precipitator system, whether it was created to treat flue gas from a heat source or to deal with particulates spilling from process streams, is likely to be far superior to any unit that could have been built 10 or 20 years ago. Among the factors underlying this superiority are these: far-more-sophisticated mathematical techniques for predicting precipitator performance; superior construction materials; computerized data banks of technical information based on 50 years or more of experience in building industrial precipitators; availability of high-quality auxiliaries such as flues, dampers and handling systems; design improvements growing both out of experience with earlier precipitators and out of accelerating research programs.

Precipitator theory

Electrostatic precipitation is a physical process by which a particulate suspended in a gas stream is charged electrically and, under the influence of the

Originally published January 17, 1977

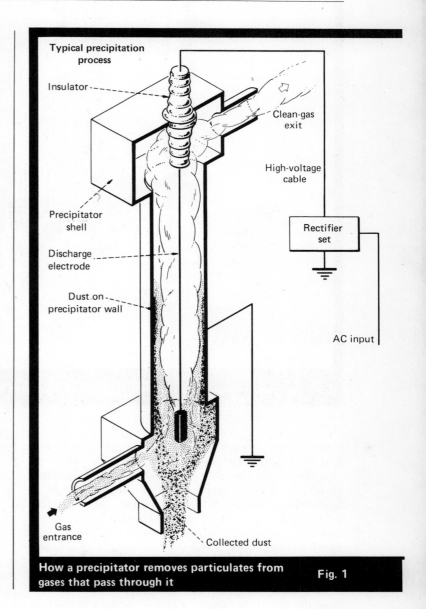

Typical precipitation process

Insulator

Clean-gas exit

High-voltage cable

Precipitator shell

Discharge electrode

Dust on precipitator wall

Rectifier set

AC input

Gas entrance

Collected dust

How a precipitator removes particulates from gases that pass through it　　Fig. 1

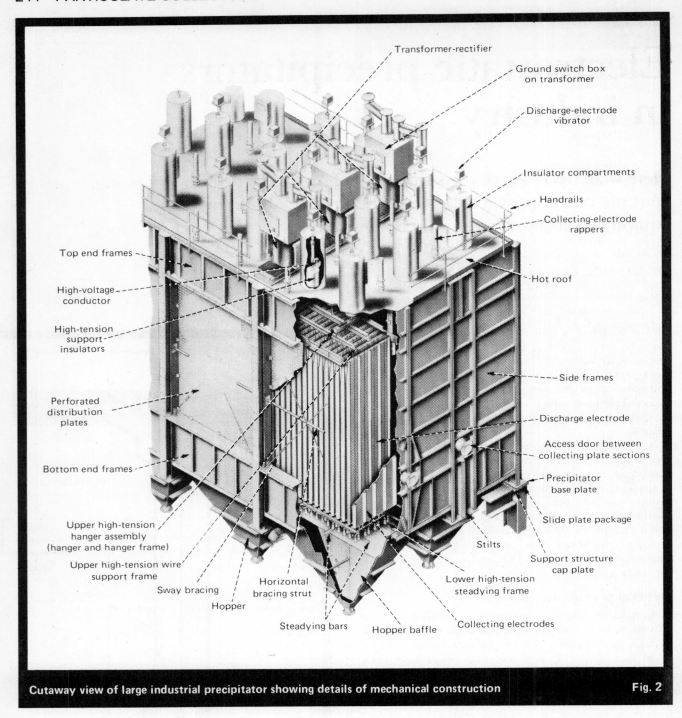

Transformer-rectifier

Ground switch box
on transformer

Discharge-electrode
vibrator

Insulator compartments

Handrails

Collecting-electrode
rappers

Hot roof

Side frames

Discharge electrode

Access door between
collecting plate sections

Precipitator
base plate

Slide plate package

Support structure
cap plate

Lower high-tension
steadying frame

Collecting electrodes

Hopper baffle

Stilts

Steadying bars

Hopper

Sway bracing

Horizontal
bracing strut

Upper high-tension wire
support frame

Upper high-tension
hanger assembly
(hanger and hanger frame)

Bottom end frames

Perforated
distribution
plates

High-tension
support
insulators

High-voltage
conductor

Top end frames

Cutaway view of large industrial precipitator showing details of mechanical construction   **Fig. 2**

electrical field, separated from the gas stream. The system that does this (Fig. 1) consists of a positively charged (grounded) collecting surface placed in juxtaposition to a negatively charged emitting electrode. A high-voltage dc charge is imposed on the emitting electrode, setting up an electrical field between the emitter and the grounded surface. The dust particles pass between the electrodes, where they are negatively charged and diverted to the oppositely charged collecting surface.

Periodically, the collected particles must be removed from the collecting surface. This is done by vibrating or rapping the surface to dislodge the dust. The dislodged dust drops below the electrical-treatment zone and is collected for ultimate disposal.

A commercial precipitator (Fig. 2) comprises symmetrical sections of collecting surfaces, discharge electrodes, suitable rapping devices, dust hoppers, and an enveloping casing and the necessary electrical energizing sets.

Mathematics and design

One major source of improvement in precipitator design is in the sophistication of the mathematical tech-

Information required to size a precipitator	Table I
Type of process	Fuel analysis*
Size or production rate of process	Dust analysis
Gas volume	Particle size
Temperature	Resistivity
Gas analysis	Efficiency required
Type of fuel*	Space limitations

* For particulate control on power boilers

niques employed to predict the exact size of a precipitator to handle a specific task. When emission standards were less stringent, larger margins of error were tolerable. Now that standards have risen to impose efficiencies of 99% or more, there is less room for error.

Major advances in precipitator design techniques have resulted in precipitators more precisely "tailored"—and therefore more efficient in operation—to a specific installation. The Deutsch equation, once the definitive tool for precipitator sizing, is no longer adequate to meet current demands for efficiencies well in excess of 98%. A modified Deutsch equation, now in use, factors in many of the newer practical considerations inadequately expressed in earlier theoretical approaches to sizing.

Too conservative a design produces unacceptably high equipment costs. Too "lean" a design means unacceptable operating and maintenance costs—not to mention stiff fines for out-of-compliance operation.

Sizing

The primary factors in precipitator sizing have been *face velocity* (the speed at which the gas travels through the precipitator), *migration velocity* (the speed at which the dust particle travels toward the plate under the influence of the electrical field) and *aspect ratio* (the ratio of precipitator height to its length).

In recent years, however, a more sophisticated approach to sizing has evolved. Extensive investigation of the relationship of process and operating variables to predicted performance, combined with a wealth of actual field experience correlating operational versus predicted performance, has been pulled together to create a central computerized bank of essential information.

From this data bank, which considers type of process, detailed particulate analysis, temperature, particle size and dust resistivity, a precipitator sizing program has been developed that generates a variety of acceptable options. If, for example, several process variations are possible (e.g., variety of fuels in a power boiler), the program considers all of them—in contrast to the old, often inadequate method of selecting a migration velocity for a single operating condition.

The modern result is a properly sized unit with less guesswork and more certainty of predictable and proper performance than ever before. Equally important, a way of mathematically modeling the effects of other alternatives is also easily available.

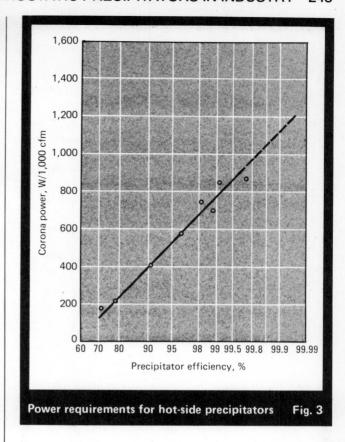

Power requirements for hot-side precipitators   Fig. 3

To obtain such a mathematical model, however, the purchaser must furnish detailed process information. Table I lists the minimum data required to size a precipitator. Without these data, assumptions would have to be made that might not clearly identify the pertinent points affecting the proper equipment selection. Also to be considered are operating and maintenance procedures, including specific schedules for process downtime and process-equipment inspection and reconditioning. These factors can influence original design parameters relating to cost and reliability.

Electrical sectionalization

In order to ensure optimization of power input and to afford the highest percentage of onstream reliability, the modern precipitator is divided into a substantially greater number of independent electrical sections than was previously done.

The efficiency of a precipitator is a direct function of the power input (Fig. 3). Any condition that adversely affects power input should be avoided in the basic design of the precipitator. Proper alignment and stability of the high-voltage system is essential.

Theoretically, the most efficient precipitator would be one in which each individual discharge electrode has its own power supply in order to maximize power input. This is highly impractical. However, it is both practical and advisable to have the precipitator divided into a number of separately energized electrical sections that can be individually isolated. This practice not only allows to some extent for variations and stratification in temperature, dust loadings and so forth, but

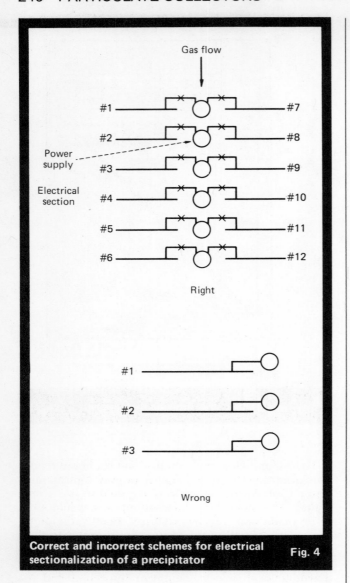

Gas flow

#1 ———— #7

#2 ———— #8

Power supply

#3 ———— #9

Electrical section

#4 ———— #10

#5 ———— #11

#6 ———— #12

Right

#1 ————

#2 ————

#3 ————

Wrong

Correct and incorrect schemes for electrical sectionalization of a precipitator   **Fig. 4**

Component reliability

Keeping the precipitator on-line requires increased reliability in the components used. To this end, research and testing is directed constantly toward upgrading existing components and developing new hardware. This includes such areas as plate and discharge-electrode designs, suspension systems, rapping processes and mechanics.

A major impetus for research in all these areas is matching component life to maintenance schedules or turnaround so that unscheduled shutdowns due to component breakdown can be avoided.

In addition, "fall-out" benefits are being realized. During one study on rapping systems, for example, cracking was noticed between rods and frames. Investigation revealed that a solid rapper rod acted as a heat sink during welding to the rapper frame and upset welding heat balance. Subsequent tests on welding techniques suggested a better way—hollow rapper rods at the welded end. Now the welded surfaces of both rod and frame respond equally to welding heat cycles. New welds exhibit 85% more resistance to cracking than those made with solid rods.

Factors affecting efficiency

In addition to the accumulation of data on precipitator sizing and design, there has been a substantial accumulation of data on operation procedures and firing practices that can cause the precipitator to lose efficiency.

Gas volume—A precipitator is a volumetric device. For example, any increase in boiler load that results in excessive flow through the precipitator will cause a loss of efficiency. A precipitator designed for 3 ft/s face velocity and an efficiency of 99% will drop to 96.5% if the velocity increases to 4 ft/s, a 33% load increase.

Temperature—A change in operating temperature may also have an effect on precipitator efficiency. Particle resistivity varies greatly in the temperature range of 200 to 400°F. Ignoring the effects of temperature on gas volume, the impact of temperature on efficiency would be, assuming 99% guarantee at 325°F (on a fly-ash application):

Temperature, °F	Efficiency, %
200	99.9+
325	99
400	99.5

Obviously, there is benefit to be derived in operating below or above the 300 to 350°F level.

Fuel—Any significant change in the type of fuel being fired will have an effect on the precipitator's performance. For example, a change from a 2%-sulfur bituminous coal to a 0.5%-sulfur, subbituminous Western coal can result in a design efficiency of 99.5% dropping to 90% or less. Other chemical constituents, such as sodium oxide, in the ash can have an effect on performance by reducing bulk resistivity.

The unit should be designed for the worst expected fuel.

Inlet loading—Since a precipitator is designed to remove a certain percentage (by weight) of the entering

it renders a smaller section of the precipitator vulnerable to external malfunctions such as dust-removal problems. When such problems occur, one section can generally be shut down for dust removal while other sections remain onstream.

For example, today a typical 99%-efficient precipitator (Fig. 4) on a 350,000-lb/h boiler would have six separate electrical sections in series and two in parallel, each with its own power supply. Ten years ago, such an approach would not even have been considered. There would have been three large sections.

In addition, it is common practice to provide some redundancy so that outage of one or more sections does not adversely impact on efficiency.

Automated controls

A relatively new field of development is the application of sophisticated electronic measuring and control devices to precipitator control circuitry. Power is now held at an optimum level automatically and dependably, despite wide variations in gas and dust conditions.

material, all things being equal, an increase of 50% at the inlet will result in the same increase at the outlet. Therefore, if an operating change involves an increase in percentage of dust, a corresponding increase at the outlet—resulting in greater opacity—can be expected.

Carbon—Variations in firing practice or coal pulverization that affect the quantity of combustible in the fly ash also impact on precipitator performance. Carbonaceous materials readily take on an electrical charge in a precipitator, but lose their charge quickly and are readily reentrained. Not only is the carbon particle very conductive, it is large and light compared with the other constituents making up fly ash.

These are the major variables to be considered if a deterioration in performance is to be avoided.

Age as a factor in performance

The question is often asked whether or not precipitator performance deteriorates with age. The answer, based on available operating experience, is "No;" however, there are two basic factors involved.

First, operating conditions that affect gas volume, temperature, gas and dust composition and so forth cannot be changed. Second, the precipitator must be maintained properly to the extent that the internals remain in good alignment and are adequately cleaned by the rapping system.

Meeting new clean-air standards

Assuming that precipitator performance has not deteriorated, there is still a major problem facing chemical plant operators. That is the ever-increasing stringency of clean-air codes, both new and revised, whether on the local, state or federal level. Collection efficiency requirements in the range of 99% and more are becoming the norm. Few precipitators installed before 1970 are rated this high.

To meet the new requirements, the plant operator has two choices. He can either upgrade existing equipment, or replace it with new collection equipment.

Upgrading existing equipment

Nothing short of additional equipment will yield 99% performance where 90% is installed. However, there are some areas where improvement can be obtained.

Gas distribution—Before efficiency requirements went to the 99% level, good gas distribution was not as critical as it is now. Therefore, it may be possible to improve the efficiency of older units by improving the flow pattern.

Ash conditioning—As mentioned earlier, a fuel change can result in a low or high resistivity situation, which can be controlled by injecting trace chemicals into the gas stream to make the dust precipitable. Such conditioning can have a markedly beneficial effect on collection efficiency.

Energization—Older precipitators are generally equipped with fewer electrical energizing sets than current practice dictates and may not be as responsive to varying operating conditions. To this extent, efficiency may be improved by increased electrical sectionalization.

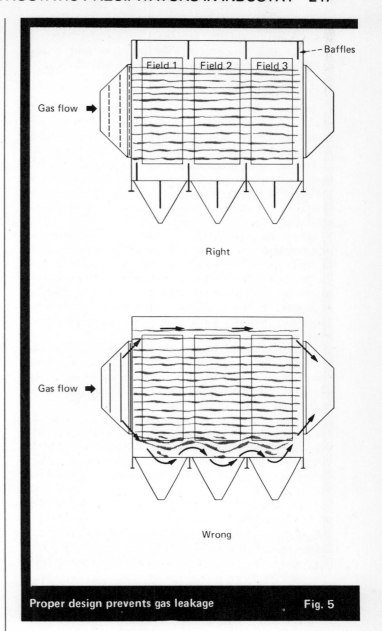

Proper design prevents gas leakage   **Fig. 5**

New equipment

The decision to install new collection equipment in series with existing equipment, or to replace the existing equipment entirely, is affected by several considerations. These include space, efficiency of the existing collection devices, their condition, pressure drop as related to possible need for a new fan and other operation considerations, and, of course, cost.

For example, assume an old, low-efficiency mechanical collector in poor condition is taking space where a new electrostatic precipitator can go. If the mechanical collector is left in place, the induced-draft fan cannot handle the pressure drop across the new equipment. In this instance, the mechanical collector should be removed and replaced with the new equipment.

In another example, a 5-yr-old precipitator is in good condition and operating at 95% efficiency. A new clean-air code calls for 99% efficiency. Since there is space downstream of the existing precipitator, the most feasible—and economical—solution is to add a new, small precipitator in series with the existing one.

Retrofitting alternatives

In some instances an existing precipitator can be made longer and higher so as to meet new requirements. However, this approach requires both a lengthy outage and the space necessary to do the modification.

Another sometimes viable approach is to "double-deck" a new precipitator over an existing unit. An alternative to this approach, assuming space is available, is to duct from the old precipitator to the new one and then double back to the existing stack.

When the precipitator is located on a building top, additional problems occur. In most cases, the building cannot support the weight of additional or new (that is, larger) equipment. This can be resolved by putting up a new support structure that penetrates the building roof and runs down to grade. An alternative solution is to, where possible, locate the new unit at grade and duct down to it.

Loss of efficiency occurs when gas bypasses the electrostatic zone in a precipitator. This can occur between the end plates and the shell, over the top of the electrical fields or in the hoppers (Fig. 5) if proper design care isn't taken.

As good flow control in a precipitator is achieved, there is a marked increase in collection efficiency. Precipitators have gone, for example, from 96% to 99.5% efficiency simply by corrections in flow control alone. Proper use of perforated plates, turning vanes and baffles is essential.

Complete and accurate information on fuel analysis and ash chemical composition is essential. The preferred data are discrete analyses, rather than merely an indication of expected ranges of constituents. With these data, the sizing program can determine the worst-case combination of constituents.

Work has also been done on cold-side sizing. Analysis of performance data on cold-side units and pilot precipitator work has indicated that there is significant deviation from the performance predicted by the commonly used Deutsch-Anderson equation at efficiency levels in excess of 98%. This led to the development of a modified equation.

In a typical example, use of the original equation would have resulted in a precipitator about 15% smaller. In addition, the program requires an identification of the boiler type, coal, mass mean particle size,* and sulfur and sodium oxide contents of the ash. Based on these inputs, the necessary collection area is indicated.

Obviously, as improvements in sizing technique develop, improvements in other design areas are necessary if precipitator efficiency is to be increased and maintained.

Gas flow distribution—Higher efficiencies demand more and more emphasis on the need for uniform gas flow. Detailed laboratory model studies are often employed to develop the most economical configuration for a new

*Diameter of a particle of average mass.

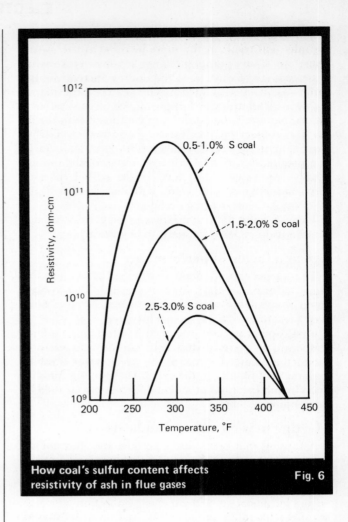

How coal's sulfur content affects resistivity of ash in flue gases   **Fig. 6**

precipitator or to identify flow problems in existing installations. In many instances, a redistribution of gas flow through the precipitator can increase collection efficiency by several percentage points.

Component reliability—The need for continuous on-line availability has required constant attention to the development of precipitator component reliability without extensive increase in equipment costs.

As a matter of fact, developments in precipitator design have advanced to the point where buyers dealing with established, reputable suppliers can be assured that they are purchasing the system and component reliability they need.

Precipitator placement—Long-range studies of precipitators in flue-gas streams indicate that gas temperature can affect precipitator performance. In certain applications, precipitators have shown that high operating temperatures are not only feasible, but economically desirable. Hot-side application can eliminate the uncertainty about dust resistivity that can result in dew-point corrosion, poor precipitator performance and low efficiency.

Ash conditioning—Either moisture or chemical treatment of flue gas can alter the electrical properties of dust particles and enhance their "precipitability." This permits greater flexibility in the choice of fuel, and more-efficient particulate collection over a wide range of operating conditions.

The first precipitator installations were cold-side operations with flue-gas temperatures seldom exceeding 300°F. In these installations, the electrical resistivity of the ash is established by a surface-conduction mechanism sensitive to the presence of minute quantities of sulfur trioxide, sodium oxide and other hydrophilic species, as well as to the partial pressure of water vapor in the system. The quantities and interreactions of these substances are not readily predictable. Therefore, low-temperature resistivity is variable and unpredictable.

However, in recent years many installations have been made ahead of the air preheater where temperatures are in the 650 to 850°F range. In these operations, resistivity is established by bulk chemical analysis of the ash, since conduction proceeds via a volume-conduction mechanism. It has been found that the predictability of fly-ash resistivity is not only much more reliable at elevated temperatures, it is virtually certain to be within a predictable range at 650 to 850°F.

Fig. 6 shows the effect of temperature on resistivity for a given low-sulfur fuel.

Because of the accompanying decrease in gas density at the higher temperatures, requirements in the corona starting potential in hot-side precipitators is greatly reduced. In addition, the elevated temperatures have been found to reduce sparkover level. The net result is a more efficient use of the electrical power input to the precipitator.

The final area of significant difference between hot- and cold-side operation is in the physical properties of the fly ash as related first to removal from the electrodes by rapping and then to removal from the precipitator dust hopper. Experience has shown that fly ash from a cold-side, high-resistivity operation is more adhesive than hot-side ash. To keep the electrodes relatively clear of this insulating layer usually requires more intense and more frequent rapping. This, in turn, limits gas velocity to about 4 ft/s to prevent excessive reentrainment of the fly ash.

Conversely, at elevated temperatures it is commonplace today to use gas velocities in the 5 to 5.5 ft/s range.

Removal of collected ash from hoppers has sometimes been a difficult and troublesome chore on cold-side installations. Hopper plugging not only reduces precipitator efficiency but, also, can cause serious damage to the internals. This may include distortion of lower high-tension framework, bowing discharge electrodes and accelerating failures. Moreover, ash buildup in the hoppers increases possibility of dust reentrainment and loss of efficiency.

Since the fly ash in high-temperature operations is almost fluid, and the gas temperature is far above the dew point, there have been essentially no problems reported with ash handling.

Ash conditioning

To combat fuel-supply uncertainties and shortages, many plants are burning lower-sulfur fuels. These low-sulfur fuels may generate more ash and gas per Btu, and the ash has an electrical resistivity several orders of magnitude greater than that from higher-sulfur coals.

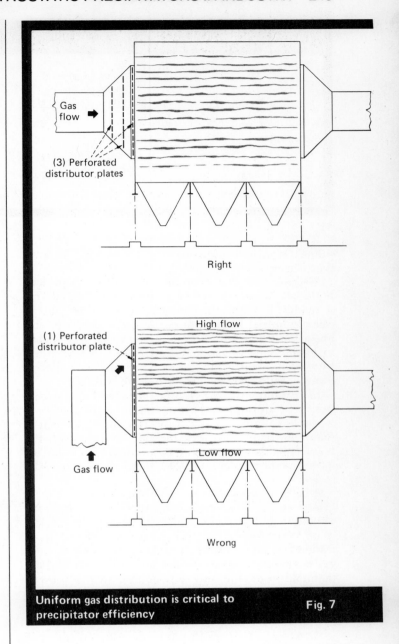

Uniform gas distribution is critical to precipitator efficiency   Fig. 7

This, too, will affect precipitator operating efficiency. For precipitators to perform at the required degree of efficiency, these ash particles can be treated chemically to enhance their charge retention.

To accomplish this, small quantities of SO_3 are injected into the flue gas. This reduces the electrical resistivity of the fly ash, making the dust more amenable to collection in the precipitator.

By improving the electrical operation, the gas-conditioning system can significantly improve precipitator performance and minimize the need to enlarge the precipitator for low-sulfur fuels.

Such a system can also be used as an "efficiency" backup system on high-sulfur-fuel operations at a reduced capital expenditure.

There are several types of commercial SO_3 gas-conditioning systems: direct injection or evaporation of liquid SO_3; catalytic conversion of SO_3; vaporization of

Economic comparison of SO₃ gas-conditioning processes						Table II
	Capital investment, thousand dollars		Operating costs*			
			¢/lb SO₃		¢/kWh	
Process	250 MW	500 MW	250 MW	500 MW	250 MW	500 MW
Molten sulfur	373	565	4.70	3.76	0.008	0.006
Liquid SO₂	320	512	8.56	7.73	0.0146	0.0132
Liquid SO₃	281	441	8.92	8.00	0.0152	0.0136
Sulfuric-acid evaporation	650	1,020	7.38	6.16	0.0123	0.0103

*Based on conditioning of a low-sulfur Western-coal flue gas to an optimum resistivity level estimated at 55 ppm SO₃.

sulfuric acid; and sulfur burning followed by the catalytic conversion of SO_2 to SO_3.

SO_3 injection—The simplest system to design and build is a liquid-SO_3 conditioning system. Liquid SO_3 is readily available in commercial quantities and is clear, colorless, stable, not particularly corrosive, and has a fairly low vapor-pressure. However, it is highly toxic and requires reasonable safeguards. SO_3 is highly hygroscopic and, when dispersed in air, immediately forms an extremely dangerous sulfuric acid mist.

In operation, liquid SO_3 is first metered into a vaporizer, then air-diluted. This maintains a constant controllable volume of gas flowing through the injection manifold (which provides adequate dispersion in the flue). The mixture is conveyed in heated lines to the injection point to prevent H_2SO_4 condensation and corrosion.

Acid vaporization—In this system, sulfuric acid is heated above its boiling point, vaporized, diluted with air and then injected into the flue ahead of the precipitator. Water vapor is always present in the acid vaporization system, so a heating system is necessary to keep gas temperature above the dew point. In addition, the manifold inside the flue must be insulated to prevent corrosion and premature condensation.

Gas distribution

As recently as 10 to 15 years ago, fully one-third of the particles treated in the precipitator were treated twice because of reentrainment caused by improper gas flow within the precipitator. Because of this, precipitators were usually sized larger than would otherwise be necessary.

However, increasingly stringent collection-efficiency requirements have tightened the criteria for what constitutes "good" gas distribution. The very high collection-efficiency levels currently required are realized with a gas-distribution pattern that permits maximum utilization of the gas-treatment zone.

Nothing will downgrade overall precipitator performance as thoroughly as maldistribution of flue gases. Careful attention to design of the flue leading to and from the precipitator (as shown in Fig. 7) can create a more uniform overall gas flow within the precipitator.

Dust reentrainment from the hoppers due to improper gas flow is a frequent cause of performance defi-

ciency in high-efficiency precipitators. Variable-porosity distribution plates, fitted at the precipitator outlet, and proper baffling inhibit those pressure gradients that would normally promote hopper sweepage. The constricting device changes resistance across the gas stream and corrects what is considered to be a little-known and often-misunderstood flow phenomenon.

Catalytic conversion—In the catalytic conversion of SO_2 to SO_3, liquid SO_2 is vaporized in a steam-heated vaporizer. This vapor is then mixed with enough air to produce a mixture containing approximately 8% SO_2 by volume. This mixture is heated in an electric heater to about 840°F, and fed to a single-stage converter. About 70 to 75% of the SO_2 can be converted to SO_3 and injected into the flue gas.

Sulfur burning—In this system, molten sulfur is pumped from a storage tank to the sulfur burner. Liquid sulfur is atomized with high-velocity air and completely burned to SO_2 in the combustion chamber. The effluent SO_2—air mixture flows out of the sulfur burner at about 1,600°F and, after cooling to 650°F in an air cooler, is converted catalytically in a one-stage bed of vanadium oxide to SO_3. Conversion efficiency is about 72%. The dilute SO_3 gas, at 1,120°F, is then transported to the precipitator distribution manifold for gas-conditioning purposes.

Table II shows an economic comparison of the processes.

In addition to economics, several other factors favor the sulfur-burning system over the others. The inert sulfur is easily handled and stored. Furthermore, the only location of a corrosive or hazardous problem is upstream of the catalytic converter.

The author

Robert Bump is product manager for industrial precipitators at Research Cottrell Inc., Box 750, Bound Brook, NJ 08805.
In this position, he analyzes customers' needs and provides technical backup. He has had 30 years of experience with precipitators and related equipment. Before joining Research Cottrell, he held several engineering positions in the field of air-quality control. He holds degrees in engineering and economics.

Electrostatic precipitators: how they are used in the CPI

The chemical process industries are a fertile ground
for precipitator applications. This article describes
some of the major ones, details typical operating conditions,
and tells how to identify and avoid situations that may adversely
affect precipitator performance.

Gilbert G. Schneider, Theodore I. Horzella, Jack Cooper and *Philip J. Striegl*, Enviro Energy Corp.

☐ In a previous article (see pp. 228-242), we explored
the many details one must be aware of in order to
select and specify precipitators intelligently. Going one
step further, we will now look into the various applications of precipitators in basic industries such as pulp
and paper, cement, and metallurgical refining. In each
case, we will attempt to identify some of the difficulties likely to arise, and how to avoid them.

The demands of air-pollution-control regulations
have brought gradual changes in the operation of the
above-mentioned industries. No one will be surprised
to hear that it is no longer possible to operate a plant at
its designed production rate if the stack effluent is not
within legal limits. Because of the precipitator's vital
role in controlling this type of pollution, the chemical
engineer must delve into each application carefully
when formulating a basic plant design.

Pulp and paper

The black-liquor recovery boiler, which gives up salt
cake in substantial amounts, has always been a source
of problems in kraft pulpmills. The resulting fumes,
laden with salt cake, are highly corrosive. Precipitators
handling the boiler offgas can be protected with
double-walled steel casings containing hot air. This
creates a "Thermos bottle" effect that minimizes corrosion.

The relatively recent emergence of the low-odor,
noncontact boiler poses an entirely different set of conditions for precipitators. So as to cut back on mercaptan odors, the boiler is not equipped with traditional
evaporators. This results in a high concentration of
high-resistivity dust fed to the precipitators, and higher
temperatures that produce a finely divided, tacky fume
having light bulk density.

For the same gas volume, such a boiler needs a
larger, more-carefully-engineered precipitator able to
work within the following conditions: gas volume, 300-

*To meet the authors, see p. 242.

Originally published August 18, 1975

Precipitator sitting atop boiler
in pulpmill saves space
and minimizes flue work

251

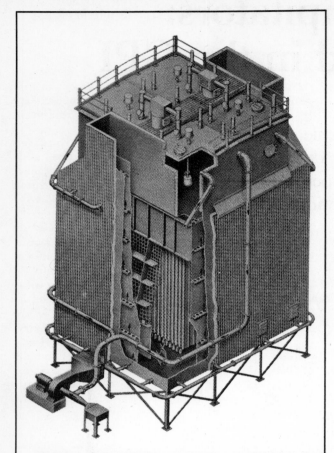

Double-walled shell through which warm air circulates protects precipitators from corrosive environments

450 ft³/(min)(ton)(d); temperature, 400-450°F; salt-cake concentration at intake, 8-12 grains/scf dry; salt-cake concentration at exit, 0.01-0.03 grains/scf dry.

In addition to using a more-conservative design, the following remedial measures are suggested: (1) employ drag conveyors in vertical-sided chambers instead of hoppers; (2) run conveyors transversely to gas flow; (3) install baffling to minimize sneakby; (4) use heavy impact rappers; (5) use blowers to keep dust moving; and (6) make model studies to ensure good gas distribution.

Precipitators can in some cases replace wet scrubbers and mechanical dust collectors in pulpmills. Effluent from the lime kilns, and from boilers can be handled effectively by precipitators.

Boilers that burn western coal need substantially larger precipitators than those using eastern (higher-sulfur) fuel. This is because of the ash-resistivity problems associated with western coal.

Sulfuric acid

In the production of acid from ore concentrates or pyrites roasting, the SO_2-laden flue gas is scrubbed down to a temperature of about 110°F and saturated. This produces small amounts of sulfuric acid mist, which must be removed before sending the gas on to the contact acid plant. The mist consists of finely divided droplets that flow freely upon precipitation.

Precipitators for this application differ from the flat-plate conventional design. A tubular pipe with a high-voltage wire centered in its middle is used as a collecting surface. Lead pipes, about 10 in dia. x 15 ft long, are commonly employed. The wires are of lead-covered steel, held taut by lead weights. The casing is generally of lead-strapped steel, with an upper and lower housing connected by the exposed tubes.

Operating conditions are: gas volumes of usually less than 100,000 ft³/min; temperature range, 95-125°F; inlet acid concentration, 100 mg/scf dry H_2SO_4; outlet concentration, less than 1 mg/scf dry H_2SO_4; efficiency, 99% +; pressure, 28-in wg below atmospheric.

It is possible to roughly estimate the number of tubes required, by assuming 100-150 ft³/(min)(pipe) for a 99% efficiency. The common practice is to employ a series/parallel orientation for the equipment—i.e., two units in series, two in parallel—so that the precipitators get two passes at mist removal.

Because of the structural problems involved, one seldom sees equipment with more than 260 pipes in a single casing. If more are required, the procedure calls for splitting the total number in half, and building two interconnecting smaller shells.

Precipitators made of lead-lined steel, with the tubes submerged in the gas stream, are now being used for mist removal. This arrangement equalizes the pressure inside and outside the pipes, and prevents their occasional collapse in situations of high-negative-pressure operation—e.g., in oil refineries. Precipitators of this type are equipped with water seals that will blow out in the event of a sudden surge of pressure.

Power generation

The three variables to consider here are: precipitator efficiency, fuel, and gas distribution.

Precipitator Efficiency—Time/cost curves in which efficiency requirements are plotted against the precipitator's purchase price are flat at the 92-97% level, but climb rapidly and become asymptotic above 99%. These days, few precipitators are specified at less than 99% removal; effluent residues must not surpass 0.02 grains/scf.

As a consequence, equipment with high-efficiency re-

Tubular array is used to precipitate mist in acid plants

quirements is derated to allow for the many operating variables it must handle while still meeting an efficiency of over 99%. This overkill syndrome explains why oversize units—e.g., 500 ft wide, 200 ft high and 80-100 ft long—are sometimes coupled with new boilers. The end-result is that cost considerations today are very different from those prevalent in the late 1960s.

Fuel—Only recently have the resistivity problems involved in the burning of low-sulfur (western) coals been widely recognized. These cause poor performance of "cold" precipitators (located downstream of the air heater).

There have been many efforts to come up with satisfactory answers for boilers that feed on a variety of coals—from using units big enough to collect almost anything, to adding SO_3 generators that are activated when the plant burns unfavorable coal.

The most popular remedy is to place the precipitator upstream of the air heater. Although the gas volume handled in this case is higher, dust precipitability obtained with these "hot" units is better, and cost considerations are consequently more favorable.

Still, data collected on the many hot precipitators in service indicate only reasonable predictability in operation. For example, there is some evidence that even at high inlet temperatures—e.g., 700°F—fly-ash chemistry is such that corona current formation is inhibited.

This points up the importance of building test units in which coal to be used is burned and precipitator performance predicted. To be sure, the results are not as reliable as in fullscale operation, but a good approximation can be reached in pilot plant facilities that are designed for handling 15,000-25,000 ft³/min at operating temperatures.

Collecting plates of pilot units should be fairly long (20 ft or longer). Three fields arranged in the direction of gas flow are enough to predict if performance will vary from field to field.

Gas distribution—As indicated in the previous article on precipitators, the best unit will function no better than the gas-flow distribution allows. Extensive problems could occur if the flow system is not carefully analyzed prior to installation. Model studies and past experience can signal obvious trouble areas to be avoided; suitable vane systems can then be designed for optimum gas distribution.

Cement

The first industrial use of precipitators was in the cement industry. There have been some changes in cement manufacturing over the years, but the calcination step and the equipment used in it remain basically the same.

Most of the early plants, built at a time of no pollution-control regulations, were conservatively sized and engineered. Consequently, the cement industry has over the years installed equipment that has performed better than specified over a variety of operating conditions.

In the U.S., most cement plants use either the wet or dry process. In the former, raw materials are ground wet and fed into a coal-, oil- or gas-fired rotary kiln for calcination into clinker. Dry processing involves dry-feed preparation, and introduction into the kiln as a powder or "raw meal." After calcination, the clinker is cooled and ground further.

Large amounts of dust arise from the rotary motion of the kiln, and there is also gas evolved by the feed. Dust loadings of 5-15 grains/scf and gas flows of 100,000-750,000 ft³/min are common. Precipitator-inlet temperatures for wet-process plants are about 450°F; for dry-process, about 650°F. As for removal efficiencies, many companies aim for a visually clear effluent, with residual concentrations of about 0.015 grains/scf.

Plants that use pelletized feed have very low dust loads (less than 2 grains/scf). But outlet-gas temperature is also low (about 250°F), and this can cause precipitator corrosion. Aluminum plates with coated steel shells are used for protection. The heavy insulation of flues, precipitator casing, hoppers and screw conveyors is also of value.

Other cement plants have preheater kilns, in which feed heats up as it falls through multiple-stage cyclones against a rising column of hot kiln gases. This type of installation has smaller kilns and smaller gas flows than conventional ones.

Part of the kiln gas can be routed through the dryers

Power plant precipitator is a typical double-deck unit

Units handling gases from rotary kilns should have CO measuring device for protection against blasts

for removing excess moisture in the feed. The moisture does not hurt precipitator performance. In fact, when handling dry kiln gas, the precipitator is usually preceded by a humidification tower that adjusts the moisture to the most favorable range (12-18% by volume).

The most common difficulties involved in operating precipitators in cement plants are:

■ Dust loading and transport. There may be large variations in the amount of dust entrained in gases leaving the kiln, which can strain the dust transport system. An analysis of the range of fluctuation of the dust concentration is a must, so that conveyors are sized to handle heavy dust loads arriving in slugs.

■ Moisture content. It is difficult to introduce moisture into dry kiln gases. Humidification towers should be designed so that they provide finely divided droplets that evaporate before the gases leave the unit.

■ Explosions. Kilns generate CO occasionally, and explosions are relatively common in the cement industry, as CO content builds up and moves into the precipitator's electrical field. Many plants have devices that measure CO and also deenergize the precipitator.

■ Sulfate deposition. At the right temperature, sublimation of sulfates can take place within the precipitator casing. The fumes precipitate on the plates, forming deposits that are difficult to remove, even with pneumatic hammers. This can be avoided by setting the inlet temperature so that sublimation takes place before the dust collection equipment. Precipitator design should be flexible enough to allow for sulfate deposition.

■ Bulk density. This can vary with certain raw materials. For instance, many Gulf Coast plants use oyster shells as a lime source. They produce a very light dust that is hard to dislodge. Some helpful techniques here include more-frequent rapping; use of larger equipment for the same gas flow; and isolation of chambers, with power off during rapping.

■ Moisture condensation. In wet-process plants, mixing cold air with wet gases causes corrosion problems, especially at points of entry around poorly sealed manhole doors, and in screw conveyors, inspection hatches and insulator compartments. When handling wet gases, it pays to install heavy insulation and make sure there is no sneakby.

■ Raw materials in dust. The outlet precipitator fields, which squeeze out the last few percentages of dust-collection efficiency, frequently trap dust that is rich in alkali content. The procedure here is to segregate the outlet hoppers and dump the collected material into a suitable disposal area.

Oil refining

Precipitators in refineries are used mainly to trap particles contained in exit gases from catalyst-regeneration systems. Regenerator offgas is usually cleaned by a series of multistage cyclones, but fine catalyst particles—all usually less than 10μ in size—succeed in escaping the cyclone system. Fine-dust concentration is about 1 grain/scf.

Gases leaving the top of the regenerator are generally at high pressure. One can either install the precipitator on the high-pressure side of the system, or reduce pressure by valving and operate the precipitator at about atmospheric pressure (but with enough positive pressure to move the gases through the rest of the system). The low-pressure approach makes for easier servicing.

Even at low pressures, there can be trouble at points where other equipment—e.g., rappers, hoppers—penetrate the precipitator shell. Tight seals to eliminate possible dust leakage are mandatory.

Catalyst dust does not present a precipitability problem. Some precipitators have had their share of resistivity problems, but these have been solved by injecting small amounts (20 ppm) of ammonia. The difficulty is thought to be associated with a low inlet temperature.

Normal operating conditions for the precipitator are: gas volumes as high as 500,000 ft³/min and inlet temperatures of 550-650°F.

Metallurgical refining

The production of nonferrous metals is one of the oldest applications for electrostatic precipitators. There are a host of problems to be dealt with here—high temperatures, cyclical operations, sublimation of salts and arsenic compounds, and corrosion.

All these are intensified by the fact that dust collected in the precipitator hoppers is a valuable product. In fact, at times the precipitator catch can contain up to 30-35% lead, zinc and arsenic. Since this inhibits precipitator performance, hoppers are usually purged when the lead or zinc concentration reaches a predetermined level. The purged material is sent to the lead or zinc smelter.

It is difficult to generalize the operating conditions of any particular smelter. It is not uncommon to find, for example, roasters and converters combined, with their effluent discharged to a sulfuric acid system. By the same token, reverberatory-furnace lines sometimes run into converter gas lines; and waste-heat boilers positioned after roasters and reverberatory furnaces are common. In short, gas-cleaning equipment must be tailored to each plant's particular mode of operation.

Copper production typically starts in a multiple-hearth roaster, although flash-, wedge- and fluid-bed roasters are also used. Roaster gas, rich in SO_2, is cleaned by a flat-plate precipitator. Normal operating conditions are: gas volume, 50,000 ft³/min; temperature, 450-550°F; and dust concentration, 1 grain/scf.

Calcined product from the roasters is further treated in a reverberatory furnace, which turns out copper matte. Reverberator gases are cooled by passing them through waste-heat boilers, and are then cleaned by plate-type precipitators. Operating conditions for this last unit are: gas volume, about 150 scf/(min)(ton of output)(d); temperature, 550-650°F; and dust concentration, 1-2 grains/scf.

Precipitators also handle the offgas from the purification of copper matte, which is done in converters. The cleanup proceeds under the following general conditions; gas volume, 80,000-150,000 ft³/min; temperature, 600-700°F; dust concentration, light.

Clean converter gas is often sent to a scrubber in preparation for sulfuric acid production. This calls for acid mist precipitators of the type described previously in our discussion of the sulfuric acid industry.

Improving electrostatic precipitator performance

The following two articles discuss ways to enhance precipitator performance. The first shows how tuning your present precipitator with a transmissometer will increase efficiency and reduce emissions. The second points out mechanical design features to look for when specifying a new precipitator.

Tuning electrostatic precipitators

Using a transmissometer to adjust operational variables can improve precipitator performance to meet EPA emissions requirements.

David I. Frenkel, Lear Siegler, Inc., Environmental Technology Division

☐ For many decades electrostatic precipitators have been collecting dust—in more ways than one. The inability to measure real-time emission levels has imposed a limit on collection efficiency. However, with the development of the transmissometer, immediate-response in-stack opacity monitoring is now possible. This allows step-by-step, trial-and-error experimentation for optimizing variable adjustments and trouble-shooting mechanical malfunctions.

A transmissometer consists of a light source and a photocell, arranged so the light passes through the gas stream before it is measured by the photocell. The change in light absorption caused by dust in the gas is a measure of its quantity, as it passes between source and photocell.

The EPA has recognized the effectiveness of opacity monitoring, and Table I lists EPA standards for such an instrument. Installed at the exit of the precipitator, the transmissometer arrangement shown in Fig. 1 is the only one that has consistently passed EPA requirements.

Before discussing transmissometer measurements, let us look at a typical precipitator pass (Fig. 2), and examine the effectiveness of stack sampling.

The basic principle of electrostatic precipitation is simple. A series of parallel plates with rows of wires between them carry high voltages of opposite polarities. As the particle-laden gas stream enters the unit, the particles are charged by the wires and are attracted to the plate. At controlled intervals, the plate is struck by a "rapper" (essentially a hammer or weight), which shakes the dust down the plate and into a hopper below. However, some dust is reentrained, and carried on to the next zone or out the stack, if it is already in zone 3 (Fig. 1).

Most precipitator acceptance tests and subsequent performance-improvement attempts use stack sampling to determine exit and entrance dust concentrations. But stack sampling is troubled by wide tolerances in accuracy and repeatability. Also, readings represent an integration of the emission levels over many minutes or hours, thereby precluding evaluation of reentrainment due to specific variables such as frequency, intensity and pattern of rapping, as well as changes in voltage level, spark rates and boiler or process load.

Without immediate response to changes in these variables, it is not possible to isolate their effects on emission levels. Trends are missed or misinterpreted, leading to incorrect adjustments.

Optical density

The transmissometer measures the optical density of the emissions stream. Relative changes in emissions can be observed directly from the spiking effect of these readings (Fig. 3).

Originally published June 19, 1978

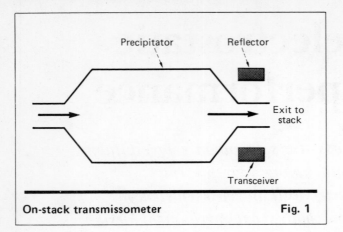

On-stack transmissometer   **Fig. 1**

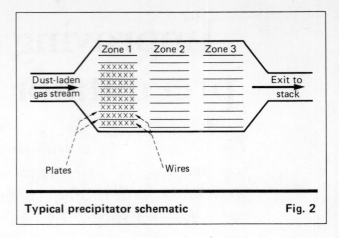

Typical precipitator schematic   **Fig. 2**

In order to correlate this information with output dust concentration, stack-sampling tests must be carried out concurrently with optical density recording. The developed correlations for a lignite-fired boiler and a bituminous-coal-fired boiler are shown in Fig. 4 and Fig. 5, respectively. As can be seen, a consistent linear relationship is established.

Having the correlations permits experimenting with the variables. The most important of these are:

1. Voltage and current levels
2. Rapping frequency
3. Rapping intensity
4. Rapping pattern
5. Mechanical condition
6. Flyash resistivity
7. Flue gas conditioning

There are many other variables involved, such as gas velocity, dust and gas distribution, and moisture content, but they will not be covered here.

Voltage and current

A common misconception is that the higher the power reading, the more effective the dust collection. This is not necessarily so. Excessive voltage levels can lead to sparking, at which time there is no dust-collecting electric field. Transmissometer data have shown that sparking actually reentrains dust, thus increasing emissions and energy consumption.

On some precipitators, a voltage level of 30 kV and 100 MA did a better job than 40 kV and 500 MA, which caused high spark rates. Power reduction led to a saving of 17 kW for one precipitator pass in a 300 MW utility plant burning bituminous coal. This came to a saving of 148,000 kWh/yr, plus far less maintenance on electrodes and other electrical equipment, which are stressed by high voltage levels. Correct voltage level adjustment can yield maximum collection efficiency at minimized operating power.

Additionally, under low boiler- or process-load con-

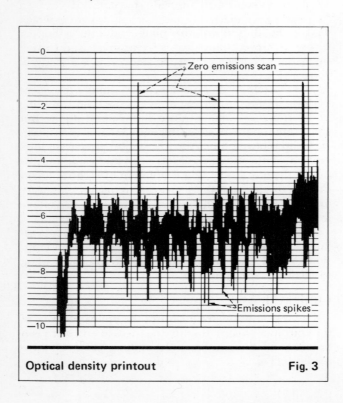

Optical density printout   **Fig. 3**

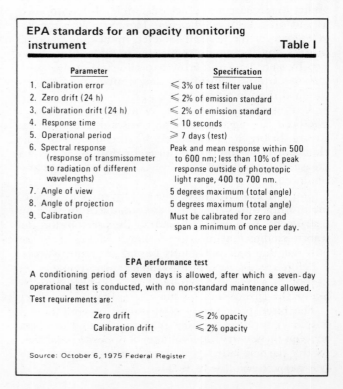

EPA standards for an opacity monitoring instrument   **Table I**

Parameter	Specification
1. Calibration error	$\leqslant$ 3% of test filter value
2. Zero drift (24 h)	$\leqslant$ 2% of emission standard
3. Calibration drift (24 h)	$\leqslant$ 2% of emission standard
4. Response time	$\leqslant$ 10 seconds
5. Operational period	$\geqslant$ 7 days (test)
6. Spectral response (response of transmissometer to radiation of different wavelengths)	Peak and mean response within 500 to 600 nm; less than 10% of peak response outside of phototopic light range, 400 to 700 nm.
7. Angle of view	5 degrees maximum (total angle)
8. Angle of projection	5 degrees maximum (total angle)
9. Calibration	Must be calibrated for zero and span a minimum of once per day.

EPA performance test

A conditioning period of seven days is allowed, after which a seven-day operational test is conducted, with no non-standard maintenance allowed. Test requirements are:

Zero drift	$\leqslant$ 2% opacity
Calibration drift	$\leqslant$ 2% opacity

Source: October 6, 1975 Federal Register

ditions, these levels can be further reduced and EPA standards still met, without power waste.

Rapping frequency

Both U.S. and European precipitators have convenient adjustments for varying rapping frequency, from intervals of minutes to hours. Every rap produces a certain amount of reentrained dust. Doubling the interval between raps does not necessarily double the dust reentrained, because accumulated dust will agglomerate. This larger particle mass lessens the probability of reentrainment.

Having transmissometer output linear with dust concentration helps determine the point where the benefit of increasing intervals between raps declines or reverses. This point is reached when plate dust-thickness brings plates and wires electrically too close, causing sparking, and in extreme cases, bridging. A change in boiler load or fuel composition will change this optimum point. Therefore different optimum frequencies are found for different fuel and load conditions.

Also, different frequencies are likely to apply for each zone. Since zone 1 tends to collect more larger particles than zone 2 or 3, it may need to be rapped more often.

In zone 3, the less frequent the rapping the better, because any reentrainment goes straight up the stack. Actually, improvements in emission levels have been achieved by merely lowering this zone's rapping frequency. Also, closing dampers at the stack exit during rapping in the rear of zone 3 improves results.

Rapping intensity

U.S.-designed precipitators have solenoid operated rappers that are convenient for changing the impact of the rapper on the plate. This is accomplished by varying the height to which the rapper is raised, or, with designs that use smaller mass and high velocity, by varying the accelerating-field size. In European designs, a fixed hammer on a rotating shaft provides the rap, and changing the rap intensity involves physically changing the rappers.

Rapping intensity can be varied on a zone-by-zone basis and will depend on the characteristics of the dust, as well as the rate of accumulation. Fuel and load changes can necessitate changes in intensity. The optimum setting is best determined by empirical tests under various conditions.

Rapping pattern

Optimizing the sequence of rapping is aided by observing the emissions reentrainment spikes.

Obviously, if all rappers are activated simultaneously, high reentrainment will occur, because there is no collecting surface active at the time. If a plate in zone 1 is to be rapped, it is better not to rap a plate in the same row in zone 2 or 3, or in adjoining, or nearby, rows.

The best settings must be experimentally determined on a unit-to-unit basis. Changes in dust characteristics may change the optimum setting.

Mechanical condition

Broken wires, damaged plates, malfunctioning rappers and other faults are often difficult to diagnose and

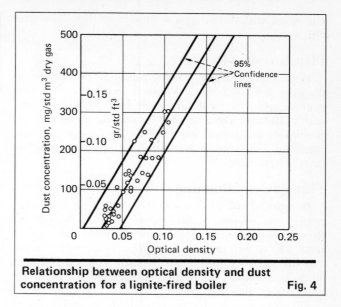

Relationship between optical density and dust concentration for a lignite-fired boiler   **Fig. 4**

isolate with the precipitator still on-line. In some instances, troubleshooting is difficult even when the equipment is off-line.

The precision transmissometer, by a process of elimination, allows the operator to locate faults. Real-time response and repeatability ensure that trends of deterioration are picked up early, thus helping to prevent catastrophic damage.

Excessive damage can be caused by acid attack, which gradually reduces plate areas and collection efficiency. By the time this problem was realized in one plant, the operator was able to walk through the holes in the plates.

Inoperative rappers are not easily detected externally. However, firing the rappers sequentially indicates rappers that do not produce a reentrainment spike, thus isolating the rapper or group of rappers that are not functioning. Similarly, damaged wires and plates can be detected by switching the power on and off in differ-

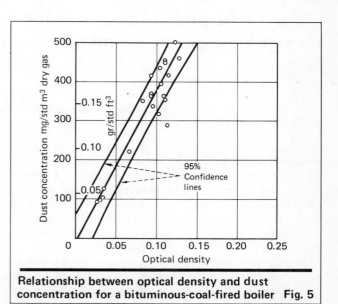

Relationship between optical density and dust concentration for a bituminous-coal-fired boiler   **Fig. 5**

ent zones, and measuring the impact on dust-collection efficiency.

Flyash resistivity

With greater emphasis being placed on coal burning, plus the development of Western coal and other new sources, resistivity will vary more frequently. Different sulfur and ash content will be encountered, and all previously mentioned adjusting tactics can provide tremendous assistance in reestablishing optimum precipitator performance.

Flue-gas conditioning

In order to lower flyash resistivity and emission levels, many precipitator owners have invested heavily in flue-gas conditioning. The chemicals used range in cost from $30 to $40 per ton for sulfur, to over $1,000 per ton for some exotic chemicals such as triethylamine. Plant investments for related equipment can be in excess of a million dollars.

How much of the chemical additive should be used? Many instances can be cited in which dosages set by manual stack-sampling data were reduced by two thirds when transmissometer readings became available. A 1600-MW utility plant had used five tons of ammonia per week, at a cost of $200 per ton, but transmissometer data halved this dosage. There are a number of instances in which the installation of gas conditioning was thought necessary, but the previously outlined tuning procedures reduced emission levels to the point where it was not.

It is highly likely that a complete precipitator control system will evolve, using transmissometer measurements and other parameters, plus a micro-processor to provide an automated supervisory function.

There are some limitations [1] to transmissometer applicability. Water droplets in the gas are measured as if they were particulate matter. Unless mist emissions are to be measured, efficient demisters must be employed after wet scrubbers, or a portion of the gas must be reheated prior to measurement in a by-pass duct.

Also, the instrument cannot correctly measure the opacity of emissions which form visible plumes after the gas exits from the stack (such as sulphuric acid mist or vapors which condense on contact with the cooler air).

References

1. Beutner, H. P., Measurement of Opacity and Particulate Emissions with an On-Stack Transmissometer, *APCA Journal*, Sept., 1974, p. 865.

The author

David I. Frenkel is marketing vice-president of the Environmental Technology Division of Lear Siegler, Inc., 74 Inverness Drive East, Englewood, CO 80110. He is responsible for all marketing activities, including new product planning, divisional planning, market research, sales support and international sales. Born in Australia, he received a B.S. in physics/math from the University of Melbourne.

Specifying mechanical design of electrostatic precipitators

Despite seventy years of development and research, design of electrostatic precipitators is still largely empirical and heavily dependent on vendor experience and reliability. Therefore, here are some key features for purchasers to consider when selecting a precipitator.

Gordon A. Lewandowski, New Jersey Institute of Technology

☐ Specifying an electrostatic precipitator (ESP) is not a straightforward procedure, because of the great complexity of ESP operation and the scarcity of precise design criteria that can be generally applied by purchasers. So many variables have an important effect on performance that neglect of any may cause the ESP to miss its design objective, resulting in months of exasperating effort to find and correct the flaw.

In addition, without precise guidance from purchasers, vendors are caught between the need for an "adequate" design and the fear of pricing themselves out of the picture. Therefore, purchasers should be as specific

as possible in detailing the options required for their particular application. At least with regard to these minimum requirements, vendor bids can be evaluated on an equal basis.

Although there are many different types of ESP, by far the most common is the parallel-plate, dry-wall type (Fig. 1); only this type will be examined here.

Gas flow

Flow maldistribution is a primary cause of poor ESP performance and efficiency. Local, high-turbulence eddies can scour particulate off the collecting plates, causing reentrainment. Once a flow maldistribution occurs, the situation continuously deteriorates, because low-velocity portions of the gas stream deposit solids in the upstream ductwork, making the velocity profile even more nonuniform.

Therefore, adjustable gas-flow modifiers should generally be included in the upstream (and sometimes downstream) ductwork. These modifiers are usually perforated plates or louvered vanes that smooth out the velocity profile in the ESP.

Flow maldistribution can be anticipated and prevented by constructing an exact scale model of the unit, particularly of the inlet and outlet duct configuration, in which velocity profiles are measured for a range of flow conditions. The model should be at least 1/16 full-scale, with flow modifiers included as needed. The goal is to obtain inlet and outlet velocity profiles in which the deviation of the peak from the mean, and the coefficient of variation (standard deviation/mean), are both less than 15% [1]. If possible, profiles should be taken at a Reynolds number of 10,000 and at the expected Reynolds number for full-scale design, where:

$$N_{\text{re}} = \left| \frac{DV\rho}{\mu} \right| = \text{Reynolds number}$$

D = plate spacing
ρ = gas density
μ = gas viscosity
V = average gas velocity between the plates

Given a uniform flow distribution, an average gas velocity of 4 ft/s, or less, in the full-scale design [2] should prevent excessive reentrainment.

Modeling studies are generally carried out by the vendor, and the resulting operating benefits can more than compensate for this added cost. In fact, correcting flow maldistributions in the field can cost ten times more than an initial modeling study [1].

Hoppers

One of the most common causes of ESP operating problems is overflowing hoppers. This not only causes dust reentrainment but also bridges the gap between discharge electrodes and the grounded frame, causing short circuiting, which burns out electrodes.

To avoid overflowing, hoppers should be oversized and have steep walls that exceed the angle of repose of the collected solids. Since moisture condensation in the hopper can lead to overflow, the walls should be heated

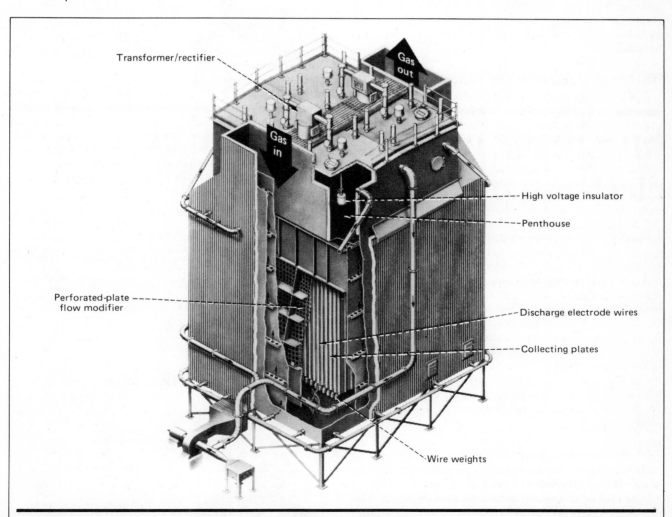

Transformer/rectifier

Gas out

Gas in

High voltage insulator

Penthouse

Perforated-plate flow modifier

Discharge electrode wires

Collecting plates

Wire weights

Parallel-plate, dry-wall electrostatic precipitator with heat jacket　　　　**Fig. 1**

(or at least insulated), and wall rappers provided. Furthermore, the solids discharge mechanism must be reliable and oversized for the expected collection rate. Emptying hoppers frequently reduces the possibility of overflow.

In addition, hoppers should be equipped with a reliable high-level alarm or with a good temperature indicator that senses the change from hot gas to cooler solids.

The integrity of the hopper shell must be carefully maintained to prevent air leakage into the hopper. In-leakage causes reentrainment of collected dust and poses a serious fire hazard if combustible materials, such as coke dust, are being collected.

Insulators

High-voltage insulators separate the discharge electrodes from the grounded frame (see Fig. 1). Cracked or short-circuited insulators will cause power loss and repair downtime. To minimize cracking, the insulators should be made of alumina instead of ceramic.

Short circuiting can be prevented by housing the insulators above the ESP in an area called the "penthouse," where a blower can provide a slight positive pressure. This prevents dust from leaking up from the main ESP body and depositing on the insulators. In addition, heating the insulators prevents moisture condensation.

Discharge electrodes and collecting plates

Discharge electrodes should be shrouded where they enter the main ESP housing. This involves placing each electrode inside a heavy-gage pipe, to prevent arcing between the electrode and grounded support frame.

Electrode wires must be centered between collecting plates, and the plates must be vertical, within a tolerance of $\pm\frac{1}{4}$ in. Off-center wires, warped plates, and plates out of alignment due to differential thermal expansion, seriously reduce ESP performance by limiting the maximum operating power. Therefore, all of the following are very important: (1) proper frame design to balance thermal expansion, (2) good quality control in plate manufacture, (3) adequate plate metal-gage, and stiffeners to prevent fatigue, (4) accurate placement of plates and wires, and (5) an experienced construction crew. The best efforts of the manufacturer to supply uniform, single-plane plates that are without warps or skews can be completely undone if a careless construction crew drops or otherwise damages the plates during unloading or erection.

Collecting plates must be manufactured with vertical baffles to reduce reentrainment of particulates. Baffle design and spacing is almost entirely a function of vendor experience.

ESP frame and housing

As mentioned, adequate consideration must be given to differential thermal expansion of electrodes and plate support frame to ensure continuous alignment. Spring-mounting the precipitator housing relieves vibrational stresses caused by the rapping mechanism. Also, insulating the entire shell minimizes internal condensation.

Rapping mechanism

European manufacturers generally specify tumbling hammer-type mechanisms for plate rapping. From an operating viewpoint, these appear to be less troublesome, although harder to adjust than electromagnetic or pneumatic devices, which can bind as a result of dust accumulation between the rapping rod and guide sleeve [2]. When hammers are used, they are normally attached in staggered fashion to a slowly rotating shaft, and only a small fraction of the collection surface is rapped at any one time. This is to avoid excessive reentrainment, or "puffing." Impact intensities greater than 60 times the force of gravity have been specified for some precipitators [3].

Choice of rapping frequency is also a matter of experience. However, the inlet sections normally collect the largest amount of solids and require a higher rapping frequency than the outlet sections.

In addition, vibrating discharge electrodes prevents dust accumulation, which reduces the effective electric field.

Miscellaneous

Allowance should be made for injection of conditioning agents into the inlet gas duct, even if such conditioning is not initially anticipated.

Where an explosion potential exists, blowout panels and an O_2 combustibles monitor should be provided as safety measures.

Finally, in order to assess ESP performance and satisfy regulatory requirements, sample ports must be located in the inlet and outlet ducts, with sampling platforms adequately sized and conveniently accessed for the large amount of equipment necessary (as specified by EPA methods [4]). Whenever possible, ports should be located in a straight run of pipe, eight stack diameters downstream and two upstream of bends and other flow disturbances.

References

1. Burton, C. L., and Smith, D. A., Precipitator Gas Flow Distribution, "Symposium on Electrostatic Precipitators for the Control of Fine Particles," EPA-650/2-75-016, Jan., 1975, pp. 191–217.
2. Greco, J., Electrostatic Precipitators—An Operator's View, "Proceedings of the Specialty Conference on Design, Operation and Maintenance of High Efficiency Particulate Control Equipment," Air Pollution Control Assn., Pittsburgh, Pa., 1973, pp. 53–69.
3. Maartmann, S., "Proceedings of the Second International Clean Air Congress," Paper EN-34F, Academic Press, New York, 1971.
4. Federal Register, Vol. 36, No. 247, Dec. 23, 1971, pp. 24882–24890.

The author

Gordon A. Lewandowski is assistant professor of chemical engineering at New Jersey Institute of Technology, 323 High Street, Newark NJ 07102. He had eight years of industrial experience with FMC Corp., Jacobs Engineering Group Inc. and Exxon Corp. Dr. Lewandowski received his B.S. and M.S. in chemical engineering from Polytechnic Institute of Brooklyn, and a D.Eng.Sc. in chemical engineering from Columbia University. He is a member of AIChE, and chairman of the North Jersey Section's Energy Committee. Also a member of ACS and the Air Pollution Control Assn., he has written articles on water pollution and particulate sampling.

Removal of Gaseous Pollutants

Equipment for controlling gaseous pollutants
Choosing a flue-gas desulfurization system
Diffusivities streamline wet scrubber design
Estimating acid dewpoints in stack gases
Automatic control of reagent feed boosts wet-scrubber efficiency
Dry scrubbing looms large in SO_2 cleanup plans
How to rate alloys for SO_2 scrubbers

Equipment for controlling gaseous pollutants

This article classifies available hardware according to the function
it performs, describes the most popular units of each type,
and, in some cases, provides basic design guidelines.

Rashmi Parekh, *Peabody Engineering Corp.*

☐ The sources of gaseous emissions in the CPI (chemical process industries) are extremely varied (see table). There is also a wide variety of equipment available to successfully clean polluted effluents to levels stipulated by environmental laws.

Partly because of this, the engineer may be confused by all the different approaches available for gas cleanup. One important factor to consider is that equipment designed for this task is highly specialized. The engineer must first analyze the physical and chemical properties of waste streams, such as concentration of pollutants, equilibrium solubility, heats of solution and reaction, and the concentration level to which the pollutant must be reduced. Only when the nature of the problem is fully understood can the engineer make an equipment selection with confidence.

Control of noxious gases in the CPI is generally done by absorbing, adsorbing or oxidizing the compounds before they reach the atmosphere. (Other techniques, such as process changes to minimize pollutants, or tall

Typical gaseous pollutants and their sources

Key element	Pollutant	Source
S	SO_2	Boiler flue gas
	SO_3	Sulfuric acid manufacture
	H_2S	Natural gas processing, sewage treatment, paper and pulp industry
	R-SH (mercaptans)	Petroleum refining, pulp and paper
N	NO, NO_2	Nitric acid manufacturing, high-temperature oxidation processes, nitration processes
	NH_3	Ammonia manufacturing
	Other basic N compounds, pyridines, amines	Sewage, rendering, pyridine base, solvent processes
Halogen:		
F	HF	Phosphate fertilizer, aluminum
	SiF_4	Ceramics, fertilizers
Cl	HCl	HCl mfg., PVC combustion, organic chlorination processes
	Cl_2	Chlorine manufacturing
C	Inorganic	
	CO	Incomplete combustion processes
	CO_2	Combustion processes (not generally considered a pollutant)
	Organic	
	Hydrocarbons — paraffins, olefins, and aromatics	Solvent operations, gasoline, petrochemical operations, solvents
	Oxygenated hydrocarbons — aldehydes, ketones, alcohols, phenols, and oxides	Partial oxidation processes, surface coating operations, petroleum processing, plastics, ethylene oxide
	Chlorinated solvents	Dry-cleaning, degreasing operations

Originally published October 6, 1975

stacks to dilute the gases, will not be covered in this article.) As such, available equipment can be classified according to the function it performs.

Wet absorbers (scrubbers)

There is a large number of devices for promoting gas absorption based on either dispersed-liquid or dispersed-gas phase.

Gas-liquid absorption processes are usually carried out in vertical countercurrent flow through packed, plate, or spray towers. The equipment falls into the following categories:

1. Dispersed gas (gas bubbles): impingement plate scrubbers
2. Dispersed liquid (film): packed towers, wetted-wall towers
3. Dispersed liquid (drops): spray towers, venturi scrubbers

For absorption of gaseous streams, good liquid-gas contact is essential. It is partly a function of proper equipment selection. Virtually any efficiency can be reached, if one is willing to pay for it in capital and operating costs.

Optimizing performance is also important. The power consumption of a modern, high-energy scrubber at its peak can be considerable because of the high pressure drop involved. Consideration should be given to using scrubbers whose pressure drop can be readily controlled while they are operating, either to deal with varying gas rates, or to maintain the efficiency at optimum value for given conditions.

Scrubbers can usually be classified by type and group according to operating principles. Some are primarily designed for particulate collection (and are covered elsewhere in this Deskbook issue), others for gaseous collection. Some accomplish both jobs simultaneously, and can heat or cool the gas stream as well.

Choosing the right scrubber for the job requires an understanding of process requirements, an analysis of the relationship between collection efficiency and cost, and an appreciation of alternative methods. Following are brief descriptions of the most common types employed as gaseous collection devices.

Packed-bed scrubbers—The conventional packed tower, familiar to most chemical engineers, is an empty shell filled with commercially available packing materials. Liquid flows over the packing in a film, and exposes a large surface area to the gas flowing upward through the interstices.

An important consideration in designing gas-absorption columns is the determination of the right liquor flowrate and the amount of exposed surface necessary to accomplish the desired degree of absorption. Tower design is based on determining the necessary volume or number of equilibrium trays and then determining, from the known gas and liquid rates, the minimum tower diameter capable of handling these flowrates.

Equations to determine the amount of packing or the number of theoretical trays can be found in most chemical engineering handbooks and will not be examined here. Actually, although the science of scrubbers has developed considerably over the past 30 years, much of the work is still empirical in nature. The best results are often obtained with units developed through years of trial-and-error experience.

A recent article published in *Chemical Engineering* (Apr. 14, 1975, pp. 70–76), which develops design data for packed towers and investigates specific types of packing materials, underscores the empirical approach.

The author, John S. Eckert, observes: "The design of absorption and regeneration towers always presents a challenge to the engineer because of the many variables . . . The complexity of the absorption/regeneration process is such that it does not lend itself to design by

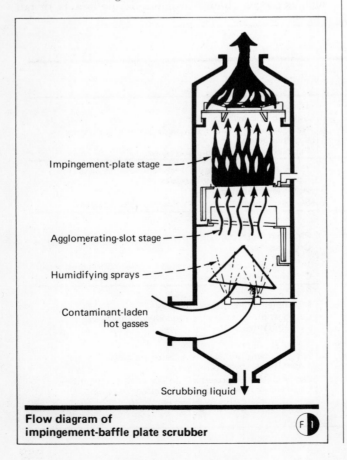

Flow diagram of impingement-baffle plate scrubber   F 1

Impingement-plate stage

Agglomerating-slot stage

Humidifying sprays

Contaminant-laden hot gasses

Scrubbing liquid

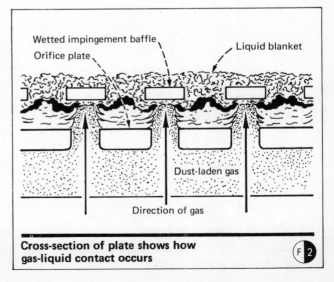

Cross-section of plate shows how gas-liquid contact occurs   F 2

Wetted impingement baffle

Orifice plate

Liquid blanket

Dust-laden gas

Direction of gas

either the concepts of theoretical plates or transfer units, because these will change both in number required and in height within the same tower and over the same packed bed. Because of this complexity, it is necessary to invent a 'variable constant' to arrive at a design procedure—so we have the mass-transfer coefficient, or $K_g a$, which is:

$$K_g a = \frac{N}{V \ln \Delta p} \qquad (1)$$

$$\ln \Delta p = \frac{P_1 - P_2}{\ln (P_1/P_2)} \qquad (2)$$

where $K_g a$ = mass-transfer coefficient, gas lb-moles/(h)(ft^3)(atm)

N = number of lb/moles of gas transferred/h

P_1, P_2 = net partial pressures, atm

$\ln \Delta p$ = natural log-mean of net drive between bottom and top of bed, atm

V = volume of bed, ft^3

$K_g a$ may be approximated using concepts of diffusional mass transfer and reaction kinetics, but coefficients in common use today have been taken from operating equipment, either in plant or engineering laboratory." The author provides a table of mass-transfer coefficients for various common packings.

$K_g a$ may then be used in the following equation to find the height of packing for a specified performance.

$$h = \frac{G_m}{K_g a P} \ln \Delta p \qquad (3)$$

where h = packed depth, ft

G_m = moles-in-gas (mass) velocity based on tower cross section, lb-mole/(h)(ft^2)

P = total pressure, atm

Spray towers—Scrubbing liquid is atomized into droplets to capture gaseous pollutants. These units have a low pressure drop and a high liquid rate, and are generally the least expensive type of scrubbing equipment. Pressure drops are in the range of 2–4 in wg; operating costs are low.

Scaling and plugging, which can be a problem in certain processes using scrubbers—e.g., limestone-slurry removal of sulfur dioxide from power-plant flue gases—don't present difficulties when a spray tower is used in a chemically balanced system. Such units need very high liquid rates to increase gas absorption efficiency, and there is very little possibility of forming highly supersaturated solutions, with attendant scaling problems. Spray towers also boast further advantages in handling large volumes of flue gases. They have the highest turndown ratio and the lowest pressure drop of all alternatives. Further, they have minimal internal surfaces on which scaling can deposit.

Ejector venturis—High-pressure spray collects gaseous pollutants in these devices, and moves the gas stream. Collection is affected by the co-current flow of the gas-liquid. Energy consumption per unit volume is high because of scrubber-liquor pumping costs. These units are often used as the first stage of more complex systems.

Venturi scrubbers—Gas and liquid are atomized in a moving gas stream. Venturi-type scrubbers use converging-diverging sections, with the liquid-gas contact at a maximum in the venturi throat. For gaseous collection, the pressure drop is in the range of 10–25 in wg. Collection efficiency for particulates increases with pressure drop, so these units often operate at 10–60 in wg.

Adjustable venturis can handle varying gas volume at constant pressure drop, or a varying pressure drop at constant gas volume. They are designed for maximum pressure drop and gas volume.

Plate scrubbers—These are vertical towers divided by plates of various configurations. As the contaminated gas passes upward through orifices in the plates, scrubbing liquor flows down across them to produce the desired gas-liquid contact.

Plate scrubbers are named for the type of plates they contain—bubble caps, sieve plates, impingement plates, etc. The units have been widely used for mass transfer, and have the ability to remove gaseous pollutants to any desired concentration if enough plates are employed.

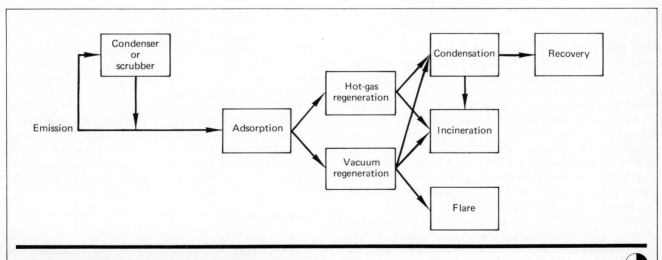

Pollution control is based on activated-carbon unit

F 3

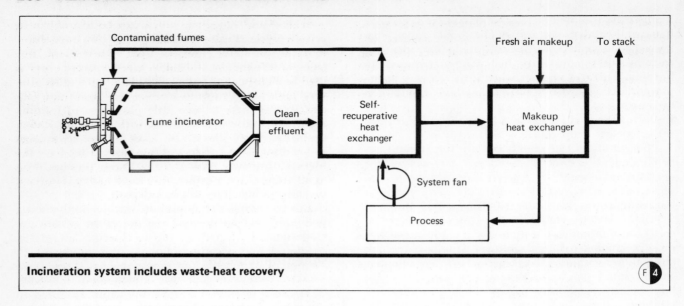

Incineration system includes waste-heat recovery   F 4

Scrubbers and corrosion

A major factor in the successful operation of a scrubber is the proper selection of materials of construction. The units are subjected to harsh environments created by industrial gases such as SO_2, SO_3, HCl, HF, and chlorine, which form sulfurous, sulfuric, hydrochloric, and hydrofluoric acids as they enter the liquid phase. In addition, chlorides and fluorides may be present.

Most high-efficiency scrubbers quench the hot inlet gas to its saturation temperature, forming acids and other corrosive materials that can easily attack steel, stainless steel, and special-purpose alloys. The attack is accelerated by the presence of high inlet temperatures, abrasive solids, high gas velocities, and corrosive recycle scrubbing-liquors.

Materials selection must be based on specific application requirements. And here, past experience or pilot-plant testing is often the best guide. The most important factors to consider when selecting materials center on the nature of the gas stream to be treated—its temperature, pressure, chemical composition and velocity, and the nature and rate requirements of the scrubbing liquor.

Dry adsorbers

Activated carbon is the most important dry adsorbent, while molecular sieves have reached the commercial stage in several new applications in the gaseous emission control area. Alumina, bauxite, and silica gel are more-polar adsorbents that find use in drying operations and in liquid-phase adsorption processes.

Activated carbon has the ability to adsorb gaseous hydrocarbons, hydrogen sulfide, sulfur dioxide, mercury vapors, carbon monoxide, and certain odors from a few ppm to as much as 40% by volume. The material can remove compounds from very dilute streams and, through regeneration, produce highly concentrated streams. This ability can increase the value of many pollutants because, once concentrated, they can be used as fuels or incinerated with minimum amounts of auxiliary fuel.

By impregnating specific reactive chemicals over the internal surface area of granular carbon, various air pollutants that might otherwise be difficult to handle can be removed from gas streams. Activated carbon impregnated with sulfur, for example, is being used at several U.S. chlorine plants to remove mercury vapor from byproduct hydrogen. When the carbon is spent, mercury can be recovered by thermal oxidation in a retort. Metal-oxide-impregnated carbon has also been employed to remove hydrogen sulfide from gas streams that do not contain oxygen.

Carbon adsorption is generally carried out in large, horizontal fixed beds with depths of 3–25 ft. Ratings of these units vary from 2,000 to about 50,000 ft^3/min, and they are often equipped with blowers, condensers, separators, and controls. A typical installation usually consists of two carbon beds; one is onstream while the other is being regenerated. Continuous-adsorption units involve a constant cycling of a doughnut-shaped bed that turns past a loading point and then past a regeneration point.

Molecular sieves, like activated carbon, usually require two fixed-bed absorbers with sequenced valves for switching the beds from adsorption to regeneration. The key element in these systems is molecular sieve/adsorbent blends—synthetic crystalline metal aluminosilicates that are highly porous adsorbents for water, some liquids, and gases.

In the past few years molecular-sieve technology has received broader chemical industry use for controlling gaseous pollutants such as sulfur dioxide, nitrogen oxides and mercury. These systems have recently been used commercially to meet gaseous emission codes at a sulfuric acid plant, a nitric acid plant, and a chlorine caustic plant.

Incineration

Many processes produce gas streams that have no recovery value, so adsorption or absorption may not be economically warranted. Incineration may be the simplest route when the gas streams are combustible.

There are two methods in common use today: direct flame and catalytic oxidation. The former usually has lower capital-cost requirements, but higher operating costs because an auxiliary fuel is often required for burning. This cost is offset when heat recovery is employed. The catalytic approach has high capital cost because of expensive metal catalysts, but needs less fuel. Either method will provide a clean, odorless effluent if the exit-gas temperature is sufficiently high.

Flame incineration—Highly combustible streams can be eliminated by direct flame incineration without the need for auxiliary fuel. These streams can also be employed as a source of fuel to thermally incinerate other less combustible, contaminated effluents.

When gaseous pollutants are not concentrated enough to support combustion directly, additional fuel must be supplied to heat the fumes to a sufficiently high temperature with adequate reaction time for complete combustion. Fume incinerators may be designed on the basis of pilot runs in mobile test units. This approach establishes design guidelines, such as oxygen requirements, mixing turbulence, combustion temperature, residence time and materials of construction. Contaminated air usually serves as the combustion air. If oxygen is limited, supplemental combustion air is provided.

Sizing an incinerator is based on the gas flowrate at incineration temperature. Design volume is determined with the following simple equation when the desired residence time is known (usually 0.5 s with hydrocarbons):

$$V_d = \text{acfm}_d(t_r) \qquad (4)$$

where V_d = design volume, ft^3
  acfm_d = design gas flowrate, actual ft^3/min
  t_r = residence time, min

The four following factors can reduce the costs of operating a fume incinerator:

- Having an oxygen content of at least 16% by volume can ensure stable combustion without the need for an additional separate source of combustion air.
- Use of heat-recovery devices of the shell-and-tube type can serve to preheat contaminated air prior to incineration, thereby greatly reducing the heat input required.
- Burning combustible waste liquids through a center-fired waste gun can supplement fuel requirements while getting rid of unwanted liquid wastes. Complete conversion of the liquid with no subsequent air pollution is ensured by proper equipment design.
- The heat contained in the incinerator exhaust can be used to meet process requirements and/or plant heating needs.

Flame incinerators operate at a low pressure drop, and require little maintenance if properly engineered. The major operating cost is fuel required to sustain appropriate temperature levels.

Catalytic combustion—When additional-fuel cost for flame incineration of very-low-heat-value pollutants is not economically feasible—or heat recovery would result in only marginal savings—catalytic combustion can convert gaseous pollutants into harmless combustion products under reasonable reaction conditions. However, catalytic units are not adaptable to gas streams containing appreciable amounts of heavy metals, liquids, or solids. Chlorinated hydrocarbons, phosphorus, arsenic, lead, boron and zinc are particularly poisonous to most catalysts.

The combustion occurs on the catalyst. Most available ones use a platinum metal electrodeposited on a heat-resistant Ni-Cr alloy, platinum metal dispersed on a high-surface-area support, or a combination of these on a ceramic honeycomb. To maintain the catalyst in an active state and supply sufficient oxygen to the process for complete combustion of organic compounds, about 1% excess oxygen, based on stoichiometry, is required.

Some thoughts on costs

Each technique for removing pollutants from process gas streams is economically feasible under certain conditions. And each specific instance must be carefully analyzed before a committment is made to any type of approach.

It is difficult to keep abreast of rising equipment costs for the various processes. In general terms, adsorption is a high-initial-cost process, but may be warranted when a product is recovered for reuse. Capital costs for package units are $5–$13/(ft^3)(min) for carbon steel, and $6–$16/(ft^3)(min) for alloy construction. Installation costs may range from 20–50% of equipment cost because of the wide variety of gas contaminants and exhaust streams involved. In the case of absorption, mild-steel-packed towers cost on the order of $2–$6/(ft^3)(min) and installation costs may run from 40–150% of the equipment costs.

On a comparative basis, flame incineration can handle a variety of process gases with constant, high-removal efficiencies. The cost of these units is about $6–$14/(ft^3)(min) installed. If process loads fluctuate, the slow response time of the incinerator could be a disadvantage. And if catalyst poisons are present, it may not be feasible to use the catalytic approach. Since oxidation conducted over a catalyst happens at a lower temperature than thermal combustion, it is often possible to use less fuel. On the other hand, today's high fuel costs are fostering the use of heat-recovery systems, and of the combination of incineration with process-steam production to take advantage of the additional heat input from the waste stream.

The author

Rashmi Parekh is product manager, wet collectors, for Peabody Engineering Corp., 835 Hope St., Stamford, CT 06907, and has been with Peabody in various engineering functions since 1968. He received an M.S. degree in chemical engineering from the University of Colorado in 1967. In 1966, he completed undergraduate work at the University of Bombay, India, with a B.S. degree in chemical engineering.

Choosing a flue-gas desulfurization system

Practical guidelines are presented here on how to evaluate scrubber operation and efficiency, reagent recycling, solid-wastes disposal, and online availability of SO_2 removal systems for industrial boilers.

James E. McCarthy, Research-Cottrell, Inc.

☐ The large number of sulfur-dioxide-removal systems now in successful operation in the U.S., Europe and Japan demonstrate that sufficient technology is available to deal with most sources of SO_2 contamination of the atmosphere.

However, no single system or single chemical process can deal with every SO_2 problem. Fortunately, for every problem or combination of problems, it is almost certain that a system can be found that offers a solution.

Among the factors influencing choice of an SO_2-removal system:

■ Amortized capital costs of various systems.

■ Operating costs of various systems, including cost of reagents, power, labor, maintenance, etc.

■ Disposal of waste products either through recovery for reuse (in the scrubber itself or some nearby industrial process) or via disposal in an environmentally acceptable way.

■ Efficiency required of the system in terms of particulate removal and SO_2 removal. If efficiency requirements are high, some systems will automatically be eliminated from consideration.

■ Availability required of the system. If scrubber outage will force plant shutdown, a high-reliability, high-availability system must be chosen.

Choosing a system

It seems likely that U.S. standards, based on the concept of Best Available Control Technology (BACT), will soon call for the removal of approximately 85% of the SO_2 in stack emissions. Hence, there will probably be greater interest in high-efficiency processes that can meet this standard. In the past, simple systems designed to achieve only 50 to 60% removal have often been acceptable.

Hand-in-hand with stricter codes covering SO_2 emissions are more-rigid standards for particulate emissions. All wet scrubbers have some ability to remove particulates. Not all, however, can meet typical air-quality codes calling for emissions containing no more than 0.02 to 0.03 grains/standard ft³. The capabilities of wet

Modular scrubbers handle 10,000 to 270,000 scfm fluegas.

scrubbers vary with factors such as grain loading (which can be reduced by precleaning the gas stream with a mechanical collector or precipitator); pressure drop in the scrubber (which affects the cost of operating fans); and size of the particulate matter to be removed. Naturally, some flue-gas-desulfurization (FGD) systems have higher particulate-removal capacity than others.

FGD systems are often installed downstream of particulate-removal systems such as precipitators, filters or mechanical collectors. Removal of particulates before they reach the FGD scrubber, of course, reduces the load on the scrubber and minimizes the volume of wet waste to be disposed of.

Another factor influencing final choice of a system is the exact reagent, or range of reagents, to be used in the scrubbing process. For example, some systems require precise amounts of specific reagents containing no sub-

Originally published March 13, 1978

stantial impurities. Other systems can use a wide range of reagents, including alkaline waste streams from other plant processes, and are virtually immune to impurities. Such versatility is particularly attractive if the reagent sources are likely to be varied or intermittent.

Present status

The widespread need for industrial SO_2-removal systems is seen by the increased number of units in operation. In mid-1977, more than 60 such systems were in operation at industrial sites (as opposed to utilities) in the U.S., and represented about a dozen different SO_2-removal processes [1]. During 1978, the total will probably reach 100, or more. At least 10 additional installations are in the planning, construction or modification stage. Many different industries are being monitored for air-quality control. Hence, revisions, improvements and modified designs are constantly called for to satisfy the SO_2-removal needs dictated by various fuels, regions and industrial processes.

New environmental laws (The Clean Air, Water Pollution Control, Safe Drinking Water, and Resource Conservation and Recovery Acts) further complicate the issue. Not only must SO_2 be removed but the wastes generated during the removal process must be disposed of in such a way as to meet regulations.

Thus, two basic challenges face the designer and user of an SO_2-removal system: (1) efficient and reliable removal of SO_2 to prevent its reaching the atmosphere, and (2) provision of some means whereby the captured sulfur or its compounds can be recovered for use in an industrial process (e.g., manufacture of sulfuric acid), or disposed of in an environmentally acceptable way.

Let us now discuss the commercially available SO_2-removal systems that have been designed for industrial use, as an aid to evaluating them. We will include a discussion of some of their differences and similarities.

Flue-gas desulfurization

There are four types of FGD systems [2] available for industrial boilers fired with fossil fuel:

1. Large, self-contained FGD units that use a mixture of water and chemical reagent to generate solid wastes. Basically, these are similar to FGD systems installed at some electric utilities, but are smaller.

2. Self-contained units of moderate size that use a mixture of water and chemical reagent to generate a liquid waste for disposal.

3. Self-contained units that use water alone to generate a liquid waste from which SO_2 can be recovered.

4. FGD systems linked to a waste stream from some industrial process to permit recovery of water and/or the alkaline reagent present in that stream.

Solid-waste units

Solid-waste FGD systems are of particular interest to industries burning high-sulfur fuels in quantity, because the solid sludges formed by these systems are inherently easier to dispose of in an environmentally acceptable manner than are liquid wastes. Let us examine the principal features for three such units:

- Dilute double-alkali system (Zurn Industries Inc.).
- Concentrated double-alkali system (FMC Corp.).

- System that normally uses lime as a reagent but can also operate with other reagents, including limestone and various waste streams (Research-Cottrell/Bahco).

Solutions containing concentrations greater than 0.15 molar of active sodium ion (Na^+) are referred to as concentrated; those less than 0.15 molar, as dilute. Double-alkali systems are designed to protect against the possibility of forming scale inside the scrubber, where it could interfere with scrubber efficiency or even cause complete shutdown.

In the double-alkali processes, SO_2 is reacted to form only soluble compounds (e.g., sodium sulfite and sodium sulfate) in the scrubber proper. These compounds in solution are subsequently pumped to a holding tank where they are further reacted to form insoluble compounds such as calcium sulfite and calcium sulfate. By restricting the formation of the insoluble compounds to nonscrubber areas, rather than the scrubber proper, protection against scaling and plugging is obtained.

In the Research-Cottrell/Bahco system, formation of calcium sulfate and calcium sulfite is encouraged in the scrubber proper, providing a one-module process. Scaling and plugging are guarded against by intermittent flushing of the scrubber with a strong flow of clear water. This flow scours insoluble compounds that may form on scrubber surfaces and prevents supersaturation of the scrubbing liquor to minimize precipitation of solids in those areas.

All of these systems are now operating on coal-fired boilers in the U.S. The Research-Cottrell/Bahco system operates at twelve sites in Europe and Japan, on gas streams from oil-fired boilers, on incinerators, and with processes for sintering metals and for producing glass and paper products.

As shown in the accompanying table, other FGD systems are also operating on industrial boilers. These include five units at various plants of General Motors Corp. (each of different design), and six units mostly at paper plants, where sulfur compounds are sometimes recovered and reused in the production process.

Other flue-gas installations

At a textiles plant in Canton, Ga., the scrubber designed by FMC's Environmental Equipment Div. uses a highly alkaline waste stream from the dye house as the scrubbing liquor. This has the twin advantage of eliminating the need for special alkaline reagent and of neutralizing the waste stream that might otherwise present a disposal problem.

Another FMC installation is at an FMC-owned soda ash plant for treating the flue gas from a 150-MW generating facility. A third installation, for which little operating information is available, is at the Mossville, Ill., plant of Caterpillar Tractor Co.

Capital and operating costs

Cost comparisons among FGD systems are quite difficult. The only sure method is to work out the detailed cost estimates for alternative systems at a specific site, and to take into account local costs of land, reagent, water, waste-disposal requirements, probable installation costs, and so on.

Data for flue-gas-desulfurization systems in operation on industrial boilers

User/location	Supplier	Capacity MW	Capacity Gas volume	Waste-product disposal
Canton Textile Mills, Inc.[1] Canton, Ga.	FMC Environmental Equipment Div.	10	42,000 acfm	Na_2SO_3/Na_2SO_4 liquor to lined holding ponds
Caterpillar Tractor Co. Joliet, Ill.	Zurn Industries, Inc.	18[2]	103,500 acfm	More than 60% solids. $CaSO_4$ sludge pond-dried for landfill.
Caterpillar Tractor Co. Mossville, Ill.	FMC Environmental Equipment Div.	57[3]	240,000 acfm	$CaSO_3$ (more than 60%) sludge to landfill.
FMC Corp. (soda-ash plant) Green River, Wyo.	FMC Environmental Equipment Div.	150[3]	—	Liquor (Na_2SO_3/Na_2SO_4) to collection pond for future landfill.
GM/Delco Moraine Div. Dayton, Ohio	Entoleter, Inc.	24[2]	55,440 scfm	Onsite clarification and dewatering facility receives Na_2SO_3/Na_2SO_4 liquor at 35 gpm.
GM Chevrolet Parma Plant Parma, Ohio	GM Environmental Design; Koch Engineering Co., fabrication	32[3]	60,000 acfm	$CaSO_3/CaSO_4$ (more than 50% solids) sludge to drying pond and eventual landfill.
GM Truck & Coach Plant Pontiac, Mich.	Peabody Process Systems	26	55,000 scfm	Landfill receives dewatered 5% solids slurry.
GM Corp. St. Louis, Mo.	Combustion Equipment Associates, Inc.	25[2]	—	Na_2SO_3 is oxidized to Na_2SO_4, then pH is neutralized; eventually to city sewer.
GM (Chevrolet Motor Div.) Tonawanda, N.Y.	FMC Environmental Equipment Div.	32	146,800 acfm	Sanitary landfill receives flyash and dewatered Na_2SO_3/Na_2SO_4.
Georgia-Pacific Paper Co. Crossett, Ark.	Neptune AirPol, Inc.	96	320,000 acfm	pH neutralized for discharge to city sewer.
Great Southern Paper Co. Cedar Springs, Ga.	Neptune AirPol, Inc.	100[4]	700,000 acfm	Clarification, ponding, eventual discharge to river.
ITT Rayonier, Inc. Fernandina Beach, Fla.	Neptune AirPol, Inc.	120[3]	—	Clarification, ponding, eventual discharge to river.
Mead Corp. Stevenson, Ala.	Neptune AirPol, Inc.	60[2]	175,000 acfm	Pulp mill uses recovered solution as makeup to cooking liquor.
Nekoosa Edwards Paper Co. Ashdown, Ark.	Neptune AirPol, Inc.	52	55,000 lb/h steam	SO_2 recovered from Na_2SO_3/Na_2SO_4 for recycling in pulping process.
Rickenbacker Air Force Base Columbus, Ohio	Research-Cottrell/ Bahco	20	55,000 scfm	Lined pond receives unstabilized calcium sulfite/ calcium sulfate sludge.
St. Regis Paper Co. Cantonment, Fla.	Neptune AirPol, Inc.	60	175,000 acfm	Onsite sewage-treatment plant receives discharged sludge.

All scrubbers in this table use an aqueous scrubbing liquor with an alkaline reagent. All are capable of removing both SO_2 and particulates from flue gas that is generated by burning fossil fuels.

Capital costs have not been included because they are often incomplete. For example, some data include costs of site acquisition and preparation while others do not. Some data include developmental or construction cost while others cover only the cost of hardware.

Capacity ratings in MW (megawatt) equivalent.

All systems in this table are operating satisfactorily, and have availability rates (when reported) in the range of 90% to 100%.

1. This installation, in addition to removing SO_2 and particulates, uses a very-high-pH effluent from the dye house as a scrubbing liquor. It neutralizes that effluent in the acid environment of the scrubber, and thereby reduces disposal problems.
2. Total of two boilers.
3. Total of four boilers.
4. Total of two boilers and two scrubbers.

As a general guide, costs for self-contained solid-waste FGD systems can be estimated at about $7 to $13/ton of coal burned. This value includes operating costs as well as amortized capital costs. If an industrial waste stream can be used for the reagent, costs can drop sharply. In most systems, the cost of reagent amounts to 25 to 35% of operating expenses.

Looking at capital investment alone, it seems fair to say that industrial FGD systems can cost as little as $60/kW for an installation rated at 35 MW equivalent, or larger, or may cost as much as $160/kW for a small unit in the range of 5 MW.

In general, the cost of FGD systems over the next decade or more can be expected to drop—for at least four reasons:

1. Capital cost of industrial units is certain to decline as both suppliers and operators gain experience and confidence. Demand for duplicate units will lead to economies of modular design and shop fabrication, cutting R&D costs associated with custom-designed units.

2. Reagent cost is likely to decline as scrubber operators take advantage of waste streams from their own and nearby plants. The use of a process-waste stream as a reagent in scrubber operation not only eliminates the cost of purchasing another reagent but can also eliminate the cost of treating or otherwise disposing of the original waste stream.

3. Alkaline process-waste streams are commonly available in industries such as metals production, textiles and papermaking. The use of such streams not only cuts operating costs but also permits economies in the capital costs of FGD systems because it may be possible to eliminate equipment otherwise needed for storing and mixing purchased reagents.

4. Many FGD systems in the future are likely to be incorporated into the original design of an industrial plant, rather than retrofitted to existing plants. This should lead to lower overall costs.

Comparisons of scrubber processes

All scrubbers use substantial quantities of liquid, usually water mixed with some alkaline reagent. The most common reagents are compounds of calcium or sodium, which are available in quantity at reasonable cost. In some cases, waste streams from other processes can be used as scrubbing liquors.

In all scrubbers, these alkaline liquors are intimately mixed with SO_2 or SO_3 in the flue gases. Such mixing promotes the formation of sulfur-containing compounds that can be subsequently removed from the scrubber liquor and disposed of.

In most cases, the chemistry of SO_2 scrubbing is familiar and straightforward, so that very few problems arise.

Rather, the most common problems are: plugging and scaling within the scrubber; possibility of corrosion, which shortens the life of the scrubber; and disposal of sulfur-bearing waste products. These problems must be carefully considered when selecting an FGD system.

Plugging and scaling

The formation of insoluble compounds such as calcium sulfate or calcium sulfite within a scrubber interferes with the flow of gas or scrubbing liquor. If such deposits become excessive, it may be necessary to shut down the scrubber and clean it out. Usually, the affected area is the mist eliminator, which prevents escape of entrained droplets from the scrubbed gases. Passages through these mist eliminators are usually narrow, and prone to plugging. Other areas vulnerable to deposits are the "wet-dry" zone, and any surfaces not thoroughly washed by scrubber liquor.

Several techniques are used to minimize scale formation. These are (a) constant flooding of the vulnerable areas, and (b) the use of a double-alkali system so that insoluble compounds do not form in vulnerable areas. Properly operated, either system is effective.

The basic advantage of dual-alkali processes is that they prevent the formation of scale inside the scrubber. These processes use an alkali solution that will form a soluble salt upon absorption of SO_2. The soluble salt is then regenerated with limestone or lime, with the resulting formation of an insoluble disposable byproduct.

The term dual-alkali can be applied to systems that scrub with solution containing ammonia, sodium salts or methylamine. However, the greatest amount of work has been done with solutions of soluble sodium salts.

When compared on an overall basis, limestone or lime scrubbers are essentially similar to dual-alkali scrubbers. Both consume limestone or lime and both produce a calcium sulfite/sulfate sludge. Some reports indicate that scaling can occur in dual-alkali scrubbing systems, with carbonate and gypsum scales forming in the absorber towers. Care must be taken to control process conditions to prevent scaling.

There were some early predictions that dual-alkali processes could use less-expensive materials of construction. This is not necessarily so, because the regeneration stage must still handle calcium sulfite/sulfate sludges as in a lime or limestone process. In addition, dual-alkali systems may be subjected to erosive effects if they are used to collect flyash.

Let us now describe the basic details of eight commercial sulfur-dioxide removal systems. These systems will be reviewed in alphabetical order by firm name.

Davy Powergas Inc.

This firm's system has a major advantage in enabling the reuse of sulfur recovered from gas streams. In some cases (including sulfuric acid or Claus plants), concentrated SO_2 is recycled directly back to the production process. In others, recovered SO_2 is converted to some other commercial form, such as elemental sulfur, liquid SO_2, or sulfuric acid.

Key to proper operation of the process is adequate pretreatment of the incoming gas. Flue gas from boilers must be cooled and cleaned of particulates before it enters the scrubber. At a Claus plant, tail gas must be incinerated to oxidize the various sulfur compounds to SO_2.

Depending on the particular installation, a single scrubber may be used, or multiple scrubbers may be installed leading to a central regeneration system. The reagent is Na_2SO_3, which combines in the scrubber with SO_2 and water to give $NaHSO_3$. Cleaned gas exits to the atmosphere while the sodium bisulfite solution passes through a surge tank to a heated, forced-circulation evaporator/crystallizer. In this recovery stage, the sodium bisulfite is split into its original components—sulfur dioxide, water and sodium sulfite. Sulfur dioxide is removed and diverted to a recovery system, while the sodium salt and water are recirculated to the scrubber.

Surge tanks permit the absorber to continue in operation while the regeneration section undergoes periodic maintenance.

FMC Corp., Environmental Equipment Div.

Essentially, the FMC concentrated dual-alkali system uses two loops. In the first, SO_2 in the flue gas is reacted with scrubber liquor, according to:

$$Na_2SO_3 + SO_2 + H_2O \rightarrow 2NaHSO_3$$

In the second loop, this compound is precipitated by calcium ion, according to:

$$2NaHSO_3 + Ca(OH)_2 \rightarrow CaSO_3 \cdot \tfrac{1}{2}H_2O\downarrow + Na_2SO_3 + 1\tfrac{1}{2}H_2O$$

If the SO_2-control process is preceded by a particulate collector, the system utilizes a low-energy disk-contactor scrubber. To achieve the required efficiency, multiple stages are used. If the system also collect particulates, a dual-throat venturi is used, followed by a cyclone mist-separator, and a meshed mist-eliminator. In either case, the same absorption chemistry applies.

Scrubbing liquor is recirculated, except for a bleed stream that is pumped to a regenerator tank in order to recover the sodium bisulfite by reacting it with slaked lime, according to:

$$2NaHSO_3 + Ca(OH)_2 \rightarrow CaSO_3 \cdot \tfrac{1}{2}H_2O\downarrow + 1\tfrac{1}{2}H_2O + Na_2SO_3$$

Residence time required for this reaction is low.

Regenerated sodium sulfite is now returned to the scrubber liquor, while calcium sulfite is fed to a thickener, where solids are increased to about 60%. Entrained sodium salt is minimized by washing.

As with the FMC dual-alkali system, the equipment components for the FMC sodium system depend on whether or not particulates are removed from the gas stream before it enters the scrubber. In either case, the chemistry in the scrubber is:

$$Na_2SO_3 + SO_2 + H_2O \rightarrow 2NaHSO_3$$

Most of the scrubbing liquor is recirculated. A portion is bled off and piped to a neutralization tank where soda ash (Na_2CO_3) is added to neutralize the sodium bisulfite. This prevents the release of sulfur dioxide in the disposal pond. The reaction is:

$$Na_2CO_3 + 2NaHSO_3 \rightarrow 2Na_2SO_3 + CO_2 + H_2O$$

The resulting sodium sulfite may be oxidized to sulfate in order to reduce chemical oxygen demand. Disposal may be in a lined pond.

Institut Français du Pétrole (IFP)

The IFP-1500 process, used primarily in Claus tail-gas cleaning, was developed in France. Gas containing a mixture of hydrogen sulfide and sulfur dioxide is injected into the bottom of a packed tower. No blower is necessary. The gas flows countercurrent to a low-vapor-pressure polyethylene glycol solvent containing a proprietary carboxylic-acid-salt catalyst in solution.

Through a series of reactions, the sulfurous gases are converted to elemental sulfur. Continuous injection of a small amount of catalyst maintains activity.

According to IFP, the system operates at 250° to 270°F—hot enough to maintain sulfur in its molten state but not so hot that large quantities of sulfur or glycol are driven off. Liquid sulfur accumulates in a sump and is continuously withdrawn.

IFP also states that the process is insensitive to changes in the gas flowrates and has operated successfully on flows as low as 30% of design level without adverse effect. The system is said to operate without maintenance for some two years, after which it is shut down for a wash with water to remove the accumulations of catalyst and sulfates from the packing.

Neptune AirPol, Inc.

One scrubber system offered by this firm for scrubbing stack gas for power boilers uses caustic or soda ash as a reagent. This system provides a choice of one or two stages. A single-stage venturi can achieve removal of 80 to 90% of the SO_2 present in the stack gas. A second (absorber) stage allows higher removal rates (to 99%, according to AirPol). The absorber section recycles caustic or soda ash with pH control, permitting recovery and recycling of the chemical reagent for maximum economy. If appropriate, sodium sulfate and sodium bisulfite are recycled to the pulping process in paper mills.

A second system offered by Neptune AirPol uses an adjustable-throat venturi through which flue gas is propelled by a booster fan. Intimate mixing of the gas and the recirculating scrubbing liquor promotes removal of both particulates and SO_2. This venturi is designed for an L/G (liquid to gas) ratio of 15 to 1. The basic chemical reaction is:

$$Na_2SO_3 + SO_2 + H_2O \rightarrow 2NaHSO_3$$

Cleaned gas passes through a mist eliminator to a stack for discharge to the atmosphere. Slurry falls to a sump, where a small bleed-stream is pumped to a regeneration area. Here, slaked lime is added to generate calcium sulfite and calcium sulfate. Next, a vacuum filter removes sulfite, sulfate and flyash for disposal, while the liquids undergo sodium-ion makeup and calcium-ion removal before being returned to the reservoir of recirculating liquor.

Peabody Process Systems

The Peabody Process Systems limestone-scrubbing system is in use at sulfuric acid plants in Florida to control sulfur dioxide emissions. Basically, the system consists of a number of spray towers, each having six spray levels, and a pumping system to recirculate spray liquor. Other major elements include an unlined settling pond and storage for limestone.

Dry flue gas (8 to 10% O_2) enters the scrubber at the top of a Peabody/Lurgi radial venturi, where it encounters slurry from a holding tank in which limestone has been dissolved. At the scrubber inlet, the SO_2 concentration is 1,000 to 3,000 ppm. Most particulates fall to the bottom of the venturi for removal, while partially cleaned gas passes to a second tank, where it moves upward from the bottom in countercurrent flow to sprays from a series of heads. The gas then goes through a radial-vane mist eliminator and a reheater before entering the stack.

All of the slurry returns to a holding tank, from which it is recirculated to the scrubber section. The recirculating limestone slurry is 70% calcium sulfite and

30% calcium sulfate, having a suspended-solids content of 15% by weight. A portion of the slurry is bled off to a nearby settling pond.

Primary problems have involved scaling and scrubber erosion. The erosion problem has been solved by relining the bottom quarter of the scrubbers to eliminate trouble from spray impingement.

Research-Cottrell, Inc.

A process developed in Sweden by Bahco and supplied in North America by Research-Cottrell uses an inverted-venturi construction that keeps pressure drop low and is capable of high efficiency for the removal of SO_2 and particulates. This process has been applied to flue gases from the combustion of coal, oil, municipal sludge, and other fuels. Lime, limestone, sodium salts, or alkaline-waste streams can be used as a reagent.

The process operates by countercurrent flow in two scrubber stages. Incoming hot flue gas enters at the bottom of the scrubber and is directed downward at the surface of the slurry, which is in a sump at the base of the tower. As the flue gas strikes the surface and changes direction, it creates a vigorous cascade of droplets that are swept upward into the first venturi chamber. Here, the gas and droplets are well mixed, the flue gas is cooled to its adiabatic saturation temperature, and SO_2 and particulates are scrubbed. The partially cleaned gas passes through an inertial drop collector, and then enters the second venturi stage, where the process is repeated. The cleaned gas passes through a mist eliminator and is then expelled to the atmosphere.

The alkaline solution is pumped from the reagent mixing tank to the top of the second-stage scrubber, through which it flows downward. A portion of the spent slurry containing solid particles of calcium sulfite, calcium sulfate and flyash is passed to a sludge-concentration and sludge-dewatering system. After more reagent is added, much of the reagent-rich water from this system is recycled to the scrubber. The concentrated sludge (about 65% solids) is relatively stable and is removed from the system for disposal as landfill.

Scaling and plugging are controlled by directing an intermittent stream of fresh water on the mist eliminator and the interior walls.

Removal rates for SO_2 and particulates of better than 98% can be achieved. Turndown capability of the system is unlimited because any desired quantity of makeup air can be drawn into the scrubber to compensate for low boiler loads.

Zurn Industries, Inc.

The Zurn dual-alkali system uses a dilute solution of sodium hydroxide as the scrubbing liquor. Reaction of the liquor with SO_2 produces soluble sodium sulfite/bisulfite/sulfate. After the reaction, this liquor passes to a regeneration subsystem. Here, sodium hydroxide is regenerated for eventual return to the scrubber system.

The addition of calcium hydroxide in a mix tank containing the reacted liquor produces insoluble calcium sulfite and calcium sulfate. These insoluble salts are, in turn, pumped to a thickener, from which they are removed in the form of a slurry. The slurry passes to rotary-drum filters where flyash and calcium sulfite/sulfate are removed. After washing to remove sodium compounds, the solids are disposed of as landfill.

Liquor recovered from the thickener and from the rotary-drum filters passes to a surge tank, from which it is pumped to a recirculating tank. Here, soda ash is added to clarify the liquid by further reducing calcium concentration by precipitating calcium carbonate, before the regenerated sodium hydroxide is returned to the scrubber.

This system is reported to have a removal efficiency of 90% and 95% for flyash and SO_2, respectively. Untreated sludge (65% solids) is relatively stable. The system must be relatively large because the circulating solution is dilute. Therefore, a relatively large volume of liquor must be circulated to achieve a high percentage of SO_2 removal.

A look at the future

Experience at industrial and utility installations in the U.S. has shown that the problems of SO_2 scrubbing can be overcome by applying basic principles of engineering. Waste-disposal problems have been largely solved, and work continues on methods of recycling the wastes. Erosion and corrosion, which seem to pose major difficulties have been overcome. The cost of reagent can be kept within reasonable limits. Scrubbers now operate routinely, with an availability close to 100%. Almost any current air-quality standard can be met at reasonable cost.

In the U.S., confidence in the availability and efficiency of SO_2-removal equipment has led to proposed standards that call for removal of 85%, or more, of SO_2 from industrial flue gas. Many installations now in service routinely achieve 90% removal or more and are capable of removing 99%. Other systems now operate at lower efficiencies but could achieve 85%, or more.

With increasing pressure on industry to switch to coal (often high-sulfur coal), it is inevitable that FGD systems will become common in the U.S. The scrubber industry is growing in size and maturity and it is clear that the day of the scrubber prototype has passed.

Bibliography

1. "Survey Report on SO_2 Control Systems for Nonutility Combustion and Process Sources," prepared for U.S. Environmental Protection Agency by PEDCo Environmental, Inc., Cincinnati, Ohio, May 1977.
2. Much of the information in this article is adapted from "Survey of the Application of Flue Gas Desulfurization Technology in the Industrial Sector," prepared for the Federal Energy Administration by Energy and Environmental Analysis, Inc., Arlington, Va.

The author

James E. McCarthy is manager of sales development for Research-Cottrell, P. O. Box 750, Bound Brook, NJ 08805. A specialist in SO_2 removal for industrial applications, he has been with the firm since 1972. Earlier, he held several industrial positions and worked as a consultant in solid-waste pollution and for a manufacturer of noise-pollution-control equipment. He has a B.S. in chemistry from St. Louis University, and an M.B.A. from the University of Buffalo.

Diffusivities streamline wet scrubber design

For eight common organic pollutants, the author gives data useful for estimating the height of a transfer unit (HTU) and the total column height.

Alex C. Mottola, U.S. Dept. of Agriculture

☐ In recent years, there has been growing concern over a number of organic air-pollutants common to industrial and food-processing operations [1,2,3]. Absorption via wet scrubbers is the usual means of abatement. In order to design such units effectively, it is valuable to have reliable data on the diffusivities of the gaseous contaminants. This article will present correlations for eight commonly encountered organic pollutants: methyl ethyl ketone, butyl acetate, allyl and ethyl sulfide, acetic acid, acrylonitrile, hexane, and isopropanol.

Packed-column design and diffusivity

The resistance to mass transfer in an absorber can be broken into two parts, a gas-phase resistance $(1/k_g)$ and a liquid-phase resistance $(1/k_l)$. The additivity of these film resistances may be expressed as [4]:

$$\frac{1}{K_g a} = \frac{1}{k_g a} + \frac{1}{H k_l a} = \frac{1}{H K_l a} \qquad (1)$$

Since it is physically impossible to measure solute concentrations at the gas-liquid interface, there is no direct way of measuring the individual gas-film coefficients (k_g) and the liquid-film coefficient (k_l). Also, because the "effective" interfacial area between the gas and liquid cannot be measured directly, the transfer rates can only be determined as "overall" coefficients. The resulting experimental value is the product of the solute flux and the total interfacial area $(K_g a$ or $K_l a)$.

Because of the complex nature of mass transfer, it is common practice for engineers to lump the many variables involved into dimensionless groups. These dimensionless groups greatly simplify the correlation of experimental data. In this article, we will cover only a few of the equations that were developed to represent experimental data for gas-film coefficients.

Van Krevelen and Hoftijzer [5] have derived for packed columns an expression for the dimensionless

Nomenclature

a	Interfacial area of packing/unit volume of tower, ft²/ft³
K_g	Overall mass transfer coefficient, lb-moles/(h)(ft²)(atm)
K_l	Overall mass transfer coefficient, lb-moles/(h)(ft²)(lb-mole)/ft³
k_g	Individual gas-film coefficient, lb-moles/(h)(ft²)(atm)
k_l	Individual liquid-film coefficient, lb-moles/(h)(ft²)(lb-mole)/ft³
H	Henry's Law constant
R	Gas constant, 82.06 (cm³)(atm)/(g-mole)(K)
T	Absolute temperature, K
L	Length of diffusion path, cm
$(\Delta P)_{lm}$	Natural log-mean partial pressure of contaminant in the gas film, mm Hg
D_{AB}	Molecular diffusivity, cm²/sec
μ	Viscosity, g/(cm)(s)
p, p_m	Density, g/cm³; molar density, g-moles/cm³
H_{OG}	Height of an overall gas phase transfer unit, ft

H_G	Height of an individual gas-film transfer unit, ft
H_l	Height of an individual liquid-film transfer unit, ft
L_M	Superficial molar liquid velocity lb-moles/(h)(ft²)
G	Superficial molar gas velocity, lb-moles/(h)(ft²)
G'	Superficial gas velocity, lbs/(h)(ft²)
e_o	Operating void in packed bed, ft³/ft³
d_s	Equivalent dia. of a sphere with the same surface area as a single particle, ft
P_A	Partial pressure of pollutant in air
P_B	Partial pressure of air
M_A	Molecular weight of pollutant
r^2	Correlation coefficient [17]
Sh	Sherwood number
Sc	Schmidt number
Re	Reynolds number
U	Time-averaged velocity of contaminant, ft/h

Subscripts

A	Contaminant
B	Air

Originally published December 19, 1977

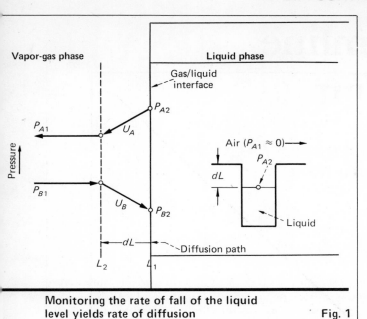

Monitoring the rate of fall of the liquid level yields rate of diffusion   **Fig. 1**

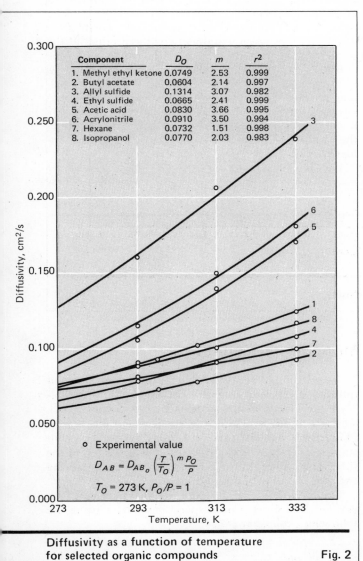

Component	D_O	m	r^2
1. Methyl ethyl ketone	0.0749	2.53	0.999
2. Butyl acetate	0.0604	2.14	0.997
3. Allyl sulfide	0.1314	3.07	0.982
4. Ethyl sulfide	0.0665	2.41	0.999
5. Acetic acid	0.0830	3.66	0.995
6. Acrylonitrile	0.0910	3.50	0.994
7. Hexane	0.0732	1.51	0.998
8. Isopropanol	0.0770	2.03	0.983

o   Experimental value

$$D_{AB} = D_{AB_o}\left(\frac{T}{T_O}\right)^m \frac{P_O}{P}$$

$$T_O = 273\,K,\ P_O/P = 1$$

Diffusivity as a function of temperature for selected organic compounds   **Fig. 2**

Sherwood number (Sh), which they hold to be valid over a wide range of Reynolds numbers (Re).

$$Sh = 0.2Re^{0.8}Sc^{0.33} \qquad (2)$$

where

$$Re = \frac{DU\rho}{\mu} \qquad (3)$$

$$Sh = \frac{k_g \Delta P_{lm} RTL}{PD_{AB}} \qquad (4)$$

$$Sc = \frac{\mu}{\rho D_{AB}} \qquad (5)$$

If the Reynolds number and the Schmidt number (which depends upon the diffusivity of component A into a medium B) are known, the Sherwood number can be calculated from Eq. (2). This number can then be used to estimate the gas-phase mass transfer coefficient, k_g, from Eq. (4), or from other equations that have been used to correlate the contacting efficiency of packing devices. This coefficient can be used to estimate the height of a transfer unit, and ultimately the column height.

Shulman and others [6,7,8] have established the nature of gas-phase coefficients and the interfacial areas for various-size Raschig rings and Berl saddles. Their work correlates the gas film coefficient as follows:

$$\frac{k_g(\Delta P)_{lm}}{G} \cdot (Sc)^{2/3} = 1.195\left[\frac{d_s G'}{\mu_G''(1 - e_o)}\right]^{-0.36} \qquad (6)$$

Further refinements of the Sherwood number were developed by Thoenes and Kramers [9], who showed that the transfer of solute between a fluid and a packed bed is controlled by different flow regimes. Experimental rate determinations for a single sphere in a regular packing have shown:

$$Sh = 1.26Re^{1/3}Sc^{1/3} + 0.054Re^{0.8}Sc^{0.4} + 0.8Re^{0.2} \qquad (7)$$

where the first term on the right represents the contribution due to laminar flow; the second term, the contribution due to turbulence; and the final term, the effect due to the stagnant film.

The widely used Colburn [10] method yields the total height of a countercurrent absorption column from the packing's height of a transfer unit. From Eq. (1), the overall height of a transfer unit (H_{OG}) can be derived in terms of the individual transfer units (H_G and H_l).

$$H_{OG} = H_G + H_l\left(\frac{M_A G}{L_m}\right) \qquad (8)$$

where

$$H_G = \frac{G}{k_g aP} \qquad (9)$$

and

$$H_l = \frac{L_m}{\rho_m k_l a} \qquad (10)$$

Eq. (8) is often used in practical engineering design work [4,11]. Mass transfer resistance is expressed in terms of an experimentally determined number which has the dimension of height only (H_{OG}).

Correlation values for diffusivity follow experimental results closely   Fig. 3

With this technique, diffusivities play an important role. For gas-phase controlled absorption operating under a given set of conditions, the experimental values of H_G and H_l can be corrected for different gas mixtures by assuming that the height of a transfer unit [4,12] is proportional to a function of the Schmidt number $(\mu/\rho D_{AB})^m$. Values for m have been determined to be 2/3 for the gas-phase film and 1/2 for the liquid-phase film [4].

The above discussion illustrates that the diffusivity coefficient is an important design parameter for mass transfer equipment such as wet scrubbers. For many air pollutants, diffusion coefficients are not available in the literature. The purpose of this article is to present diffusivity data for a few organic compounds known to be troublesome.

Measuring the diffusivity

Several methods are available for the measurement of diffusion coefficients in gases. The simplest, and probably most widely used, was developed by Stefan [4,12]. The test liquid is placed in a small glass tube, almost filling the tube. The tube and liquid are held at constant temperature, and air is passed over the top of the tube. The rate of diffusion is determined by measuring the rate of fall of the liquid level.

The coefficient of diffusion as defined by Fick's Law can be determined from experimental data using the following expression [12]:

$$D_{AB} = \frac{RT}{P \ln (P_{B_2}/P_{B_1})} \cdot \frac{\rho_A}{M_A} \cdot \frac{L_2^2 - L_1^2}{2\theta} \quad (11)$$

Diffusivity estimates

The following experimental data for methyl ethyl ketone illustrate the correlation technique used to determine diffusivities.

$T = 20°C$ (293K) $\theta = 65.33$ h
$L_1 = 4.55$ cm $P = 754.9$ mm Hg (0.993 atm)
$L_2 = 6.01$ cm

In the absence of experimental values, the Clausius-Clapeyron equation was used to evaluate the vapor pressure of all components. From the literature [14,15]:

$\rho_A = 0.805$ g/cu cm at 20°C
$\Delta H_v = 8149.5$ cal/g-mole
$T = 25°C$ (298K); $P = 90.6$ mm Hg
$M_A = 72.12$ g/g-mole

Using the Clausius-Clapeyron equation, $P_{A_2} = 71.6$ mm Hg. Substituting into Eq. (12) and converting the time interval from hours to seconds:

$$D_{AB} = \frac{RT}{P \ln\left[\dfrac{P_{B_2}}{P_{B_1}}\right]} \cdot \frac{\rho_A}{M_A} \cdot \frac{L_2{}^2 - L_1{}^2}{7,200\theta}$$

$$D_{AB} = 0.0889 \text{ cm}^2/\text{s}$$

In Fig. 1, L_1 is the position of the liquid surface, L_2 the top of the film thickness. The gentle stream of air across the tube removes the vapor, thereby enhancing the vapor pressure gradient $(P_{A_2} - P_{A_1})$.

The variation of D_{AB} with respect to temperature (T) and pressure (P) on the system can be expressed by the empirical Loschmidt-von Obermayer equation [13], where:

$$D_{AB} = D_{AB_0}\left(\frac{T}{T_0}\right)^m \frac{P_0}{P} \tag{12}$$

Analytical Procedure

The diffusion tubes used in these experiments were made from true bore (2.64 mm) tubing with an expanded bulb section capacity of about 5 ml. The test chemical was placed in the bulb so that the surface of the liquid extended into the straight section of the tube. The tube with its contents was immersed in a circulating hot-water bath.

At least three temperatures (20, 40, 60°C) were used, and all were maintained within ±0.1°C. The rate of evaporation was monitored with a cathotometer. The barometric pressure was recorded with each liquid level reading. A schematic representation of the measurements recorded during the experiment is shown in Fig. 1.

Results show good fit

For the eight pollutants, Fig. 2 plots diffusivity in units of cm² per second as a function of temperature, using Eq. (12). Experimental values are given over the temperature range of 273 to 333 K at atmospheric pressure. The reference temperature (T_o) in the Loschmidt-von Obermayer equation is 273 K.

As can be seen from Fig. 2, the Loschmidt-von Obermayer equation provides a good empirical fit for the measured values of diffusivity. Correlation coefficients (r^2) calculated for the curves [17] are quite high, and close to one.

In Fig. 3, diffusivities obtained by experiment are compared with diffusivities calculated from the correlations for the various contaminants. Observed and predicted values are very close over the range of the correlations. The greatest departure of computed diffusivities from experimental ones occured with allyl sulfide, which shows a 3% maximum deviation. This deviation may have resulted from dimerization of the unsaturated molecule.

Acknowledgement

The author is indebted to Dr. Zoltan A. Schelly, University of Texas at Arlington, for many helpful discussions on this work.

References

1. "National Goals in Air Pollution Research—Report of the Surgeon General's Ad Hoc Task Group on Air Pollution Research Goals," U.S. Dept. of Health, Education and Welfare, PHS Publication 804, Aug. 1960, p. 23.
2. "Air Pollution from Alfalfa Dehydrating Mills," U.S. Dept. of Health, Education and Welfare, Bureau of State Services, 1960.
3. "Report on Interstate Air Pollution in the Selbyville, Delaware–Bishop, Maryland Area," U.S. Dept. of Health, Education and Welfare, Div. of Air Pollution Abatement Branch, Nov. 1965.
4. Sherwood, T. K., and Pigford, R. L., "Absorption and Extraction," McGraw Hill Book Co., 1952, pp. 16, 56, 131, 287.
5. Van Krevelen, D. W., and Hoftijzer, P. J., *Chem. Eng., Progr.*, Vol. 44, 1948, p. 529.
6. Shulman, H. L., and DeGouff, J. J. Jr., *Ind. Eng. Chem.*, Vol. 44, 1952, p. 1915.
7. Shulman, H. L., Ullrich, C. F., Proulx, A. Z., and Zimmerman, J. O., *A.I.Ch.E.J.*, Vol. 1, 1955, p. 253.
8. Zimmerman, J. O., Master of Chemical Engineering Thesis, Clarkson College of Technology, 1953.
9. Thoenes, D. Jr., and Kramers, H., *Chem. Eng. Sci.*, Vol. 8, 1958, p. 271.
10. Colburn, A. P., *Trans. Am. Inst. Chem. Eng.*, Vol. 35, 1939, p. 211.
11. Perry, J. H., "Chemical Engineers' Handbook," McGraw Hill Book Co., 3rd ed., 1950, p. 687.
12. Norman, W. S., "Absorption, Distillation and Cooling Towers," John Wiley and Sons, Inc., 1961, p. 21.
13. Quinn, E. L., and Jones, C. L., "Carbon Dioxide," Reinhold Publishing Corp., Stamford, Conn., 1936, p. 45.
14. Weast, R. C., "Handbook of Chemistry and Physics," Chemical Rubber Co., 52nd ed., 1972, pp. C-222, D-155.
15. Riddick, J. A., and Bunger, W. B., "Organic Solvents–Physical Properties and Methods of Purification," Wiley-Interscience, Vol. 2, 1970, p. 243.
16. Burgwald, T. A., I.I.T. Research Institute, Project No. C8172, Chicago, Ill., Jan. 1971.
17. Volk, William, "Applied Statistics for Engineers," 2nd ed., McGraw-Hill, New York, 1969, p. 266.

The author

Alex C. Mottola is a research chemical engineer serving as project leader in the U.S. Dept. of Agriculture's Environmental Engineering Laboratory, P.O. Box 5677, Athens, GA 30604. Currently involved in air-pollution control research, he received the USDA Superior Service Award for his part in developing an alfalfa wet-fractionation process. A member of the AIChE, he holds a B.S. in chemistry from Syracuse University and a B.S. in chemical engineering from the University of New Mexico.

Estimating acid dewpoints in stack gases

Finding the dewpoints of sulfuric acid is critical
to the design and operation of stacks handling
combustion gases. Here is a summary of data
and a new correlation for estimating such dewpoints.

Robert R. Pierce, Pennwalt Corp.

☐ All fossil fuels contain a certain quantity of sulfur—e.g., natural gas, as much as 50% hydrogen sulfide
(which is removed); and organic matter in sewage
sludge, ½–1%. When such fuels are burned, the sulfur is
oxidized to sulfur dioxide. While sulfur dioxide is recognized as a pollutant, it is not in itself a source of
trouble in furnaces, boilers and chimneys. But when
sulfur dioxide is further oxidized to sulfur trioxide,
material problems may develop.

Sulfur trioxide and water have a tremendous affinity
for each other; when temperatures are lowered to the
dewpoint, the two combine rapidly—in less than one
second—to form sulfuric acid. The sulfuric acid molecules also have a powerful affinity for water, so that the
concentrations of sulfuric acid occurring at elevated
dewpoint temperatures are very corrosive to steel and
almost all plastics, as well as hydraulic cement composites such as concrete, castables, gunites, and mortar.

Sulfuric acid solutions have low vapor pressures and,
consequently, high boiling temperatures. A 98.2% concentration of sulfuric acid forms a constant-boiling
azeotrope at 333°C (631°F). As the concentration is
reduced, the proportion of sulfuric acid in the vapor
drops drastically to as little as 1% by weight in the
vapor phase for 85% acid in the liquid phase. Therefore,
1% sulfuric acid in a gas has the corrosive properties of
85% sulfuric acid in the liquid. At a liquid concentration of 70% by weight, which liquid has a normal
boiling point of 165°C (329°F), the concentration in
the vapor phase drops to about 0.01%, so that this acts
as corrosively as 70% liquid sulfuric acid at 165°C.

So, only a very small amount of sulfur trioxide in
combustion gas is required to draw water from the gas
and form a fairly concentrated acid that has a high
boiling-point temperature even under negative pressure.

For example, if a combustion gas containing 10%
water vapor by volume also contains as little as 40 ppm
sulfur trioxide by volume, that gas has 0.22 weight
percent of sulfuric acid (based on sulfur trioxide and

water alone). The liquid phase in equilibrium with this
gas has a sulfuric acid concentration of 82.5% by weight
at a boiling point of 148°C (298°F) and 0.1 atmospheric pressure. As soon as the temperature of such a
mixture drops below 148°C, liquid must form; and this
powerfully acidic liquid will condense on any solid
surface that is cool enough.

Further, if the gas is cooled below this dewpoint by
radiation, etc., a mist of acid droplets is formed. That is,
a gas containing a mixture corresponding to only 0.22%
by weight of sulfuric acid will form a droplet or film
containing 82.5% by weight of sulfuric acid. Of course,
only a very small amount of the water in the gas is
condensed out with the acid.

Equilibrium data available

Because of its importance to the power generation
industry, the subject of acid condensation has received a
great deal of attention over the years. The classic study
by J. S. Thomas and W. F. Barker in 1925 [1] on the
vapor pressures of sulfuric acid solutions in the 89–99%
range is still quoted in the handbooks, and for good
reason. However, experimental difficulties have delayed
development of accurate data on more-dilute solutions.
F. H. Verhoff and J. T. Banchero have recently made a
critical review of the available data and developed an
equation from which the dewpoint can be calculated
when the percentage of water vapor and sulfur trioxide
in the gas is known [2]. The agreement of their equation with some of the better experimental work is remarkably good, the maximum deviation being on the
order of 7°C. Transposed, the equation is:

$$\frac{1,000}{T_{DP}} = 1.7842 + 0.0269 \log P_{H_2O} -$$
$$0.1029 \log P_{SO_3} + 0.0329 \log P_{H_2O} \log P_{SO_3}$$

where T_{DP} is the dewpoint in °K (273 + °C) and the
pressures (P) are in atmospheres.

As an example, in a gas at 760 mm pressure, con-

Originally published April 11, 1977

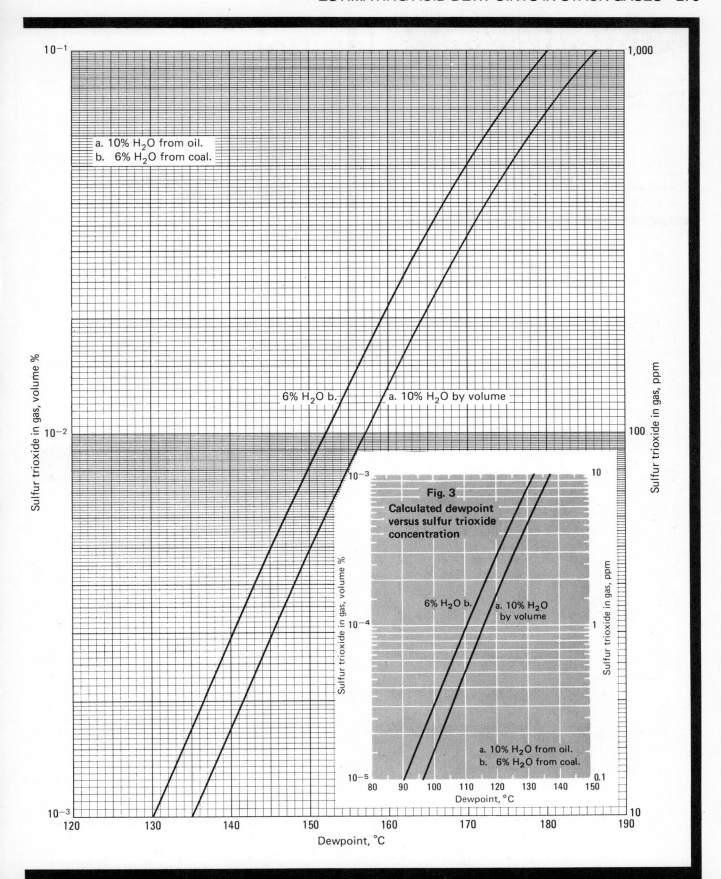

Dewpoint versus sulfur trioxide concentration   Fig. 1

Estimate of sulfur trioxide in combustion gas

Oil fired units (Table 1)

Sulfur in fuel, %		0.5	1.0	2.0	3.0	4.0	5.0
Excess air, %	Oxygen in gas, %	Sulfur trioxide expected in gas, ppm					
5	1	2	3	3	4	5	6
11	2	6	7	8	10	12	14
17	3	10	13	15	19	22	25
25	4	12	15	18	22	26	30

Coal-fired units (Table 2)

25	4.0	3-7	7-14	14-28	20-40	27-54	33-66

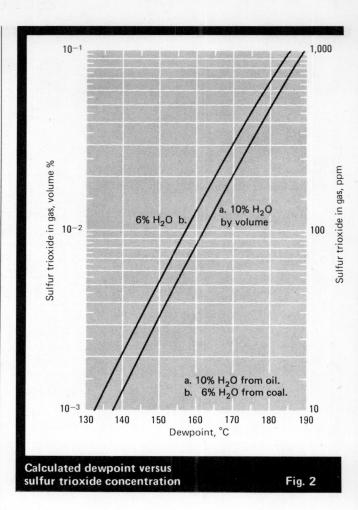

Calculated dewpoint versus sulfur trioxide concentration   **Fig. 2**

taining 10% water vapor and 100 parts per million sulfur trioxide, P_{H_2O} would be 10^{-1} and P_{SO_3} would be 10^{-4} atm. (Logs are base 10.) Using these values in the equation, we get T_{DP} equal to $435°K = 162°C$ ($324°F$). Fig. 2 presents curves of dewpoints versus the log of volume-percent sulfur trioxide calculated from this equation at two values of the percentage of water vapor. The 10% curve is for a typical combustion gas when burning oil, and the 6% curve is for a typical combustion gas when burning coal. Since the curves are only about 5°C apart, interpolation is easy.

We have independently made a correlation of the Thomas and Barker data with that of R. Haase and H. W. Borgmann, resulting in the curves shown in Fig. 1 [3]. At the low end of the scale (10 ppm sulfur trioxide), our curves read about 2°C lower and at 100 ppm about 5°C lower than in Fig. 2. In the range of 120 to 140°C (248 to 284°F), Verhoff and Banchero report positive deviations on the order of 4°C from the determinations of E. S. Lisle and J. D. Sensenbaugh [4] and W. E. Francis [5], so the curves of Fig. 1 are probably more accurate than those of Fig. 2.

We did not attempt to cover the range below 10 ppm of sulfur trioxide in our correlations. The form of Verhoff and Banchero's equation is such that it does not reduce to the dewpoint of water at zero parts per million of sulfur trioxide. This means that at some point the equation must start yielding low results. This is apparent from a comparison of Verhoff and Banchero's predicted dewpoints with Lisle and Sensenbaugh's experimental points. In the range of 121 to 100°C (250 to 212°F), the predicted points are 2.5–4.0°C low. Nevertheless, Fig. 3 has been prepared from Eq. 1 to give an indication of what might be the case in the 0.1 to 10.0-ppm sulfur trioxide range.

Use of curves—oil

Because the percentage of water in the gas is readily estimated, its determination should not present a problem. However, determination of the sulfur trioxide is not so easy and all plants may not have the equipment and techniques available. In the absence of an actual determination of the sulfur trioxide, Table I may be used as a guide for oil-fired units, but many factors, such as firing rates and the presence of vanadium in the oil, may change the figures drastically. One article

shows 40 ppm sulfur trioxide at high load and only 5 ppm at low load. Also, additives such as magnesium oxide may greatly reduce the sulfur trioxide concentration. But fortunately, the curves of Fig. 1 and 3 are very steep, and doubling the sulfur trioxide concentration only increases the dewpoint by 6.5°C.

We have no information on the effect of electrostatic precipitators on the sulfur trioxide concentration, but some increase could be expected from the extra detention time in the 371–538°C (700–1,000°F) range, the extra metal contacts, and the electrical effects.

When high-efficiency wet scrubbers are used, much of the sulfur trioxide is removed; but the temperatures are drastically reduced, and the percentage of water increased, so that any sulfuric acid mist getting through the scrubbers would require considerable reheat to be dissipated. For example, if only 1.0 ppm of sulfur trioxide remained, and the percentage of water were increased to 15.5% by the adiabatic quench from 149°C to 55°C (300 to 131°F), the dewpoint would be 121°C (249°F). So a reheat of $121 - 55 = 66°C$ ($151°F$), plus enough extra reheat to take care of cooling in the chimney, would be necessary.

Fortunately the oxidation of sulfur dioxide to sulfur trioxide is slow in the absence of a catalyst. At furnace temperatures such as 1,000°C (1,832°F), the percentage of conversion is low. Otherwise, the dewpoint problem would be much worse than it is, because at 150°C

(302°F) the theoretical conversion is very high and a high dewpoint temperature would result.

Agreement with practice

It is believed that any significant disagreement of actual dewpoints with Fig. 1 will be due to faulty determination of the percent of sulfur trioxide. The agreement of Verhoff and Banchero's equation with data by Lisle, Sensenbaugh, and Francis is so good (and in the direction of Fig. 1) that we should have the dewpoints bracketed to plus or minus 3°C at least. Below 10 ppm sulfur trioxide, there could be significant disagreements, as noted before.

Many published results show 130–150°C (266–302°F) dewpoints, which are predicted by the curves for 5–50 ppm sulfur trioxide. Such temperature measurements are always in question where combustion gases are concerned. Unless shielded and well ventilated with the gas, thermocouples may record an integrated temperature composed of the gas temperature and the adjacent stack-wall temperature. What is more important from the corrosion standpoint, however, is the skin temperature of the interior of the stack. Measuring this temperature poses its own problems. Unless suitable precautions are taken, conduction along a metallic thermocouple may cause it to report a significantly lower temperature than the actual inside wall.

Use of curves—coal

In one respect, combustion gas from coal is in a more favorable position than gas from oil as regards dewpoint. The percentage of water will be 6 or less, unless there is a large amount of inherent moisture; and this yields a 5°C advantage right away. On the other hand, reduction of the percentage of excess air is much more difficult, and the availability of coals with 1% or less sulfur is limited.

The ash in coal, especially ash that is high in iron oxide from pyrite, may increase the formation of sulfur trioxide through the catalytic effect of uncombined iron oxide. On the other hand, the ash from some coals contains sufficient calcium and magnesium oxides to absorb much of the sulfur trioxide. These effects make it difficult to reach valid generalizations.

J. D. Piper and H. Van Vliet report 14–39 ppm sulfur trioxide (by volume) when burning coal containing 2.9% sulfur on a dry basis [6]. These concentrations correspond to 1–2% of the sulfur dioxide content. The ad hoc panel on Control of Sulfur Dioxide from Stationary Combustion Sources (National Research Council) mentions exactly the same range in its report [7], while S. Katell of the U.S. Bureau of Mines used 2% in his calculations [8]. Table II is drawn up on this basis for estimating dewpoints in the absence of actual sulfur trioxide determinations.

However, several industry people say that they design on the basis of maintaining the temperature of a steel stack from 135°C (275°F) to 149°C (300°F) all the way to the top. The lower temperature would give protection only up to 2.5% sulfur in the coal, according to Table II used in conjunction with Fig. 1. Some coals in the U.S. run considerably above this 2.5% sulfur content. Another source cites acid dewpoints of 107–116°C

(225–240°F), which correspond to only 0.7–1.5 ppm sulfur trioxide in Fig. 3. Even with a 1% sulfur coal, these concentrations correspond to only 0.1–0.2% of the sulfur in the coal going to sulfur trioxide, by contrast to the 1–2% cited above. We plan to make an extensive survey of the coal-burning power plants in the U.S. to try to resolve such disagreements.

Other contaminants

Although hydrochloric acid has some of the water-holding characteristics of sulfuric acid, its constant boiling mixture contains 20% hydrochloric acid at 110°C (230°F). This is below the sulfuric acid azeotrope, so the sulfuric acid controls. When flue gases are scrubbed, both sulfuric acid and hydrochloric acid are condensed, and some of these acids move together into the chimney, drastically increasing corrosion rates.

With scrubbed gases from coal, hydrofluoric acid has also been measured at under 2.0 ppm, which is not sufficient to endanger brick linings but which may (along with sulfuric acid) increase the corrosion rates of metals. As mentioned, sulfur dioxide is not a problem in itself; its saturated solution in the presence of about 2,000 ppm sulfur dioxide by volume in the gas has practically the same vapor pressure as pure water.

Hydrofluoric acid and nitric acid (from nitrous oxides plus air in the combustion gases) also form constant boiling mixtures; but like hydrochloric acid, these exist in low concentrations (50 ppm or less) in the gas phase and do not lower the vapor pressure of the water significantly.

References

1. Thomas, J. S. and Barker, W. F., *J. Chem. Soc.*, Vol. 127, p. 2820 (1925).
2. Verhoff, F. H. and Banchero, J. T., *Chem. Eng. Prog.*, Vol. 70, p. 71 (1974).
3. Haase, R. and Borgmann, H. W., *Korrosion*, Vol. 15, p. 47 (1961).
4. Lisle, E. S. and Sensenbaugh, J. D., *Combustion*, Vol. 36, p. 12 (1965).
5. Francis, W. E., "The Measurement of the Dew Point and H_2SO_4 Vapor Content of Combustion Products," Gas Research Board, Communication GRB 64 (1952).
6. Piper, J. D. and Van Vliet, Hazen, *Trans. ASME*, Aug. 1958, p. 1251.
7. Ad Hoc Panel on Control of Sulfur Dioxide from Stationary Combustion Sources, Committee on Air Quality Management, National Research Council, 1970. "Abatement of Sulfur Oxide Emissions from Stationary Combustion Sources."
8. Katell, S., *Chem. Eng. Prog.*, Vol. 62, Oct. 1966, p. 67.

Acknowledgements

The author is grateful for the assistance of Elliott J. Roberts, who conceived the approach used for the presentation of the calculated data.

The author would also like to thank James A. Conover of Pennwalt Corp. for his assistance in the securing and evaluation of available engineering data.

The author

Robert R. Pierce is Technical Manager of the Industrial Chemical and Chemical Specialties Div., Pennwalt Corp., Pennwalt Bldg., Three Parkway, Philadelphia, PA 19102. He has worked for Pennwalt since 1941, when he started as a laborer in the Industrial Chemicals Dept. in Portland, Ore. Since 1947, when he moved to Pennwalt's Specialty Chemicals Div. in Philadelphia as Product Manager, he has become well known as a materials specialist who exploited chemical engineering principles for corrosion control. An active member in many professional societies, he holds a B.S. in chemical engineering from Oregon State University.

Automatic control of reagent feed boosts wet-scrubber efficiency

A constant reagent feedrate often results in excessive chemicals consumption or inadequate removal of odorous contaminants from exhausted process gas.

R. Puri, The Pillsbury Co.

☐ Oxidation is gaining wider acceptance over incineration for the control of odorous compounds in process exhaust gases because of its low operating cost, and the higher cost of fuel for incineration.

Particle and odor pollutants in exhaust gases are normally controlled by a combination of Venturi and packed-tower scrubbers. This combination collects liquid and particles by the mechanisms of inertial impaction and interception, and removes gaseous pollutants by liquid-gas diffusional processes.

Water in the Venturi scrubber washes particles from the exhausted process gas, and cools and absorbs a major portion of the odorous compounds that are soluble in water. In the packed-tower scrubber, the gas is scrubbed with a solution of a chemical reagent that absorbs and oxidizes most of the odorous compounds.

A properly designed packed-tower scrubber having the required number of transfer units provides not only the residence time needed for the diffusional or absorption process but also for the reactions between the constituents of the solute gas and the scrubbing reagent. Of course, a substantial quantity of the oxidizer must be recycled so that the concentration of the solute gas in it does not change appreciably during the absorption process.

Multistage scrubbing may be required

Typical odor contaminants present in process emissions are ammonia and organic amines, sulfides and disulfides, aldehydes, mercaptans, fatty acids, hydrocarbons and phenolic compounds. Most can be effectively oxidized by such agents as an aqueous solution of sodium hypochlorite and potassium permanganate.

Frequently, multistage scrubbing systems are installed when single scrubbing reagents will not reduce the odor as desired. A multistage system is also needed when a gas contains two (or more) contaminants, such as amines and fatty acids, and the scrubbing reagent cannot react with both. In such a case, the first-stage solution may be a weak caustic, which saponifies the fatty acids, and the second stage sulfuric acid, which reacts with the amines to produce sulfates.

For adequate mixing, the reagent may be injected into either the recycle line or the recycle tank of the packed-tower scrubber. Generally, the solution is injected at a constant rate determined from experiences with other applications, from pilot tests of the system, or from tests performed on the system at startup.

A constant chemical feedrate may not cope with variations, caused by process fluctuations, in the odorous content of the exhaust gases. If the feedrate is higher than necessary, excessive consumption of the reagent will run up operating costs. Additionally, particles suspended in the gas stream will tend to settle on the packing of the scrubber, and finally collect in the sump. These particles will be oxidized, and so increase consumption of the reagent, or decrease odor reduction.

For these reasons, automatic control of the solution feedrate is essential. Also, continuous recording will provide data for air-pollution-control authorities, as well as indicate malfunctions in the system.

Control of pH is often critical

The concentration of reagent in the scrubber can be controlled and monitored continuously by measuring solution properties that vary with its concentration—such as oxidation-reduction potential or electrolytic conductivity. Because there is usually an optimum pH

Originally published October 23, 1978

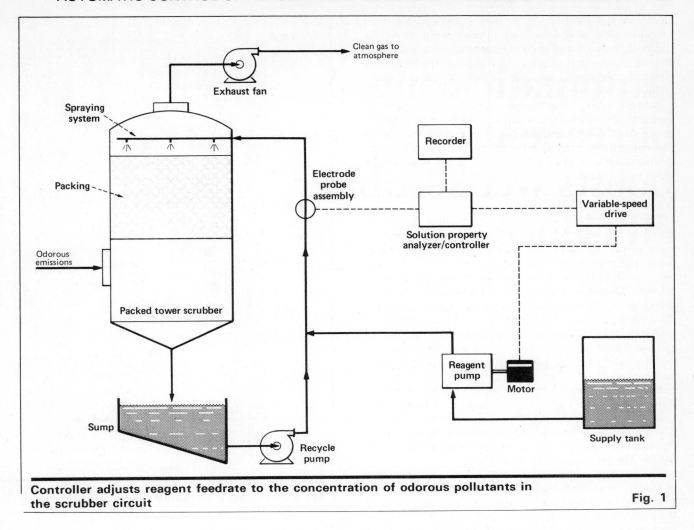

Controller adjusts reagent feedrate to the concentration of odorous pollutants in the scrubber circuit

Fig. 1

level of a reagent for oxidation in odor control, pH level, and either oxidation-reduction potential or solution conductivity, should be controlled.

When aqueous sodium hypochlorite is the scrubbing solution, pH should be maintained at 10 or higher. Chlorine exists in the water as hypochlorous acid ($HOCl^+$) and hypochlorite ion (OCl^-), and it is the latter that oxidizes the pollutants to reduce odor. The proportion of hypochlorite ion to hypochlorous acid changes radically with the pH of the scrubbing solution. For example, at a pH of 7, only 20% of the chlorine is in the hypochlorite ion form, but at a pH of 8, the proportion rises to 70%. Maintaining the pH of the solution at 10 or higher prevents the evolution of free chlorine.

When potassium permanganate is the reagent, the pH should be maintained from 8 to 10 because the hydroxyl ions work as catalysts for the oxidation of organic compounds. Moreover, any fatty acids formed by the oxidation of aldehydes or alcohols react with the alkaline solution to produce odorless alkali metal salts. A pH above 10 may result not only in excessive secondary oxidation of primary oxidation products, but also in the loss of the oxygen formed in the primary reaction by the reduction of potassium permanganate in the presence of hydroxyl ion.

Oxidation-reduction potential measurement is critical in many chemical processes, such as the chemical oxidation of cyanide waste, the chemical reduction of chromate, and wastewater treatment. In the manufacture of commercial bleach solution (sodium hypochlorite), oxidation-reduction potential measurement and control is vital to achieving optimum stability of bleach liquor.

The measuring assembly consists of a metal electrode and a reference electrode. The metal electrode creates an electric potential proportional to the oxidizing or reducing potential of the solution. The reference electrode produces a constant potential in any solution, which serves as a reference point from which to measure the variable potential created by the other electrode.

Electrolytic conductivity measurement

The conductivity of a solution is a function of its concentration, and may be determined by its electrolytic conductance. Electrolytic conductivity is measured in terms of the specific conductance (mhos/cm) of the solution by a pair of electrodes whose dimensional configuration and spacing are precisely fixed.

Temperature compensation is usually included in the conductivity cell, to create a resistance change that is related to the temperature coefficient of the solution.

Localized polarization effects, such as electrolysis, must be minimized in conductivity measurement to reduce the back voltage and enhance measurement accuracy. Porous carbon or platinum black may be applied to the electrode surfaces to increase surface area and thereby reduce current density and back voltage.

Several types of conductivity meters are available to measure the electrolytic conductivity of a solution in terms of specific conductance, percent hydroxide or percent sulfuric acid. For strong caustic solutions, conductivity cells of glass should be avoided because the glass may corrode.

Methods of controlling

The automatic feed-control system may be pneumatic or electric. Whether the means is the measurement of pH, oxidation-reduction potential, or conductivity, the purpose is to adjust the feedrate so that the concentration of the scrubbing liquid remains approximately constant. Measurement of potential is best suited to the control of sodium hypochlorite or potassium permanganate concentration, and of conductivity to monitor potassium permanganate, sulfuric acid or sodium hydroxide reagents. For sodium hydroxide or sulfuric acid solutions, pH measurement is usually most suitable.

Fig. 1 shows such a control system. A sample point should be provided ahead of the point of reagent injection, in either the scrubber's recycle line or its sump. The electrode probe assembly is installed so as to sense the reagent concentration of the solution. It sends signals to the analyzer/controller.

The analyzer determines whether the concentration of reagent is above or below the set point. The controller, in turn, changes the speed of the electric motor of the solution feed pump via a variable-speed drive. The concentration of the scrubbing solution may be continuously logged by a recorder.

In the electronic control system shown, the solution flow is controlled by a variable-speed drive, whereas in a pneumatic one, the flow would be regulated by an orifice positioner (metering orifice). The air signal is the standard 3 to 15 psi, and the analog signal is available in 1 to 5, 4 to 20, or 10 to 50 mA. Both the electronic and pneumatic controllers provide the same functions, e.g., proportional band, dead band and reset.

For best results, the injection point should be close to the sampling point so that dead time will be short. Of course, this can also be achieved by appropriately adjusting the proportional band and reset action.

Developing calibration curves

A number of simple titrations on the actual process can be performed to develop calibration curves that relate solution concentration to its oxidation-reduction potential or conductivity. Amperometric titration is useful for spot-checking solution concentration when developing calibration curves.

(In amperometric titration, the current passing through the titration cell between an indicator and reference electrode is measured as a function of the volume of the titrating solution. This current diminishes until all the oxidant is neutralized, finally ceasing at the

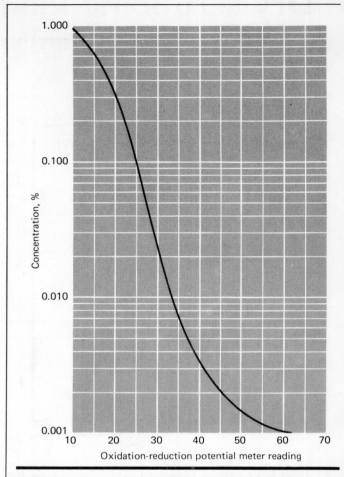

Calibration curve for sodium hypochlorite   Fig. 2

equivalence point, indicating the titration endpoint.)

A typical calibration curve (this one of the concentration of sodium hypochlorite vs. oxidation-reduction potential) is shown in Fig. 2. Any concentration of solution can be achieved by fixing the set-point value proportional to the concentration desired.

If, for a particular application, one need be only concerned with the oxidizing power of the solution, the solution's temperature may be allowed to vary, as long as control of solution concentration accomplishes the desired degree of oxidation. This will save the cost of thermal control.

The author

Rajaish Puri is a staff engineer with The Pillsbury Co. (608 Second Ave. So., Minneapolis, MN 55402), involved in the design of systems for air pollution control, heat transfer, energy conservation, refrigeration, and materials handling. He previously worked for Environmental Research Co. of St. Paul (with which he was affiliated when he wrote this article). Holder of an M.S. in mechanical engineering from Washington State University and an M.B.A. from University of Minnesota, he is an associate member of ASME and a registered engineer in Minnesota.

Dry scrubbing looms large in SO$_2$ cleanup plans

The latest standards on SO$_2$ emissions from new coal-fired utilities favor the use of dry scrubbing, which, according to its supporters, boasts lower investment and operating costs and avoids sludge-disposal problems.

☐ The newest fashion on the SO$_2$-control scene is the emergence of nonregenerable, dry-scrubbing systems for flue-gas desulfurization (FGD). The first U.S. commercial installation based on this technique went onstream last month at Strathmore Paper Co.'s pulpmill in Strathmore Mass.; four more such facilities will start up within the next three years (see table); and several other projects are expected to be announced by the end of this year.

The underlying technology, based on SO$_2$ removal either by injection of a dry alkaline sorbent or by an alkaline slurry or solution in a spray dryer (see box), has been available for a few years. But potential users, noting a lack of commercial operating experience, and perhaps wary of the extravagant claims made years ago for such cleanup methods as wet scrubbing,

have been slow in switching to dry scrubbing—despite what its proponents say are impressive advantages over wet scrubbing (lower investment and operating costs, no sludge-disposal problems).

Nonregenerable dry scrubbing, however, does have one important limitation: it appears to be economic only at low SO$_2$ concentrations in flue gas. This is because the dry methods (dry injection, spray dryer) use expensive sorbents—lime, sodium carbonate, or such naturally occurring carbonates as trona (a hydrous sodium carbonate) and nahcolite (sodium bicarbonate). Wet scrubbing, the most widespread FGD method, employs cheaper limestone, so it has an operating-cost edge at high SO$_2$-removal levels.

GOVERNMENTAL BOOST—The U.S. Environmental Protection Agency (EPA) had dry-scrubbing technology very much in mind when it issued

tougher standards in June, covering emissions of SO$_2$, particulates and nitrogen oxides from new coal-fired power plants (350 will be built in the U.S. between now and 1995).

Until now, utilities could lower SO$_2$ emissions by either burning low-sulfur coal or using scrubbers to remove the emissions given off by high-sulfur coal. But after the June regulations, all new coal-burning facilities (except those using low-sulfur anthracite) will have to install scrubbers.

Observing that it would probably be more economical to employ a wet scrubber when the coal's sulfur content is above 1.5%, the agency devised a sliding-scale emissions-removal requirement aimed at fostering the growth of dry-control methods. For example, new facilities emitting more than 0.6 lb of SO$_2$/million Btu must install scrubbers able to remove 90% of the emissions. But those that use low-sulfur coals (with emissions under 0.6 lb/million Btu) can employ scrubbers only 70% efficient. This rule is expected to generate additional interest in dry scrubbing among utilities planning new construction.

DRY VS. WET—Among the strongest proponents of dry scrubbing is the Basin Electric Power Cooperative, which plans to install units of the spray-dryer type at plants near Beulah, N.Dak., and Wheatland, Wyo. According to Kent E. Janssen and Robert L. Eriksen, two company spokesmen who presented a paper at a recent EPA symposium on FGD (March 5-8, at Las Vegas, Nev.): "Wet scrubbers are complex and cumbersome, cause a heavy capital burden, are costly to operate, difficult to maintain, and consume a sizable part of the energy generated by the plants they serve. The byproduct, sludge, produces additional environmental problems by taking up otherwise useful land for settling ponds."

As Janssen and Eriksen see it, dry scrubbing comes out ahead in the following aspects of an FGD system:

■ Waste handling. Dry control yields a dry waste, so there is no need for sludge-handling equipment.

■ Wet/dry interface. Scaling and plugging is common in wet scrubbers at wet/dry interfaces and in scrubber packing materials and demisters. In the spray-drying method of dry scrubbing, only dry powder comes into contact with the scrubber walls.

Guide to nonregenerable flue-gas desulfurization

In contrast with regenerable FGD methods, in which the scrubbing agent is recovered for reuse and the waste is usually converted into a useful product—e.g., elemental sulfur, sulfuric acid—nonregenerable techniques do not recycle the scrubbing medium and must dispose of either a sludge or a dry waste.

Wet scrubbing and dry scrubbing are the two available nonregenerable FGD methods. In the first, an aqueous slurry or solution, frequently of lime or limestone, is brought into direct contact with flue gas. The waste product is usually a sludge of calcium sulfite.

Dry scrubbing can be done either by dry injection, in which a dry sorbent (soda ash, trona, nahcolite) comes into contact with flue gas, or by spray drying, in which the heat from flue gas is used to evaporate the water off a sprayed alkali (lime or soda ash) slurry or solution. The outcome in either case is a dry-powder mixture of fly ash and sulfates, which is collected by filter bags or in an electrostatic precipitator.

Originally published August 27, 1979

Roundup of U.S. commercial-size dry-scrubbing projects

User/location	Developer/licensor	Plant/fuel	System	Status and comments
A. Industrial				
Strathmore Paper Co. Strathmore, Mass.	Mikropul Corp.	Pulp mill Flue-gas flow: 60,000 acfm Existing pulverized coal boiler burns 2.5% sulfur coal	Lime-based Spray dryer (1 unit—15 ft dia.) Two-fluid nozzle Baghouse collector (pulsed air)	Started up July 1979. SO_2 removal efficiency 75%.
Celanese Corp. Cumberland, Md.	Rockwell International and Wheelabrator-Frye in a joint venture	Acetate-fiber plant Flue-gas flow: 65,000 acfm Stoker-fired boiler (from storage) burns 1.5-2% sulfur coal	Lime-based Spray-dryer (1 unit) Baghouse collector (4 compartments)	Slated for completion in January 1980.
B. Utility				
Otter Tail Power Co. leads group of five utilities Coyote Station, near Beulah, N.Dak.	Rockwell International and Wheelabrator-Frye in a joint venture	410 MW (under construction) Flue-gas flow: 1.89 million acfm Fuel: lignite, 0.78% sulfur (avg.)	Sodium-based (soda ash) Spray dryer (4 units) Baghouse collector	Construction to be completed in late 1981. Contract for $36 million awarded in December 1977.
Basin Electric Power Cooperative (Bismarck, N.Dak.) Laramie River Station-Unit 3 Wheatland, Wyo.	Babcock & Wilcox	500 MW (under construction) Flue-gas flow: 2 million acfm Fuel: Wyoming sub-bituminous coal, 0.81% sulfur (max.)	Lime-based Spray dryer—horizontal reactor Electrostatic precipitator	The first of the utility's three units, now under construction, is expected to start up in 1980. Units 1 and 2 are wet-scrubber systems. Identical electrostatic precipitators will be used with both the wet and dry systems.
Basin Electric Power Cooperative (Bismarck, N.Dak.) Antelope Valley, near Beulah, N.Dak.	Joy Mfg. Co. and Niro Atomizer in a joint venture	440 MW (under construction) Flue-gas flow: 2 million acfm Fuel: lignite, 1.22% sulfur (max.)	Lime-based Spray dryer Baghouse collector	Onstream date for the first dry scrubber is targeted for April 1982. Bids for a second dry-scrubbing system will be going out this year.

■ **Materials of construction.** Wet scrubbers require corrosion-resistant alloys or coatings; dry-control systems can use low-carbon steel for vessels.

■ **Maintenance.** In wet systems, maintenance for the slurry-handling equipment is high, because of the need to recirculate corrosive materials at high pressures and volumes. Dry systems operate at low material-volumes and low pressures; liquid/gas ratios are about 0.2-0.3 gal/1,000 acf, compared with about 40-100 gal/1,000 acf for a wet scrubber.

■ **Energy requirements.** Dry systems require about 25-50% of the energy needed by a wet system.

■ **Water consumption.** This is much less for dry systems. For instance, the dry scrubber at Basin Electric's Laramie River Station Unit #3 will use about 50% as much water as is needed for the wet scrubber in Unit #1 or #2. Cooling-tower blowdown or ash water may be used in the spray dryer.

WHO'S ACTIVE—The list of firms doing research on spray-dryer systems includes such names as Carborundum Co. (Niagara Falls, N.Y.), Joy Mfg. Co. (Montgomeryville, Pa.), Niro Atomizer Inc. (Columbia, Md.) and Rockwell International Corp. and Wheelabrator-Frye Inc.'s Air Pollution Control Div., both in Pittsburgh. Among those studying dry-injection systems are Carborundum, and Energy & Pollution Controls, Inc. (Bensonville, Ill.).

The Electric Power Research Institute (EPRI) also is interested in dry scrubbing, mainly in dry-injection systems. According to project manager Navin Shah, dry injection is simpler than spray drying, in terms of both equipment and processing. He notes that, because the latter employs a solution or slurry, the flue gas may need reheating after evaporation of the sorbent liquid. Should flue-gas temperature drop for some reason, clogging may occur in the subsequent collection equipment (a baghouse or electrostatic precipitator).

In an EPRI-sponsored program, KVB Inc. (Tustin, Calif.) tested a sodium-based dry-injection system for one and a half years on a bench-scale basis. Among the parameters studied: sodium/sulfur stoichiometric ratios, injection methods and temperatures, baghouse operation temperatures, incoming SO_2 concentrations, and sorbent particle size.

EPRI is also supporting a Bechtel Corp. economic-comparison study involving a dry-injection system (based on nahcolite and trona) and wet scrubbing. According to Shah, a draft report that may be available by the end of the year indicates that dry-injection economics are highly dependent on reagent costs.

The Tennessee Valley Authority (TVA) too is in the throes of completing an economic assessment of dry scrubbing—under contract with the EPA. The goal is to compare a conventional wet-limestone scrubbing system with lime-based spray-drying.

Another EPA-funded study has been awarded to Radian Corp. (Austin, Tex.), which will review dry-scrubbing technology; a draft report was due at the end of July.

In addition, EPA is funding a few small demonstration-scale projects on dry scrubbing. Envirotech Corp. (Lebanon, Pa.) is about to start a dry-injection study at the Martin Drake Station in Colorado Springs, Colo., using a baghouse collector.

Looking ahead, an industry expert predicts that use of additives to enhance FGD at high SO_2 concentrations will be tried in dry-scrubbing applications, as has been the case in wet-scrubbing systems.

How to rate alloys for SO₂ scrubbers

New data on several stainless steels will help

you specify materials that will

stand up to acid scrubbing liquors

carrying chlorides and other contaminants.

H. T. Michels and E. C. Hoxie,
The International Nickel Co.

☐ Many power plants and industrial facilities that have switched to coal as a boiler fuel rely on wet scrubbers to remove pollutants from combustion gases. A major hindrance to reliable operation of these devices is the severe corrosion caused by contact with SO_2-containing gases and acid scrubbing liquors.

Scrubbers for this difficult service have been fabricated from a variety of materials. Such stainless steels as Type 316L have found widespread use. However, at certain pH levels and chloride concentrations even this alloy will suffer localized attack. At more-stringent conditions, alloys containing more chromium, molybdenum and nickel may be required. Although these alloys are more expensive initially, their extended life may economically justify their selection.

Materials requirements

Wet scrubbing liquors range from locally available water to slurries and solutions that can be either acid or alkaline. These liquors contact the flue gas, absorb SO_2 (plus any trace amounts of SO_3), and become more acidic. Chloride ions, which are always present, are usually introduced with the coal, but may also accompany the water and chemicals dissolved in the scrubbing liquor. Closed-loop recirculating systems, which minimize water pollution, can build up the chloride level, resulting in an extremely aggressive acid-chloride solution.

Fig. 1 shows a schematic highlighting trouble spots observed in various scrubber installations. Where the flue gas enters and is quenched by the spray nozzles, localized attack and flyash erosion are likely to occur. In the recirculating pump, erosion-corrosion can present a problem. Inside the absorber, pitting and crevice corrosion may crop up, whereas stress-corrosion cracking has been observed in reheater tubes. Dewpoint corrosion takes place downstream of the mist eliminator, and fatigue cracking has caused the failure of exhaust fans.

Factors that enhance performance

Kopecki and McDaniel [1] have identified several factors that contribute to the good performance of Type 316L and other stainlesses in a wet limestone scrubber (of

Originally published June 5, 1978

La Cygne design) owned by Kansas City Power and Light:

1. Regular, weekly cleaning of each scrubber module to minimize scale buildup preserves scrubber efficiency and removes deposits that contribute to localized attack.

2. Careful regulation of the recirculating scrubber solution to a pH within a 5.5 to 5.7 range controls scaling and cuts down on corrosion problems.

3. Maintaining a low chloride level favors good materials performance.

These authors concluded that Type 316L was highly suited for scrubber construction because of its good resistance to corrosion, low tendency toward scale adhesion, and ease of cleanability. The low carbon of the "L" grades minimizes detrimental carbide precipitation in the heat-affected zones of welds. This property explains the high resistance of these grades to intergranular corrosion attack.

Coatings, despite their temperature limitations, have been widely used in SO₂ scrubbers because of their low initial cost. They are especially attractive in low-tempera-

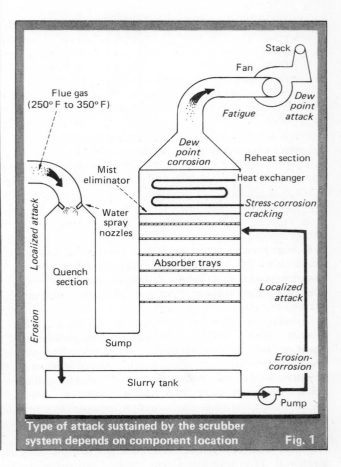

Type of attack sustained by the scrubber system depends on component location   **Fig. 1**

Nominal compositions of some alloys that have been used for wet-scrubber construction   Table I

Alloy	Cr	Mo	Ni	Fe	Cu	C	Other
Wrought							
Inconel 625[1]	22	9	Balance	3	—	0.05	3.6Cb + Ta
Hastelloy B[2]	1*	28	Balance	5	—	0.05*	2.5Co*
Hastelloy C-4[2]	16	16		3*	—	0.015*	2.0Co*, 0.7Ti*
Hastelloy C-276[2]	16	16	Balance	5	—	0.02*	2.5Co*, 4W
Hastelloy G[2]	22	6.5	Balance	20	2	0.05*	2.5Co*, 1W*
Incoloy 825[1]	22	3	Balance	30	2	0.03	2.1 Cb + Ta
Allegheny Ludlum AL-6X[3]	20	6	24	Balance	—	0.03	
Jessop JS-700[4]	21	4.5	25	Balance	—	0.05	0.3Cb
Jessop JS-777[4]	21	4.5	25	Balance	2.4	0.05	?
Carpenter 20 Cb-3[5]	20	2.5	35	Balance	3.5	0.07*	
Haynes 20 Mod[2]	22	5	26	Balance	—	0.05*	
Type 304	19	—	9	Balance	—	0.08*	
Type 216	20	2.5	7	Balance	—	0.08*	0.3N, 8.5Mn
Type 316	17	2.5	12	Balance	—	0.08*	
Type 316L	17	2.5	12	Balance	—	0.03*	
Type 317	19	3.5	13	Balance	—	0.08*	
Type 317L	19	3.5	13	Balance	—	0.03*	
Type 317LM[6]	19	4.25	14	Balance	—	0.03*	
Type 317L Plus[4]	19	4.25	14	Balance	—	0.03*	
Type 409	12	—	—	Balance	—	0.05*	0.4Ti
Type 410	12	—	—	Balance	—	0.15*	
Type 430	17	—	—	Balance	—	0.12*	
E-Brite 26-1[3]	26	1	—	Balance	—	0.005*	
Titanium	—	—	—	—	—	0.10*	Balance Ti
Carbon steel	—	—	—	Balance	—	0.20*	
Cast							
CF-3M	19	2	11	Balance	—	0.02*	
CF-8M	19	2	9	Balance	—	0.08*	
CG-8M	19	3	11	Balance	—	0.08*	
CD4 MCu	26	2	5	Balance	3	0.04*	
CN 7M	20	2	29	Balance	3	0.07*	
IN-862	21	5	24	Balance	—	0.07*	
CW-12M-2	18	18	Balance	6	—	0.07*	

[1] Trademark of the International Nickel Co.
[2] Trademark of Stellite Div. of Cabot Corp.   *Maximum
[3] Trademark of Allegheny Ludlum Steel Corp.
[4] Trademark of Jessop Steel Corp.
[5] Trademark of Carpenter Technology Corp.
[6] Trademark of Eastern Stainless Steel Corp.

ture areas, even though a temperature excursion can have a devastating effect. Scrubber ductwork and stacks present special problems to coatings, in that the underlying steel sheet is thin, and prone to rapid thermal shocks and severe flexing [2].

Maintenance poses an additional complication when coatings are used, since the scrubber must be isolated for repair. In one instance, a scrubber was severely damaged when a welder tried to weld a patch onto the steel shell from the outside. The coating on the inside surface ignited and the natural chimney effect spread the flames.

Tests on alloys

No single material seems universally suitable for SO₂-scrubbers from the viewpoints of both cost and performance. In order to gain insight into the performance of the iron-nickel-chromium-molybdenum family of alloys in wet scrubbers, The International Nickel Co. has undertaken an extensive corrosion-testing program. The nominal chemical compositions of the alloys examined in this program are listed among those in Table I.

Samples of the alloys were mounted on galvanically insulated test spools of about 2.25 in. dia. These were installed in various locations within the scrubber unit. Duplicate samples contained in the same spool provided several data points at each location.

Previous experiments using such test spools [4] indicated that plain carbon steel and Type 304 stainless steel were generally inadequate in wet scrubbing environments. This work also showed that stress-corrosion cracking is generally not a problem below 140°F (60°C). Other test work and industrial experience [5] indicate that resistance

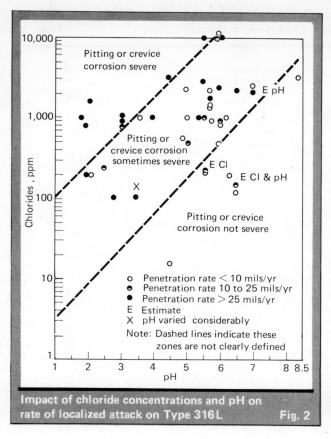

Impact of chloride concentrations and pH on
rate of localized attack on Type 316L   Fig. 2

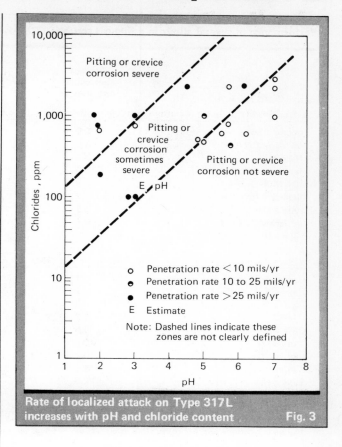

Rate of localized attack on Type 317L
increases with pH and chloride content   Fig. 3

to stress-corrosion cracking can be achieved with higher nickel contents.

Influence of pH and chloride content

Several important trends were discovered in this work. As shown in Fig. 2, there is a rough correlation between the degree of localized attack on Type 316L and the pH and chloride contents of the scrubbing liquor. Additional corrosion-spool data, mostly at temperatures of 120 to 140°F, have confirmed this.

As may be seen in Fig. 2, the rates of localized attack (which encompasses both pitting and crevice corrosion) fall into three regions, occurring at roughly 10, 10 to 25, and more than 25 mils/yr, respectively.

Localized corrosion increases as the chloride level increases and the pH of the scrubbing liquor decreases. At a given pH, the environment becomes more aggressive as the chloride content is increased, and at a given chloride level as the pH is decreased.

Deposition of solids can lead to crevice formation,

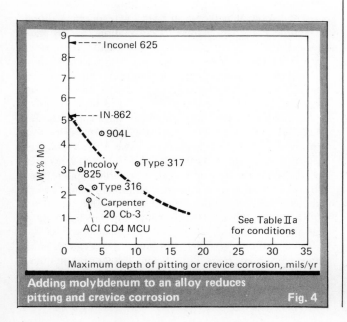

Adding molybdenum to an alloy reduces
pitting and crevice corrosion   Fig. 4

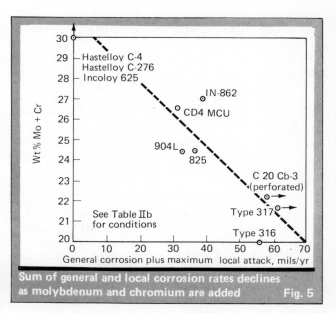

Sum of general and local corrosion rates declines
as molybdenum and chromium are added   Fig. 5

Performance of several alloys in a flue-gas scrubber. Corrosion rates varied with sample location			Table II
Alloy	**Average corrosion rate, mils/yr**	**Maximum pit depth, mils**	**Maximum crevice corrosion, mils**
a. Samples placed in spray pattern between venturi rods and spray header*			
Inconel 625	2.6	None	Incipient
Hastelloy C-276	3.0	None	None
IN-862	3.2	None	None
Hastelloy C-4	3.3	None	Incipient
CD4 MCu	3.8	3	None
Incoloy 825	4.2	2	Incipient
Type 317	4.3	10	Incipient
Carpenter 20 Cb-3	4.5	2	Incipient
Type 316	4.5	4	Incipient
b. Samples placed between mist eliminater vanes and exhaust stack†			
Hastelloy C-4	0.1	Incipient	None
Hastelloy C-276	0.1	Incipient	Incipient
Inconel 625	0.3	Incipient	Incipient
IN-862	1.3	37	14
CD4 MCu	8.5	24	13
Type 317	10.0	51 (perforated)	25
Inconel 825	10.3	27	20
Carpenter 20 Cb-3	14.2	45 (perforated)	21
Type 316	22.9	33	33

*Flue gases from coal-fired boiler were scrubbed with boiler water (pH = 1.7 to 2.2; Cl = 200 ppm; avg. gas velocity = 2,400 to 3,000 ft/min; avg. temperature = 130°F). Exposure time was 94d.

†Same conditions and exposure time as above, except avg. temperature = 150°F.

resulting in localized attack as dissolved oxygen is depleted and chlorides concentrate. Oxygen-poor, chloride-rich regions break down the passive films that give stainless steels corrosion resistance. Even though pitting or crevice corrosion may not be favored under design operating conditions, they may be initiated at a lower pH during an upset. Once initiated, such attack may continue even after the unit is returned to normal operation, since pits can propagate by an autocatalytic mechanism involving the hydrolysis of dissolved metal ions.

For Type 317L (3.17% Mo), a similar relationship between corrosion rate and pH and chloride content was determined (Fig. 3). The data in the figure came from 26 test spools that were exposed at a median temperature of 130°F (34°C). Higher temperatures of 199°F (93°C) and 277°F (136°C) were reached during excursions. Although essentially the same behavior pattern as for Type 316L was observed, Type 317L showed some improvement in corrosion resistance. Additional data should confirm that higher molybdenum contents generally impart higher resistance to pitting and crevice corrosion.

Adding molybdenum and chromium

The beneficial effects of molybdenum were identified in the previous study [4], which indicated that alloys containing more than 3.5% Mo, such as Type 317L stainless steel, would be highly resistant to localized attack in most scrubbing liquors.

The protection afforded by molybdenum was also observed under the severe conditions of low pH (approxi-

mately 2) and 200 ppm chloride (Fig. 4). However, it should be noted that a second test spool exposed at a different location in the same scrubber at seemingly the same conditions showed more-severe attack. (Exceptions: the high-nickel alloys such as Inconel 625 and Hastelloy C-276 and C-4.)

Curiously, when the sum of the wt% Mo and Cr in the alloys of this test spool was plotted against the sum of the general corrosion rate plus the maximum localized attack, a trend such as that shown in Fig. 5 was found. Although no simple explanation for this correlation can be made, molybdenum content alone does not ensure high resistance. Chromium content is obviously also a consideration. In high-chromium, high-molybdenum alloys, nickel content has to be increased to stabilize the austenitic structure.

An examination of the two spools shows that the first (Table IIa) was directly in the spray pattern between the venturi rods and the spray header, where the average gas velocity was 2,400 to 3,000 ft/min (732 to 914 m/min), and the temperature was 150°F. The second spool (Table IIb) was placed between the mist eliminator vanes and the exhaust stack, where the average gas velocity was 700 to 800 ft/min (213 to 244 m/min), and the temperature was 130°F.

These facts suggest that the lower temperature of the latter spool may have been below the dewpoint during operation or upset periods, and that an aggressive acid condensate severely attacked the stainless steels. The chromium and molybdenum contents of the high-nickel

Corrosion rates observed in a regenerative sodium sulfite scrubber				Table III
	Location 1*		Location 2†	
Alloy	Corrosion rate, mils/yr	Maximum depth of localized attack, mils	Corrosion rate, mils/yr	Maximum depth of localized attack, mils
Hastelloy G	0.7	None	0.0	None
Type 317	1.2	Incipient	0.1	13
Type 316	1.0	7	0.1	18

*Venturi exit gas plus entrained, scrubbing liquid at 126°F, pH 1.5, containing 1,200 ppm chlorides. Light discontinuous solids deposit on samples.
†Scrubbing liquid leaving venturi at 124°F, pH 1.5, and containing 1,200 ppm chlorides. Continuous solids deposit on samples.

stainless-steel alloys was sufficient to provide protection from this environment.

Higher corrosion rates measured on the more-resistant alloys at higher velocity (Table IIa) can be laid to erosion-corrosion in the high-velocity, particulate-laden liquor. Even under these conditions, where general corrosion is the dominant mode of attack, the benefit of higher Cr, Mo and Ni content in minimizing localized attack is apparent. It appears that nickel in particular accelerates repassivation of damaged corrosion films.

Additional observations—deposits

Limited data from two spools in a regenerative Na₂SO₃-type scrubber unit are presented in Table III. Although deposits were present on each spool, they were light and discontinuous in the first location, and localized attack was restricted to Type 316. In the second location, where deposits were continuous, localized attack was evident on both Types 316 and 317. Other data, taken from a power-plant scrubbing system operating with a chloride level of 800 ppm and a pH of 1, suggest that areas of severe pitting and crevice corrosion crop up beneath "black deposits." Under these deposits, chlorides build up to higher levels than in the scrubbing liquor.

The progress of pitting, and to a lesser extent crevice corrosion, can be divided into an incubation period and a propagation period. Higher chloride levels and lower pH levels appear to reduce the incubation period. In addition, the incubation period appears to be shortened if deposits are present.

Alloy selection

The test-spool data should be of assistance to those charged with selecting a particular alloy for a given pH and chloride content. One may use Fig. 2 and 3 as approximate guidelines for the performances of Types 316L (2.3% Mo) and 317L (3.17% Mo). However, local variations in temperature, pH and chloride content may be disruptive factors. Process upset conditions that cause these variables to fluctuate widely must be taken into account during design.*

Note has already been taken of the influence of deposits on materials performance. These deposits create crevices and act as a host for chlorides. Pitting and crevice corrosion may initiate under these oxygen-poor deposits even though the overall chloride content and pH value would tend to rule out localized corrosion. A sudden rise in temperature or drop in pH, as a consequence of a process upset, will further enhance localized corrosion.

The following design and operating guidelines will help to diminish the severity of corrosive attack in scrubbers that handle SO₂:

1. Maintain the pH of the scrubbing solution as high as other process conditions will permit. (Tendency toward scaling deposits at higher pH will pose a restriction.)

2. Maintain chloride levels as low as possible.

3. Minimize deposition of solids.

4. Maintain dry conditions downstream of the scrubber mist-eliminator.

5. Keep operating temperature low in the wet areas of the scrubber system.

6. Reserve more-resistant alloys for the above-mentioned trouble spots within the scrubber system.

*Inco is presently conducting a program at its Paul D. Merica Research Laboratory in Sterling Forest, Suffern, N.Y., in which a broad range of Fe-Ni-Cr-Mo alloys are being evaluated for pitting resistance by an electrochemical technique. Immersion tests at various pH and chloride levels are being run to verify the trends developed via electrochemical techniques. It is expected that this information will be published in the near future.

References

1. Kopecki, E. S., and McDaniel, C. F., Corrosion Minimized, Efficiency Enhanced in Wet Limestone Scrubbers, *Power Engineering*, April 1976, pp. 86-89.
2. Berger, D. M., Coatings for Power Plants, *Power Engineering*, April 1976, p. 90.
3. *The Wet Scrubber Newsletter*, The McIlvane Co., No. 37, July 31, 1977, p. 7.
4. Hoxie, E. C., and Tuffnell, G. W., A Summary of Inco Corrosion Tests in Power Plant Flue Gas Scrubbing Processes, prepared for the symposium, Resolving Corrosion Problems in Air Pollution Control Equipment, sponsored by APCA, IGCI and NACE, 1976.
5. Copson, H. R., Effect of Composition on Stress-Corrosion Cracking of Some Alloys Containing Nickel, "Physical Metallurgy of Stress Corrosion Fracture," Rhodin, T. N., ed., Interscience Publishers, New York, 1959, pp. 451-456.

The Authors

Harold T. Michels is a program manager for sales development with The International Nickel Co., 1 New York Plaza, New York, NY 10004, where he is concerned with the use of nickel-containing alloys for corrosive services in the power industry. A former corrosion research scientist, he holds an M.S. in metallurgy and a Ph.D. in materials science from New York University and a B.E. in mechanical engineering from the City College of New York. He is registered as a professional engineer in New York and California.

Earle C. Hoxie, a program manager for sales development with Inco, is concerned with the use of nickel-containing alloys for corrosion-resistant applications in the chemical process industries. He has more than 25 years experience in the field of corrosion. He holds a B.S. in metallurgical engineering from Lehigh University. Certified as a corrosion specialist by the National Assn. of Corrosion Engineers, he is a registered professional engineer (corrosion engineering) in California.

INDEX